BIOMASS (Biological Investigations of Marine Antarctic Systems and Stocks) was a unique, large-scale, long-term international research programme established in 1977 to investigate the ecology of the Southern Ocean. Its main aim was to gain a greater understanding of the biological systems and stocks in the marine Antarctic environment, thereby providing a sound basis for the future management of the living resources within this immense ocean. The programme was drawn to a close in 1991 and its completion marked by a colloquium which brought together key researchers to summarize and review the results obtained.

This volume, arising from that colloquium, provides a succinct, up-to-date account of the ecology of the Southern Ocean and serves as a comprehensive record of this unique and successful international project.

Southern Ocean ecology: the BIOMASS perspective

Southern Ocean ecology: the BIOMASS perspective

Edited by

SAYED Z. EL-SAYED
Department of Oceanography
Texas A&M University

CAMBRIDGE
UNIVERSITY PRESS

Published by the Press Syndicate of the University of Cambridge
The Pitt Building, Trumpington Street, Cambridge CB2 1RP
40 West 20th Street, New York, NY 10011–4211, USA
10 Stamford Road, Oakleigh, Melbourne 3166, Australia

First published 1994

Printed in Great Britain at the University Press, Cambridge

A catalogue record for this book is available from the British Library

Library of Congress cataloguing in publication data

Southern Ocean ecology: the BIOMASS perspective/edited by Sayed Z. El-Sayed.
p. cm.
Includes indexes.
ISBN 0-521-44332-6
1. Marine ecology – Antarctic Ocean – Congresses. 2. Marine biology – Antarctic Ocean – Congresses. 3. Biomass – Antarctic Ocean – Congresses. 4. Marine resources – Antarctic Ocean – Congresses. 5. BIOMASS Program – Congresses. I. El-Sayed, Sayed Z. (Sayed Zakaria), 1926–
QH95.58.S68 1994
574.5′2636′09167 – dc20 92–40608 CIP

ISBN 0 521 44332 6 hardback

Contents

Glossary of acronyms

AAAS	American Association for the Advancement of Science
AABW	Antarctic Bottom Water
ACC	Antarctic Circumpolar Current
ACMRR	Advisory Committee on Marine Resources Research
ADBEX	Antarctic Division BIOMASS Experiment
AZ	Antarctic Zone
BAS	British Antarctic Survey
BIOMASS	Biological Investigations of Marine Antarctic Systems and Stocks
BOFS	Biogeochemical Ocean Flux Study
BSBBW	Bransfield Strait Basin Bottom Water
BWPFE	BIOMASS Working Party on Fish Ecology
CCAMLR	Commission for (*or* Convention on) the Conservation of Antarctic Marine Living Resources
CCSZ	Coastal and Continental Shelf Zone
CDW	Circumpolar Deep Water
CPUE	catch per unit effort
CSW	Continental Shelf Water
DEE	Daily Energy Expenditure
DMS	dimethylsulphide
DMSP	β-dimethylsulphoniopropionate
DOM	dissolved organic matter
EASIZ	Ecology of the Antarctic Sea-Ice Zone
EEZ	Exclusive Economic Zone
ENSO	El Niño – Southern Oscillation
EPOS	European *Polarstern* Study
FAP	fluorescent age pigment
FIBEX	First International BIOMASS Experiment
FRAM	Fine Resolution Antarctic Model
GEOSECS	Geochemical Ocean Sections Study
GLOBEC	Global Ocean Ecosystem Dynamics Programme

GOSEAC	Group of Specialists on Environment and Conservation
HNLC	high-nutrient, low-chlorophyll (waters)
IABO	International Association for Biological Oceanography
ICES	International Council for the Exploration of the Sea
IFZ	Ice-Free Zone
IGBP	International Geosphere–Biosphere Programme
IGY	International Geophysical Year (1956–7)
IOC	Intergovernmental Oceanographic Organization
ICSU	International Council of Scientific Unions
ISAS	International Survey of Antarctic Seabirds
IUCN	International Union for the Conservation of Nature
IWC	International Whaling Commission
IWSOE	International Weddell Sea Oceanographic Expedition
JARE	Japanese Antarctic Research Expedition
JGOFS	Joint Global Ocean Flux Study
LOICZ	Land–Ocean Interactions in the Coastal Zone
LME	large marine ecosystem
MIZ	Marginal Ice Zone
MVBS	mean volume backscattering strength
NMDS	non-metric multidimensional ordination
NOAA	National Oceanic and Atmospheric Administration
PF	Polar Front
PFZ	Polar Frontal Zone
POM	particulate organic matter
POOZ	Permanently Open Ocean Zone
PPIZ	Permanent Pack-Ice Zone
PSU	Practical Salinity Unit
RACER	Research on Antarctic Coastal Ecosystem Rates
SACW	South Atlantic Central Water
SAF	Sub-Antarctic Front
SAZ	Sub-Antarctic Zone
SCAR	Scientific Committee on Antarctic Research
SC-CAMLR	Scientific Committee for the Conservation of Antarctic Marine Living Resources
SC-CCAMLR	Scientific Committee of the Commission for the Conservation of Antarctic Marine Living Resources
SCOR	Scientific Committee on Oceanic Research
SIBEX	Second International BIOMASS Experiment
SIBW	South Indian Basin Bottom Water
SIZ	Seasonal Ice Zone
SL	standard length
SPIZ	Seasonal Pack-Ice Zone
STF	Subtropical Front
SSW	Summer Surface Water
SW	Surface Water *or* Shelf Water
TAAF	Terres Australes et Antarctiques Francaises
UML	upper mixed layer
UNCLOS	United Nations Convention on the Law of the Sea
WCRP	World Climate Research Programme
WDW	Warm Deep Water
WEBW	Weddell–Enderby Basin Bottom Water

WG-CEMP	Working Group for the CCAMLR Ecosystem Monitoring Programme
WG-FSA	Working Group on Fish Stock Assessment
WG-Krill	Working Group on Krill
WOCE	World Ocean Circulation Experiment
WSC	Weddell–Scotia Confluence
WW	Winter Water

Foreword

The BIOMASS Colloquium, held in Bremerhaven, Germany, in September 1991, was the culmination of the major programme of Biological Investigations on Marine Antarctic Systems and Stocks, formulated in 1976, and directed to improving understanding of the ecology of the vast Southern Ocean. The origins of the programme lie in the establishment, in 1972, of a Subcommittee of the SCAR Working Group on Biology, on Marine Living Resources of the Southern Ocean. In 1975 this Subcommittee was upgraded by SCAR to a Group of Specialists on Living Resources of the Southern Ocean, which was later renamed the SCAR Group of Specialists on Southern Ocean Ecosystems and their Living Resources, in co-sponsorship with SCOR (as Working Group 54), IABO and ACMRR.

A keystone species of the Antarctic marine ecosystem is a small shrimp-like organism called krill (*Euphausia superba*), potentially a major human food source that may provide an important supply of protein for the future. Krill are crucial to the Antarctic pelagic food chain, providing food for the higher species including whales, seals, penguins, fishes and squid.

The BIOMASS Programme was a major collaborative effort of scientists from many nations; one of its main objectives was to provide an adequate body of knowledge for the wise management of the living resources of the Southern Ocean surrounding the continent of Antarctica. It grew out of the scientific community's concern for maintaining the underlying ecological relationships between the living organisms in this ocean. The BIOMASS Programme facilitated the achievement of an international convention – the 1980 Convention on the Conservation of Antarctic Marine Living Resources (CCAMLR) – as part of the Antarctic Treaty System. It has been an important source of information for the Scientific Committee of CCAMLR; the BIOMASS investigators were commissioned to produce two key reports for CCAMLR, on the *Biology and Status of Exploited Antarctic Fish Stocks*, and on the *Biology and Ecology of the Antarctic Krill*, both published as BIOMASS Scientific Series 6 and 9, respectively.

BIOMASS was one of the first international collaborations of scientists attempting to provide information for the formulation and application of multispecies management policies, before an ecosystem is seriously threatened or destroyed by commercial exploitation. Because of the immensity of the circumpolar Southern Ocean – some fifty million square kilometres – and its participating membership of eleven countries, the programme was truly international in character.

The overall programme initiated, promoted and drew together marine scientific studies by the participating nations. Scientists standardized methodology, planned programmes, exchanged information in workshops, coordinated the work of joint expeditions and made provision for data archiving and management by setting up the BIOMASS Data Centre. Two multiship, multinational cruises – FIBEX and SIBEX (First and Second International BIOMASS Experiments) – took place in 1981 and 1984–5. Some 32 international workshops were held around the world. The BIOMASS Programme also led to an impressive number of publications, including many valuable contributions in scientific journals, in addition to 68 Reports in the BIOMASS Report Series, 23 BIOMASS Handbooks, 10 volumes in the BIOMASS Scientific Series, and 25 issues of the BIOMASS Newsletter. The Programme transformed knowledge of the ecology and oceanography of the Southern Ocean, but has also had an impact on scientific research far beyond that region.

In the process the BIOMASS Programme trained and encouraged many young scientists from a variety of nations, disciplines and backgrounds, bringing them together in workshops, as invited participants in the Group of Specialists' meetings, and at the Colloquium.

The administrative costs of this major programme extending over fifteen years totalled less than US$600 000, that is less than $40 000 a year, contributed predominantly by the Antarctic Operating Agencies, but also by the International Council of Scientific Unions (ICSU) and its Scientific Committees on Antarctic (SCAR) and Oceanic (SCOR) Research. The leadership of this intensive and extensive international programme was in the hands of Professor S. Z. El-Sayed (with a tiny Secretariat at Texas A&M University) and the members of the BIOMASS Executive, to whom, and to the many Chairmen, Convenors and contributing scientists, we offer our appreciation and thanks.

In this volume, arising from the contributions to the BIOMASS Colloquium, some of the many accomplishments of the Programme are described, including aspects of physical oceanography; estimates of krill stocks, advances in krill biology, physiology and biochemistry; contributions to the knowledge of organisms at higher trophic levels, particularly fish and birds; and the operation and achievements of the BIOMASS Data Centre. There was remarkable unanimity among those attending the Colloquium that the BIOMASS Programme had been a resounding success and that related programmes, such as the International Geosphere–Biosphere Programme (IGBP) and its core programme the Joint Global Ocean Flux Study (JGOFS), now under way, as well as others now being planned (e.g. the Global Ocean Ecosystem Dynamics Programme, GLOBEC), could build on the BIOMASS data and experience. SCAR and SCOR are proud to have been closely involved in this success story.

Richard M. Laws
President, SCAR

Jarl-Øve Stromberg
President, SCOR

Preface

This book is the proceedings of the BIOMASS Colloquium held at the Alfred Wegener Institute, Bremerhaven, Germany, in September 1991. It is a record of accomplishments of the international BIOMASS programme during the decade of the 1980s. It traces the historical context and circumstances that led to the establishment in 1972 of the subcommittee on the Marine Living Resources of the Southern Ocean (of the SCAR Working Group on Biology), and the subsequent upgrading, in 1976, of that subcommittee to the SCAR Group of Specialists on Southern Ocean Ecosystems and Their Living Resources. That group was charged with planning, developing and implementing the BIOMASS Programme.

The book consists of 22 peer-reviewed chapters in six sections. The chapters' arrangement follows closely the organizational structure of the Colloquium. The opening chapter gives a short narrative of the history of BIOMASS, which is followed by description of the physical and chemical settings of the two regions of the Southern Ocean that were the main foci of the BIOMASS field work, namely, the Southwest Atlantic Ocean and the Prydz Bay area (in the Indian Sector). This is followed by chapters on the phytoplankton and zooplankton (not including krill). Because of the great attention given to krill research during BIOMASS, the several papers presented at the Colloquium underscore the preoccupation of the BIOMASS investigators with the Antarctic's keystone organism. The chapters on higher trophic levels, namely fish and birds, document BIOMASS accomplishments in ichthyo- and avifauna research. The interactions among these trophic levels are highlighted in the chapters presented on the subject.

The final chapters are intended as a look to the future. Speakers were asked to respond to the challenge presented by the Convenor, who asked 'Is there life after BIOMASS?' Predictably, the speakers rose to the challenge and provided thought-provoking and stimulating commentaries.

For each of the Colloquium's six sessions, discussants were invited to present their impressions on the papers presented, and to evaluate their strengths and shortcomings. They were later asked to summarize their evaluations, which we are pleased to include under the heading 'Discussant's report' at the end of each section.

As is the case in similar multinational, multidisciplinary programmes whose lifespans extend more than a decade, and in which hundreds of researchers are involved, it is natural to expect that most of the chapters in the book are essentially review papers of the state-of-the-art knowledge in the various disciplines at the end of BIOMASS. This was the charge given beforehand to the authors, all of whom made special effort to comply with that charge.

It is indeed a pleasure to acknowledge with deep gratitude the valuable contributions to this volume made by the authors and discussants. I am very much indebted to the several reviewers for their diligence and thoroughness, and for making helpful suggestions and thoughtful comments for improvement of the manuscripts.

Grateful acknowledgements should be extended to those members of the BIOMASS community at large, to the members of the SCAR Group of Specialists on Southern Ocean Ecosystems and Their Living Resources, to the Chairmen and members of the BIOMASS Technical Groups, Working Parties and *ad hoc* Groups. Without their interest, enthusiasm and devotion to the cause of BIOMASS, the programme would not have become the reality and the success we believe it was.

Finally, The British Antarctic Survey deserves our appreciation for housing and supporting the operational aspects of the BIOMASS Data Centre. Special thanks are extended to the two main sponsors of BIOMASS, SCAR and SCOR, and to the contributors to the BIOMASS Special Fund, who over the long duration of the programme have given generously to finance almost all aspects related to BIOMASS: its workshops, meetings and publications, including subsidy of the present book.

Sayed Z. El-Sayed
College Station, Texas, U.S.A.
July 1992

Contributors

Dr F. Brandini
Centro de Biologia Marinha, Universidade Federal do Parana, Trav. Alfredo Bufren, 140-terrer, Curitiba, Parana 80 020, Brazil.

Professor Andrew Clarke
British Antarctic Survey, High Cross, Madingley Road, Cambridge CB3 0ET, UK.

Dr John Cooper
Percy FitzPatrick Institute of African Ornithology, University of Cape Town, Rondebosch 7700, South Africa.

Dr John P. Croxall
British Antarctic Survey, High Cross, Madingley Road, Cambridge CB3 0ET, U.K

Dr Guy Duhamel
Museum National d'Histoire Naturelle, 43 rue Cuvier, F-75231 Paris Cedex 05, France.

Dr Sayed Z. El-Sayed
Department of Oceanography, Texas A&M University, College Station, TX 77843-3146, U.S.A.

Dr Inigo Everson
British Antarctic Survey, High Cross, Madingley Road, Cambridge CB3 0ET, U.K.

Dr E. Fahrbach
Alfred-Wegener-Institut für Polar- und Meeresforschung, Am Handelshafen 12, Columbusstrasse, D-2850 Bremerhaven, Germany.

Dr G. E. Fogg
School of Ocean Sciences, University College of North Wales, Marine Science Laboratories, Menai Bridge, Anglesey, Gwynedd LL59 5EY, U.K.

Dr Mitsuo Fukuchi
Associate Professor of Marine Ecology, Physiology and Ecology Department, National Institute of Polar Research, 9–10 Kaga 1-chome, Itabashi-ku, Tokyo 173, Japan.

Dr G. Hempel
Zentrum für Marine Tropenökologie, Universitätsallee GW1/A, 2800 Bremen 33, Germany.

R. B. Heywood
British Antarctic Survey, High Cross, Madingley Road, Cambridge CB3 0ET, U.K.

Dr G. Hosie
Krill Research Group, Australian Antarctic Division, Channel Highway, Kingston 7050, Tasmania, Australia.

Dr Gerd Hubold
Institut für Seefischerei, Bundesforschungsanstalt für Fischerei, Palmaille 9, D-2000 Hamburg 50, Germany.

Dr George Hunt
Department of Ecology and Evolutionary Biology, School of Biological Sciences, University of California, Irvine, CA 92717, U.S.A.

Dr Jean-Claude Hureau
Museum National d'Histoire Naturelle, 43 rue Cuiver, F-75231 Paris Cedex 05, France.

Dr Guy Jacques
Ministere des Universités, Université Pierre et Marie Curie, Laboratoire Arago, 66650 Banyuls-sur-Mer, France.

Dr Janusz Kalinowski
Centro Nationale di Ricerca, Instituto di Ricerca sulla Pesca Maritima, Molo Mandracchio, 60100 Ancona, Italy.

Dr Adi Kellermann
Geschäftsstelle Ökosystemforschung, Wattenmeer, c/o Umweltbundesamt, Bismarckplatz 1, 1000 Berlin 33, Germany.

Dr Karl-Hermann Kock
Institut für Seefischerei, Bundesforschungsanstalt für Fischerei. Palmaille 9, D-2000 Hamburg 50, Germany.

Dr Richard M. Laws
Master, St Edmund's College, Mount Pleasant, Cambridge CB3 0BN, U.K.

Dr M. Lipski
Department of Polar Research, Institute of Ecology, Polish Academy of Sciences, Dziekanow Lesny, 05-092 Lomianki, Poland.

Dr Harvey J. Marchant
Australian Antarctic Division, Channel Highway, Kingston 7050, Tasmania, Australia.

Mr Denzil Miller
Sea Fisheries Research Institute, Private Bag X2, Rogge Bay 8012, Cape Town, South Africa.

Dr Armando Mujica
Facultad de Ciencias de Mer, Universidad Catolica del Norte, Coquimbo, Chile.

Dr Eugene J. Murphy
British Antarctic Survey, High Cross, Madingley Road, Cambridge CB3 0ET, U.K.

Dr A. W. North
British Antarctic Survey, High Cross, Madingley Road, Cambridge CB3 0ET, U.K.

Dr Julian Priddle
British Antarctic Survey, High Cross, Madingley Road, Cambridge CB3 0ET, U.K.

Dr L. Quetin
Marine Science Institute, University of California, Santa Barbara, CA 93106, U.S.A.

Dr Robin M. Ross
Marine Science Institute, University of California, Santa Barbara, CA 93106, U.S.A.

Dr Egil Sakshaug
Trondhjem Biological Station, University of Trondheim, Bynesveien 46, N-7018 Trondheim, Norway.

Dr Sigrid Schnack-Schiel
Alfred-Wegener-Institut für Polar- und Meeresforschung, Am Handelshafen 12, Columbusstrasse, D-2850 Bremerhaven, Germany.

Dr Kenneth Sherman
National Oceanic and Atmospheric Administration, National Marine Fisheries Services, Northeast Fisheries Center, Narragansett Laboratory, South Ferry Road, Narragansett, RI 02882-1199, U.S.A.

Dr Yasu Shimadzu
Senior Research Coordinator, Research Division, National Research Institute of Fisheries Science, Fisheries Agency, Ministry of Agriculture, Forestry and Fisheries, 5-1 Kachidoki 5-chome, Chuo-ku, Tokyo 104, Japan.

Dr Volker Siegel
Institut für Seefischerei, Bundesforschungsanstalt für Fischerei, Palmaille 9, D-2000 Hamburg 50, Germany.

Dr Neville Smith
Bureau of Meterology Research Centre, Box 1289K, Melbourne, Victoria 3001, Australia.

Mr Manfred Stein
Institut für Seefischerei, Bundesforschungsanstalt für Fischerei, Palmaille 9, D-2000 Hamburg 50, Germany.

Dr J.-Ø. Stromberg
Kristineberg Marine Biological Station, Kristineberg 2130, S-45034 Fiskebackskil, Sweden,

Mr Mark R. Thorley
BIOMASS Data Centre, British Antarctic Survey, High Cross, Madingley Road, Cambridge CB3 0ET, U.K.

Dr Phil N. Trathan
British Antarctic Survey, High Cross, Madingley Road, Cambridge CB3 0ET, U.K.

Dr Paul Tréguer
Institut d'Etudes Marines, Université de Bretagne Occidentale, 6 avenue V. Le Gorgeu, F-29287 Brest Cedex, France.

Dr R. Williams
Australian Antarctic Division, Channel Highway, Kingston 7050, Tasmania, Australia.

Mr Eric J. Woehler
Australian Antarctic Division, Channel Highway, Kingston 7050, Tasmania, Australia.

(a)

(b)

(c)

(d)

(e)

(f)

(g)

(h)

(i)

(j)

(k)

(l)

(m)

(n)

(o)

Members of the BIOMASS fleet: (a) ARA *Almirante Irizar* (Argentina); (b) MV *Nella Dan* (Australia); (c) RV *Professor W. Besnard* (Brazil); (d) RV *Capitan Alcazar* (Chile); (e) *Marion Dufresne* (France); (f) *Meteor* (Germany); (g) *Polarstern* (Germany); (h) *Umitaka Maru* (Japan); (i) RV *Kaiyo Maru III* (Japan); (j) *Professor Siedlecki* (Poland); (k) RV *Africana* (South Africa); (l) MV *S.A. Agulhas* (South Africa); (m) *Academic Knipovich* (U.S.S.R.); (n) RRS *John Biscoe* (U.K.); (o) *Melville* (U.S.A.).

History, organization and accomplishments of the BIOMASS Programme

S. Z. EL-SAYED

Background and history of the BIOMASS Programme

IN JULY 1968, on the occasion of holding the Second SCAR Biology Symposium at Scott Polar Research Institute, Cambridge, U.K., two of the Symposium's fourteen sessions dealt with plankton and the pelagic resources of the Southern Ocean. It was obvious from the six krill review papers presented (mostly by Russian scientists) that little was known of the biology, ecology and population dynamics of what was perceived as one of the most significant creatures in the Antarctic marine ecosystem. A few in the audience took note of this deficiency in our knowledge, not only with regard to krill, but also the Southern Ocean ecosystem as a whole. This lack of knowledge spurred a group of U.S. scientists to organize in 1972 the first multidisciplinary oceanographic cruise on board the USNS *Eltanin*. The objective of the cruise was to study the structure and function of the Ross Sea ecosystem (El-Sayed, 1973).

Between the time the SCAR Symposium was held and the *Eltanin* Cruise 51 was under way, the krill fishery was still in its infancy, though showing every sign of potential expansion. On the other hand, the harvest of other Antarctic marine living resources (e.g. finfish) was in full swing (Kock & Shimadzu, this volume). It was then that concern over the need for proper management and conservation of these resources was voiced by members of the scientific community and national and international agencies and organizations. The concern of scientists for the conservation of Antarctic marine living resources, and in particular krill, stemmed from the fact that despite past research activities, notably those of the *Discovery* Investigations in the late 1920s and 1930s, there was no adequate information on the stocks of these resources. However, effective conservation and management depended upon a continuing flow of data and information regarding these stocks.

In this regard, one has to recall that in the early 1970s a fierce debate was raging between the preservationists and the scientific community. On one hand, the preservationists perceived development in the Antarctic as a potential threat to the ecosystem, or the manifestation of the 'boom and bust' syndrome which has occurred throughout much of the world fisheries during the past several decades. The scientific community, on the other hand, has a vested interest in the Antarctic in that it would like to retain this almost 'pristine ecosystem' as an exclusive 'test tube environment' for future study. Beyond these publicly discussed issues are the peripheral ones associated with the balance of power in the Antarctic jurisdictional matters, and the potential relationship between utilization of living resources and the non-living resources (oil and minerals) of the Antarctic continent and its continental shelf.

Realizing that unwise and unregulated past exploitation had decimated the Southern Fur Seal and baleen whale populations, and recognizing krill's key position in the Southern Ocean food web and its impending exploitation, the Scientific Committee on Antarctic Research (SCAR) foresaw a need for substantial expansion of scientific research on the Antarctic marine ecosystem. Thus, in August 1972, SCAR established a subcommitee (of the Biology Working Group) on the Marine Living Resources of the Southern Ocean. The charge of the subcommittee was to assess the present state of knowledge of the Antarctic marine ecosystem and to stimulate investigations of the ecology and population dynamics of the organisms at different trophic levels. The subcommittee held its first meeting in May 1974 at McGill University (Montreal, Canada) in conjunction with the SCOR/SCAR Polar Oceans conference. At this meeting the subcommittee developed a series of recommendations, one of which stated:

> We realize the enormous difficulties of undertaking the formidable task of studying the marine ecosystem of the vast expanses of the Southern Ocean when only one or two oceanographic ships are utilized. We, therefore, recommend that a well co-ordinated International Biological Investigation of the Southern Ocean (IBISO), utilizing multiple ships, be undertaken . . .

The Montreal recommendations were approved by the Thirteenth Meeting of SCAR in September 1974, and by the International Coordination Group for the Southern Ocean of the Intergovernmental Oceanographic Organization (IOC) at its meeting in Buenos Aires, Argentina, in July 1974. Thus, in March 1975, the Executive Council of IOC invited the SCAR subcommittee:

> . . . to prepare practical proposals for collaborative investigations in biological oceanography in the area and to prepare proposals for working toward the organization of multiship studies for submission to SCAR and SCOR and through them to the IOC International Co-ordination Group . . .

At the Eighth Antarctic Treaty Consultative Meeting held in Oslo, Norway, in June 1975, it was recommended that the governments:

> . . . invite SCAR to continue its scientific work on these matters and to consider convening, as soon as practicable, a meeting to discuss current work and report on progress for the study of conservation of Antarctic marine living resources . . .

In recognition of these new responsibilities, the SCAR Executive at its meeting in June 1975 upgraded the status of the marine living resources subcommittee to a group of specialists. The group was later reorganized as the SCAR Group of Specialists on Southern Ocean Ecosystems and Their Living Resources. In November of the same year, the Executive Committee of SCOR (Scientific Committee on Oceanic Research) agreed to co-sponsor the group and adopted it as Working Group 54. Other sponsors included the International Association for Biological Oceanography (IABO) and the Advisory Committee on Marine Resources Research (ACMRR) of the Food and Agriculture Organization (FAO).

In order to initiate action in relation to IOC and Antarctic Treaty invitation, the first meeting of the reconstituted Group of Specialists was convened at the Scott Polar Research Institute, Cambridge, 6–8 October 1975. During this meeting, the group agreed that its first task should be to review the existing information and to bring together knowledge of ongoing and presently planned programmes of marine biological investigations. The group also undertook the preparation of practical proposals for long-term cooperative investigations of the Southern Ocean. The group welcomed the offer of the United States to host a scientific meeting on the living resources of the Southern Ocean.

Thus, in August 1976, the U.S. National Academy of Science hosted in Woods Hole, Massachusetts, the First International Conference on Living Resources of the Southern Ocean. The chief objective of the Conference was to review present knowledge of the living resources of the Southern Ocean and to develop a proposal for future cooperative studies in the area. This proposal became known as the Biological Investigations of Marine Antarctic Systems and Stocks (BIOMASS).

In the proposal, members of the Group of Specialists outlined in great detail several research programmes which they envisaged to be significant in contributing to a better understanding of the physics, chemistry and biology of the Southern Ocean. These included: hydrography, nutrient chemistry, phytoplankton, zooplankton, krill, cephalopods, fish, seaweeds, benthic invertebrates, birds and modelling. Although there were sections in the BIOMASS document dealing with marine mammals (i.e. seals and whales), it was felt that since research on these animals came under the purview of the Seal and Whaling Conventions, they should be left out of the BIOMASS programme.

BIOMASS implementation

The BIOMASS programme was originally conceived as a ten-year research programme during which two large-scale oceanographic surveys were conducted. The austral summer of 1980–1 was chosen for the First International BIOMASS Experiment (FIBEX) in which 13 ships from 11 nations participated in the largest biological oceanographic expedition ever mounted in the Southern Ocean. The Second International BIOMASS Experiment (SIBEX), Phase I (1983–4) and Phase II (1984–5), were the final collaborative field efforts of this ambitious programme.

For nearly ten years, until its demise in 1985, the Group of Specialists on Southern Ocean Ecosystems and Their Living Resources, with substantial assistance from its network of Technical Groups, Working Parties and *ad hoc* Groups, provided the guidance and coordination needed to ensure the success of the programme. The Group succeeded in organizing numerous meetings and workshops, and was instrumental in establishing (in 1985) the BIOMASS Data Centre.

Flushed with a great sense of accomplishment after the successful completion of the last of the BIOMASS field exercises, the Group of Specialists held in June 1985, in Dammarie-les-Lys, France, what turned out to be its last meeting. For in August of that year, at its annual meeting, the then SCAR Executive unexpectedly proposed to the other cosponsors a move to disband the Group of Specialists! Thus the Group was effectively disbanded along with its Technical Groups, Working Parties and *ad hoc* Groups. However, to soften this rash act, the SCAR Executive agreed to create a five-man BIOMASS Executive to manage the newly established BIOMASS Data Centre, to plan and organize future data analysis workshops and to oversee the last function of the programme, i.e. the holding of the BIOMASS Colloquium.

Objectives of BIOMASS

The principal objective of the BIOMASS Programme is *to gain a deeper understanding of the structure and dynamic functioning of the Antarctic marine ecosystem as a basis for the future management of potential living resources* (El-Sayed, 1977). The two main components of that objective are: (i) to contribute to man's understanding of the Southern Ocean and its biota; and (ii) to develop a sound ecological strategy on which to base future management decisions, thereby ensuring rational exploitation and the continued maintenance of the important ecological relationship between exploited and unexploited stocks.

From the outset it was recognized that part of the BIOMASS programme would be carried out by individual investigators or groups of investigators studying topics which they found particularly interesting and challenging. However, it was also recognized that, to achieve the overall objective, it was essential to establish a firm framework as to the structure of the investigations. The ultimate objective would be achieved by the production of a predictive model of the whole system. It was clear that in this model krill would play a central role, both as the major herbivore and as the dominant food source for most of the higher organisms.

A general model would be developed in two stages. First would be models describing the population dynamics of the main species in the system. These would require estimation of the biomass of each population, its rates of change and, as far as possible, estimation of the parameters of growth, mortality, and reproduction. Such models are well developed for some mammals and birds but need to be extended, especially for krill. A predictive submodel of krill production at different levels of exploitation should be developed, first for the Atlantic sector, taking into account fluctuations and stock identity. At the second stage, multispecies models would be developed requiring quantitative information on the interactions between the dominant species (food consumption, etc.).

BIOMASS Programme priorities

In developing the BIOMASS Programme, the Group of Specialists agreed that without ignoring

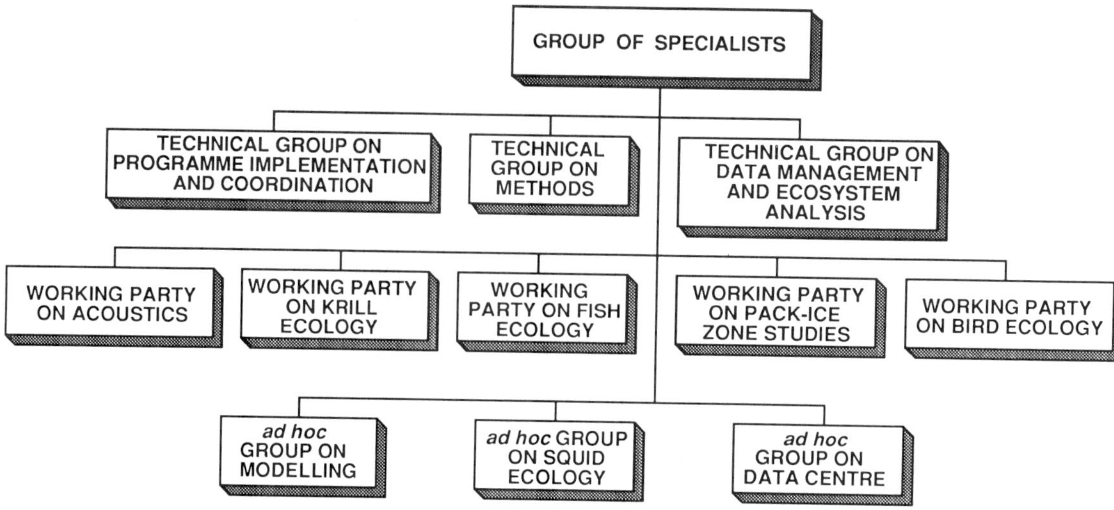

Fig. 1. BIOMASS organizational structure.

other problems, there was an urgent need to address the problem of resource evaluation of krill, and proposed that apart from the study of the life history of krill, the following questions should be given priority.

How much krill is there?
What are the trends and fluctuations in krill abundance?
What is the population identity of krill?
What is the swarming behaviour of krill?
To what extent are krill food-limited?
How are the total populations of krill consumers subdivided into unit populations?
How much krill is consumed by each predator species?
What are the present trends in the stocks of consumers?
What will be the impact of krill harvesting at different levels on the dynamics of the populations of krill consumers?

Coordination of the BIOMASS Programme

As mentioned earlier, the coordination of BIOMASS was achieved by: a network of **Technical Groups**, which had a continuous role for the duration of BIOMASS; **Working Parties**, which were needed for a particular phase of the programme; and ***ad hoc* Working Groups**, whose tasks were quite specific and normally of short duration. Fig. 1 shows the organizational structure of the BIOMASS Programme.

It is interesting to note in this respect that, although the Technical Group on Programme Implementation and Coordination had only seven months to plan FIBEX, all vessels were in place on schedule using recommended equipment for detecting and sampling krill. This was a singular achievement since all the coordination was done without the aid of a Secretariat, and long before the BIOMASS Data Centre was created!

The success of FIBEX in describing the distribution of krill on a large scale, and in estimating its abundance, is due in large measure to the Working Party on the Acoustic Estimation of Krill. Although all the FIBEX ships did not ultimately adopt the standard acoustic methodology, the Acoustic Working Party was at least able to ensure that there was sufficient standardization in such matters as instrumentation and data reporting to enable the data from all ships to be compared and combined. The degree to which every participating ship adopted the same sampling strategy and adhered to the main unifying theme was no small achievement.

First International BIOMASS Experiment (FIBEX)

The prime objective of FIBEX was to address the question 'How do we determine how much krill there is in the Antarctic?' This was the main unifying theme of the FIBEX project.

As a result of the Woods Hole Conference, it was known that the distribution of krill is very patchy and that the species sometimes occurs in large local concentrations ('swarms'). Thus, two further objectives were identified for FIBEX: 'How much of the total stock of krill occurs in swarms?' and 'What is the structure of krill swarms?'

It was also recognized that the habitat of Antarctic krill is so vast that acoustic methods were probably the only feasible way to obtain a satisfactory answer to the central question, and then only in relation to that segment of the stock which aggregates in swarms. It should be realized that the acoustic technique sees krill in terms of target strength, i.e. the ratio of the intensities of reflected to incident sound at one metre from the target. This parameter is influenced by three factors in particular, whose combined effect could lead either to an underestimate or to an overestimate of abundance. The first of these is the circulation aspect of animals in the water column; the second is carapace density, which could become important if there is moulting synchrony within a population; the third is echo-sound calibration (see Everson and Miller, this volume, for details).

It was also apparent that it was beyond the capacity and resources of FIBEX to survey the entire habitat of Antarctic krill, and so a regional approach was developed. Thus efforts were concentrated in conducting four survey areas reflecting, in the main, the availability of ships (Fig. 2). The most intensive survey area was the west Atlantic sector, where ships from Argentina, Chile, F.R.G., Poland, U.S.A. and U.S.S.R. were deployed. The other survey areas were the western Indian Ocean (France and South Africa); the eastern Indian Ocean (Australia and Japan); and the Pacific sector (Japan). The primary objectives of the krill acoustic survey were twofold: to examine the methodology for future assessments of total krill abundance, and to provide instantaneous pictures of the abundance of krill in the S.W. Atlantic and S.W. Indian sectors of the Southern Ocean.

Second International BIOMASS Experiment (SIBEX)

The success and shortcomings of FIBEX provided a useful basis of experience for planning SIBEX. Thus, while FIBEX provided a synoptic picture of the distribution of krill, SIBEX's aim was to obtain a deeper understanding of the dynamics of the krill-dominated part of the Antarctic ecosystem. Further, a prominent feature of the Antarctic marine ecosystem is the pack-ice. Its seasonal advance and retreat determine nutrient cycling and algal growth in large parts of the Southern Ocean. Pack-ice covers much of the krill population in winter, provides a platform for birds and seals, and acts as a barrier for most species of whales. It shelters the water column from the wind, cuts off surface aeration, and reduces the light available for photosynthesis.

Compared with the frequency of open-water studies, relatively little was done during FIBEX in terms of multidisciplinary study of pack-ice, ice-edge and adjacent open areas. This was mainly because so few of the participating vessels were adequately ice-strengthened. As a consequence, FIBEX investigators missed an important segment of the krill population. It is for these reasons that SIBEX planners gave special attention to the questions related to the pack-ice.

Another lesson learned from FIBEX was the detection of a number of 'superswarms' such as the one that was encountered by the U.S. vessel *Melville* (Macaulay *et al.*, 1984). As it appears that a substantial proportion of the krill stock could be contained within these swarms, the SIBEX planners undertook to study them more thoroughly during SIBEX. In particular, information was needed on the composition and identity of the swarms, and whether there is a control process of formation and dissolution of swarms. Information was also sought on the proportion of the total population contained in the swarms and how it varies.

Also as a result of FIBEX, it was realized that the acoustic estimates of krill biomass needed to be refined. This meant that calibration and intercalibration problems needed to be sorted out, and more reliable target strength information was required (Everson *et al.*, 1990). Moreover, it was noted that detailed analysis of the oceanographic data gathered during FIBEX was required to identify the water masses associated with regions of high krill abundance. In addition, further assessments were needed

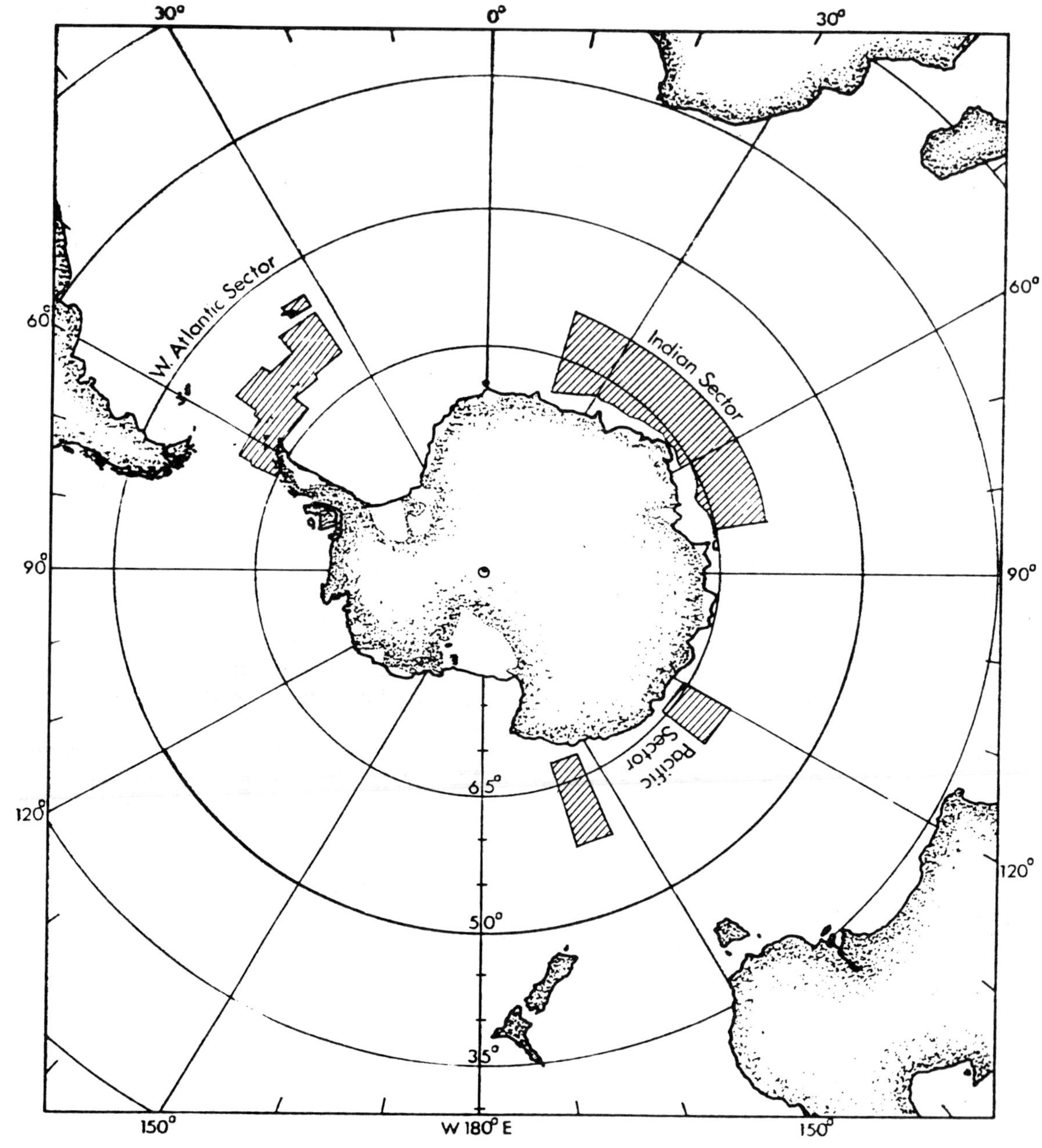

Fig. 2. The four study sites (shaded areas) where krill hydroacoustic surveys were carried out during FIBEX.

of the sexual maturity stages, particularly spawning status of krill populations encountered during FIBEX; and these distributions needed to be reconciled with data on their larval distribution.

Although krill studies still constituted the backbone of SIBEX investigations, it was appreciated that the research thrust of SIBEX must seek to balance the wider, long-term ecosystem perspective with more immediate management in research oriented, for example, on stocks of Antarctic fish which already showed signs of overfishing. It is for this reason that fish studies formed an integral part of SIBEX.

Thus, the primary objectives of SIBEX were to enhance our understanding of the dynamics of the krill-dominated part of the Antarctic marine ecosystems and to obtain a broad picture of krill-swarm distribution in relation to the prevailing mesoscale environmental features, particularly the advance and retreat of the sea-ice.

Although the field exercises of BIOMASS came to an end with the completion of SIBEX II, the post-SIBEX years were noted for the fish-stock assessment surveys by several nations, and the successful mounting of the European *Polarstern* Study (EPOS), and more recently the CCAMLR target-oriented cruises and surveys.

Accomplishments of BIOMASS

A careful appraisal of the accomplishments of the BIOMASS programme must await the completion of the analysis of the massive data sets that were collected during the tenure of the programme, and in particular the synthesis of the results obtained. Like any other 'historic event', one is still too close to that event to pass judgement on its success or shortcomings. Only time will tell whether the planners and architects of BIOMASS have failed or succeeded in their efforts. However, at least for the time being, all indications are that the programme has been a great success.

There are several yardsticks by which one can measure the success of the programme. These fall into two basic categories: scientific accomplishments, and international cooperation. As Professor Hempel (this volume) mentions, one needs to compare the scene before and after BIOMASS in order to appreciate the magnitude and impact this programme has played in transforming our knowledge of the Southern Ocean and its biota. Thus no effort will be made here to catalogue the accomplishments of BIOMASS as the 20 papers that follow will give a fairly reasonable account of the programme's successes and shortcomings. Further elaboration on BIOMASS achievements and failures is to be found in the six Discussants' Reports (this volume) as well as in Professor Fogg's (this volume) thoughtful evaluation of the programme.

Regarding the role BIOMASS has played in fostering international scientific cooperation, suffice it to say that within its short lifespan, BIOMASS has rekindled great international interest in Antarctic affairs reminiscent of the International Geophysical Year (IGY) in 1956–7. The era of individual national expeditions (e.g. *Belgica, Pourquoi pas?, Gauss, Meteor, Discovery, Ob* and *Eltanin*) has been replaced by well-coordinated, multiship, multinational expeditions. This is attested to by the mounting of the largest biological expedition ever undertaken in the Southern Ocean and by the numerous scientific expeditions to the Antarctic during the past few years by such newcomer nations as India, China, Italy, the Republic of Korea, and most recently, Spain, Peru, and Uruguay. This tradition of international cooperation has continued in the post-SIBEX years as exemplified by the EPOS cruise in 1988.

Further, the 16 BIOMASS workshops, symposia, and colloquia held in recent years in such places as Cambridge (U.K.), Tokyo (Japan), Bremerhaven (F.R.G.), Bariloche (Argentina), Wilmington (North Carolina, U.S.A.), Rio de Janeiro (Brazil), Ravello (Italy) and College Station (Texas, U.S.A.), where BIOMASS data and results have been discussed by scientists from the BIOMASS community, attest to the high degree of scientific cooperation, goodwill and camaraderie that has developed among the Antarctic community as a result of the BIOMASS programme. Perhaps the height of this cooperative spirit is best exemplified by the highly successful data analysis workshops where, in the best tradition of scientific cooperation, scientists from varied political persuasions and backgrounds have agreed to pool their unpublished data and subject them to communal analysis, and to collaborate later in joint publications.

It was in this atmosphere of international collaboration, and with the growing interest in the conservation of the Southern Ocean ecosystem and its living resources, that the *Convention for the Conser-*

vation of Antarctic Marine Living Resources (CCAMLR), of the Antarctic Treaty, was ratified in 1982. Everson and Miller (this volume) quite correctly pointed out that CCAMLR has its roots in BIOMASS. Indeed, both BIOMASS and CCAMLR are rooted in their concern over the development of a sound ecological strategy for the exploitation of living marine resources and for the conservation of the living resources of the Southern Ocean. Early on, BIOMASS planners realized that in developing successful fishery research programmes the entire ecosystem would need to be considered. The 'species-by-species approach' has proven inadequate and faulty (Sherman, this volume). Thus, any attempt at harvesting a single species without regard to possible effects on other components in the system may set up irreversible reactions that will result in major changes in the ecosystem. This ecological approach struck a responsive chord in CCAMLR and was put into immediate use by the drafters of the CCAMLR convention, who foresaw that harvesting levels should take into account the effect on the whole ecosystem, not only on the large but also on the dependent species (Croxall, this volume).

Is there life after BIOMASS?

Although the SCAR/SCOR Group of Specialists on Southern Ocean Ecosystems and Their Living Resources was disbanded in 1985, it became clear that some new international scientific group was required to continue to facilitate international cooperation in the Southern Ocean ecology and related fields. This led SCAR, with its extensive experience of fruitful coordination of such research, to establish (jointly with SCOR) in 1986 the SCAR Group of Specialists on Southern Ocean Ecology. The Group was charged with identifying important fields of research in Antarctic marine ecology and proposing cooperative studies. Another SCAR group is the Group of Specialists on the Antarctic Sea-Ice Zone, whose ten-year programme for an international collaborative study includes a biological component.

The third group is the governmental organization, CCAMLR, which has the mandate to conserve the living resources of the Antarctic within the context of the ecosystem. The fledgling CCAMLR has already made good progress towards implementing a well-designed predator-based monitoring study and has adopted several conservation measures in connection with fish resources around South Georgia. It has reached agreement regarding the protection of immature fish and has recently set a precautionary limit to krill catch (Everson and Miller, this volume).

Perhaps the greatest challenge facing any such new group or organization is, as Kellerman and North (this volume) point out, how to apply the scientific and managerial knowledge gained through the BIOMASS programme to enhance future international scientific work towards the understanding and conservation of the Antarctic marine ecosystem as an integral part of the Earth's biogeosphere. One hopes that the claimant heirs to BIOMASS would live up to this challenge, and would maintain the tradition that has been the essence of BIOMASS, namely of pushing forward the frontier of knowledge, and fostering international cooperation in Southern Ocean research.

Acknowledgements

This article is dedicated to the memory of the late Dr Mary Alice McWhinne, one of the pioneers in krill biology research, and Chief Scientist on USNS *Eltanin* Cruise 51, the first multidisciplinary oceanographic cruise in the Antarctic. It is also dedicated to Dr George A. Llano for the role he has played, while at the National Science Foundation, in funding both the *Eltanin* cruise and the First International Conference on Living Resources of the Southern Ocean, that gave the impetus to the BIOMASS programme and led to its development.

References

El-Sayed S Z (1973) *Biological Oceanography. Antarctic Journal of the U.S.*, Vol. VIII, No. 3, pp. 93–100. [Special issue commemorating *Eltanin*'s accomplishments.]

El-Sayed S Z (ed) (1977) *Biological Investigations of Marine Antarctic Systems and Stocks*, Vol. I. Scott Polar Research Institute, Cambridge, England. 79 pp.

Everson I, Watkins J L, Bone D G, Foote K G (1990) Implications of a new acoustic target strength for abundance estimates of Antarctic krill. *Nature* **345**:338–40.

Macaulay M D, English T S, Mathisen O A (1984) Acoustic characterization of swarms of Atlantic krill (*Euphausia superba*) from Elephant Island and Bransfield Strait. *J Crust Biol* **4** (Spec. No. 1): 16–44.

Antarctic marine environment: physical oceanography

Antarctic environment – physical oceanography: the Antarctic Peninsula and Southwest Atlantic region of the Southern Ocean

M. STEIN & R. B. HEYWOOD

Introduction

THE *DISCOVERY* Investigations (1925–39) clearly indicated that the biogeography of Antarctic marine organisms was largely governed by the dynamics of the water circulation (Deacon, 1937; Hart, 1934, 1942; Hardy & Gunther, 1935; Baker, 1954). Subsequent work has continued to reveal the considerable spatial and temporal variability occurring in the physical and chemical parameters of Antarctic waters. This variability must have a profound effect on the biota. Consequently, the BIOMASS Programme recognized the need for a concurrent study of the marine environment if the short-and long-term fluctuations in the distribution, abundance, productivity and behaviour of the organisms were to be understood (SCAR/SCOR, 1977). The environments of the seas off the Antarctic Peninsula and of the Scotia Sea are particularly complex. Waters of the Antarctic Circumpolar Current, the Weddell Gyre, and the Eastwind Drift intermingle, within an area of complex topography, which includes Drake Passage, Bransfield Strait and the Scotia Arc, to produce complicated mixtures of water masses.

This paper seeks to collate for the biologist descriptions of those features of these environments which will probably assist in interpreting biological observations. A detailed description of the physical processes involved has not been thought necessary; in most instances these are at best imperfectly understood (SCOR, 1985). Ship-time and resources restricted BIOMASS physical oceanography to a monitoring regime. Consequently, full use has been made of published data from physical oceanography projects outwith the BIOMASS programme to present as complete a picture as possible.

Antarctic Circumpolar Current

On leaving the confines of Drake Passage and entering the Scotia Sea the Antarctic Circumpolar Current (ACC) is deflected north and west by the submerged ridge of the Scotia Arc, before it turns east again to flow around South Georgia and across the Atlantic Ocean sector of the Southern Ocean (Deacon, 1933, 1937; Gordon & Goldberg, 1970). It consists of sub-Antarctic and Antarctic Surface waters from the Pacific Ocean sector of the Southern Ocean and from the Bellingshausen Sea (Clowes, 1934; Nowlin & Clifford, 1982). Some mixing with Weddell Sea Surface Water occurs in the Scotia Sea. However, the southern arm of the Scotia Arc effectively restricts the flow of water out of the Weddell Sea (Carmack & Foster, 1975) and most enters the Scotia Sea across the Weddell–Scotia Confluence east of the South Orkney Islands (Gordon, 1966, 1967). Consequently there is a marked gradient of mixing across the Scotia Sea with the influence of

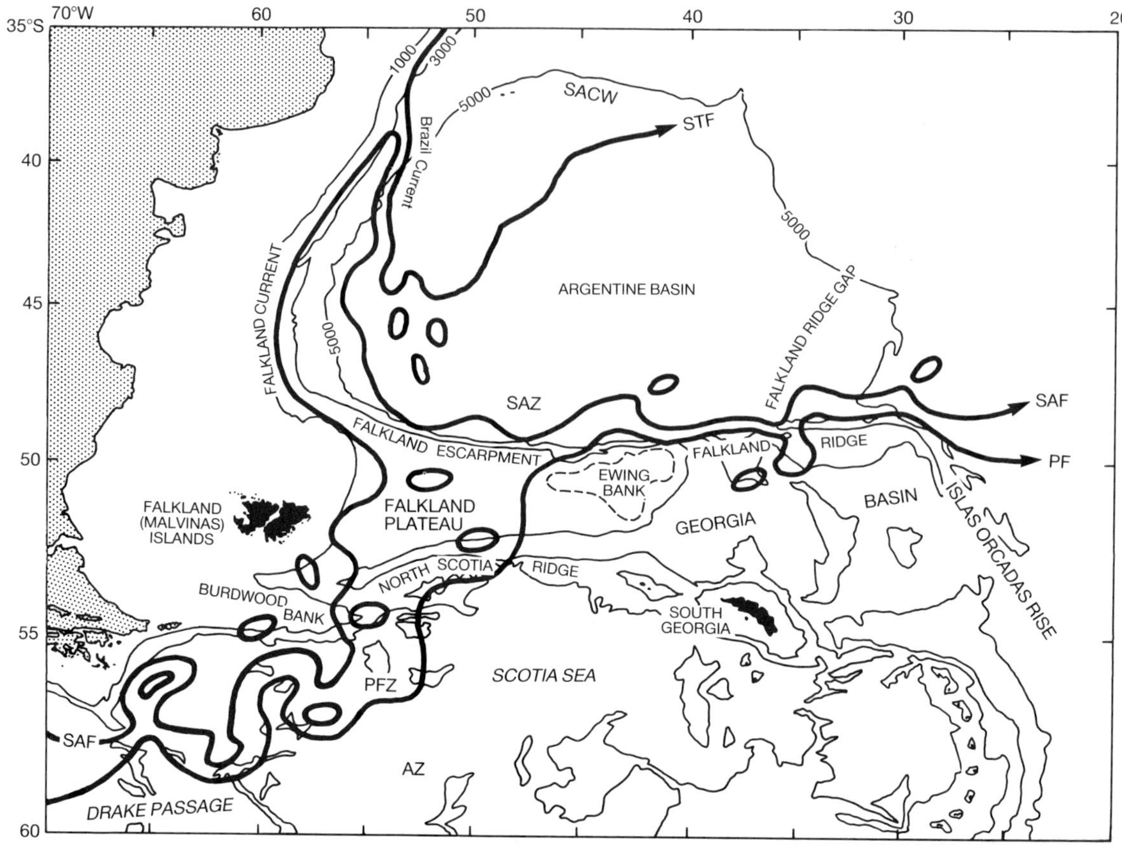

Fig. 1. The upper ocean water masses and fronts in the western South Atlantic (from Peterson and Whitworth, 1989).

SACW	South Atlantic Central Water	PF	Polar Front
		SAZ	Sub-Antarctic Zone
STF	Subtropical Front	PFZ	Polar Frontal Zone
SAF	Sub-Antarctic Front	AZ	Antarctic Zone

The broken line around Ewing Bank is the 2000 m isobath.

Weddell Sea Surface Water being greatest to the south and east of South Georgia (Anon, 1983; Heywood, 1985). The core salinities and silica concentrations of the original waters mixing in the ACC are very distinctive and permit the identification of their relative influence at any oceanographic station. The salinity of the temperature minimum layer, recorded during FIBEX, was approximately 34.00 Practical Salinity Units (PSU) for south-east Pacific Antarctic Surface Water leaving Drake Passage, approximately 34.50 PSU for central Weddell Sea Surface Water, and approximately 34.25 PSU for central Scotia Sea waters (Anon, 1983). Equivalent silica concentrations recorded during British Antarctic Survey cruises are 20–30 μM, 110–120 μM and 40–60 μM SiO_3-Si, respectively, with a particularly sharp transition from 116 μM to 78 μM across the Weddell–Scotia Confluence west of the South Orkney Islands (R. B. Heywood, unpublished data).

Peterson and Whitworth (1989) have described the two principal fronts within the ACC, the Sub-Antarctic Front (SAF) and the Polar Front (PF) with respect to their spatial patterns and relations to deep water masses, from the northern Drake Passage to the western and southern Argentine Basin. East of Drake Passage, the Fronts diverge rapidly. The SAF turns sharply northward as a part of the Falkland Current, while the PF remains oriented mainly toward the north-east (Fig. 1). The wide intervening Polar Frontal Zone (PFZ) is a complex of mesoscale eddies, the upper layers of which consist of Sub-Antarctic Mode Water from the Pacific. The SAF remains part of the Falkland Current until the confluence with the Brazil Current. The SAF then turns southward with the return flow, and then eastward along the northern edge of the Falkland Escarpment and Ewing Bank. Here the SAF and PF tend to merge to form a single intense current core strongly influenced by bottom topography. At times the eastward current loops southward through a gap in the Falkland ridge into the Georgia Basin, thereby supplying relatively warm and salty sub-Antarctic water to the Antarctic Zone. East of the gap, the two fronts become separated and are distinct features again in the vicinity of the Islas Orcadas Rise.

At depth, within Drake Passage and the western Scotia Sea, the SAF and PF are not lateral boundaries between distinct water masses, but are instead identified by enhanced vertical displacements of property isopleths.

Weddell–Scotia Confluence

The southern ridge of the Scotia Arc forms an effective barrier between the waters of the Weddell and Scotia Seas. Model (1958) named the boundary between the water regimes the 'Bellingshausen Front', but Gordon (1967) renamed it the 'Weddell–Scotia Confluence' because there was little evidence of either convergence or divergence. The zone varies from 10 to 100 km in width. Gordon (1989) suggests that the water is best considered as continental margin water, advected in the region along the extreme western edge of the Weddell Gyre. The water is relatively cold and low in salinity throughout the water column. Convective events tend to extend to considerable depths. Temperature profiles obtained during the International Weddell Sea Oceanographic Expedition (IWSOE) (1975), FIBEX (1981) and the Anglo-German Joint Biological Expedition (1982) show a very marked discontinuity extending throughout almost all the water column (Deacon & Moorey, 1975; Stein, 1981, 1983; Heywood, 1985).

Gordon (1967), in his review of data for Antarctic waters between 20°W and 170°W, showed the boundary extending outside the South Shetland Islands in the southern Drake Passage. Schröder (1983) used data from RV *Meteor* (1980) to identify a sharp salinity front between Weddell Sea Surface Water and warmer, less saline waters, which he considered to be the Weddell–Scotia Confluence within the Bransfield Strait. Heywood (1985) obtained similar evidence during the Anglo-German Cruise in 1982. SIBEX (Anon, 1990) provided considerable evidence to show that the position of the western arm of the Confluence could vary considerably over a brief period of time as well as annually.

The seas off the west coast of the Antarctic Peninsula

The physical environment is complex and very variable in this region, where several major water masses meet and interact with a continental shelf cut by numerous underwater valleys and basins (Heywood, 1985; Popkov, 1987; Solyankin & Spiridonov, 1987). Surface water from the South-East Pacific Basin and the Bellingshausen Sea approaches the coast in the ACC. It is turned southwestwards under the influence of the easterly winds and the submarine topography to reform the current of the Eastwind Drift, which has been interrupted by the Antarctic Peninsula and the South Shetland Islands. This westward countercurrent, which extends throughout the depth of the water column in the abyss off the continental shelf, has been described by Gordon (1966), Hollister & Elder (1969) and Whitworth *et al.* (1982). Heywood (1985) showed that the north-easterly coastal movement of the cold Bellingshausen (winter) Surface Water was stopped by the shallow sill at the entrance to Bismarck Strait in February 1982 (Fig. 2). Local accumulations of icebergs, pack-ice and glacial melt cooled and diluted the surface water, and the influence of the Warm Deep Water varied in general with the depth

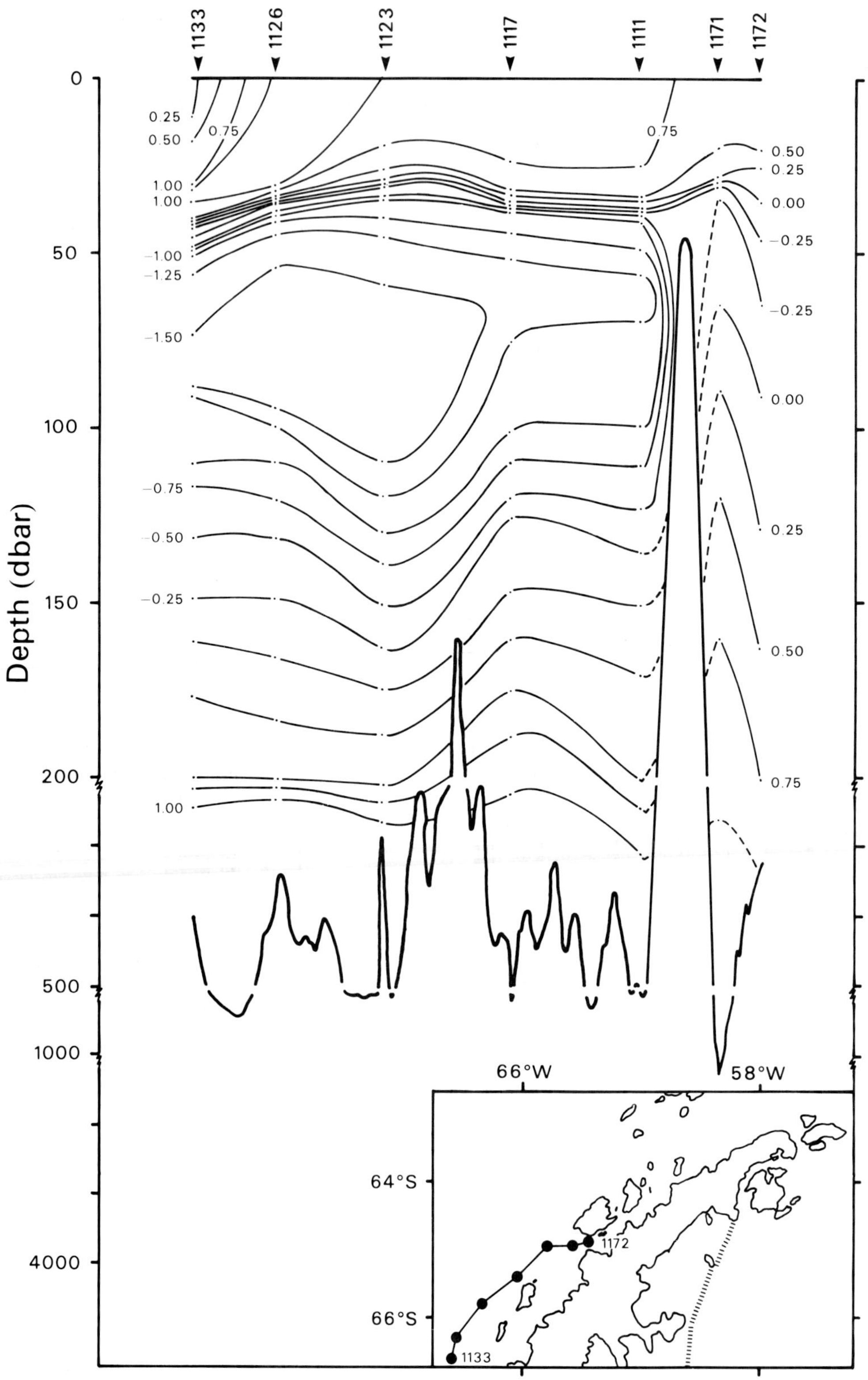

Fig. 2. Potential temperature isotherms along a transect running northwards out of the Bellingshausen Sea along the coast of the Antarctic Peninsula, February 1982. (From Heywood, 1985.)

of water. Movement of water into some straits was restricted by high sills, and cold saline Bottom Water had formed in the small but deep basins. Stein (1992) investigating upper ocean geostrophic shear patterns off the Antarctic Peninsula region, obtained evidence for the existence of cyclonic gyres, which might indicate places of permanent local upwelling.

Bransfield Strait

Antarctic Surface Water from the Weddell Sea enters Bransfield Strait around Joinville and d'Urville Islands, and flows in geostrophic balance southward along the northern coast of the Antarctic Peninsula. At about 61°W, the main basin of the Strait is blocked by a chain of islands and associated shelf, stretching from Brabant Island to Livingston Island. The southward-flowing waters are deflected to circulate cyclonically within the Strait. Some of the Antarctic Surface Water, flowing out of the South-East Pacific Basin and the Bellingshausen Sea through Drake Passage, also enters the Strait, passing between the islands stretching across the southern entrance. The contraflow of Weddell Sea Surface Water restricts the penetration of this water and confines it to the northern side of the Strait. Consequently, the dominant current within the Bransfield Strait is towards the north-east, and the outflow passes around Elephant and Clarence Islands to enter the Scotia Sea. Shallow sills prevent the movement of deeper water. Deep basins within the Strait contain only Bottom Water, which is colder and more saline than the Antarctic Bottom Water of Drake Passage and the Scotia Sea, and which is formed *in situ* during the seasonal freeze of Surface Water.

The general physical oceanography was well documented (Clowes, 1934; Gordon & Nowlin, 1978; Heywood, 1985; Stein, 1986) but the SIBEX data sets revealed the continuous and considerable movement of the boundaries between the major water masses which occurs under atmospheric forcing (Anon, 1990).

South Georgia

Only limited work was carried out around South Georgia during the 1981 FIBEX period because the U.K. vessel had to withdraw through mechanical failure. However, British Antarctic Survey (BAS) scientists carried out intensive investigations during the following austral summer (1981–2) and during the winter of 1983. A core salinity for the temperature minimum layer of approximately 34.00 PSU and relatively low dissolved reactive silicate concentrations of 20–50 μM recorded at all the stations indicated that the surface water around the island was originally Antarctic Surface Water from the South-East Pacific Basin. Water of Weddell Sea origin was not present during this period. However, at least six water masses could be distinguished within the research area, the result of local mixing with sub-Antarctic water carried in eddies within the PFZ and locally modified coastal waters, and other local processes (Priddle *et al.*, 1986).

In the Scotia Sea the maximum displacement of the Polar Fronts is limited by the bottom topography (Gordon & Goldberg, 1970) but may be as much as 100 nautical miles (186 km). This is sufficient to affect the composition of the water flowing around South Georgia. Evidence of a major shift of the Polar Frontal Zone southwards to encompass South Georgia was obtained during the BAS austral Winter Cruise in 1983. The surface isotherms present a general picture of the conditions in the Surface Water Layer at that time (Fig. 3). The waters were homogeneous to depths of 100–150 m throughout the Scotia Sea. The Polar Front was less than 20 nautical miles wide in Drake Passage, and was sharply defined by the 2.0 and 0.0 °C isotherms. The 2.0 °C isotherm extended almost vertically to below 760 m (the depth limit of the expendable bathythermographs used). Mixing over the ridge of the Scotia Arc and the South Georgia shelf broadened the Polar Front as the water flowed eastwards. The northern boundary of the Polar Front was ill-defined here, and the position given for the surface 2.0 °C isotherm is where it was first detected on passage from South Georgia to the Falkland Islands after the study period. Salinities and concentrations of dissolved reactive silicate recorded for the Surface Water layer in the Polar Front ranged from 34.10 to 34.05 PSU and from 21 to 32 μM, respectively. A tongue of water at *ca.* 2 °C intruded under the Surface Water Layer to occupy the depths between 300 and 1100 m as far south as the 0.0 °C surface isotherm. Here the heavier Surface Water layer (σ_t 27.50 to 27.46) of the Antarctic Zone sank below that of the Polar Front (σ_t 27.31 to 27.28).

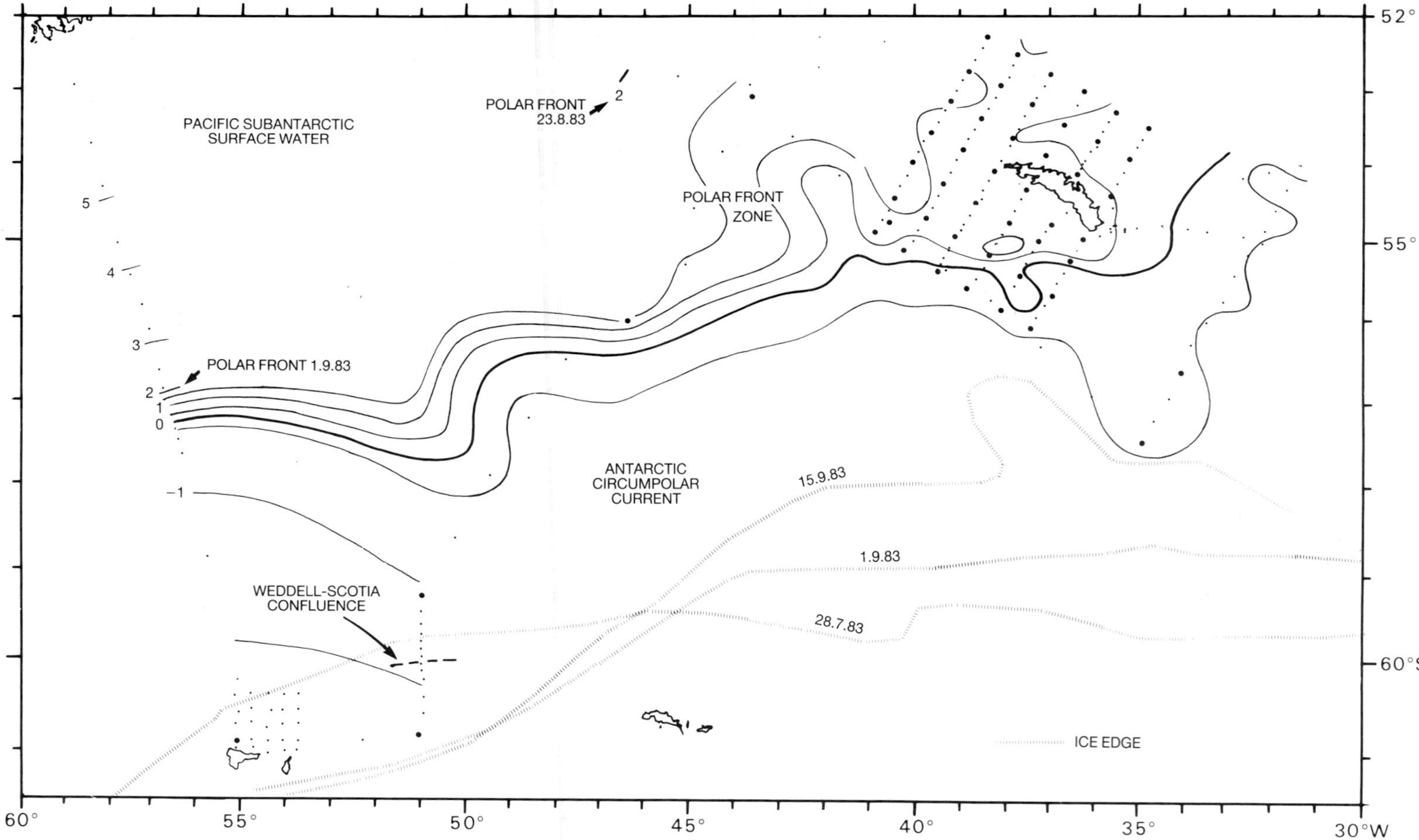

Fig. 3. Major water masses within the Scotia Sea, August–September 1983. Also shown are surface isotherms and the approximate northerly limit of pack ice occurrences. (From Heywood *et al.*, 1985.)

Currents

The kinematic structure of the Antarctic Circumpolar Current between the tips of South America and the South Shetland Islands is given by Nowlin *et al.* (1977), based on moored current meter measurements and geostrophic computations. They report current bands of high eastward speeds separated by regions of lower speed or even westward flow in the central Drake Passage. Off the north coast of the South Shetland Islands they derive an eastward flow, which amounts to about 10 cm s^{-1}. Long-term current measurements in the bottom waters at locations over the outer shelf and slope of the South Shetland Islands and along the continental rise show persistent flow from Atlantic to Pacific along isobaths at speeds of 10–20 cm s^{-1} (Nowlin & Zenk, 1988).

At the eastern edge of the South Shetland Archipelago a steady flow of about 12 cm s^{-1} heading 65° (true) was recorded by moored instruments between mid-November 1977 and the end of January 1978 (Stein & Cornus, 1979). Six year-long records from moorings deployed in Drake Passage revealed mean flow directions consistent with those from geopotential anomaly charts (Pillsbury *et al.*, 1979). The mean deep flow (2667 m) off the South Shetland Islands was almost directly to the east (east component 5.5 cm s^{-1}, north component 0.4 cm s^{-1}). Rattray (1985) compared the orientation of the deep mean flow at 20 sites in Drake Passage to the geostrophic contour distribution. The directions of the mean and low-frequency flows coincide on average, although there are several individual large differences between them. Stein (1988) found similar results for the area to the north-west of Elephant Island and between Elephant and Clarence Islands.

Madejski & Rakusa-Suszczewski (1990) measured the direction and velocity of iceberg drift during 1987 by photogrammetric methods. Drift along the southern coast of King George Island followed the NE and ENE directions of wind and water currents. The highest mean velocities (0.6–0.8 ms^{-1} were recorded in March–April; the lowest velocities (0.1–0.2 ms^{-1}) in May–July. August–October velocities (0.4–0.6 ms^{-1}) were higher than those in November–December (0.4–0.5 ms^{-1}).

Atmospheric forcing in the Antarctic Peninsula region

The geographical positions of the boundaries (fronts) between sub-Antarctic, Antarctic and Weddell Sea Surface Waters vary almost continuously under the influence of a series of atmospheric depressions that drive the water circulation (Fig. 4). Satellite imagery elucidates such a scenario (Fig. 5). The data originate from the Soviet Meteor satellite S40 (visible spectrum). The situation is shown on November 14, 1987, 10.45 UTC (Fig. 5*a*) and on November 26, 1987, 09.13 UTC (Fig. 5*b*). In Fig. 5*b* the archipelago is clearly visible as far east as Elephant and Clarence Islands. On both days wind speeds were well above 30 ms^{-1} or 12 Bft. As on the cloud layer, the gale must have acted on the upper surface water layers.

The energy transfer from atmosphere to ocean is via friction, which drives the pure drift current. Although the main force for convection is through cooling, cyclonic motion leads to doming of isopycnals and subsequently enhances convection in the surface layer. This might disturb the stratification of the upper layers far beyond the depth of frictional influence, D, which is 39 m for the Antarctic Peninsula region (Stein, 1986).

Perturbation of the surface layer then leads to deep convection, which is believed to be the process responsible for deep water formation in the Bransfield Strait area (Gordon & Nowlin, 1978). FDRAKE 75 tritium data confirm deep convection in the eastern Bransfield Strait basin. Oxygen measurements performed during spring 1987 by RV '*Polarstern*' in the central basin south of King George Island reveal a profile that indicates high oxygen values in the deep basin water (500m: 9.50 mg l^{-1}; 1000m: 9.66 mg l^{-1}; 1500m: 9.81 mg l^{-1}; 1900m: 9.98 mg l^{-1} (data: Rutgers van der Loeff, personal communication)). These oxygen values require fast convective overturn rates which mix the highly oxygenated surface waters (> 11 mg l^{-1}) deep into the bottom water domain. According to Gordon & Nowlin (1978) the winter surface water contributes by convective means to the deep and bottom waters of the basins. Weddell Sea inflow, being the most saline surface water component observed, is the most likely candidate for a surface source water. A further process initiated by air–sea interaction is the intensified inflow of Circumpolar

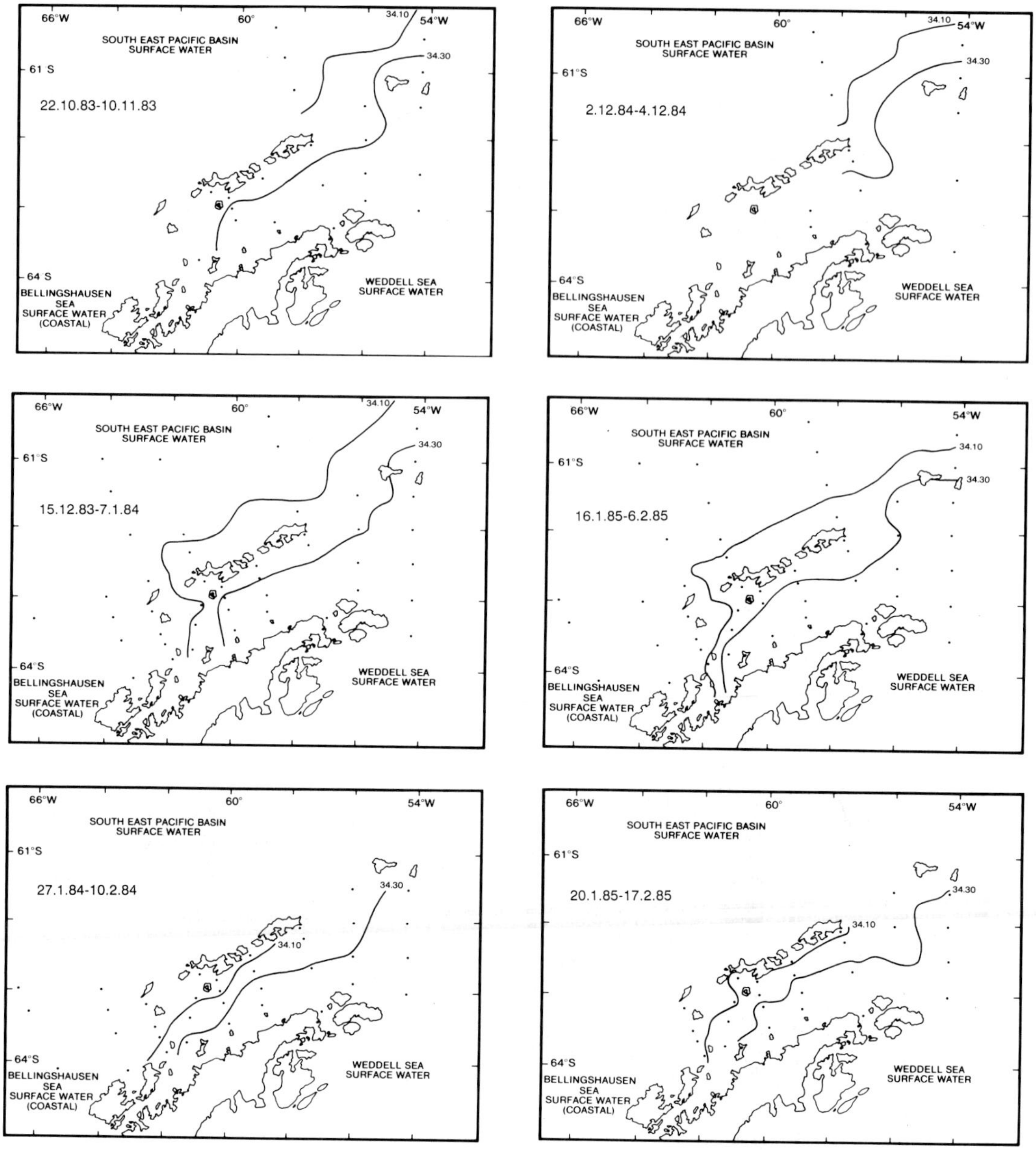

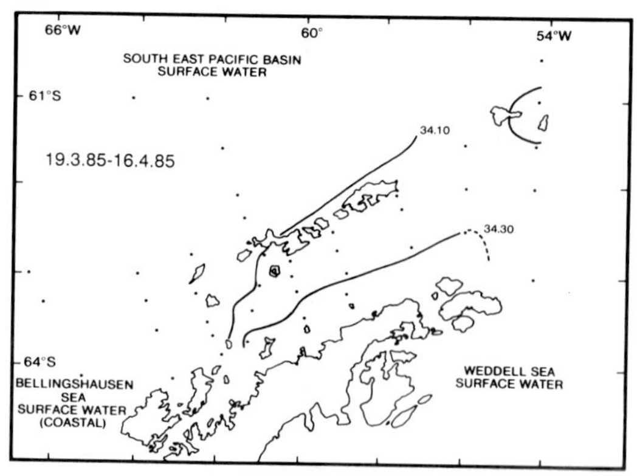

Fig. 4. Changes in the boundaries between main water masses within the 100 m layer, as depicted by selected isohalines of mean salinity (PSU). (From Anon, 1990.)

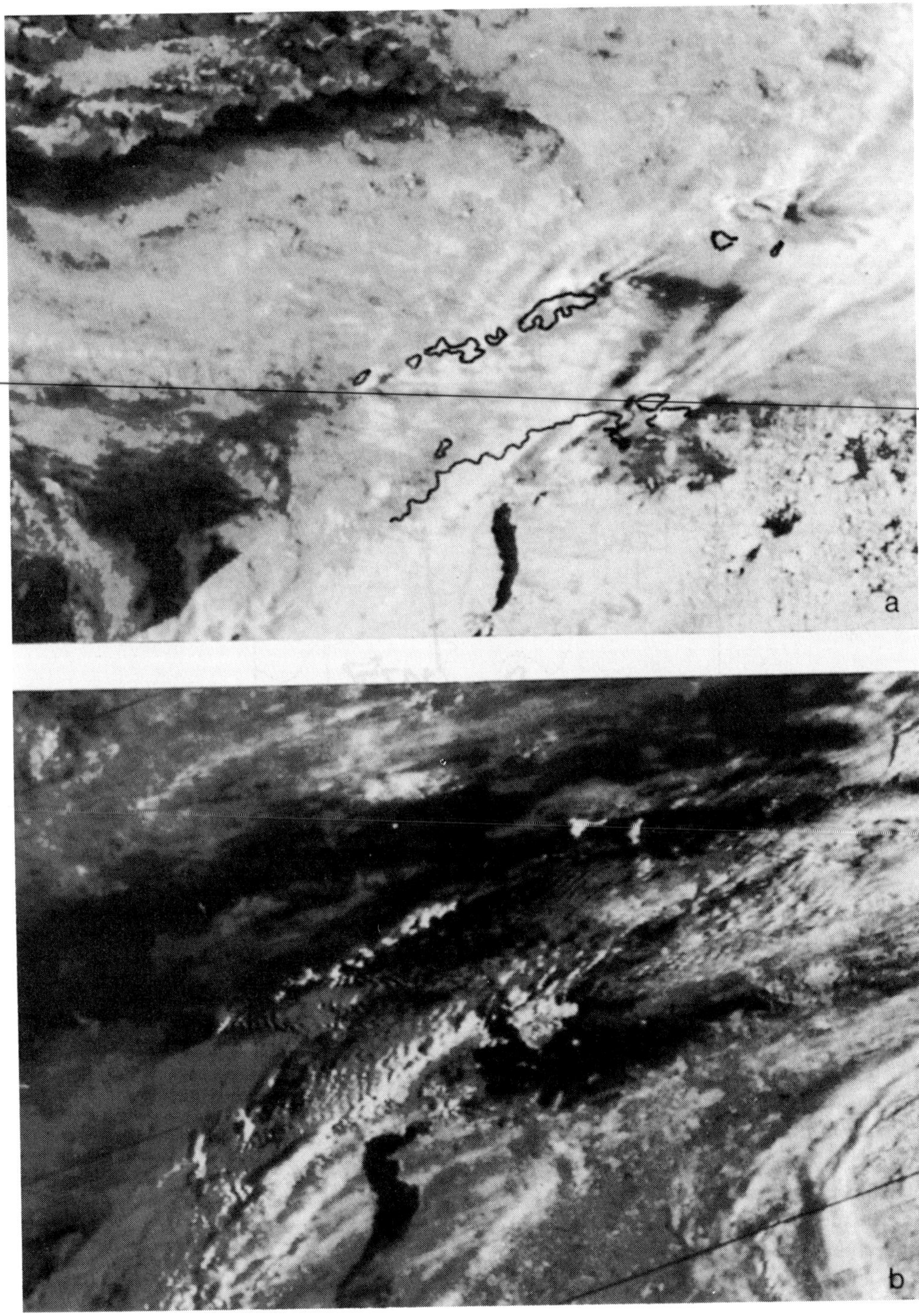

Fig. 5. Satellite imagery from Bransfield Strait (visible spectrum). (a) 14 November 1987, 10.45 UTC; (b) 26 November 1987, 09.13 UTC. (From Stein, 1989.)

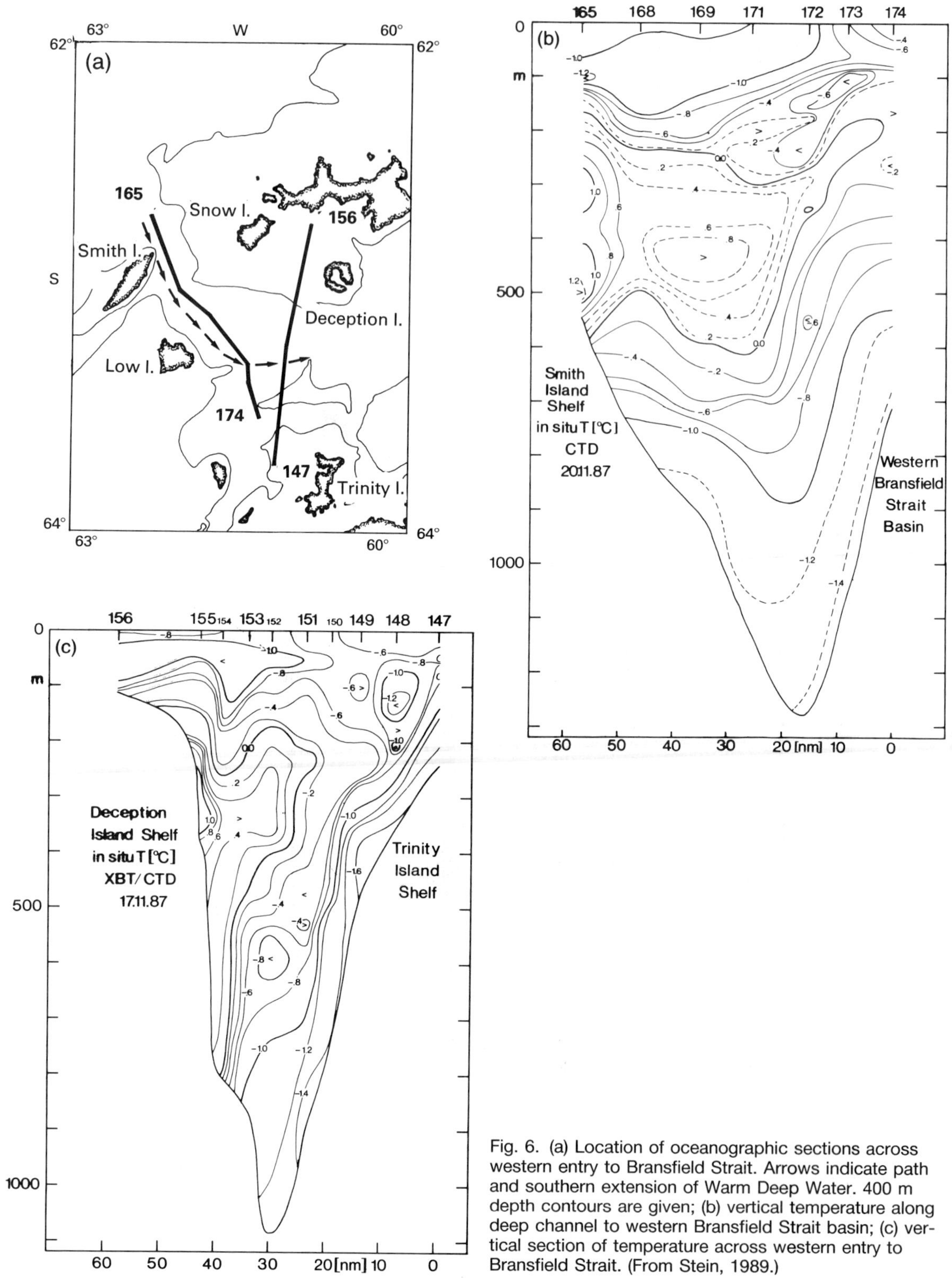

Fig. 6. (a) Location of oceanographic sections across western entry to Bransfield Strait. Arrows indicate path and southern extension of Warm Deep Water. 400 m depth contours are given; (b) vertical temperature along deep channel to western Bransfield Strait basin; (c) vertical section of temperature across western entry to Bransfield Strait. (From Stein, 1989.)

Deep Water (CDW) through the western boundary of the Bransfield Strait. Clowes (1934) and Gordon & Nowlin (1978) state the existence of eastward advection either of an isolated bolus, or of a current which enters the central basin from the west, passing to the south of Deception Island. FDRAKE 75 data show for the western basin a broad salinity maximum and oxygen minimum within the upper kilometre, but this signal is much less intense and shallower than within the waters of the adjacent open ocean.

During the SIBEX Physical Oceanography Workshop held in Cambridge, U.K., 9–27 March 1987 (Anon, 1990) it was revealed that during strong northerly winds South-East Pacific Basin surface water penetrated farther into the Bransfield Strait, especially along a deep channel between Snow and Low Islands. This action blocked water of Bellingshausen Sea origin entering the strait from the southwest. During the workshop this phenomenon was called a 'flip-flop' mechanism.

Based on these assumptions Stein (1989) initiated a mesoscale study during spring 1987 to follow the path of water masses entering the Bransfield Strait across its western boundary. The approximate southern extension and path of the CDW can be traced from the arrows in Fig. 6, if one takes the 0 °C isotherm as characteristic water mass boundary. From the regional wind field an inflow situation was favoured. On 20 November 1987 wind speed peaked at 24 m/sec from WNW.

The homogeneous surface layer might reflect frictional influence and turbulent mixing due to prevailing gales. Below the surface layer the core of the CDW inhabits the depth range 250–500 m with temperatures above 1 °C. Further east in the deep channel south of the Deception Island shelf this water mass is found from about 250 to 400 m depth. As shown in a temperature–salinity (T–S) analysis, both CDW masses have the same characteristics (Fig. 7). For the heat budget of the Bransfield Strait, it is therefore essential whether inflow conditions from Drake Passage are favourable or not.

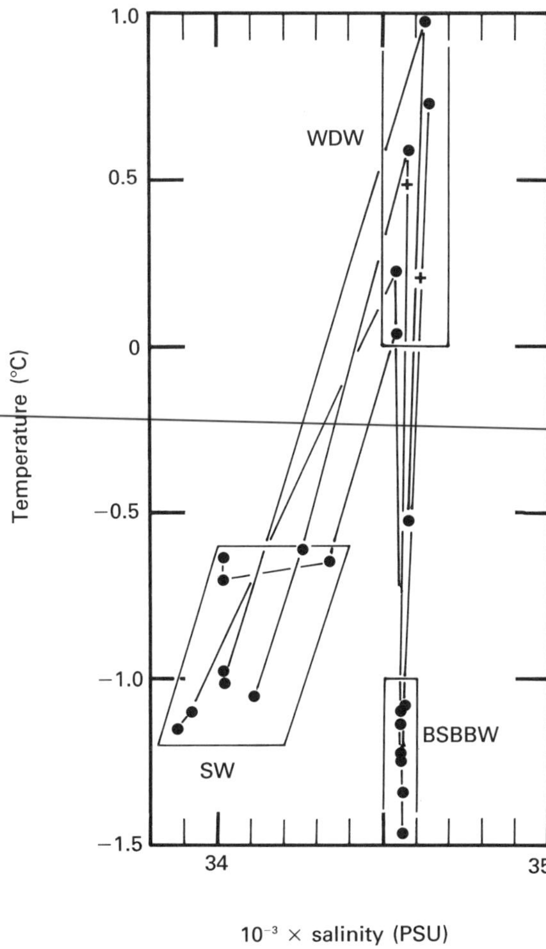

Fig. 7. T–S diagram of section along inflow channel to western Bransfield Strait basin. WDW, Warm Deep Water; SW, Surface Water; BSBBW, Bransfield Strait Basin Bottom Water. (From Stein, 1989.)

Interactions between biota and environment

The BIOMASS Programme recognized the need for the close coupling of environmental studies with studies of the biota if the distribution, abundance, productivity and behaviour of the Southern Ocean marine organisms were to be fully understood (SCAR/SCOR, 1977). Little use has been made of the extensive physical and chemical data sets during the biological workshops, although it was recommended at the first Post-FIBEX Hydrographic Data Interpretation Workshop (Anon, 1982*a*) that a physical oceanographer should be present at all biological workshops and meetings to discuss the implications of hydrographic phenomena. Fortunately, integration and analysis of environmental and biological data has occurred at national level.

Variation in community structure, biomass and levels of productivity has been related to fronts, eddies and even subtle differences in water bodies for phytoplankton (Priddle *et al.*, 1986; Heywood & Priddle, 1987), zooplankton (Atkinson & Ward, 1988; Bölter *et al.*, 1988; Rakusa-Suszczewski, 1988), krill (Maslennikov & Solyankin, 1980; Poluiaktov *et al.*, 1983; Solyankin, 1984; Stein & Rakusa-Suszczewski, 1984), fish (Kellerman & Kock, 1988), squid (Rodhouse, 1989) and even seabirds (Hunt *et al.*, 1990).

Of particular significance has been the discovery of major redistributions of krill associated with movement of the Polar Front Zone (Heywood *et al.*, 1985; Makarov *et al.*, 1988). A southward movement of the Front during the winter of 1983 led to a low abundance of krill throughout a large area of the Scotia Sea, which persisted into the 1983–4 summer. This resulted in very low survival rates for offspring of those bird and seal species that are dependent on krill during the breeding season. The long-term records of such survival rates, available for species breeding on Bird Island (off South Georgia), indicate that a drastic reduction in local krill abundance, and therefore major shift southwards of the Polar Front, is a frequent but not regular occurrence. The phenomenon will have to be accommodated in any ecosystem-based, krill fisheries management plan.

Conclusion

The physical phenomena regulating ecosystems range in spatial and temporal scale of interaction over many orders of magnitude. To date, this has presented unsurmountable difficulties to the researcher. The process-orientated approach of the Second BIOMASS Experiment (SIBEX; Anon, 1982*b*) required the investigation of physical processes on small spatial and temporal scales that related to the biological observations being made. Fortunately, the past two decades have also seen a resurgence of interest in the physical processes of the Southern Ocean generally. The information gained from these parallel programmes has made it possible to relate the BIOMASS observations to such basic-scale features as the Antarctic Circumpolar Current, the Weddell Sea Gyre, etc., and to atmospheric forcing. The SIBEX discovery that the waters of the southern Bransfield Strait were particularly sensitive to the effects of local atmospheric depressions may prove to be a valuable tool in understanding the population dynamics of this particularly complex area. One notable gap in the knowledge required of the Southern Ocean environment by biologists is a detailed qualitative and quantitative description of mass transport, particularly of the Surface Water.

However, expansion of knowledge of the physicochemical environment of the Southern Ocean will be exponential during the next decade, which will see major international programmes such as the World Ocean Circulation Experiment (WOCE). Shipboard observations will be made on an hitherto unprecedented scale, linked to satellite observations and utilized in powerful numerical models such as the U.K. Fine Resolution Antarctic Model (FRAM). Nevertheless the need will remain for physical and chemical measurements to be made during biological programmes at scales which reflect, and are therefore meaningful to studies of, the interactions between biotic and abiotic processes (Murphy *et al.*, 1988). BIOMASS was the first major international programme to attempt such a multidisciplinary approach. Most future studies will require finer detail, but the BIOMASS data set should not be forgotten or ignored; it still contains much information of value.

References

Anon (1982*a*) First Post-FIBEX Hydrographic Data Interpretation Workshop. Hamburg, F.R. Germany, 20–26 September 1982. *BIOMASS Rep Ser* **30**: 1–10.

Anon (1982*b*) BIOMASS Scientific Advisory Group for SIBEX-1. *BIOMASS Rep Ser* **23**: 1–24.

Anon (1983) Second Post-FIBEX Hydrographic Data Interpretation Workshop. Hamburg, F.R. Germany, 16–20 May 1983. *BIOMASS Rep Ser* **31**: 1–26.

Anon (1990) SIBEX Physical Oceanography Workshop. Cambridge, U.K., 9–27 March 1987. *BIOMASS Rep Ser* **62**: 1–38.

Atkinson A, Ward P (1988) Summer-winter differences in copepod distribution around South Georgia. *Hydrobiologia* **167/168**: 325–34.

Baker A de C (1954) The circumpolar continuity of Antarctic plankton species. *Discovery Rep* **27**: 201–18.

Bölter M, von Bodungen B, Liebezeit G, Meyer M (1988) The pelagic ecosystem of the Bransfield Strait, Antarctica: An analysis of microbiological, planktological and chemical characteristics by multivariate analysis. In: Sahrhage D (ed) *Antarctic Ocean and Resources Variability*. Springer, Berlin, pp. 160–8.

Carmack E C, Foster T D (1975) On the flow of water out of the Weddell Sea. *Deep-Sea Res* **22**: 711–24.

Clowes A J (1934) Hydrology of the Bransfield Strait. *Discovery Rep* **9**: 1–64.

Deacon G E R (1933) A general account of the hydrology of the South Atlantic Ocean. *Discovery Rep* **7**: 171–238.

Deacon G E R (1937) The hydrology of the Southern Ocean. *Discovery Rep* **15**: 1–24.

Deacon G E R, Moorey J A (1975) The boundary region between currents from the Weddell Sea and Drake Passage. *Deep-Sea Res* **22**: 265–8.

Gordon A L (1966) Potential temperature, oxygen and circulation of bottom water in the Southern Ocean. *Deep-Sea Res* **13**: 1125–38.

Gordon A L (1967) Structure of Antarctic waters between 20°W and 170°W. *Antarctic Map Folio Series* **6**: 10 pp.

Gordon A L (1989) Physical oceanographic setting of the *Siedlecki* January 1987, South Shetland Island data set. In: Scientific Committee for the Conservation of Antarctic Marine Living Resources. *Selected Scientific Papers 1988*. Part 2. CCAMLR, Hobart, Tasmania, pp. 161–84.

Gordon A L, Goldberg R D (1970) Circumpolar characteristics of Antarctic waters. *Antarctic Map Folio Series* **13**: 5 pp.

Gordon A L, Nowlin W D Jr (1978) The Basin Waters of the Bransfield Strait. *J Phys Oceanogr* **8**: 258–64.

Hardy A C, Gunther E R (1935) The plankton of the South Georgia whaling grounds and adjacent waters, 1926–27. *Discovery Rep* **11**: 1–456.

Hart T J (1934) On the phytoplankton of the south-west Atlantic and the Bellingshausen Sea, 1929–31. *Discovery Rep* **8**: 1–268.

Hart T J (1942) Phytoplankton periodicity in Antarctic surface waters. *Discovery Rep* **21**: 261–356.

Heywood R B (1985) Environmental conditions in the Antarctic Peninsula area of the Southern Ocean during the Anglo-German Joint Biological Expedition, February 1982. *Meeresforschung* **30**: 220–39.

Heywood R B, Priddle J (1987) Retention of phytoplankton by an eddy. *Continental Shelf Res* **7**: 937–55.

Heywood R B, Everson I, Priddle J (1985) The absence of krill from the South Georgia zone, winter 1983. *Deep-Sea Res* **32**: 369–78.

Hollister D, Elder R B (1969) Contour currents in the Weddell Sea. *Deep-Sea Res* **16**: 99–101.

Hunt G L Jr, Heinemann D, Veit R R, Heywood R B, Everson I (1990) Distribution, abundance and community structure of marine birds in southern Drake Passage and Bransfield Strait, Antarctica. *Continental Shelf Res* **10**: 243–57.

Kellermann A, Kock K-H (1988) Patterns of spatial and temporal distribution and their variation in early life stages of Antarctic fish in the Antarctic Peninsula region. In: Sahrhage D (ed) *Antarctic Ocean and Resources Variability*. Springer-Verlag, Berlin, pp. 147–59.

Madejski P, Rakusa-Suszczewski S (1990) Icebergs as tracers of water movement in the Bransfield Strait. *Antarct Sci* **2**(3): 259–63.

Makarov R R, Maslennikov V V, Solyankin E V, Spiridonov V A, Yakovlev V N (1988) Variability in population density of antarctic krill in the western Scotia Sea in relation to hydrological conditions. In: Sahrhage D (ed) *Antarctic Ocean and Resources Variability*. Springer, Berlin, pp. 231–6.

Maslennikov V V, Solyankin E V (1980) Role of water dynamics in the maintenance of a *Euphausia superba* Dana population in the Weddell Sea. *Oceanology* **20**: 192–5.

Model F (1958) Ein Beitrag zur regionalen Ozeanographie der Weddellsee. Deutsche Antarktika Expedition 1938–39. *Wiss Ergebn* **2**: 63–96. (In German.)

Murphy E J, Morris D J, Watkins J L, Priddle J (1988) Scales of interaction between Antarctic krill and the environment. In: Sahrhage D (ed) *Antarctic Ocean and Resources Variability*. Springer-Verlag, Berlin, pp. 120–30.

Nowlin W D, Clifford M (1982) The kinematic and thermohaline zonation of the Antarctic Circumpolar Current at Drake Passage. *J Mar Res* (Suppl) **40**: 481–507.

Nowlin W D, Whitworth T, Pillsbury R D (1977) Structure and transport of the Antarctic Circumpolar Current at Drake Passage from short-term measurements. *J Phys Oceanogr* **7**: 788–802.

Nowlin W D, Zenk W (1988) Westward bottom currents along the margin of the South Shetland Island Arc. *Deep-Sea Res* **35**: 269–301.

Peterson R G, Whitworth T III (1989) Subantarctic and polar fronts in relation to deep water masses through the southwestern Atlantic. *J Geophys Res* **94**(C8): 10817–38.

Pillsbury R D, Whitworth T III, Nowlin W D, Sciremammano F Jr (1979) Current and temperatures as observed in Drake Passage during 1975. *J Phys Oceanogr* **9**: 469–82.

Poluiaktov V F, Solyankin E V, Khvathii N V, Khomutov A V (1983) Hydrological conditions as factors responsible for krill distribution in the southern section of the Bransfield Strait. In: Liubimova T G, Makarov R R, Kamenskaia E A (ed) *Antarktcheskii kril'. Osobennosti raspredeleniia i sreda*. Legkaia i Pishchevaia Promyshlennost', Moscow, pp. 99–102. (In Russian.)

Popkov V V (1987) Characteristics of oceanographic conditions in the antarctic Pacific sector (Kharakteristika okeanograficheskikh uslovii tikhookeanskogo sektora Antarktiki). In Makarov R R (ed) *Biologo-okeanograficheskie Issledovaniia Tikhookeanskogo Sektora Antarktiki (Biological and Oceanographic Investigations of the Pacific Sector of the Antarctic Ocean).* VNIRO, Moscow, pp. 41–61. (In Russian.)

Priddle J, Heywood R B, Theriote E (1986) Some environmental factors influencing phytoplankton in the Southern Ocean around South Georgia. *Polar Biol* **5**: 65–79.

Rakusa-Suszczewski S (1988) Differences in the hydrology, biomass and species distribution of plankton, fishes and birds in the Bransfield Strait and the Drake Passage during FIBEX 1981 and SIBEX 1983/84. In: Sahrhage D (ed) *Antarctic Ocean and Resources Variability*. Springer, Berlin, pp. 214–18.

Rattray M Jr (1985) Effect of bathymetry on the deep flow in Drake Passage. *Deep-Sea Res* **32**(2A): 127–47.

Rodhouse P J (1989) Cephalopods in the diet of wandering albatrosses and sea-surface temperatures at the Sub-Antarctic Front. *Scient Mar* **53**: 277–81.

SCAR/SCOR (1977) *Biological Investigations of Marine Antarctic Systems and Stocks (BIOMASS)*. Vol 1: *Research Proposals*. University Library, Cambridge, 79 pp.

SCOR (1985) General circulation of the Southern Ocean: Status and recommendations for research. Report by SCOR Working Group 74. WCP-108: 50 pp.

Schröder M (1983) Characteristics of a salinity front in the eastern Bransfield Strait, the source region of the Weddell–Scotia Confluence. *Deut Hydrogr Zschr* **36**(6): 217–35.

Solyankin E V (1984) Oceanographic conditions affecting bioproductivity in the Atlantic sector of the Antarctic (Osobennosti okeanologicheskikh uslovii formirovaniia bioproduktivnosti vod v atlanticheskom sektore Antarktiki). In: Gershanovich D E, Bogdanov M A, Sapozhnikov V V (ed) *Promyslovo-okeanograficheskie Issledovaniia Produktivnykh zon Morei i Okeanov. Sbornik Nauchnykh Trudov (Fishery and Oceanography of Marine Productive Areas. Collected Papers)*, VNIRO, Moscow, pp. 71–91. (In Russian.)

Solyankin E V, Spiridonov V A (1987) Qualitative aspects of the distribution of *Euphausia superba* Dana in the Antarctic Peninsula region (Kachestvennye aspekty raspredeleniia *Euphausia superba* Dana v raione Antarkticheskogo poluostrova). In: Makarov R R (ed) *Biologo-okeanograficheskie Issledovaniia Tikhookeanskogo Sektora Antarktiki (Biological and Oceanographic Investigations of the Pacific Sector of the Antarctic Ocean)*. VNIRO, Moscow, pp. 151–76. (In Russian.)

Stein M (1981) Thermal structure of the Weddell–Scotia confluence during February 1981. *Meeresforschung* **29**(1): 47–52.

Stein M (1983) The distribution of water masses in the South Shetland Islands area during FIBEX. *Mem Nat Inst Polar Res Spec Issue* No. 27 (*Proceedings of the BIOMASS Colloquium in 1982*): 16–23.

Stein M (1986) Variability of water masses and currents off the Antarctic Peninsula during SIBEX. *Arch FischWiss* **37**: 25–50.

Stein M (1988) Variation of geostrophic circulation off the Antarctic Peninsula and in the southwest Scotia Sea. In: Sahrhage D (ed) *Antarctic Ocean and Resources Variability*. Springer-Verlag, Berlin, pp. 81–91.

Stein M (1989) Seasonal variation of water masses in Bransfield Strait and adjacent waters. *Arch FischWiss* **39**: 15–38.

Stein M (1992) Variability of local upwelling off the Antarctic Peninsula, 1986–1990. *Arch FischWiss* **41**: 131–58.

Stein M, Cornus H P (1979) Ozeanographische Untersuchungen. In: *Antarktis-Expedition 1977/78 der Bundesrepublik Deutschland. Arch FischWiss* **30** (Beih 1): 1–119. (In German.)

Stein M, Rakusa-Suszczewski S (1984) Meso-scale structure of water masses and bottom topography as the basis for krill distribution in the SE Bransfield Strait February–March 1981. *Meeresforschung* **30**(2): 73–81.

Whitworth T, Nowlin W D, Worley S J (1982) The net transport of the Antarctic circumpolar current through Drake Passage. *J Phys Oceanogr* **12**: 960–71.

Physical and chemical oceanography in the vicinity of Prydz Bay, Antarctica

N. SMITH & P. TRÉGUER

Introduction

THE CIRCULATION and physical and chemical characteristics of water masses in the different sectors of the Southern Ocean are generally well documented (e.g. Gordon *et al.*, 1978; Gordon & Molinelli, 1982; Mantyla & Reid, 1983). Both the Weddell Sea (Atlantic sector) and Ross Sea (Pacific sector) have been the subject of numerous physical and chemical studies. The data collected through these studies and the subsequent analyses provide the basis for much of our knowledge of the oceanography of Antarctic waters. The third significant, though less extensive, embayment of Antarctica, Prydz Bay, is located in the Indian sector. It is not situated as far south as either the Weddell Sea (76°S) or the Ross Sea (77°S); open waters of Prydz Bay proper (Fig. 1) reach only to about 69°S. Compared with the Weddell and Ross Seas this region, and the adjacent waters to the north (occasionally referred to as the Amery Sea), have received relatively little attention.

Although most studies in this area have primarily been focused on biological aspects of the Antarctic ecosystem, physical and chemical data have often been collected during these cruises from which the coastal and offshore circulation can be inferred. This review examines hydrological and chemical data collected during French and Australian cruises in the region (Table 1). We do not attempt to give a comprehensive account of all studies and data collected in this region, either before or during BIOMASS, but instead concentrate on building a general picture of the currents and water properties, assuming this more restricted data set is sufficiently representative of the long-term mean circulation pattern. This review makes obvious the need for improved analysis and interpretation of the existing data. As will become evident, the sampling provided by any single cruise is usually not sufficient to unambiguously resolve the circulation and water mass characteristics of these waters.

The layout of this review is as follows. After a brief introduction to the geographical and topographical characteristics of the region our present understanding of the general circulation in the vicinity of Prydz Bay is discussed, based primarily on the data from the Australian cruises. The principal water masses in the study region are identified and French cruise data are used to estimate open ocean entrainment into the winter water layer, and to assess variability and formation mechanisms of deep water masses.

General features of the study area

The study area is shown in Fig. 1. We are concentrating our attention on the region south of 60°S and between 60°E and 90°E. Prydz Bay is characterized

(a)

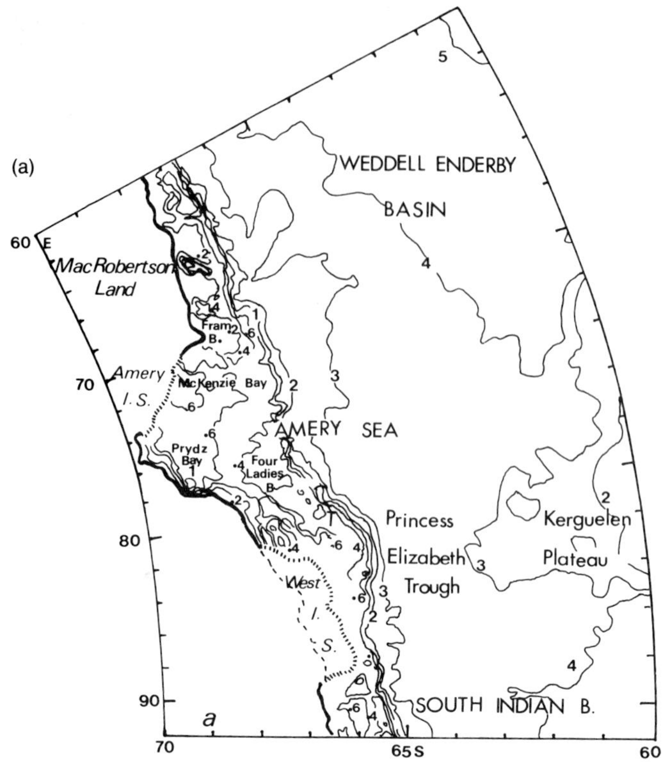

(b)

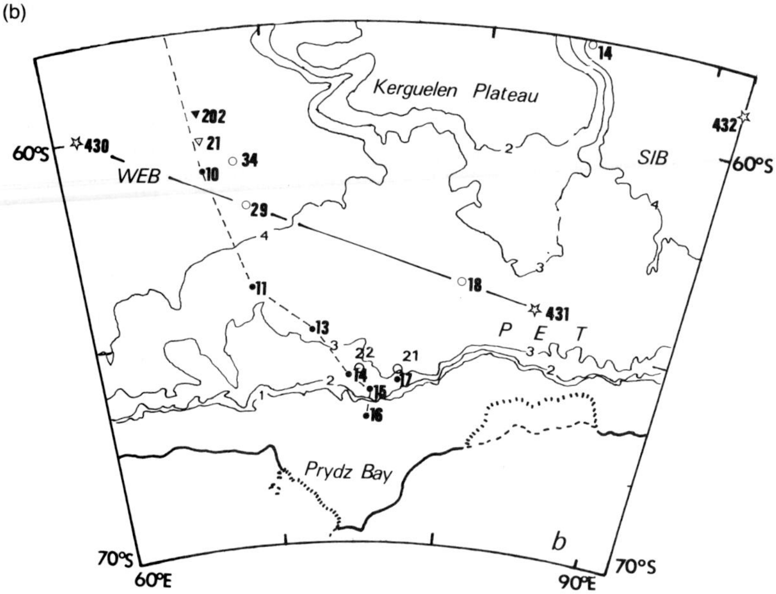

Fig. 1. (a) Geography and bathymetry of Prydz Bay, the adjacent continental shelves, and the seas near Prydz Bay. The isobaths are marked in kilometres. (b) Location of selected stations occupied by the French observation programme (Table 1): ANTIPROD cruises 1 (△), 2 (▲) and 3 (●), and SIBEX (○). Three stations occupied during the GEOSECS programme (430–2) are also marked (*).

Table 1. *A list of the Australian and French cruises undertaken during the BIOMASS Programme*

The references are to the data reports, which summarize the aims of, and detail the data collected during, each survey.

Cruise	Date	Stations	References
Territoire des Terres Australes et Antarctiques Françaises Cruises			
ANTIPROD 1	Jan 1977		Jacques, 1978
ANTIPROD 2	Mar 1980		Jacques, 1982; Tréguer *et al.*, 1982
ANTIPROD 3	Mar–May 1985		Fiala & Fontugne, 1987
SIBEX	Feb–Mar 1985		Souchu *et al.*, 1990
INDIGO 3	Jan 1987		Mantisi *et al.*, 1991
Australian Antarctic Division Cruises			
FIBEX	Jan–Mar 1981	52	Dong *et al.*, 1982; Kerry *et al.*, 1987*a*
ADBEX 1	Nov–Dec 1982	79	Kerry & Woehler, 1987
ADBEX 2	Jan–Feb 1984	22	Kerry *et al.*, 1987*b*
SIBEX 2	Jan 1985	64	Woehler *et al.*, 1987
ADBEX 3	Sep–Dec 1985	50	Woehler *et al.*, 1987
AAMBER	Feb–Mar 1987	96	Woehler & Williams, 1988

by depths near 600 m, with Fram Bank and Four Ladies Bank (depths typically around 400 m) providing a partial barrier to exchange with the deep ocean (for this review we will assume Prydz Bay includes the MacKenzie Bay region). The southern boundary of Prydz Bay is provided by the Amery Ice Shelf. East and west of Prydz Bay the continental shelf is characterized by relatively deep channels shoreward of the outer shelf break, with occasional deep trenches providing direct access for coastal waters to the deep sea. The presence of the West Ice Shelf near 85°E is a significant factor in the circulation and characteristics of the waters over the shelf in this region.

Sea-ice covers much of the embayment during autumn, winter and spring. The four-year mean ice coverage for 1973–6 was greater than 70% from April to December according to passive microwave measurements (Zwally *et al.*, 1983). Recurrent polynyas have been detected in the Prydz Bay region (Knapp, 1969; Kupetskii, 1958, 1959; Leont'yev, 1973) and have been associated with anomalously warm waters found in the central part of the Bay. The ice breaks out in early summer and by January open waters predominate. The greatest seasonal and interannual variability in East Antarctic ice coverage tends to occur in the vicinity of Prydz Bay, although this variability is considerably less than experienced in the Weddell and Ross Seas (Lemke *et al.*, 1980; Allison, 1989*a*). The interannual variability is exemplified by the comparatively harsh conditions experienced by Denisov & Myznhikova (1978) compared with earlier Soviet expeditions (e.g. Zverev, 1959). Streten (1983) has linked the ice conditions in this region to cyclogenesis in the Indian sector of the Southern Ocean.

The waters to the north (sometimes referred to as the Amery Sea) are delimited by the Weddell–Enderby Basin in the west and the South Indian Basin in the east; the Kerguelen Plateau to the north-east provides a substantial barrier to the eastward flow of the Antarctic Circumpolar Current and to the east–west passage of deeper water masses. This region is out of the direct influence of the Weddell Sea Gyre, whose eastern extension reaches to about 25–30°E (Whitworth, 1988). The abyssal flow is partly fed from the Weddell–Enderby Basin but is prevented from penetrating further eastward by the Kerguelen Plateau; it is diverted south and contributes to the westward-flowing arm of a deep gyre similar to that of the Weddell–Enderby Basin (Foster & Carmack, 1976; Jacobs & Georgi, 1977). The Princess Elizabeth Trough which separates the Kerguelen Plateau and the Antarctic continent is a relatively deep channel which allows restricted exchange between the Weddell–Enderby Basin and the South Indian Basin.

Horizontal circulation in the vicinity of Prydz Bay

Smith *et al.* (1984), Middleton & Humphries (1989) and Nunes Vaz & Lennon (1993) have mapped the horizontal circulation of Prydz Bay and the adjacent waters of the South Indian Ocean, based primarily on data collected as part of Australia's contribution to BIOMASS (refer to Table 1 for details of the individual cruises and references to the cruise reports). Some cruises resolved the flow within Prydz Bay, while others concentrated on the waters off the continental shelf, or maintained a broadly spaced sampling pattern over both the continental shelf and the deeper waters. In no case was the sampling sufficiently dense in space and time to compile a 'snapshot' of the horizontal circulation, and no attempt has been made to incorporate all available data (both chemical and physical) in a more complete inverse study of the time-mean circulation. Prior to these surveys the most complete picture of the horizontal circulation was established by the 1967 *Ob* hydrographic survey (Grigor'yev, 1967, 1970; see also the discussion in Smith *et al.*, 1984).

Geostrophic maps

The general circulation pattern within the study region (Fig. 2) is characterized by a broad, deep, eastward-flowing Antarctic Circumpolar Current north of the Antarctic Divergence (around 64°S), and clockwise-circulating gyres to the south connecting the shelf and continental shelf waters with the offshore flow (Smith *et al.*, 1984; Nunes Vaz & Lennon, 1993). The northward outflows near the West Ice Shelf (around 90°E) and off MacRobertson Land near 60°E are typical of the Antarctic zone (Zverev, 1963; Smith *et al.*, 1984), and have been partly corroborated by direct measurements (Nunes Vaz & Lennon, 1993). The Prydz Bay circulation shows a clockwise gyre fed by a broad inflow from the north-east, and a stronger coastally confined outflow toward the west.

The flow within Prydz Bay and the outflow past Fram Bank toward the west are at least partially controlled by bathymetry and, in the case of the outflow, are thought to influence biological distributions, particularly that of krill, *Euphausia superba* (Smith *et al.*, 1984; Lennon & Nunes Vaz, 1988). However, it is difficult to draw strong inferences concerning the role of physics in biological production: the existing surveys do not have the spatial resolution (horizontal and vertical) to resolve the frontal features which are probably important in biological activity; there is too little information on the circulation outside the summer period (particularly upwelling rates and vertical mixing); and the existing analyses have been unable to usefully combine disparate data types (physical, chemical) and the long historical record to give a coherent picture of the mean flow in the region.

Direct current estimations

The limited direct measurements that are available (Hodgkinson *et al.*, 1988; Nunes Vaz & Lennon, 1993) generally support the circulation patterns inferred indirectly from hydrographic measurements. Current meter measurements westward of Fram Bank over the continental shelf support the dynamic height inference of a strong littoral westward current between Fram Bank and approximately 60°E, and subsequent movement of this flow toward the north (Hodgkinson *et al.*, 1988; Nunes Vaz & Lennon, 1993). A current meter deployed west of Four Ladies Bank (near 67°S, 73°E) showed a mean westward residual flow, but with both southward and northward components, so giving somewhat ambiguous support to the geostrophic flow patterns suggested by Smith *et al.* (1984) and Middleton & Humphries (1989).

Middleton & Humphries (1989) argue that variable winds adjacent to the continental rise could excite shelf waves superimposed on the westward wind-driven currents. Moreover, current meter measurements over the shelf indicate mostly diurnal frequencies (Hodgkinson *et al.*, 1988) and, since these too propagate as continental shelf waves at these latitudes, provide a further mechanism for inducing Circumpolar Deep Water onto the shelf. Subsequent tidal enhancement of mixing through vertical shearing and intrusions provides a mechanism for mixing high-salinity shelf water with warm deep water, and the resultant mixture is sufficiently dense to flow out and down the continental slope.

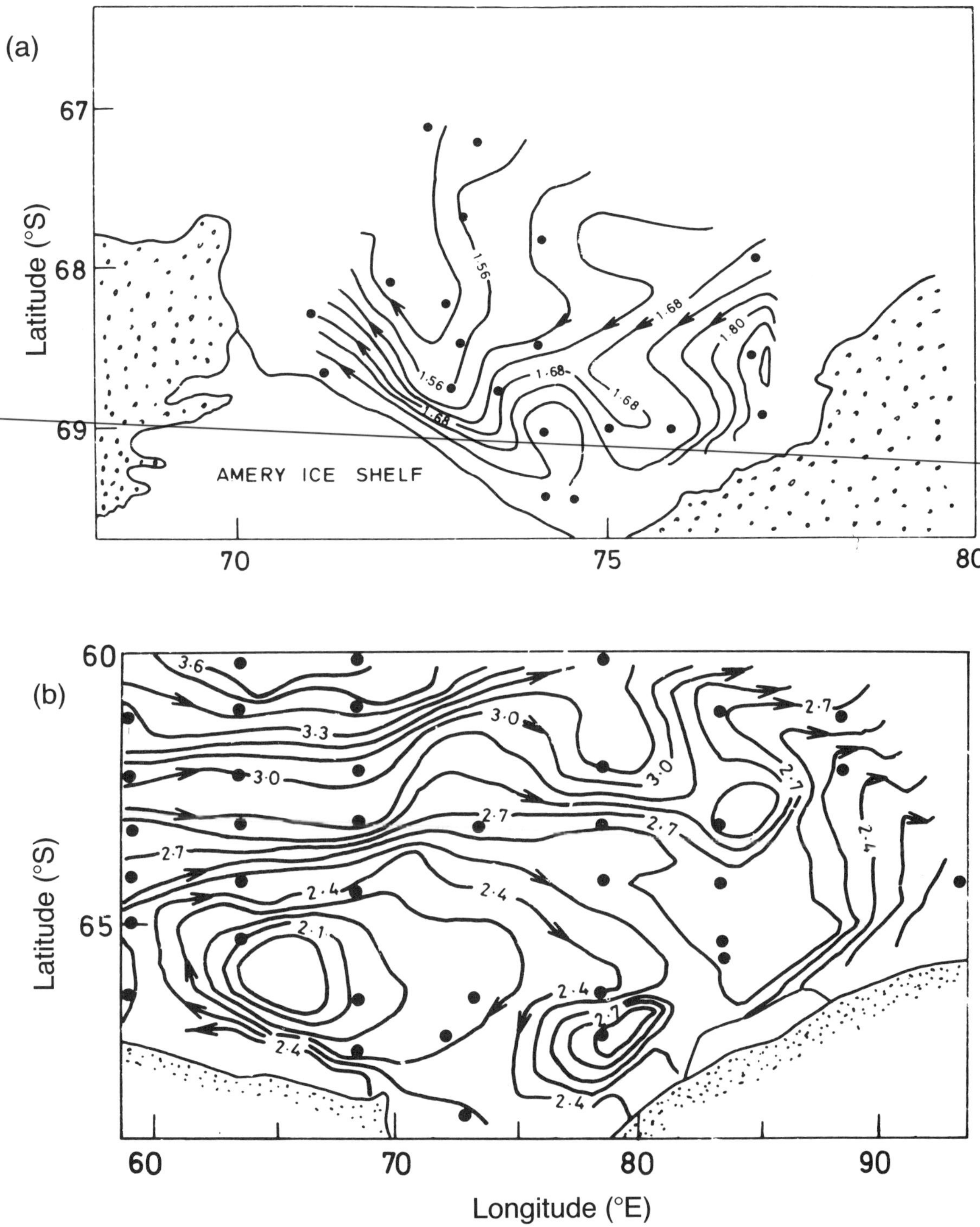

Fig. 2. Circulation of the surface waters in the vicinity of Prydz Bay in summer 1985 (after Middleton & Humphries, 1989) estimated by using the dynamic method. (a) Geopotential anomaly in Prydz Bay relative to 500 m, based on ADBEX 1 data. The contour interval is 0.04 $m^2\ s^{-2}$. (b) Geopotential anomaly relative to 600 m for the seas adjacent to Prydz Bay based on Australian SIBEX data. Contour interval is 0.1 $m^2\ s^{-2}$.

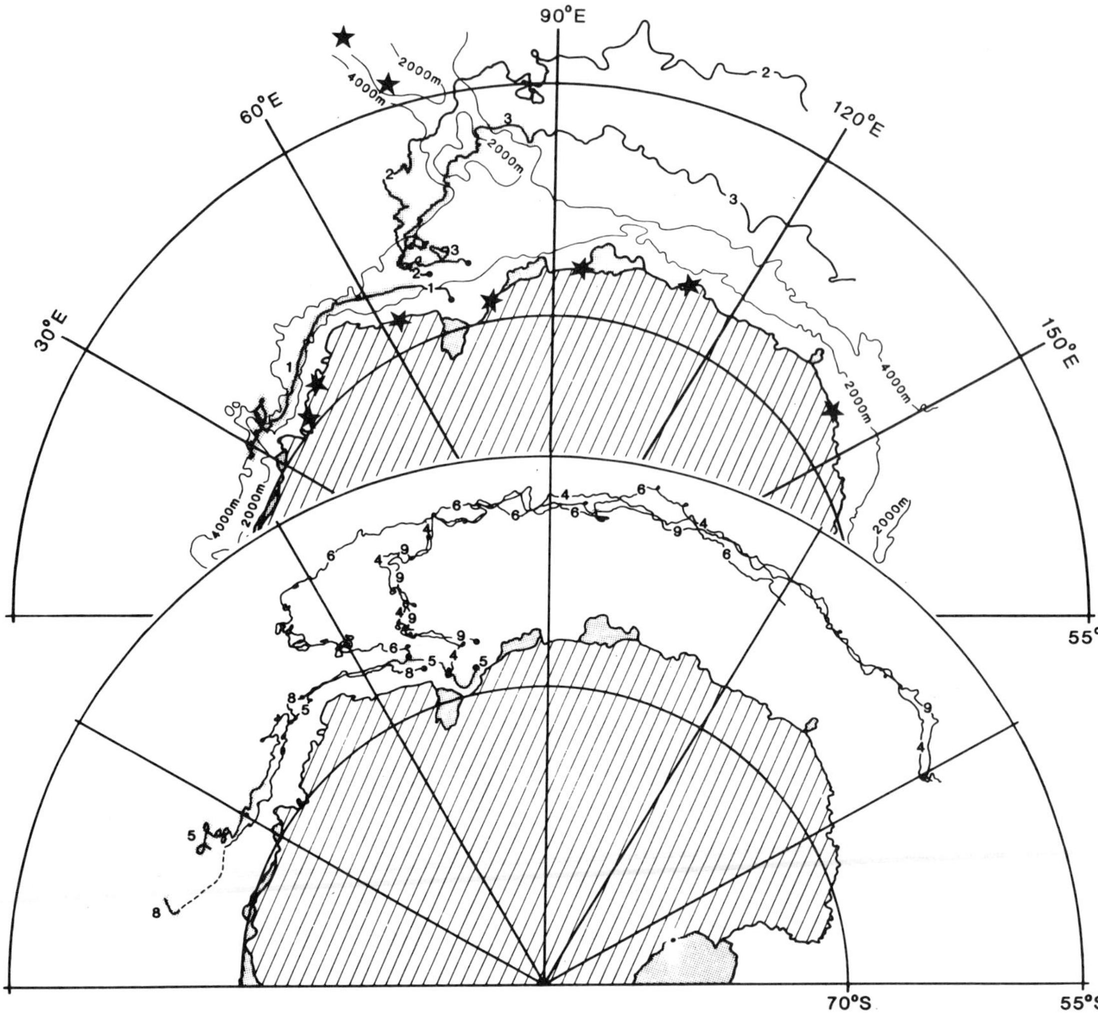

Fig. 3. Drift of data buoys deployed in the vicinity of Prydz Bay in February/March 1985 (top) and March 1987 (bottom). Numbers on the tracks are the last digit of the buoy identification (e.g. 1=ICEX41). Stippled tracks indicate those periods when the buoys were within the sea-ice edge (reproduced from Allison, 1989*a*). The filled stars are meteorological stations used by Allison (1989*a*) in the interpretation of the buoy motions.

Table 2. *Physical (potential temperature θ and salinity S) and chemical characteristics of the principal water masses in the vicinity of Prydz Bay as employed in this study*

The definitions are derived from Foster and Carmack (1976) and from results of both the French and Australian stations occupied in the region. Acronyms are as in the text. CSW refers to the high-salinity variety of continental shelf water.

Water mass	θ(°C)	S (PSU)	O_2	NO_3	PO_4	H_4SiO_4
			(μmol kg^{-1})			
WW	−1.87	34.34	311	31.8	2.18	67.8
CDW	0.61	34.68	210	31.9	2.24	99.2
CSW	−1.85	34.56	—	—	—	—
Ice Shelf Water	−2.00	34.50	—	—	—	—
AABW	−0.51	34.66	242	32.5	2.21	139.0

Middleton & Humphries (1989) termed this mass Prydz Bay Bottom Water. Current meter measurements over the shelf are broadly consistent with this scenario.

Further information on the horizontal circulation comes from natural drifters (e.g. icebergs; Tchernia & Jeannin, 1980) and purpose-built buoys (Allison, 1989*b*). A total of nine buoys were deployed in late summer of 1985 and 1987 in the vicinity of Prydz Bay (Fig. 3). The buoys were subsequently trapped in the advancing seasonal sea-ice and drifted with the ice pack. The data collected by the buoys show the complex energy exchange involved with advancing and retreating pack-ice and the important role of bathymetry in steering both ocean currents and the surface-drifting ice. The drift pattern in the vicinity of Prydz Bay is generally consistent with the patterns inferred from hydrographic data. In particular we note the northward breakout of several buoys west of Fram Bank, while other buoys deployed in similar regions continued to drift westward with the littoral current.

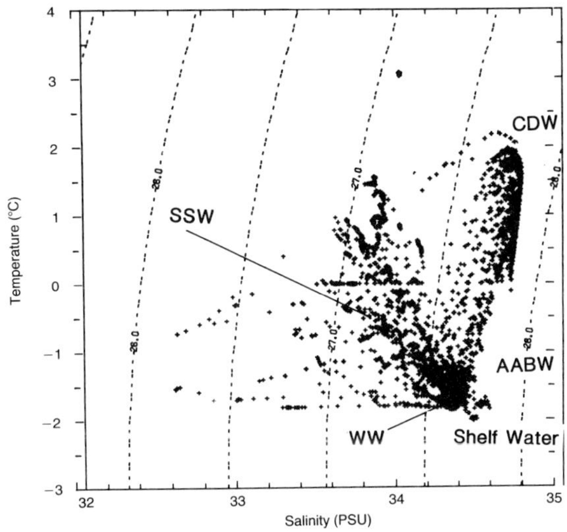

Fig. 4. An illustrative distribution of potential temperature and salinity determinations for the Prydz Bay study region (from Smith *et al.*, 1984). Also shown are the regions of the temperature–salinity plane associated with the water masses discussed in the text and defined in Table 2. The Shelf Water samples comprise both Continental Shelf Water and Ice Shelf Waters. Dashed lines mark lines of constant σ_T.

Water masses and vertical structure

The study area lies south of the Polar Front Zone, so it is convenient to adopt Deacon's (1937) water mass classifications (see Foster & Carmack, 1976, and Table 2): Antarctic Surface Water which, depending on season, can be subdivided into Summer Surface Water (SSW) and Winter Water (WW); Circumpolar Deep Water (CDW; we are specifically interested in the warm deep water component); Antarctic Bottom Water (AABW); and Shelf Water (SW; usually broken down into specific components, for example Continental Shelf Water (CSW) and Ice Shelf Water). The definition of these water masses is shown schematically in Fig. 4, together with data from an early Australian cruise (Dong *et al.*, 1982), and are tabulated in Table 2.

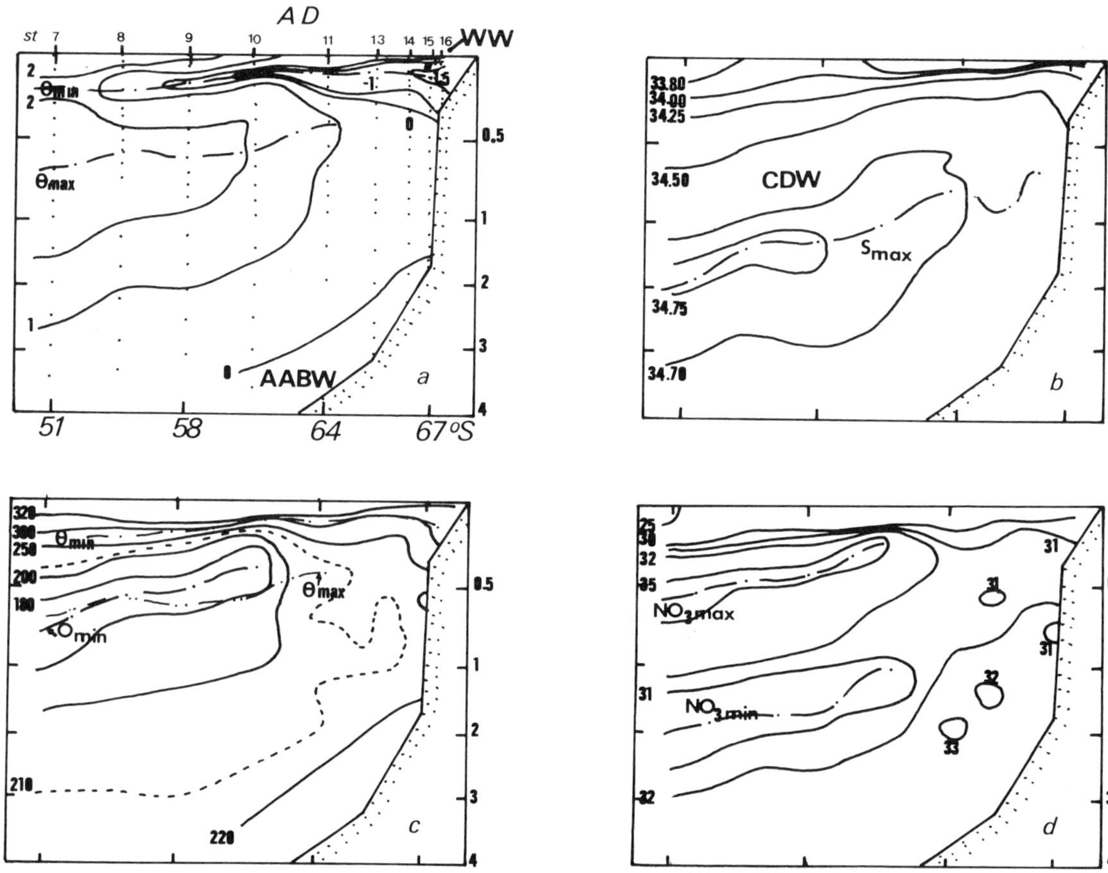

Fig. 5. Transect through the eastern part of the Weddell–Enderby in January–February 1984 (see cruise track marked in Fig. 1b). Station (St) 16 is close to Four Ladies Bank. The vertical axis is marked in kilometres. Note the change in scale after 1000 m. (a) Potential temperature, θ, in °C; θ_{min} and θ_{max} mark the temperature minimum (i.e. WW) and temperature maximum core layers. AD marks the approximate location of the Antarctic Divergence; WW and AABW are defined in Table 2 and in the text. (b) Salinity, S, in PSU (Practical Salinity Units). Note the position of the S_{max} core. (c) Oxygen concentration in μmol kg^{-1}. Note that the oxygen minimum (O_{min}) is similarly located to the temperature maximum. (d) Nitrate (NO_3) in μmol kg^{-1}. Both the minimum and maximum core layers are marked.

Typical hydrological and chemical spatial characteristics of these water masses are shown in Fig. 5 along a summer 1984 transect of a French cruise (Fiala & Fontugne, 1987), along the eastern perimeter of the Weddell–Enderby Basin.

Over the continental shelf the WW is no longer well defined as a single mode, although the temperature remains near the freezing point, so the waters are referred to generically as Shelf Waters. The salinities can vary over a large range depending upon brine rejection during freezing, convection and mixing with subsurface water, and horizontal advection and interleaving of water masses. Following Carmack (1977), Smith *et al.* (1984) accordingly differentiated between low-salinity shelf water, originating from the east near the West Ice Shelf, and high-salinity continental shelf water due to freezing and convection within Prydz Bay proper (the boundary between high- and low-salinity CSW used here and in Smith *et al.* (1984) is 34.5 compared with 34.6 PSU in Carmack (1977)). The formation of the latter type is well illustrated by data collected through a complete annual cycle in the fast ice zone (Fig. 6; Allison *et al.*, 1982; Allison, 1989*a*). The salt rejection from the freezing of sea-ice drives deep mixing of the waters, transporting properties of surface origin into the deep ocean. The column is rapidly restratified in early summer as ice melts, adding a fresh surface layer which is subsequently heated by solar radiation. It is this relatively high-salinity shelf water mass which is linked to CDW modification adjacent to the shelf. Allison *et al.* (1985) point out that the local budget also requires net advection of salt to and from the site at different times of the seasonal cycle.

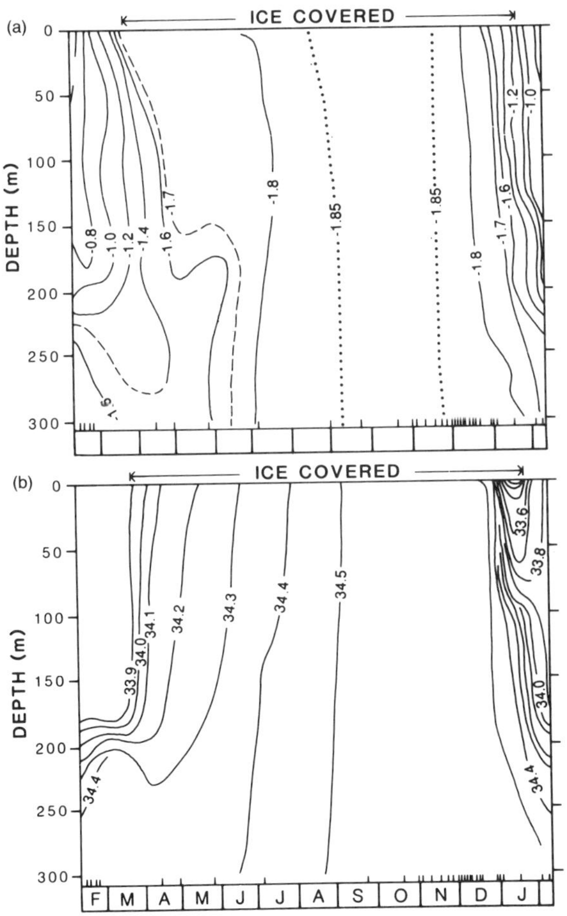

Fig. 6. Annual cycle of temperature (°C) (a) and salinity (PSU) (b) in the fast-ice zone over the continental shelf near Mawson station (near 61°E). (From Allison, 1989*a*.)

A further near-freezing high-salinity variety of SW is present in the vicinity of the Amery Shelf (see Fig. 4) and is termed Ice Shelf Water (Smith *et al.*, 1984). Although this mode has very distinctive thermohaline properties, it is unlikely to play a major part in off-shelf water mass formation. It is found at depth in the Prydz Bay region and most likely accrues over long time scales owing to its continued isolation from outside waters.

Summer Surface Water

The SSWs do not exist as a distinct mode but are represented by a relatively wide range of temperatures and salinities (<4 °C and 33.5–34.0 PSU; see Figs 4 and 5). This variation results from the seasonal influence of warming and melting at the ocean surface and intermittent cooling and mixing events. Complex upper ocean structures with double haloclines, intrusive layers, and intervening mixed layers have often been reported (e.g., Smith *et al.*, 1984; Middleton & Humphries, 1989). The SSW layer is typically around 30 m deep but may be as shallow as 10 m and can occasionally extend to 60 m depth within the shelf zone (Smith *et al.*, 1984; Tréguer *et al.*, 1987; Middleton & Humphries, 1989; Souchu *et al.*, 1990). As noted previously, the conditions in the surface layer are closely linked to the prevailing ice conditions (Grigor'yev, 1970; Denisov & Myznhikova, 1978; Smith *et al.*, 1984). In the oceanic domain the SSW layer may extend beyond 100 m (see, for example, Fig. 5). While SSW has no direct role to play in the formation of deep water masses, the buoyancy accrued in the SSW due to melting and heating in the summer months provides a barrier which must be broken down before penetrative convection can take place in the winter months.

Winter Water

From spring to autumn, below the SSW and above the relatively warm CDW, there lies a temperature minimum water mass which is generally considered to be the remnant of the winter mixed layer, and is thus known as Winter Water (WW) (for example, see Figs 4 and 5). This water mass has distinct modal and core properties: temperatures less than −1.5 °C and salinity close to 34.3 PSU (Smith *et al.*, 1984; Tréguer *et al.*, 1987; Middleton & Humphries, 1989; Souchu *et al.*, 1990).

The 1984 and 1985 French cruises (see Table 1) focused on the characteristics of WW, especially for temperature and oxygen concentration (Fig. 7). Temperatures in the WW layer were typically in the range −1.56 °C to −1.87 °C. During these cruises a calibrated temperature/oxygen probe was used to get continuous vertical profiles (Fig. 7a); the results clearly indicated that the WW layer was oxygen undersaturated. For stations located off the continental shelf during summer, but still near the pack-ice, the oxygen concentrations were 81% of saturation in 1984, and 85–89% of saturation in 1985; that is 40–67 μmol kg^{-1} below saturation. Such values for oxygen undersaturation have been reported for the WW layer of the Weddell Sea by Gordon *et al.* (1984) who measured values averaging 48 μmol kg^{-1} below saturation, just at the beginning of the seasonal ice retreat.

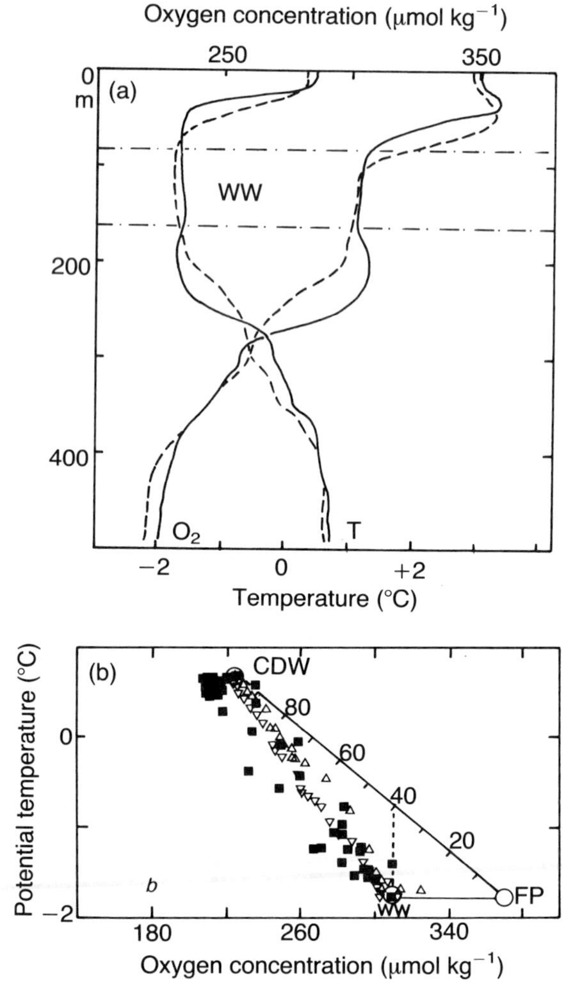

Fig. 7. A demonstration of the oxygen under-saturation of Winter Water in the vicinity of Prydz Bay. (a) Profiles of oxygen (O_2) and temperature (T) from $CTDO_2$ measurements in the upper 500 m layer on 25–26 January 1985 for stations 21 (continuous line) and 22 (dashed) of the French SIBEX cruise (see Fig. 1b and Table 2). (b) Potential temperature versus oxygen for ANTIPROD 3 (■) and SIBEX (△) stations located in the open ocean off Prydz Bay in the summer of 1984. FP marks the freezing point, WW the Winter Water. The dashed line indicates the required proportions of warm CDW and fully oxygenated winter mixed layer water required to get the measured oxygen levels in the WW.

Estimation of warm deep water entrainment

Gordon *et al.* (1984) discuss possible causes of the undersaturation of WW. The two main possibilities are: (i) local non-conservative processes (oxidation of organic matter produced in the surface layer, and heterotrophic respiration); and (ii) local entrainment of low-oxygenated warm CDW into the mixed layer during winter when atmospheric exchange is suppressed by ice cover. This entrainment continues as long as penetrative convection is being driven by surface brine rejection (Fig. 6, although based on data in the fast-ice zone, is probably a reasonable representation of this process). Many authors (for example, Weiss *et al.*, 1979; Edmond *et al.*, 1979; Gordon *et al.*, 1984) have noted the persistence of the relatively low oxygen concentration in the WW layer through different seasons and concluded that non-conservative processes are probably not the major mechanism controlling the oxygen concentrations in this layer. For the time being we will assume that these conclusions for the Weddell Sea are also appropriate for the present study region.

Following Gordon *et al.* (1984) the degree of oxygen undersaturation in the WW can be used to develop an estimate of the percentage of warm deep water entrained into the winter mixed layer during the period of ice cover, and thus infer the deep upwelling rate. This calculation is based on the assumption that at the beginning of winter (just prior to ice coverage) the well-mixed upper layer is in equilibrium with the atmosphere; that is, oxygen-saturated. Unfortunately there are little data to corroborate this assumption, nor are there measurements to indicate whether non-local processes (e.g. advection) may be important, or whether oxygen concentrations in the upper waters are surface-modified at other times of the year. Also, these measurements were not taken immediately after the pack ice retreated, so it is possible that some WW oxygen enrichment may have taken place in the intervening two-month period. However, estimates with a typical diffusion rate of 0.5 $cm^2\ s^{-1}$ suggest that such effects are probably quite small.

With these assumptions in mind, and the attendant uncertainty in the estimation, Fig. 7 suggests a mixture of about 40% warm CDW (0.65 °C and oxygen concentration 230 μmol kg^{-1}) and 60% winter surface mixed layer water (−1.87 °C and 370 μmol kg^{-1}) is required to form a WW mass whose oxygen concentration is near 310 μmol kg^{-1}. On an annual basis, for a homogeneous WW layer 100 m thick, entrainment of 40 m of warm deep water corresponds to an upwelling rate of 1.1×10^{-4} cm s^{-1}. This estimate is of the same order of magnitude as that obtained by Gordon *et al.* (1984) and by Schlosser *et al.* (1987), who estimated entrainment rates of 15–35 m per year for the Weddell Sea. These results tentatively suggest that deep water entrainment is stronger in this region than in the Weddell Sea. However it should be borne in mind that the method does depend on several assumptions, which remain largely unsubstantiated, and that, in any case, we are comparing just a few estimates from a very large region; neither the Weddell Sea nor the present region estimates may be truly representative of the large-scale, long-term mean entrainment. On a more positive note, it would seem that neither the WW nor the CDW oxygen concentration fields are particularly noisy and that the signal (the oxygen concentration difference) is clear, so this returns some degree of confidence in the results.

As a final point, it is worth while returning to the question of non-conservative processes. There is no doubt that biological processes play a role in diminishing the oxygen content in subsurface layers. According to Nelson & Smith (1986), in the marginal ice zone, the total flux of matter exported from the surface layer (that is, new production) should be no more than 0.4 g C $m^{-2}\ d^{-1}$. Total mineralization would require an oxygen consumption of approximately 40 mmol $m^{-2}\ d^{-1}$. To produce an oxygen consumption as large as 50 mmol m^{-3} in a layer 100 m thick, we would need to assume a residence time of about 4 months for the particulate organic matter (that is, a mean sinking rate of about 25 m per month). Such a long time seems unrealistic, at least for the majority of organic matter, which comprises large particles whose sinking rates have been demonstrated to be as large as 250 m per day (e.g. Tsunogai *et al.*, 1986; Bodungen *et al.*, 1986). Although nothing is known about the rate of oxygen consumption for the mineralization of dissolved organic matter in this layer, according to previous studies (Packard *et al.*, 1971; Skopintsev, 1976; Knauer *et al.*, 1979) we could expect low values in such cold Antarctic waters. Thus we may safely assume that the contribution of non-conservative processes to low oxygen content of the WW layer is small.

Continental Shelf Water

Relatively salty and dense waters are often found over the continental shelf. Smith *et al.* (1984) identified both low- and high-salinity varieties, but for the present discussion, it is the high-salinity variety which is of most interest (temperatures less than −1.5 °C, salinity greater than 34.50 PSU). In contrast to the open oceans, where entrainment of deeper, less saline waters restricts the salinity increase during winter-time mixing, mixing over the shelves often involves the whole water column (Fig. 6; Allison *et al.*, 1982, 1985). From March through to May convection driven by surface ice formation and brine rejection erodes the halocline, after which there is a steady increase in the salinity and decrease in temperature until a homogeneous vertical structure is achieved. It is this process which we identify with high-salinity shelf water formation. Note that there is a prolonged period of production and that presumably the horizontal circulation is involved with the annual cycle of redistribution of these waters (Grigor'yev, 1970; Smith *et al.*, 1984).

While CSW appears to be confined to Prydz Bay proper and the adjacent continental shelves, it seems likely that at least some of this water is forced off the shelf in a modified form, possibly through the episodic coastally trapped wave process described by Middleton & Humphries (1989). They termed this modified water mass Prydz Bay Bottom Water (perhaps modified Prydz Bay shelf water would have been a more appropriate name), and noted that it would be insufficiently dense to reach abyssal depths, but that it might interleave with Circumpolar Deep Water at intermediate depths in the manner described by Carmack & Killworth (1978). Indeed each of the physical oceanography programmes listed in Table 1 have sought evidence of CSW involvement in a locally sourced variety of AABW. To date, none has found definitive unambiguous evidence for such a role, although the indirect evidence for a role in intermediate depth water mass formation is strong. This lack of success is due at least in part to the lack of near-bottom current measurements in the key shelf slope region west of Fram Bank, the likely exit route for any Prydz Bay modified, dense CSW.

Circumpolar (warm) Deep Water

In the west Indian sector of the Southern Ocean the CDW usually comprises two components (Fig. 5; see also Le Jehan & Tréguer, 1983). At depth the major input is provided by a water mass related to North Atlantic Deep Water, easily detectable by a salinity maximum (34.70–34.75 PSU), almost coinciding with a NO_9 ($O_2 + 9 \times NO_3^-$; Broecker, 1974) minimum (468–480 mmol kg^{-1}). Following the core layer southwards it rises along the 32.43–32.45 isopycnals between 3000 m and 1000 m, and along the 27.80–27.83 isopycnals above 1000 m. In the upper layers the influence of the Indian and Pacific Oceans is stronger (Jacobs & Georgi, 1977; Le Jehan & Tréguer, 1983), as evidenced here by the oxygen minimum and nitrate maximum core layers (less than 180 μmol O_2 kg^{-1} and greater than 35 μmol NO_3 kg^{-1}, respectively) (Fig. 5c, d) which rise with the 27.6–27.7 isopycnals. This general scheme is modified in the region south of the Antarctic Divergence. There, along the continental slope, a single core layer is evident. The salinity maximum is encountered at levels between 600 and 1000 m, with temperatures near 0.65 °C.

Using data collected by French cruises we suggest that this modified warm CDW south of the divergence is the end product of the CDW–CSW mixing process described previously, and improperly termed Prydz Bay Bottom Water by Middleton & Humphries (1989) (Jacobs (1991) prefers the terminology Antarctic Slope Front waters). We have already noted that current meters deployed between Four Ladies and Fram Banks indicate an episodic exchange of waters between the shelf and the continental slope. Fig. 8 suggests that along-isopycnal mixing of 1/3 CSW and 2/3 Indian sector CDW (1.65 °C and 34.76 PSU at the level of the S_{max} core layer near the latitude of the Antarctic Divergence) would form a mixture with the characteristics of the modified CDW found south of the divergence and west of Prydz Bay. Hydrological and chemical data (oxygen and freons) collected by a French cruise in January 1987 off MacRobertson Land confirms that this modified water mass contains freons in quantities consistent with input of surface water.

Antarctic Bottom Water

AABW is found in almost all sections that penetrate to the bottom in the deep ocean off Prydz Bay. However, with the data collected during BIOMASS we are unable to clearly define the role played by local (that is, Prydz Bay and the adjacent continental shelves) deep mixing and water formation processes in maintaining the AABW mass, other than to suggest mechanisms whereby high-salinity shelf water might mix with relatively warm and salty deep water to form a denser variety. We restrict our discussion to two aspects. First, by comparing AABW samples taken up to eight years apart, we look for possible temporal variability which, if it existed, might suggest a local bottom water source. Then possible connections and mixing between bottom waters of the Weddell Enderby Basin (which presumably are dominated by Weddell Sea AABW formation) and the South Indian Basin (where Ross Sea AABW is dominant) are investigated.

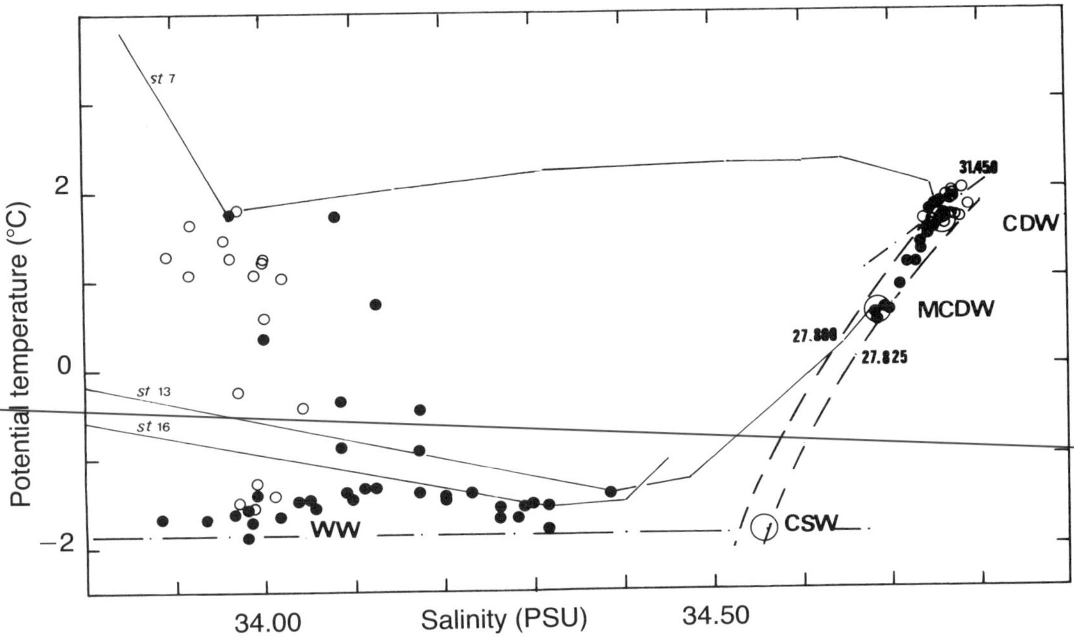

Fig. 8. Diagram of potential temperature versus salinity for the temperature minimum Winter Water mass layer and the salinity maximum layer (as shown in Fig. 5) for stations distributed S of the Polar Front (data from ANTIPROD 1 (open circles) and French SIBEX (filled circles) cruises (see Table 1). Solid lines are for ANTIPROD 3 data; station 7 was located N of the Atlantic Divergence (AD), and stations 13 and 16 were situated S of the AD. The dashed line demarcates the waters of the S_{max} core layer, and suggests a mixing path for CSW and warm deep water south of the Antarctic Divergence. CDW_1 and CDW_2 mark Circumpolar Deep Water north and south of the Antarctic Divergence (South Indian and modified forms, respectively).

AABW variability

To examine possible temporal variations in the physical and chemical characteristics of deep and bottom waters in the study region we compare data collected at similarly located stations in the eastern part of the Weddell–Enderby Basin from the first four French cruises (see Table 1):

1. 16 March 1977, 60°26′S, 66°27′E, 4670 m;
2. 23 February 1980, 59°58′S, 66°28′E, 4450 m;
3. 24 January 1984, 61°04′S, 66°21′E, 4475 m;
4. 6 February 1985, 61°00′S, 68°18′E, 4380 m.

The maximum distance between these stations is 240 km. The methods for measuring physical and chemical parameters during these cruises were similar on all cruises (see Annex 1: Methods). The vertical profiles of temperature, salinity, oxygen and nutrient outside the surface layers are similar (not shown).

Fig. 9 shows scatter diagrams of various physical and chemical parameters plotted against potential temperature for the stations above. In general, the plots reveal only relatively small variations in any of the water properties with time, and mostly within the range of instrumental and spatial sampling error. A nearby station occupied during GEOSECS (at 59°59′S, 60°58′E on 10 February 1978) yields relationships similar to those depicted in Fig. 9. Using this station as the absolute reference, and considering only samples in the AABW class (refer to

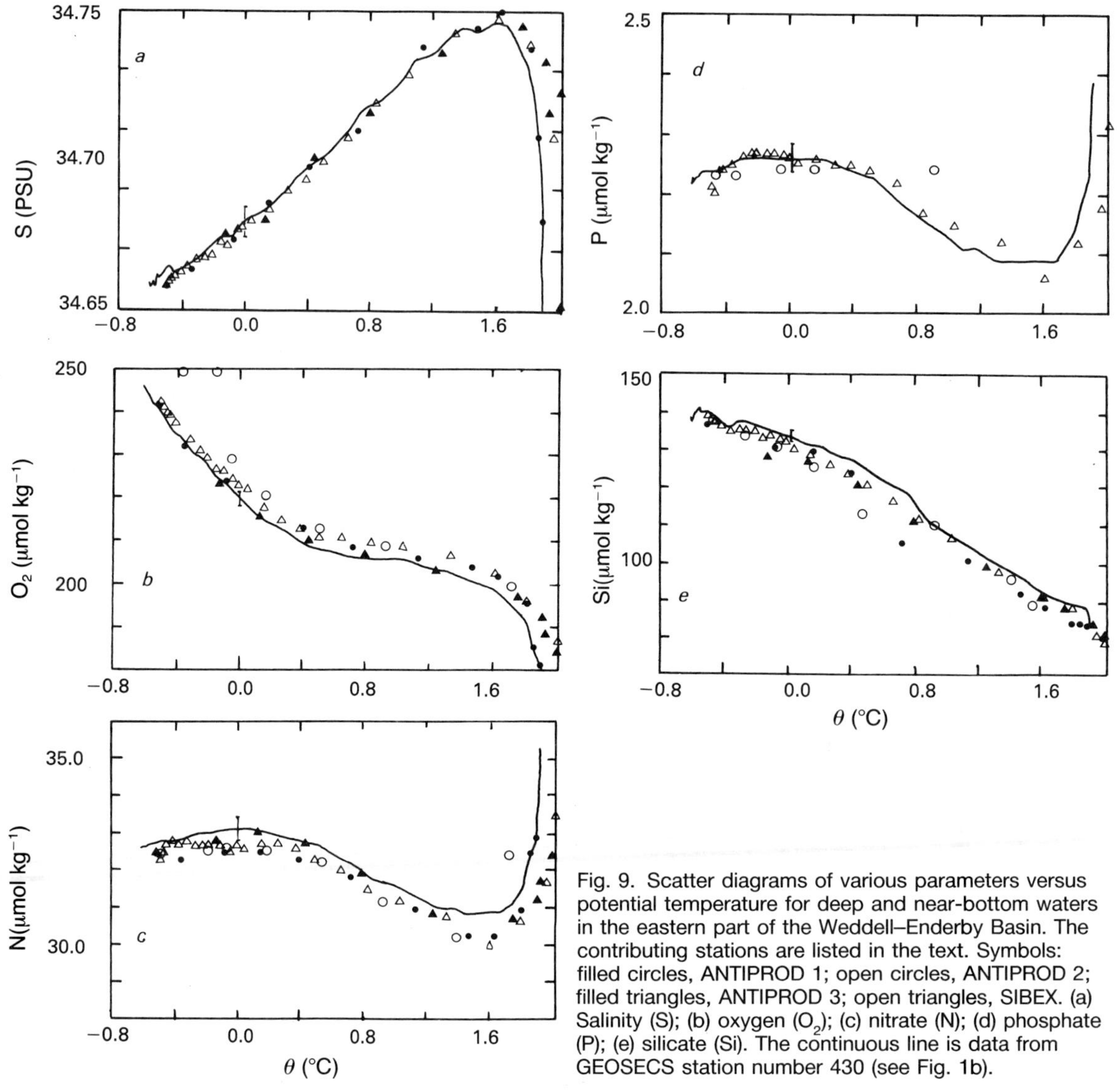

Fig. 9. Scatter diagrams of various parameters versus potential temperature for deep and near-bottom waters in the eastern part of the Weddell–Enderby Basin. The contributing stations are listed in the text. Symbols: filled circles, ANTIPROD 1; open circles, ANTIPROD 2; filled triangles, ANTIPROD 3; open triangles, SIBEX. (a) Salinity (S); (b) oxygen (O_2); (c) nitrate (N); (d) phosphate (P); (e) silicate (Si). The continuous line is data from GEOSECS station number 430 (see Fig. 1b).

Table 2), the scatters were typically 0.000 to 0.005 PSU for salinity, −0.5 to 2.6 μmol kg^{-1} for oxygen, 0.0 to −0.7 μmol kg^{-1} for nitrate, −0.03 to 0.04 μmol kg^{-1} for phosphate, and 0 to −10 μmol kg^{-1} for silicate. This level of consistency in physical and chemical properties over a period of eight years (a significant fraction of the estimated renewal time of Southern Ocean deep waters; see, for example, Gordon & Taylor, 1975) leads to the conclusion that the bottom waters at the eastern boundary of the Weddell–Enderby Basin are nearly constant. This is consistent with the conclusion presented previously, that Prydz Bay itself seems unlikely to contribute directly to the AABW mass. This suggests that the distinctive AABW identified by Jacobs & Georgi (1977) and Mantyla & Reid (1983) might arise from a continual process of interleaving of Enderby Land/Prydz Bay coastal waters with dense deep water, but this process has not yet been resolved.

AABW exchange between the Weddell–Enderby and South Indian Basins

Since the temporal variability in the Weddell–Enderby Basin data is small, we can safely composite the entire data set (including the aforementioned GEOSECS data; see Fig. 1b) in order to estimate interaction between the various varieties of AABW filling the basin in the vicinity of the Kerguelen Plateau. The physical data have been used to construct the θ-S diagram in Fig. 10a (refer to the figure legend for an explanation of the symbolism). The transition from deep to bottom water varieties is continuous for the Weddell–Enderby Basin data (solid circles). However, for the South Indian Basin (open circles) an inflection point appears near −0.25 °C and 34.675 PSU. The distribution of the data looks like an inverted 'Y' (according to Carmack, 1977).

Referenced against the σ_{4000} isopycnals this illustrates the fact that: (i) waters at the Prydz Bay shelf break and along the continental slope do not directly participate in the bottom water formation process (as concluded previously); and (ii) the characteristics of the bottom waters in the vicinity of the Princess Elizabeth Trough are consistent with mixing of bottom waters from the Weddell–Enderby and South Indian Basins (AABW of Weddell Sea and Ross Sea origins, respectively). A detailed comparison of stations near the Princess Elizabeth Trough with stations occupied in the Weddell–Enderby Basin proper shows that the influence of South Indian Basin AABW drops off sharply with distance from the Princess Elizabeth Trough.

Despite a somewhat wider dispersion, the oxygen salinity diagram (Fig. 10b) reveals a similar 'Y'-like distribution. This dispersion may be the result of non-conservative influences, especially in the relatively young slope waters. The silicate–salinity distribution (Fig. 10c) reveals major differences compared with the previous two plots. The 'crook-

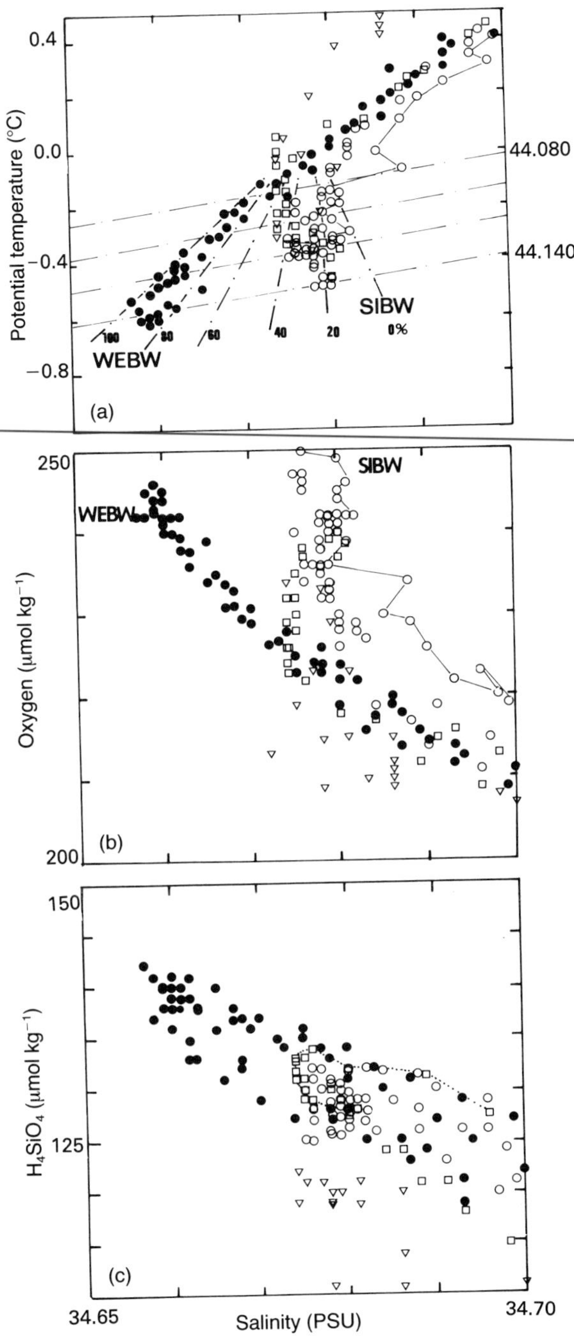

Fig. 10. Scatter diagrams for various parameters versus salinity for near-bottom waters of the Weddell–Enderby Basin (WEB; filled circles, data from stations GEOSECS 430, SIBEX 34 and SIBEX 29), South Indian Basin (SIB; open circles, data from stations GEOSECS 432, SIBEX 14 and SIBEX 7: 54°35S, 81°15E, 4025 m, not shown on Fig. 1b), the Princess Elizabeth Trough (PET: squares, data from GEOSECS 431 and SIBEX 18), and continental slope off Enderby Land (triangles, stations 14, 15 and 17 of ANTIPROD 3). (a) Potential temperature; (b) oxygen; (c) silicate. The dot-dashed lines indicate the percentages of Weddell-Enderby Basin Bottom Water (WEBW) in the blend of WEBW and South Indian Basin Bottom Water (SIBW) along the σ_{4000} lines.

shaped' distribution for stations located in the South Indian Basin and Princess Elizabeth Trough (the dotted line in Fig. 10c) indicates a silicate maximum some 100–200 m above the bottom, as found by Edmond *et al.* (1979). The scatter suggests that silicate is at least partly controlled by non-conservative behaviour in the deep and bottom waters. In the Weddell–Enderby Basin, for a given salinity, the silicate concentrations can vary by as much as 10 $\mu mol\ kg^{-1}$, well outside the range of analytical uncertainty (see Annex 1: Methods). This special behaviour for silicate in deep and bottom waters is in agreement with previous studies in the Southern Ocean (Le Jehan & Tréguer, 1983; Jacobs and Georgi, 1977), and suggests that silicate is only a semi-conservative tracer.

Summary and concluding remarks

This review has briefly examined the circulation and physical and chemical water properties over the shelf and in the deep ocean in the vicinity of Prydz Bay, relying on data collected by French and Australian cruises during the BIOMASS period to delineate key processes.

The dynamic method has been used to resolve the horizontal circulation within Prydz Bay and in the region adjacent to the bay. The general pattern is consistent with direct measurements from several current meters and with natural and purpose-built buoys. However, the accumulated knowledge from these data has not vastly improved upon knowledge of the circulation derived from earlier work, mainly by Soviet scientists, excepting that there is now greater confidence in the constancy of the circulation pattern and somewhat better detail in what are thought to be important areas of the shelf. There remains a great deal of information in the accumulated data set for the region (both physical and chemical) that has yet to be exploited. For example, thermodynamic numerical models and inverse models have not played a significant role in the interpretation of the data thus far.

From a biological perspective, information on the vertical circulation, frontal features, and the long-term mean annual cycle of currents will be critical. It is easy to be critical of the dominant role played by biology in setting the design of the physical and chemical ocean observation programmes, but it would seem that knowledge of the key biological processes is still not sufficiently advanced to provide the motivation, justification and resources for a more intensive physical and chemical survey. The physics and chemistry of the region are apparently not sufficiently interesting and distinctive to warrant such action on their own.

There is both direct and indirect evidence that Prydz Bay high-salinity continental shelf water does get mixed with off-shelf warm circumpolar deep water, probably through the agency of continental shelf waves, and that the resultant mixture is dense enough to flow out and down the continental slope, at least to intermediate depths. The properties of a modified form of warm deep water found off Enderby Land seems to support this. However, at this point, there is no evidence of Prydz Bay continental shelf water being directly involved in bottom water formation, although speculation does continue (Nunes Vaz & Lennon, 1993; T. Foster, personal communication).

In general the water masses in the vicinity of Prydz Bay do not have distinctive characteristics compared with Weddell and Ross Sea varieties. The waters over the shelf are interesting (anomalously high temperatures; dense ice shelf water; concentrated wintertime convection and formation of high-salinity shelf water) but there is no strong evidence of these waters playing a significant role in the global scheme of deep water formation. The low oxygen concentrations of winter water are similar to those elsewhere about the continent, although there is a suggestion of a slightly stronger entrainment process.

The properties of Antarctic Bottom Water in the region have remained constant over the period of BIOMASS according to French cruise data. In view of the known interannual variability in Prydz Bay, this again suggests the bottom waters found in the region are not formed by local processes. The bathymetry of the deep ocean region, characterized by the Weddell–Enderby Basin in the west, the South Indian Basin in the east, and the Kerguelen Plateau and Princess Elizabeth Trough in between, does seem to be significant in the movement and mixing of bottom waters. The present analysis demonstrated that the Princess Elizabeth Trough is an important conduit in the functioning of the deep ocean system, and that the Kerguelen Plateau is important not only for its influence on the Antarctic Circumpolar Current but also as a barrier to deep water movement. Perhaps the role of bathymetry in facilitating mixing of Ross Sea and Weddell Sea AABW masses may be more critical and of more interest for the global physical and chemical water budgets than any local water formation processes.

Acknowledgements

P.T. wishes to thank Guy Jacques, Michel Fontugne, Michel Fiala and Patrick Mayzaud, chief scientists of the ANTIPROD, SIBEX and APSARA cruises. L. Gamberoni provided the CTD data and continuous oxygen measurements (SIBEX cruise). P. Souchu made valuable suggestions on an earlier draft of this review. We are especially grateful to TAAF/Mission de la Recherche and CNRS (JE 034692) for logistic and financial support.

The difficult task of organizing, collecting and recording the Australian cruises was undertaken by many Antarctic Division scientists and technical staff, and for that N.S. is extremely grateful! The assistance of Richard Nunes Vaz (who kindly supplied a draft of an as yet unpublished manuscript), Bill Budd and Jason Middleton is gratefully acknowledged. N.S.'s contribution is on behalf of all the above.

The criticism and helpful suggestions by two anonymous reviewers greatly improved the manuscript. Thank you for your patience and input.

References

Allison I (1989*a*) The East Antarctic sea ice zone: Ice characteristics and drift. *GeoJournal* **18**: 103–15.

Allison I (1989*b*) Pack-ice drift off East Antarctica and some implications. *Ann Glaciol* **12**: 1–8.

Allison I, Tivendale C M, Akerman G J, Tann J M, Wills R H (1982) Seasonal variations in the surface energy exchanges over Antarctic sea ice and coastal waters. *Ann Glaciol* **3**: 12–16.

Allison I, Tivendale C M, Copson G R (1985) Annual salt and energy budget beneath an Antarctic fast ice cover. *Ann Glaciol* **6**: 182–6.

Bodungen B von, Smetacek V S, Tilzer M M, Zeitzschel B (1986) Primary production and sedimentation during spring in the Antarctic Peninsula region. *Deep-Sea Res* **33**: 177–94.

Broecker W S (1974) "NO", a conservative water mass tracer. *Earth Planet Sci Lett* **23**: 177–94.

Carmack E C (1977) Water mass characteristics of the Southern Ocean south of the Polar Front. In: M V Angel (ed) *A Voyage of Discovery*. *Deep-Sea Res* **24** (Suppl): 15–41.

Carmack E C, Killworth P (1978) Formation and interleaving of abyssal waters off Wilkes Land, Antarctica. *Deep-Sea Res* **25**: 357–69.

Carrit D E, Carpenter J H (1962) Comparison and evaluation of currently employed modifications of the Winkler method for determining dissolved oxygen in sea-water, NASCO report. *J Mar Res* **24**: 286–318.

Deacon G E R (1937) The hydrology of the Southern Ocean. *Discovery Rep* **15**: 1–124.

Denisov A S, Myznhikova M N (1978) Osobennosti gidrologisheskogo rezhima v zalive Priuds (fevral' 1973 g). (Features of the oceanographic regime in Prydz Bay, February 1973.) *Trudy Sov Antarkt Eksped* **68**: 100–5. (In Russian.)

Dong Z, Kerry K R, Shearer J D, Wright S W (1982) *First International BIOMASS Experiment (FIBEX) Voyage MV Nella Dan, January–March 1981. Data Report – Oceanography*. Antarctic Division Publication, Kingston, Tasmania, 166 pp.

Edmond J M, Jacobs S S, Gordon A L, Mantyla A W, Weiss R S (1979) Water column anomalies in dissolved silica over opaline pelagic sediments and the origin of the deep silica maximum. *J Geophys Res* **84**: 7809–26.

Fiala M, Fontugne M (1987) Résultats de la campagne ANTIPROD 3 – APSARA 2 (été 1984). *Rapports Campagnes Mer TAAF* **84**: 54–74. (In French.)

Foster T D, Carmack E C (1976) Frontal zone mixing and Antarctic Bottom Water formation in the southern Weddell Sea. *Deep-Sea Res* **23**: 301–17.

Gordon A L, Chen C T A, Metcalf W G (1984) Winter layer entrainment of Weddell Deep Water. *J Geophys Res* **89**: 637–40.

Gordon A L, Molinelli E M (1982) *Southern Ocean Atlas. Thermohaline Chemical Distributions and the Atlas Data Set*. Columbia University Press, New York, 111 pp. 233 plates.

Gordon A L, Molinelli E M, Baker T (1978) Large-scale relative dynamic topography of the Southern Ocean. *J Geophys Res* **83**: 3023–32.

Gordon A L, Taylor H W (1975) Heat and salt balance within the cold waters of the world ocean. In: *Numerical Models of Ocean Circulation*. National Academy of Sciences, Washington DC, pp. 54–6.

Grigor'yev Y A (1967) Circulation of the surface waters in Prydz Bay. *Sov Antarct Exped* **7**: 74–6.

Grigor'yev Y A (1970) Hydrological investigations in Prydz Bay. *Trudy Sov Antarkt Eksped* **54**: 180–99.

Hodgkinson R P, Colman R S, Kerry K R, Robb M S (1988) *Water Currents in Prydz Bay, Antarctica, during 1985*. Australian National Antarctic Research Expedition Notes No. 59. Antarctic Division, Kingston, Tasmania, Australia, 127 pp.

Jacobs S S (1991) On the nature and significance of the Antarctic Slope Front. *Mar Chem* **35**: 9–24.

Jacobs S S, Georgi D T (1977) Observations on the southwest Indian Antarctic Ocean. In: M V Angel (ed) *A Voyage of Discovery*. *Deep-Sea Res* **24** (Suppl): 43–84.

Jacques G (1978) *Campagne Antiprod 1 (1–28 mars 1977)*. Résultats Campagnes Mer CNEXO, Publication 16, Paris, 150 pp. (In French.)

Jacques G (1982) Production pélagique dans le secteur Antarctique de l'Océan Indien. *J CNFRA* **53**: 141 pp. (In French.)

Kerry K R, Woehler E J (1987) *Oceanographic Data:*

Prydz Bay Region. Antarctic Division BIOMASS Experiment I, MV Nella Dan, November–December 1982. Australian National Antarctic Research Expedition Notes No. 52, Antarctic Division, Kingston, Tasmania, Australia, 166 pp.

Kerry K R, Woehler E J, Robb M S (1987*a*) *Oceanographic Data: Prydz Bay Region. FIBEX, MV Nella Dan, January–March 1981*. Australian National Antarctic Research Expedition Notes No. 49, Antarctic Division, Kingston, Tasmania, Australia, 114 pp.

Kerry K R, Woehler E J, Robb M S (1987*b*) *Oceanographic Data: Prydz Bay Region. SIBEX II, MV Nella Dan, January 1985*. Australian National Antarctic Research Expedition Notes No. 53, Antarctic Division, Kingston, Tasmania, Australia, 122 pp.

Knapp W W (1969) A satellite study of large stationary polynyas in Antarctic coastal water. PhD thesis, University of Wisconsin, 118 pp.

Knauer G A, Martin J H, Bruland K W (1979) Fluxes of particulate carbon, nitrogen and phosphorous in the upper water column of the northeast Pacific Ocean. *Deep-Sea Res* **26**: 97–108.

Kupetskii V N (1958) Stationarnye polyn'i v zamerzayushchikh moryakh (Recurring polynyas in freezing seas). Vestnik Leningradskogo Gosudarstvennogo Universiteta (LGU), No. 12. (In Russian.)

Kupetskii V N (1959) O prichinakh anomalii gidrologicheskikh uslovii zaliva Olaf Priuds (On the causes of anomalous hydrological conditions in Prydz Bay). *Izvestiya Vsesoyuznogo Geographicheskogo Obshchestva (VGO)* **91**: 356–7. (In Russian.)

Le Jehan S, Tréguer P (1983) Uptake and regeneration $\Delta Si/\Delta N/\Delta P$ ratios in the Indian sector of the Southern Ocean. Originality of the biological cycle of silicon. *Polar Biol* **2**: 127–36.

Lemke P, Trinkl E W, Hasselman K (1980) Stochastic dynamic analysis of polar sea ice variability. *J Phys Oceanogr* **10**: 2100–20.

Lennon G W, Nunes Vaz R A (1988) *Towards a Coordinated Program of Australian Antarctic Oceanography*. Unpublished manuscript for the Australian Antarctic Division, Kingston, Tasmania, Australia.

Leont'yev Y B (1973) Ice conditions in Prydz Bay (December 1971–February 1972). *Sov Antarct Exped* **8**: 557–8.

Mantisi F, Beauverger C, Poisson A, Metzl N (1991) Chlorofluoromethanes in the western Indian sector of the Southern Ocean and their relations with geochemical tracers. *Mar Chem* **35**: 1–4, 151–8.

Mantyla A W, Reid J R (1983) Abyssal characteristics of the World Ocean. *Deep-Sea Res* **30**: 805–33.

Middleton J H, Humphries S E (1989) Thermohaline structure and mixing in the region of Prydz Bay. *Deep-Sea Res* **36**: 1255–66.

Nelson D M, Smith W O Jr (1986) Phytoplankton bloom dynamics of the western Ross Sea ice-edge. II. Mesoscale cycling of nitrogen and silicon. *Deep-Sea Res* **33**: 1389–412.

Nunes Vaz R A, Lennon G W (1993) Physical oceanography of the Prydz Bay region of Antarctic waters. *Antarct Sci* (submitted).

Packard T T, Healy M L, Richards F A (1971) Vertical distribution of the activity of the respiratory electron transport system in marine plankton. *Limnol Oceanogr* **16**: 60–70.

Schlosser P, Roether W, Rohardt G (1987) Helium-3 balance of the upper layers of the northwestern Weddell Sea. *Deep-Sea Res* **34**: 365–77.

Skopintsev B A (1976) Oxygen consumption in the deep waters of the ocean. *Oceanology* **15**: 556–61.

Smith N R, Zhaoqian D, Kerry K R, Wright S (1984) Water masses and circulation in the region of Prydz Bay, Antarctica. *Deep-Sea Res* **31**: 1121–47.

Souchu P, Percelay L, Tréguer P (1990) Caractéristiques chimiques des eaux antarctiques dans la région du Plateau des Kerguelen en été 1985. *Rapports Campagnes Mer TAAF* **85–01**: 25–42. (In French.)

Streten N A (1983) Antarctic sea ice and related atmospheric circulation during FGGE. *Arch Met Geophys Biokl* **32**: 231–46.

Strickland J D H, Parsons T R (1972) A practical handbook of sea water analysis. *Fish Res Bd Can Bull* **1**: 155–64.

Tchernia P, Jeannin P F (1980) Observations on the Antarctic East Wind drift using tabular icebergs tracked by satellite Nimbus F (1975–1977). *Deep-Sea Res* **27**: 467–74.

Tréguer P, Appriou P, Bordin G, Cuet P, Le Jehan S (1987) Caractéristiques hydrologiques et chimiques des eaux du secteur Indien de l'Océan Austral. *Rapports Campagnes Mer TAAF* **84**: 54–74. (In French.)

Tréguer P, Delmas R, Le Corre P, Le Jehan S, Masson A (1982) Caractéristiques chimiques des eaux antarctiques et subantarctiques dans le secteur Indien de l'Océan Austral (mars 1980). *J CNFRA* **63**: 55–80. (In French.)

Tréguer P, Le Corre P (1975) Manual d'analyses des sels nutritifs dans l'eau de mer. Utilisation de l'AutoAnalyzer II. Document UBO, Brest, 110 pp. (In French.)

Tsunogai S, Noriki S, Harada K, Kurosaki T, Wanatabe Y, Maeadaa M (1986) Large but variable particulate flux in the Antarctic Ocean and its significance for the chemistry of Antarctic water. *J Oceanogr Soc Jap* **43**: 83–90.

Weiss R F, Östlund H G, Craig H (1979) Geochemical studies of the Weddell Sea. *Deep-Sea Res* **26**: 1093–120.

Whitworth T III (1988) The Antarctic Circumpolar Current. *Oceanus* **21**: 53–8.

Woehler E J, Williams R (1988) *Oceanographic Data: Prydz Bay Region. Australian Antarctic Marine Biological Ecosystems Research 86/87, MV Nella Dan, February–April 1987*. Australian National Antarctic Research Expedition Notes No. 63,

Antarctic Division, Kingston, Tasmania, Australia, 136 pp.

Woehler E J, Kerry K R, Humphries S E, Robb M S (1987) *Oceanographic Data: Enderby Land Region. Antarctic Division BIOMASS Experiment III, MV Nella Dan, September–December 1985*. Australian National Antarctic Research Expedition Notes No. 54. Antarctic Division, Kingston, Tasmania, Australia, 111 pp.

Zverev A A (1959) Anomalous sea water temperatures in Olaf Prydz Bay. *Sov Antarct Exped* **1**: 269–71.

Zverev A A (1963) Currents in the Indian sector of the Antarctic. *Trudy Sov Antarkt Eksped* **17**: 144–55.

Zwally H J, Campbell W J, Carsey F D, Gloersen P (1983) *Antarctic Sea Ice, 1973–1976: Satellite Passive-Microwave Observations*. National Aeronautics and Space Administration Report, NASA SP-459, Washington, D.C. 206 pp.

Annex 1: Methods used for the ANTIPROD and French SIBEX cruises

Water samples: Collected by means of classical hydrocasts (2–5 l Niskin samplers) except for SIBEX, which used a rosette sampler operating in parallel with a Neil Brown CTD.

In situ temperature: determined using reversing thermometers and/or calibrated CTD: accuracy ±0.01 °C.

Salinity: measured aboard ship by means of either Beckman or Guidline salinometers and/or calibrated CTD (Neil Brown system); accuracy better than ±0.005 PSU. Data acquired with Neil Brown CTD accuracy better than ±0.005 PSU.

Oxygen: During the ANTIPROD 1 cruise oxygen was determined by Carrit & Carpenter's (1962) method (accuracy ±0.5 μmol kg^{-1}). Otherwise determined by Strickland & Parson's (1972) method (accuracy ±1.1 μmol kg^{-1}). For SIBEX a continuous calibrated oxygen determination was also available.

Orthosilicic acid (called silicate): analysed aboard ship using unfiltered samples stored in polythylene flasks and following the automated method described by Tréguer & Le Corre (1975) (accuracy ±1%).

Nitrate: unfiltered samples stored in polythylene flasks with immediate determination on board ship during ANTIPROD 1 and 2, otherwise frozen samples stored at −20 °C analysed 4–6 months later. In both cases nitrate was determined by following Tréguer & Le Corre's (1975) automated method (accuracy ±1%).

Phosphate: unfiltered samples collected in polyethylene flasks during ANTIPROD 1, and in glass flasks during other cruises. The phosphate concentration was determined according to Tréguer & Le Corre's (1975) method. For ANTIPROD 1 and 2 the determination was done immediately on board ship (accuracy ±1%); otherwise the samples were processed as for nitrate (accuracy ±1.5%).

Discussant's report: Antarctic marine environment – physical oceanography

E. FAHRBACH

TWO PAPERS were presented in the Colloquium Session on *Antarctic marine environment physical oceanography*. The first one, by M. Stein & R. B. Heywood, reported on the Antarctic Peninsula and Southwest Atlantic Region of the Southern Ocean and the second, by N. R. Smith & P. Tréguer, on the Amery Sea and the Region of Prydz Bay. The authors gave comprehensive overviews of the sea areas where the major part of the BIOMASS activities took place.

Both areas have common features which distinguish them from other parts of the Southern Ocean. They are each a part of one of the subpolar cyclonic gyres: the Weddell Gyre or the somewhat smaller Amery Gyre. Those gyres are of particular interest because they represent areas of significant meridional advective transports in the Southern Ocean, which is normally much more determined by zonal advection in the Antarctic Circumpolar Current. Another common feature of the two study areas is that they include the boundary between two oceans. At these boundaries, water masses of significantly different origin converge and form new water masses. In the Southwest Atlantic Region, water masses of the Antarctic Circumpolar Current and the outflow from the Weddell Sea form the Weddell–Scotia Confluence. In the Amery Sea, waters off the Weddell–Enderby Basin meet those of the South Indian Basin.

In spite of these similarities, significant differences are obvious. The Amery Sea is bound to the south by the Amery Shelf Ice, which exerts a strong influence on the shelf water masses by subsurface cooling and melting. This leads to the formation of deep water which, even if it is not dense enough to sink to the bottom, plays a significant role in the ventilation process. In contrast, the Southwest Atlantic Region is not bounded by a significant shelf-ice area and the water mass formation processes are dominated by the mixing processes in the Weddell–Scotia Confluence.

In addition to the natural differences between the two areas, the direct interests of the authors made them focus their views on different aspects. Stein & Heywood concentrated on the description of the near-surface features, giving detailed descriptions of ocean currents and water masses, possible forcing mechanisms and implications for biota. Smith & Tréguer, on the other hand, emphasized the water mass formation processes and interbasin exchanges in the deep and bottom waters.

As a result of the two presentations, we can say that we know the relevant features and processes in the two areas in much detail and we know that there can be significant changes in time. However, some questions arise:

> How representative are these two areas? Can we use the knowledge derived from these special areas in the subpolar gyres to understand observations from later or earlier periods?

How important is the time change? Can we use the present data to understand observations from other areas in the Southern Ocean?
Did we observe normal conditions?
Can we establish quantitative relationships between physical and biological processes?
Can we use observed correlations between physical and biological processes for prediction purposes?

The answers to these questions cannot be given on the basis of the present data sets, owing to their limitations in space and time. Processes at all temporal and spatial scales are superimposed on the observations, but the contribution of the different scales cannot be determined from local and small-scale observations. This makes it rather difficult to establish causal relations between physical and biological processes and calls for tremendous future efforts, if one takes into account the outstanding activities of the BIOMASS participants. Extensive programmes like the World Ocean Circulation Experiment (WOCE) will allow us to make a much better description of the Southern Ocean after investigation with a large-scale hydrographic grid. However, these large-scale programmes will not be able to deliver data as detailed as those from the BIOMASS cruises; such detail can only result from process-oriented studies. Ship-borne process studies will always suffer from limited time for observation, and from lack of knowledge about the initial and boundary conditions. However, progress can be expected from three sources:

Remote sensing techniques will deliver quasi-continuous data sets over vast areas. This should help to document the atmospheric forcing, as well as oceanic initial and boundary conditions.
Numerical models are becoming highly sophisticated, and refinement in horizontal and vertical resolutions should facilitate the study of detailed processes in a large-scale system.
Oceanographic mooring and drifting buoy techniques are now reliable enough to make independent measurements of long-term time series or spatial distribution.

The most promising approach for obtaining a satisfactory physical background to gain understanding of the biological processes will be to use a combination of a hierarchy of models and observations. Large-scale models which are driven by well-defined forcing conditions and controlled by long period mooring data in carefully selected areas, as well as large-scale hydrographic and drifting buoy surveys, will provide the boundary conditions for small-scale, high-resolution process models. These must be driven by actual forcing conditions, and can be run to address specific questions occurring in regional observations. Observational process studies will be necessary to calibrate these models. We are far from achieving that goal. However, there are some promising efforts in support of an optimistic point of view:

Firstly, large-scale models exist. The Fine Resolution Antarctic Model (FRAM) run by a group of researchers in the U.K. at the universities and at the Institute of Oceanographic Science's Deacon Laboratory, is a good example. Even if present computation facilities do not allow running of various experiments with such a model, the tremendous increase of computer power during the past decade means that such experiments may be possible in the near future.
Second, process models also exist. The Dutch effort during EPOS (see C. Veth, *Marine Chemistry* **35** (1–4): 63–76) showed the possibility of simulating biological observations by means of a very simple one-dimensional mixed-layer model driven by observed forcing conditions.
Third, the feasibility of combining process models into large-scale models was demonstrated by the ice–ocean models, where detailed mixed-layer and ice formation processes were successfully integrated into large-scale models.
Finally, the new generation of satellites (e.g. ERS 1) will undoubtedly improve our knowledge of the forcing conditions.

In addition to these technical improvements, there remains an important gap to close. In spite of large multidisciplinary programmes being developed under the umbrella of 'Global Change', there is still the feeling among many biological and physical oceanographers that their fields are not sufficiently appreciated or understood by the other disciplines. This requires an intensive effort on both sides to enter into meaningful dialogues and to maintain close contacts with each other if these programmes are to succeed.

Antarctic marine stocks: phytoplankton and zooplankton

Pattern and variability of phytoplankton biomass in the Antarctic Peninsula region: an assessment of the BIOMASS cruises

J. PRIDDLE, F. BRANDINI, M. LIPSKI & M. R. THORLEY

Introduction

THE BIOMASS Programme, initiated in 1977, responded to increasing concern over the consequences, both actual and potential, of a new episode of marine living resources exploitation in the Southern Ocean. Its two aims were to increase understanding of the Southern Ocean ecosystem, and to develop an ecologically based strategy for exploitation and conservation of marine species (BIOMASS, 1977). Although concentrating on exploitable species, especially krill, the BIOMASS field programme undertook measurements of a very wide range of variables (Thorley & Trathan, this volume). The field programme comprised two major campaigns: the FIBEX broad-scale survey and the SIBEX multiship repeat surveys.

In this chapter, we consider the results from cruises in the vicinity of the Antarctic Peninsula (Fig. 1a), with emphasis on the SIBEX station grid in Bransfield Strait and the southern Drake Passage. Measurements of phytoplankton biomass (as particulate chlorophyll *a*) and concentrations of inorganic nutrients, were carried out on most cruises as standard measurements at hydrographic stations. Results from most of these cruises have been published (Table 1) and the information has served to enhance and broaden our understanding of the phytoplankton biomass distribution and biogeography in an area already well studied (see, for instance, Hart, 1934; Burkholder & Sieburth, 1961; Priddle, 1985; Huntley *et al.*, 1991 for summaries). Sadly, the BIOMASS phytoplankton data were never analysed as a combined set, as was undertaken for several other topics (especially krill biology). Thus there is no information available for phytoplankton distribution and dynamics which derives from the total data set. With this in mind, we decided to carry out such an analysis as our contribution to the BIOMASS evaluation colloquium, and to concentrate specifically on an assessment of the combined data set. We have set out to identify the additional value of the BIOMASS phytoplankton data as a single coherent data set, as opposed to the analyses of individual cruises.

Analysis of variability

Our approach to the combined data set has been based on the attempt to partition the total variability into various processes and causes. First, we had to satisfy ourselves that the data from the different cruises could be combined, and that differences in technique between cruises did not prevent the data being treated as a single series of measurements. Following that, we addressed a series of potential causes for variation within the data set, and examined the data with a view to quantifying these if possible. It should be accepted that the survey

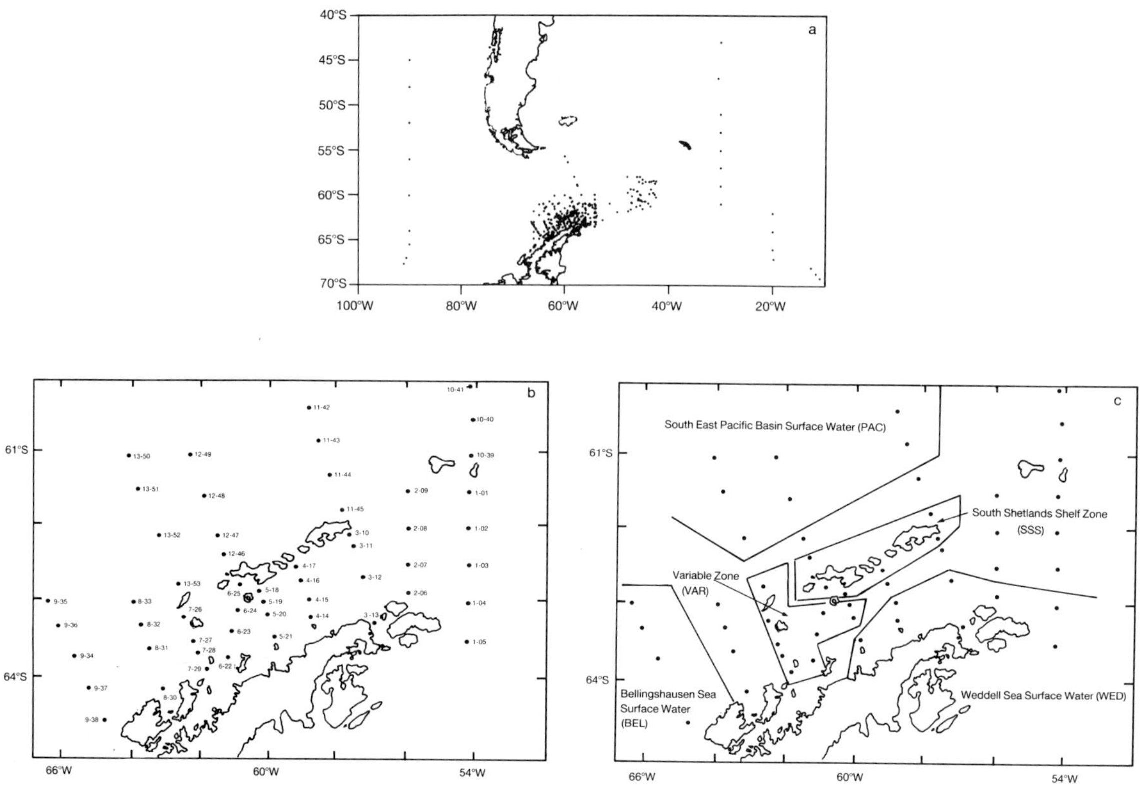

Fig. 1. (a) Positions of all BIOMASS stations in the South-east Pacific and South-west Atlantic regions of the Southern Ocean, adjacent to the Antarctic Peninsula. (b) The BIOMASS–SIBEX grid of standard stations in the Bransfield Strait – southern Drake Passage area. (c) Delineation of characteristic zones within the SIBEX grid (see text for discussion).

designs in the FIBEX and SIBEX cruise surveys were not specified with some of these questions in mind, so that significant residual variation should be expected.

Variation due to regional or water-mass differences

Measurements during the BIOMASS cruises confirmed that the study area is hydrologically and biogeographically complex. It might be expected that classification of the data into different regional regimes would account for much of the overall variation.

Variation due to nutrient availability and physical structure of the water column

At a smaller scale, variation in phytoplankton abundance can be expected to be influenced by potentially limiting features, such as the availability of inorganic nutrients, and by water-column stability.

Temporal variation: changes through the growth season

As well as spatial variation in abundance, we would expect time-related differences as phytoplankton bloom and decline through the growth season.

Table 1. *BIOMASS cruises in the Antarctic Peninsula region*

See Fig. 1.

Cruise	Date	Reference
FIBEX		
Holmberg (Argentina): HOFX[c]	19 Jan–16 Feb 1981	[a]
Itzumi (Chile): ITFX	26 Jan–23 Feb 1981	Uribe, 1982 (F, B)[b]
W Herwig (Germany): HEFX	26 Jan–1 Mar 1981	[a]
Odissey (USSR): ODFX	1 Feb–20 Mar 1981	[a]
Siedlecki (Poland): SIFX	14 Feb–13 Mar 1981	Lipski, 1982 (S, A)
SIBEX 1		
Polarstern (FRG): PSS1	23 Oct–16 Nov 1983	Nast & Gieskes, 1986
Siedlecki (Poland): SIS1	21 Dec 1983–8 Jan 1984	Lipski, 1985 (S, A)
W Besnard (Brazil): BES1	21 Jan–9 Feb 1984	MBB Kutner *et al.*, unpublished (F, B)
Alcazar (Chile): ACS1	27 Jan–11 Feb 1984	Uribe, 1987 (F, A)
SIBEX 2		
Polarstern (FRG): PSS2	21 Nov–4 Dec 1984	Weber & El-Sayed, 1987
Kayo Maru (Japan): KMS2	27 Nov 1984–27 Jan 1985	[a]
J Biscoe (UK): JBS2	16 Jan–6 Feb 1985	Mullins & Priddle, 1987 (F, B)
Alcazar (Chile): ACS2	24 Jan–12 Feb 1985	Uribe, 1987
W Besnard (Brazil): BES2	27 Jan–15 Feb 1985	MBB Kutner *et al.*, unpublished (F, B)
W Herwig (Germany): HES2	10 Mar–8 Apr 1985	[a]

[a] No paper published on chlorophyll data.
[b] Symbols in parentheses indicate methodology: F, fluorometric determination of chlorophyll; S, spectrophotometric determination of chlorophyll; A, standard depths; B, light depths.
[c] Cruise identifiers indicate ship name and the BIOMASS campaign: thus HOFX is *Holmberg*–FIBEX, ACS2 is *Alcazar*–SIBEX2, etc.

Temporal variation: interannual changes

The SIBEX programme was carried out over the summers of two successive years, which were known to have differed hydrologically and biologically (Priddle *et al.*, 1988). Such variation might explain a component of the variation within the data set.

Residual variation in the data could be attributed to a wide range of sources, especially given that the survey design was not appropriate to several key scales of phytoplankton patchiness (Murphy *et al.*, 1988).

The BIOMASS phytoplankton data set for the Antarctic Peninsula region

Analysis was undertaken on data from eight BIOMASS cruises to the Southern Ocean in the vicinity of the Antarctic Peninsula (Fig. 1a, Table 1). Three of these cruises were part of the FIBEX experiment, which comprised a near-synoptic, wide-ranging survey with little or no spatial overlap between cruises. By contrast, the SIBEX cruises were carried out on a carefully specified grid of stations, which was repeated by each cruise in turn. For the Antarctic Peninsula SIBEX area, the grid comprised two arrays of stations, one set within the Bransfield Strait and a second in the southern Drake Passage adjacent to the South Shetland Islands (Fig. 1b).

Station selection

Data analysis was carried out during June–September 1991, using the data deposited at the BIOMASS Data Centre (Cambridge, U.K.) and a subset of this database at the Alfred-Wegener-Institut (Bremerhaven, Germany). This data set did not then include information for one SIBEX cruise (ACS2), which was only deposited with the BIOMASS Data Centre after this paper was prepared. We have also excluded the cruise KMS2, which sampled stations along a series of widely spaced, long transects and was wholly different from the other cruises. For most of the analysis, we have chosen to concentrate on the SIBEX grid of 53 standard stations. Some of the FIBEX cruises sampled fortuitously at, or very near to, SIBEX standard station positions, and these data have been incorporated with the SIBEX data set. However, the number of these additional stations is low, and there was very little overlap between the two cruises which sampled near the Antarctic Peninsula (ITFX and SIFX). The other FIBEX cruise, HOFX, was carried out near the South Orkney Islands, and thus has no relation to the remainder of the data set for the purposes of this paper.

Stations have been selected for this study on the basis of the availability of sufficient values of particulate chlorophyll *a* to allow integration over the top 50 m of the water column. This criterion was adopted in order to maximize the number of stations of acceptable standard; a comparison with 100 m integration is presented below. Taking all BIOMASS cruises, this provides 343 sample sites. Of these, 235 were undertaken at SIBEX standard stations, during seven cruises. It is these stations which are used in the bulk of the analyses here, in order to use a more uniform data set and thereby remove artefacts introduced by the different spatial coverage on the various cruises. The smaller number of stations than expected ($7 \times 53 = 371$) results from several causes: the small number of FIBEX stations included; missed or misplaced stations during SIBEX; and the exclusion of parts of the grid in some cases. Data for inorganic nutrients were also integrated to 50 m depth. More stations are missing from this data set, with the result that 173 stations were available for nitrate, 172 for phosphate and 176 for silicate.

Quality control

As noted already, the phytoplankton data had not been analysed previously except as individual cruises by the scientists involved. Nutrient and physical data had been checked in the course of BIOMASS hydrographic analyses.

Obviously, our first task was to check the data and satisfy ourselves that we could use all information without concern that systematic errors could arise simply through lack of comparability between measurements from different cruises. This problem was exacerbated by the lack of documentation of methods in the BIOMASS database. Standard methods have been specified (BIOMASS, 1983), but there was no indication that these had been followed in all cases. We have established from the original published data that five cruises used fluorimetric determination of chlorophyll *a*, while at least two others used spectrophotometric methods (Table 1). Depths sampled also varied: some cruises sampled routinely at standard depths, whilst others sampled at irradiance depths (used in growth rate studies). So far as we are aware, no cruises attempted intercalibration of methods.

We can only resort to very crude quality control for this data set. Following detailed examination of the individual profiles, we have compared summary statistics for the individual cruises. Given the large number of stations for each cruise, and the considerable overlap in both space and time-of-year, we would expect that the summary statistics should be similar for each cruise, although individual stations might differ markedly. Taking all available data, it is encouraging that there is good agreement between results from different cruises. The data are markedly skewed, and the values of maxima vary widely (Table 2). However, median values for integrated chlorophyll amount are sufficiently similar for us to believe that there is no serious risk in accepting these data as being compatible. Many of the differences are explicable when cruise patterns are known. Thus the high median and minimum values for the cruise ITFX reflect the fact that this cruise did not sample in the known low-biomass waters of the South-east Pacific.

Results for inorganic nutrients for the SIBEX cruises were assessed during the course of data analysis of the physical oceanographic data. We have not attempted further data quality control on these, or the FIBEX data. In nearly all cases, the nutrient data appear reasonable although some unexpected data emerge. Summary statistics for the nutrient data are presented in Table 3.

Depth distribution and integration

As noted above, we have been forced to compromise between maximizing the station coverage, and integrating to a depth which should be representative of the water column. Ideally, we would have chosen to integrate over the top 100 m, in order

Table 2. *Chlorophyll abundances (mg m^{-2}) integrated to 50 m depth for BIOMASS cruises in the Antarctic Peninsula region*

	All stations			SIBEX Standard Stations		
Cruise[a]	*N*	Median	Range	*N*	Median	Range
HOFX	30	19.1	5.2–165.4	not sampled		
ITFX	37	64.0	12.9–327.0	9	60.5	17.4–208.5
SIFX	42	28.5	2.8–200.5	11	14.4	8.4–120.8
SIS1	62	54.4	5.9–446.0	50	60.1	5.9–391.7
BES1	44	13.5	0.45–61.9	41	14.0	0.45–61.9
ALS1	46	17.6	1.9–153.2	36	17.6	1.9–59.4
JBS2	43	18.4	1.4–288.9	43	18.4	1.4–288.9
BES2	47	18.9	2.7–1334.5	45	18.3	2.7–1334.5
'ALL BS'[b]	313	22.5	0.45–1334.5	235	25.1	0.45–1334.5

[a] For abbreviated cruise identifiers, refer to Table 1.
[b] 'ALL BS' includes all stations for cruises other than HOFX, which worked in a separate area.

Table 3. *Summary statistics for chlorophyll and nutrient amounts integrated to 50 m depth for BIOMASS–SIBEX standard stations (Fig. 1b) and for individual zones (Fig. 1c).*

	Chlorophyll (mg m^{-2})			NO_3 (mol m^{-2})			PO_4 (mol m^{-2})			Silicate (mol m^{-2})		
Region	*N*	Median	Range	*N*	Median	Range	*N*	Median	Range	*N*	Median	Range
All standard stations (53)	235	20.8	0.45–1334.5	173	1.11	0.23–1.61	172	0.09	0.02–0.12	176	2.88	0.37–6.94
WED[a](8)	34	21.3	5.5–325.8	23	1.22	0.60–1.42	24	0.09	0.03–0.11	24	3.96	1.95–4.91
BEL[b](5)	23	19.5	4.3–346.2	16	1.09	0.86–1.32	18	0.09	0.05–0.10	18	1.97	0.90–3.30
PAC[c](7)	29	6.5	0.45–61.2	22	1.02	0.58–1.31	20	0.08	0.02–0.09	22	1.09	0.37–2.04
SSS[a](6)	31	43.5	15.3–197.4	25	1.19	0.58–1.61	24	0.08	0.03–0.11	24	3.28	1.58–5.16
VAR[a](10)	43	43.5	1.36–1334.5	32	1.00	0.23–1.37	33	0.08	0.02–0.11	33	3.30	1.31–6.94

[a] No nutrient data for SIS1.
[b] No nutrient data from ITFX, SIFX, SIS1.
[c] No nutrient data from ITFX, SIS1, ACS1.

to include the entire euphotic zone and cover any chlorophyll profiles with deep maxima. However, to do so would have excluded several stations with poor depth resolution, and omitted one cruise entirely. Integration to 50 m depth proves to be a good representation of the chlorophyll profile for the great majority of stations. Chlorophyll integrated to 50 m depth was a good predictor of the 100 m integrated value for all stations and cruises (Fig. 2a). A similar result was obtained by Holm-Hansen & Mitchell (1991) for a comparison of chlorophyll biomass integrated over various depths for the RACER grid of stations in the south-western Bransfield Strait. Calculation for the present study of the ratio of 50 m to 100 m integrated values provides a crude statistic describing the shape of the chlorophyll profile, with a value of 0.5 indicating uniform mixing. For all cruises, the mean and median values of this statistic were similar (i.e. approximately normal distribution), varying between 0.52 and 0.63 and indicating a slight tendency for the chlorophyll maximum to occur in the upper 50 m (Fig. 2b). Surface values of chlorophyll concentration were in turn good predictors of the 50 m integrated values (Fig. 2), also in agreement with the findings of Holm-Hansen & Mitchell (1991).

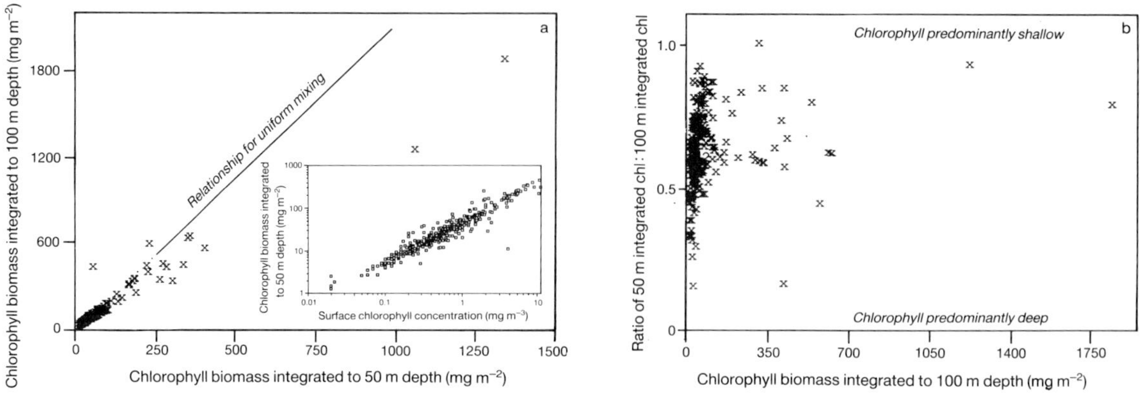

Fig. 2. (a) Relationship between depth-integrated chlorophyll to 50 and 100 m for BIOMASS stations where 100 m integration was possible; inset, relationship between 50 m depth-integrated chlorophyll and surface (< 10 m depth) concentration. (b) Relationship between the ratio of 50 m to 100 m integrated chlorophyll, and 100 m integrated chlorophyll.

Analysis of the data set

1. Regional and water-mass differences

Results of research cruises over several decades, including extensive oceanographic sampling during the BIOMASS Programme, indicate that the geographic area of the SIBEX station grid is hydrographically complex. An up-to-date summary of the hydrography of the region has been produced as the result of the analysis of the SIBEX physical oceanographic data (BIOMASS, 1990; see also Stein & Heywood, this volume). Several water masses are present in the area, deriving principally from the Weddell Sea, the South-east Pacific Ocean and the Bellingshausen Sea. Within the research area, especially in the Bransfield Strait, mixing and local modification of these waters produce a mosaic of water types. Not only is this region complex, but it is also highly variable in places as the boundaries between different water masses shift in response to atmospheric forcing (Amos, 1987; Niiler *et al.*, 1991; Stein & Heywood, this volume).

Clearly, regional and water-mass differences will influence the character and abundance of phytoplankton and no further analysis can be undertaken in this study until this is examined. However, it would be a formidable task to assign each station in our study to a particular water type on the basis of temperature–salinity and/or nutrient data. For an understanding of phytoplankton abundance, it would also probably be pointless, as such characterization provides only an instantaneous view of a continuous process, and also neglects other quasi-constant features such as proximity to the coast. Consequently, we have selected stations from the SIBEX station grid which we consider to be unequivocally associated with particular water types or environmental conditions (Fig. 1c). Three of these groups are characterized by particular water types: South-East Pacific Basin waters at the north-east edge of the station grid (PAC), Bellingshausen Sea at the southern end (BEL), and Weddell Sea waters on the eastern side of the Bransfield Strait (WED) where water enters the strait from the Weddell Sea. The other two groups have been selected to represent environmental conditions without 'allegiance' to a particular water type. Stations in shallow water close to the coast of the South Shetland Islands form one group (SSS), while a zone (VAR) situated between Smith, Deception, Trinity and Brabant Islands deliberately focuses on an area of known hydrographic variability (Niiler *et al.*, 1991; Stein & Heywood, this volume). In all cases, we have been conservative in our selection of stations, resulting in small numbers of samples in each group (Table 3).

The PAC group (7 grid points, 29 individual stations) had low values for chlorophyll and all three nutrients, according with previous observations. WED (8 grid points, 34 stations) and BEL (5 grid points, 23 stations) tended to be broadly similar in their characteristics, although WED stations had higher silicate and nitrate. The SSS group (6 grid points, 31 stations) had surprisingly low chlorophyll abundance despite the predominance of neritic habitats. The VAR group (10 grid points, 43 stations) contained the stations with the highest chlorophyll abundances, and predictably tended to be the most variable in all properties.

The BIOMASS data set provides some insight into regional variation, although nothing new has emerged from this study. In particular, high phytoplankton biomass and production in the southern Bransfield Strait–Gerlache Strait region (covered by the VAR zone in this analysis) is well documented (e.g. Hart, 1934; Burkholder & Sieburth, 1961; Holm-Hansen & Mitchell, 1991; and the BIOMASS studies cited in Table 1).

2. Correlation of phytoplankton abundance with spatial variation of nutrient concentration and water column properties

Scientific interest in the role of marine phytoplankton in global carbon cycling has reinforced interest in factors determining primary production. In Antarctic waters, the apparently paradoxical situation of low phytoplankton production and high nutrients (El-Sayed, 1984; Holm-Hansen & Mitchell, 1991; Mitchell & Holm-Hansen, 1991) has led to speculation on factors controlling phytoplankton production and nutrient utilization. The underutilized nutrient pool in the Southern Ocean has been noted as a possible indication of potentional enhancement of phytoplankton production, and hence of carbon dioxide drawdown (e.g. Martin *et al.*, 1990). Thus it seems appropriate to use the large BIOMASS data set to examine phytoplankton–nutrient relations and possible limiting factors.

Nutrient concentration and phytoplankton abundance

The spatial pattern of nutrient distribution has already been outlined in the previous section. Regional variability shown by the BIOMASS data set accords well with previous studies in the region. A noteworthy feature is the typically low nutrient status and low phytoplankton abundance in the South-east Pacific Basin waters (PAC) (Table 3). However, a simple correlative analysis of the total data set fails to indicate any overall trend (Fig. 3a–c). This is clearly a dynamic picture in which any large-scale pattern is hidden in smaller-scale spatial and temporal variability. Even repetition of the analysis for individual regions (not illustrated) does not show any relationship. The samples span a range of conditions, including both pre- and post-bloom, when the history of nutrient uptake would be very different. It is only when a localized study over a short period of time can closely follow the uptake of nutrients by a coherent bloom that the expected inverse relationship between phytoplankton biomass and nutrient concentration can be demonstrated (e.g. Heywood & Priddle, 1987). This simple approach to nutrient relations will not work with a data set such as that considered here, which embodies large-scale spatial and temporal variation.

Water-column stability

It is generally accepted that nutrient limitation in the conventional sense does not normally affect Antarctic marine phytoplankton (e.g. El-Sayed 1984; Holm-Hansen, 1985; Smith & Sakshaug, 1990), although recent suggestions that growth is limited by availability of trace elements remain equivocal (De Baar *et al.*, 1990; Martin *et al.*, 1990). Over the several decades of investigation of phytoplankton dynamics in the Southern Ocean, the strong vertical mixing of ice-free water columns has often been indicated as a major limiting factor. Thus Hart (1934, pp. 186–93) argued that physical mixing and its effect on the light climate experienced by phytoplankton was a major determinant of phytoplankton growth around South Georgia and in the Bransfield Strait. To some extent, his remarks anticipated the concept of critical depth introduced by Sverdrup (1953). Although originally intended to describe the initiation of the temperate spring bloom, this representation of the scaling of mixed-layer depth to light penetration has proved a useful concept in phytoplankton ecology (see Smetacek & Passow, 1990; Nelson & Smith, 1991). The intensity and regularity of wind-mixing typical of ice-free Antarctic waters, combined with relatively slow algal growth rates, appears likely to be one of the most important factors in phytoplankton bloom

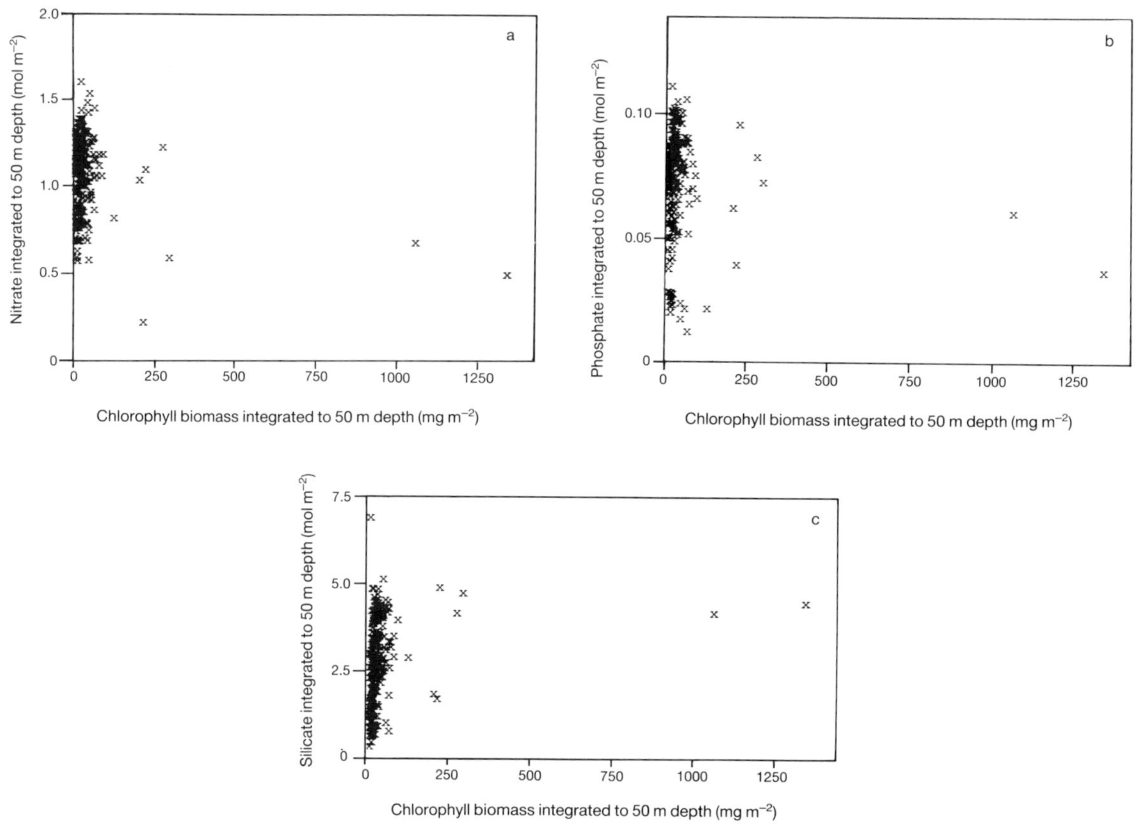

Fig. 3. Scatter plots of (a) nitrate, (b) phosphate and (c) silicate in relation to chlorophyll biomass.

formation (Sakshaug *et al.*, 1991, see also Platt *et al.*, 1991).

In a systematic study of phytoplankton dynamics in the RACER programme, Mitchell & Holm-Hansen (1991) investigated the relationship between phytoplankton abundance and the depth of the upper mixed layer (UML) in part of the Bransfield Strait. In their study, UML was defined by the depth at which seawater density σ_t changed by >0.05 over 5 m. This criterion, supported by more subjective assessment, appears to provide a reliable indication of UML (N.J. Fenton, personal communication). Mitchell & Holm-Hansen (1991) showed a clear negative relation between UML and chlorophyll biomass, and were able to describe this by a simple model of phytoplankton production and loss processes. Lipski & Rakusa-Suszczewski (1990) related vertical distribution of chlorophyll *a* over the water column to vertical stability (E) in the eastern Bransfield Strait in November. At the time of their sampling, water-column stratification was poorly developed at most stations and here phytoplankton biomass was low and distributed uniformly. Stations with higher water-column stability had higher phytoplankton abundance, and Lipski & Rakusa-Suszczewski were able to demonstrate a relation between depth of stratification and depth of chlorophyll maximum. The BIOMASS data set does not allow us to extend either approach to the wider area of the SIBEX study, as physical data are not of sufficient quality or resolution to calculate the density gradients needed to define UML. Instead, we have simply calculated the density gradient over the top 50 m of the water column as an indication of stability, without being able to identify a mixed layer depth *per se*. Most of the stations had very low,

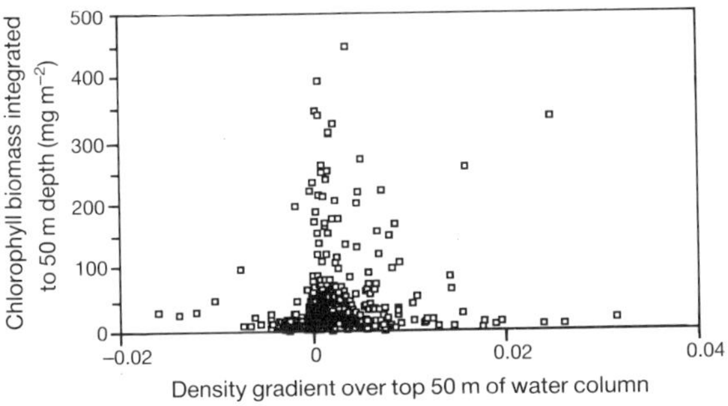

Fig. 4. Scatter plot of integrated chlorophyll biomass in relation to water column stability (density gradient).

near-zero density gradients but presented a wide range of chlorophyll abundance (Fig. 4). The only conclusions that can be drawn from these data is to observe that very unstable water columns had low biomass, and that very high biomass only occurred in stable water columns. Thus our analysis of the BIOMASS data set, although limited by data quality, endorses the findings of Mitchell & Holm-Hansen (1991), although it offers no information on the dynamics of phytoplankton growth and bloom development.

In this section, we have attempted to investigate aspects of phytoplankton dynamics in the SIBEX study area near the Antarctic Peninsula, using very simple techniques of correlation. The results are, not surprisingly, poor. The BIOMASS data set is based on a widely spaced grid of stations sampled at widely separated times, which has meant that the key processes of phytoplankton dynamics were not resolved. In considering the spatial scale of the surveys, we recognize that regional differences are represented in addition to the finer scales over which phytoplankton blooms develop. Correspondingly, the time scales of the surveys were more likely to indicate large-scale change than local dynamics. It is the temporal variation in the data set which is examined in the remainder of this study.

3. Temporal variability: changes through the growth season

One potential strength of the BIOMASS data set is the coverage of the growth season, through several consecutive cruises sampling on repeated station grids. Here we have combined data from all suitable stations to produce a single data set spanning five months (Table 1). This ignores year of sampling; an alternative treatment of the data to examine interannual variation is presented below.

Examination of the time series of chlorophyll biomass for the complete SIBEX grid shows a pattern of continuous low values punctuated by maxima (Fig. 5a). The first of these occurred at the start of the sampling period (December) and regional analysis indicates that these were associated with Weddell and Bellingshausen water. The abundance maxima in January–February were produced in the VAR and SSS groups. The spatial distribution of these maxima may have occurred purely by chance, although it appears reasonable that different processes may have prompted bloom formation in different areas. Hart (1942) offered a broad explanation of the timing of the phytoplankton bloom in the context of the seasonal change in insolation, and hence the latitudinal variation of the occurrence of sufficient light to enable phytoplankton growth to take place. However, he acknowledged that this scenario was applicable to the open ocean but that coastal areas offered more complex conditions (cf. Clarke, 1988; Clarke *et al.*, 1988). Holm-Hansen & Mitchell (1991; their figs 2 and 3) describe the progression of phytoplankton growth in the Bransfield Strait region during December–March. High phytoplankton abundance was noted over much of the RACER station grid during December, when sea-ice was still present outside the Strait. High biomass was still present sporadically in the research area during January, but declined thereafter. The later occurrences of high phytoplankton biomass appeared to be primarily neritic, for instance in Gerlache Strait, or to represent patches of high phytoplankton abundance advected from such areas (see also Heywood & Priddle, 1987). Thus we would suggest that the earlier phytoplankton maxima in the BIOMASS data represent the 'spring bloom' whereas the later peaks in the VAR and SSS zones were isolated events, probably associated either with coastal blooms or with water advected from them.

Similar time series can be constructed for the nutrients. Although lack of data from cruise SIS1 means that December data are not available (Fig. 5b–d), and taking information from all regions collectively, a decline in nitrate and phosphate amounts can be seen in the slightly shorter time series for

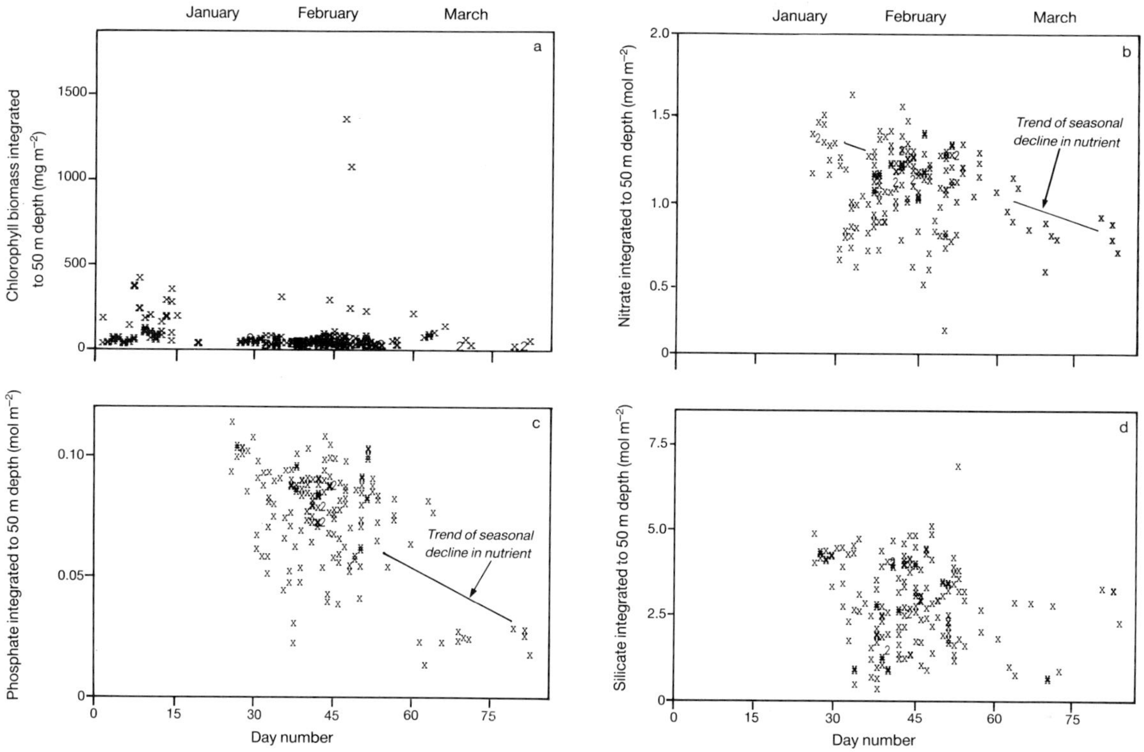

Fig. 5. Time series of chlorophyll biomass and nutrients for all SIBEX stations. Day numbers in the series start at the first date sampled (mid-December), and data from all years of BIOMASS studies in the region are combined. Lines in (b) and (c) are fitted by eye.

these nutrients (Fig. 5b,c). No statistical relation can be suggested by the data, but a decrease of the order of 0.5 mol m^{-2} for nitrate is suggested for the complete growth season, averaged over all areas. Using the conventional Redfield ratio for the elemental composition of plankton, this can be equated to 3.3 mol carbon m^{-2}. Karl *et al.* (1991) suggest that a slightly higher carbon–nitrogen ratio may be appropriate for phytoplankton particulate material for this region, which would equate the nitrate decline to nearly 5 mol carbon m^{-2}. In either case, the equivalent chlorophyll production, assuming a carbon: chlorophyll ratio of 40, of 1000–1500 mg chlorophyll m^{-2} accords with the maximum biomass recorded during the BIOMASS studies in the area. Similarly, the decline in phosphate from approximately 0.10 mol m^{-2} in January to 0.06 mol m^{-2} at the end of the study suggests a total chlorophyll production of 1200–1800 mg chlorophyll m^{-2}. These figures are in close agreement, and also correspond well to the suggested total seasonal phytoplankton particulate production for Bransfield Strait of 8.4 mol carbon m^{-2}, 0.8 mol nitrogen m^{-2} and 0.05 mol phosphorus m^{-2} obtained from five sediment trap arrays in the RACER study grid (Karl *et al.*, 1991). Karl *et al.* (1991) note that although highly productive regions in the Bransfield Strait differ markedly over short time scales from regions which do not experience intense phytoplankton blooms, overall fluxes from the epipelagic zone are similar.

Silicate did not show any recognizable depletion (Fig. 5d). This can probably be attributed to the high concentrations and systematic regional variability of this nutrient in the study area. Assuming com-

Table 4. *Median and range for chlorophyll and nutrients in the 'variable' zone (VAR) and the South Shetlands Shelf (SSS) during the two years of SIBEX*

			Chl (mg m^{-2})		NO_3 (mol m^{-2})		Silica (mol m^{-2})		PO_4 (mol m^{-2})	
Zone	Cruise	N	med	range	med	range	med	range	med	range
VAR	SB 1	14	24.1	1–46	1.13	0.7–1.4	3.21	1.3–4.6	0.09	.07–.10
	SB 2	13	51.0	4–1335	0.97	1.5–1.3	4.35	2.2–6.9	0.08	.04–.10
SSS	SB 1	10	31.4	20–62	1.30	1.0–1.6	2.87	1.6–3.8	0.09	.06–.10
	SB 2	10	45.0	13–87	1.19	0.6–1.3	3.48	3.2–5.2	0.09	.07–.11

parable diminution of the silicon pool to that seen in the nitrate and phosphate we would expect a decrease of the order of 0.5 mol m^{-2}, which would be difficult to resolve relative to depth-integrated amounts ranging from 0.1 to 6.9 mol^{-2}. Furthermore, the absence of nutrient data for December, when phytoplankton growth in the Weddell Sea waters appeared to be taking place, may have missed the highest, pre-bloom silicon concentrations.

4. Temporal variation: interannual change

The SIBEX cruises were undertaken over a period of two years, although sampling only during the summer period. These two consecutive seasons were very different from one another. The first year of SIBEX appears to have been anomalous. Water temperatures were warmer than those reported previously (Rojas, 1985), and abundance and distribution of some zooplankton were unusual. Salps dominated the zooplankton over much of the area (Mujica & Asencio, 1985; Katsuragawa *et al.*, 1989). Katsuragawa and co-workers also indicate that zooplankton biomass (measured acoustically), and hence grazing pressure on phytoplankton, would have been higher during SIBEX 1 than SIBEX 2. In a broader context, the summer of SIBEX 1 appeared to have been preceded by a period of low krill abundance in the Scotia Sea and around South Georgia (Croxall *et al.*, 1988; Priddle *et al.*, 1988). The season of SIBEX 2 appears to have been normal.

We have examined the data for the SIBEX cruises divided into the two seasons, omitting SIS1 in order to have comparable times of sampling for the two years. Again, small sample sizes militate against statistical tests, but there are indicators of differences in phytoplankton and nutrient abundances between the two years, particularly for the SSS and VAR groups (Table 4). For both zones, median chlorophyll was lower in SIBEX 1 than in SIBEX 2, consistent with the suggestion that grazing losses might have been higher in the first year. Nitrate concentrations were higher in SIBEX 1 than in SIBEX 2, presumably confirming lower phytoplankton abundance and production. Interestingly, the silicate values indicate an opposing trend, and may be indicative of differing input of Weddell Sea water to the Bransfield Strait in the two years (BIOMASS, 1990; Stein & Heywood, this volume).

Conclusions and evaluation

We have set out to evaluate the BIOMASS data set of phytoplankton-related variables for a series of cruises in the region of the Antarctic Peninsula. Our approach has been to assess the utility of the data set as a single unit, with the aim of extracting information which could not be produced by a single cruise viewed in isolation. It should of course be repeated that although the BIOMASS cruises were designed to be treated as a series in this way, the specification of individual cruises did not take the special requirements of phytoplankton studies into account.

First, we can be reasonably confident that the data can be combined. Despite the use of different methods and sampling protocols, data from the different cruises in the series are at least comparable. However, more rigorous adherence to standard methods, intercalibration, and comprehensive documentation of the database would have provided some amount of certainty that our interpretation of the data is justified.

Regional patterns in a diverse system such as that selected for the BIOMASS SIBEX study in Bransfield Strait and neighbouring waters will clearly be an important component in phytoplankton variability in large-scale surveys such as these cruises. Examination of subsets of the data, each representing clearly circumscribed environments, allows the regional pattern in the spatial variation of the data set to be described. However, apart from the opportunity to use more samples, we have not achieved a better regional description than might be obtained from summary information from one or more cruises in the BIOMASS series. Nevertheless, the groups of stations proved useful in subsequent analysis.

Use of the combined survey data to explore phytoplankton dynamics with respect to macronutrients and water-column stability was clearly inappropriate. Relative to the coarse station grid employed in the SIBEX survey cruises, the phytoplankton was highly patchy in space and time. Measurements made on station cannot be compared with observations made of laboratory or other controlled systems. The phytoplankton and their environment both have 'memories' of previous events which contribute to the current situation. Thus low chlorophyll may co-occur with high nutrient concentrations before a bloom develops, or with low nutrients afterwards. Simple correlative analyses will only provide clear results in smaller-scale, more detailed studies than the BIOMASS survey.

The BIOMASS data proved to be of greater value as time series. Although regional differences were apparent, these did not mask overall variability. Indication of both within-season and interannual changes point to interesting new information which emerges from the complete data set. Changes in nutrient concentrations during the growth season suggest that production was low overall, with isolated blooms among low phytoplankton biomass. Indications of year-to-year variation are intriguing, but hard to substantiate.

Overall, the BIOMASS data on phytoplankton abundance in the vicinity of the Antarctic Peninsula have added to the information on the large-scale variation in the region, including the seasonal cycle and possible interannual variation linked to hydrographic and biological changes. However, they do not provide data appropriate to the complete analysis of the processes controlling phytoplankton production and loss in the study area.

Acknowledgements

We thank L. P. Kurdelski, H. Pfeiffenberger, M. Reinke (and others) from the Alfred-Wegener-Institut, Bremerhaven, for their excellent cooperation for computer analysis on the BIOMASS data set. The data themselves were submitted to the BIOMASS Data Centre by scientists from several countries, and it is a pleasure to acknowledge the effort involved both in the fieldwork and in data validation and submission. F. Brandini was supported by The Alexander von Humboldt Foundation while working on this paper.

References

Amos A F (1987) RACER: Physical oceanography of the western Bransfield Strait. *Antarct J US* **22**: 137–40.

BIOMASS (1977) *Biological Investigations of Marine Antarctic Systems and Stocks*. Volume 1: *Research Proposals*. SCAR, Cambridge, U.K. 79 pp.

BIOMASS (1983) *A Handbook for the Measurement of Chlorophyll a in Netplankton and Nannoplankton* (BIOMASS Handbook 9). 44 pp.

BIOMASS (1990) SIBEX Physical Oceanography Workshop. Cambridge, U.K., 9–27 March 1987. *BIOMASS Rep Ser* **62**: 1–38.

Burkholder P R, Sieburth J McN (1961) Phytoplankton and chlorophyll in the Gerlache and Bransfield Straits of Antarctica. *Limnol Oceanogr* **6**: 45–52.

Clarke A (1988) Seasonality in the Antarctic marine environment. *Comp Biochem Physiol* **90B**: 461–73.

Clarke A, Holmes L J, White M G (1988) The annual cycle of temperature, chlorophyll and major nutrients at Signy Island, South Orkney Islands, 1969–82. *Br Antarct Surv Bull* **80**: 65–86.

Croxall J P, McCann T S, Prince P A, Rothery P (1988) Reproductive performance of seabirds and seals at South Georgia and Signy Island, South Orkney Islands, 1976–1987: implications for Southern Ocean monitoring studies. In: D Sahrhage (ed) *Antarctic Ocean and Resources Variability*. Springer-Verlag, Berlin, pp. 261–85.

De Baar H J W, Buma A G J, Nolting R F, Cadée G C, Jacques G, Tréguer P J (1990) On iron limitation of the Southern Ocean: experimental observations in the Weddell and Scotia Seas. *Mar Ecol Prog Ser* **65**: 105–22.

El-Sayed S Z (1984) Productivity of the Antarctic waters – a reappraisal. In: Holm-Hansen O, Bolis L, Gilles R (eds) *Marine Phytoplankton and Productivity*. Springer-Verlag, Berlin, pp. 19–34.

Hart T J (1934) On the phytoplankton of the south-west Atlantic and Bellingshausen Sea, 1929–1931. *Discovery Rep* **8**: 1–268.

Hart T J (1942) Phytoplankton periodicity in Antarctic surface waters. *Discovery Rep* **21**: 261–356.

Heywood R B, Priddle J (1987) Retention of phytoplankton by an eddy. *Continental Shelf Res* **7**: 937–55.

Holm-Hansen O (1985) Nutrient cycles in Antarctic marine ecosystems. In: Siegfried W R, Condy P R, Laws R M (eds) *Antarctic Nutrient Cycles and Food Webs*. Springer-Verlag, Berlin, pp. 6–10.

Holm-Hansen O, Mitchell B G (1991) Spatial and temporal distribution of phytoplankton and primary production in the western Bransfield Strait region. *Deep-Sea Res* **38**: 961–80.

Huntley M, Karl D M, Niiler P, Holm-Hansen O (1991) Research on Antarctic Coastal Ecosystem Rates (RACER): an interdisciplinary field experiment. *Deep-Sea Res* **38**: 911–41.

Karl D M, Tilbrook B D, Tien G (1991) Seasonal coupling of organic matter production and particle flux in the western Bransfield Strait, Antarctica. *Deep-Sea Res* **38**: 1097–126.

Katsuragawa M, Nonato L V, Madueira L S P, Amaral J C, Leal Ferreira S L B (1989) Analysis of krill distribution in the Bransfield Strait and Drake Passage by means of hydroacoustic survey during SIBEX I and II. *Pesq Antárt Bras* **1**(1): 11–23.

Lipski M (1982) The distribution of chlorophyll a in relation to the water masses in the southern Drake Passage and the Bransfield Strait (BIOMASS-FIBEX, February–March 1981). *Pol Polar Res* **3**(3/4): 143–52.

Lipski M (1985) Chlorophyll *a* in the Bransfield Strait and the southern part of Drake Passage during BIOMASS-SIBEX (December 1983–January 1984). *Pol Polar Res* **6**(1/2): 21–30.

Lipski M, Rakusa-Suszczewski S (1990) Early summer patterns of vertical distribution of chlorophyll *a* (Bransfield Strait, Antarctica, November 1986). *Polsk Arch Hydrobiol* **37**: 287–93.

Martin J H, Gordon R M, Fitzwater S E (1990) Iron in Antarctic waters. *Nature* **345**: 156–8.

Mitchell B G, Holm-Hansen O (1991) Observations and modelling of the Antarctic phytoplankton crop in relation to mixing depth. *Deep-Sea Res* **38**: 981–1007.

Mujica A, Asencio V (1985) Larvas de peces, eufáusidos y estructura comunitaria del zooplancton del estrecho Bransfield (Crucero SIBEX-INACH, 1984). *Ser Cient INACH* **33**: 159–86.

Mullins B W, Priddle J (1987) Relationship between bacteria and phytoplankton in the Bransfield Strait and Southern Drake Passage. *Br Antarct Surv Bull* **76**: 51–64.

Murphy E J, Morris D J, Watkins J L, Priddle J (1988) Scales of interaction between Antarctic krill and the environment. In: Sahrhage D (ed) *Antarctic Ocean and Resources Variability*. Springer-Verlag, Berlin, pp. 120–30.

Nast F, Gieskes W C (1986) Phytoplankton observations relative to krill abundance around Elephant Island in November 1983. *Arch FischWiss* **37**: 95–106.

Nelson D M, Smith W O (1991) Sverdrup revisited: critical depths, maximum chlorophyll levels and the control of Southern Ocean productivity by the irradiance/mixing regime. *Limnol Oceanogr* **36**: 1650–61.

Niiler P P, Amos A, Hu J-H (1991) Water masses and 200 m relative geostrophic circulation in the western Bransfield Strait region. *Deep-Sea Res* **38**: 943–59.

Platt T, Bird D F, Sathyendranath S (1991) Critical depth and marine primary production. *Proc R Soc Lond* B**246**: 205–17.

Priddle J (1985) Species composition of net phytoplankton from Drake Passage, Bransfield Strait and Scotia Sea during summer, 1982. *Meeresforschung* **30**: 240–50.

Priddle J, Croxall J P, Everson I, Heywood R B, Murphy E J, Prince PA, Sear C B (1988) Large-scale fluctuations in distribution and abundance of krill – A discussion of possible causes. In: Sahrhage D (ed) *Antarctic Ocean and Resources Variability*. Springer-Verlag, Berlin, pp. 169–82.

Rojas R (1985) Descripción de la estrutura térmica del estrecho Bransfield a base de observaciones de XBT (SIBEX-FASE I-CHILE). *Ser Cient INACH* **33**: 89–120.

Sakshaug E, Slagstad D, Holm-Hansen O (1991) Factors controlling the development of phytoplankton blooms in the Antarctic ocean – a mathematical model. *Mar Chem* **35**: 259–72.

Smetacek V, Passow U (1990) Spring bloom initiation and Sverdrup's critical depth model. *Limnol Oceanogr* **35**: 228–33.

Smith W O, Sakshaug E (1990) Polar phytoplankton. In: Smith W O (ed) *Polar Oceanography*. Academic Press, New York, pp. 477–525.

Sverdrup H U (1953) On conditions for the vernal blooming of phytoplankton. *J Cons Perm Int Explor Mer* **18**: 287–95.

Uribe E (1982) Influence of the phytoplankton and primary production of the Antarctic waters in relationship with the distribution and behaviour of krill. *Ser Cient INACH* **28**: 147–63.

Uribe E (1985) Distribución de clorofila "a" en el estrecho Bransfield durante el verano austral 1984 (SIBEX-Fase I). *Ser Cient INACH* **33**: 121–58.

Uribe E (1987) Distribución de clorofila "a" en el estrecho Bransfield durante el verano austral 1985 (SIBEX-Fase II). *Ser Cient INACH* **36**: 93–106.

Weber L H, El-Sayed S Z (1987) Contributions of the net, nano- and picoplankton to the phytoplankton standing crop and primary productivity in the Southern Ocean. *J Plankt Res* **9**: 973–94.

Phytoplankton of the Indian Antarctic Ocean

G. JACQUES & M. FUKUCHI

Introduction

THE ANTARCTIC Ocean is a new target area for study of the control of excess atmospheric CO_2 by the marine biosphere. As a matter of fact, the question of whether the Antarctic Ocean is a net sink or a source for CO_2 is still a matter of debate. This is due to a poor understanding of the contribution to total primary production of the different subsystems (Table 1) and to the great seasonal variability of the CO_2 exchanges at the atmosphere–water interface. Even open oceanic areas show a heterogeneity over space and time that was unsuspected several years ago. This heterogeneity is manifested first and foremost by the presence of frontal systems. The Antarctic Ocean encircles the Antarctic Continent; the specific area and the main characteristics of each oceanic sector are defined by:

the shape and position of the continent;
the bottom topography;
the position of frontal zones;
the extent of the pack-ice.

The aim of international programmes such as the Southern Ocean JGOFS is to relate the results obtained to an oceanic scale. It is clear that imbalance exists, with most results available from a limited number of study areas: the Weddell Sea, the Ross Sea, the Drake Passage and surroundings. By comparison, oceanic zones such as the Indian sector are poorly documented. In this review, we will try to characterize, for each of the four subsystems of the Indian sector, the nutrient regime (Jacques & Tréguer, 1986; Tréguer & Jacques, 1992), the level and nature of primary productivity, and the nature of the factors which control primary production.

Table 1. *Frontal zones and main subsystems in the Indian sector of the Southern Ocean*

The boundaries of the Indian sector are the 30°E and 150°E meridians (FAO classification).

Zone	Antarctic Ocean (34×10^6 km^2)		Indian sector (39%; 13.1×10^6 km^2)		Other sectors (61%; 20.9×10^6 km^2)	
CCSZ	2.5%	0.9×10^6 km^2	1%	0.1×10^6 km^2	3%	0.8×10^6 km^2
SIZ	47.5%	16.1×10^6 km^2	41%	4.7×10^6 km^2	55%	11.4×10^6 km^2
CCSZ + SIZ	50%	17×10^6 km^2	42%	4.8×10^6 km^2	58%	12.2×10^6 km^2
POOZ	41%	14×10^6 km^2	47%	6.2×10^6 km^2	38%	7.9×10^6 km^2
PFZ	9%	3×10^6 km^2	11%	2.1×10^6 km^2	4%	0.8×10^6 km^2
POOZ + PFZ	50%	17×10^6 km^2	58%	8.3×10^6 km^2	42%	8.7×10^6 km^2

Review of data and discussion

Main features of the Indian sector

In the Indian sector (Fig. 1), the continent has a regular coastline which follows the 65–67°S parallels, with the exception of the Prydz Bay and the Lützow–Holm Bay. These indentations are not as deep as the Ross Sea in the Pacific sector and the Weddell Sea in the Atlantic sector, but extend southwards to 70°S. Moving north, different fronts are encountered (Deacon, 1982, 1983; Lutjeharms *et al.*, 1985). Using oceanographic observations collected over the past 22 years from JARE (Japanese Research Expedition) icebreakers (Nagata *et al.*, 1988), we can summarize the approximate positions of these fronts in the Indian sector (Fig. 2) and their spatial–temporal variability (Table 2) as follows:

1. The *Antarctic Divergence*, a transition region between the east and west wind belts which is accompanied by upwelling. This occurs between 62°S and 63°S because the continent extends relatively far north in the Indian Ocean.

2. The *Polar Frontal area*, defined by different parameters. In this review, we will follow the nomenclature proposed by Nagata *et al.* (1988) for

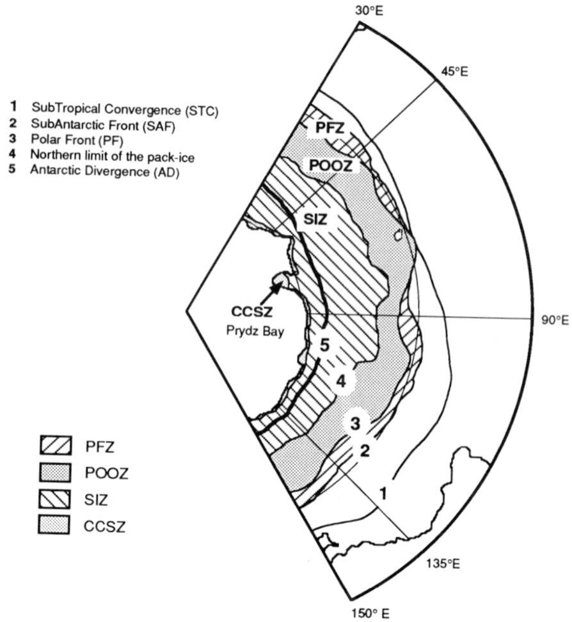

Fig. 1. General map of the Antarctic Ocean indicating the main frontal zones and subsystems. PFZ, Polar Front Zone; POOZ, Permanently Open Ocean Zone; SIZ, Seasonal Ice Zone; CCSZ, Coastal and Continental Shelf Zone; PIZ, Permanent Ice Zone.

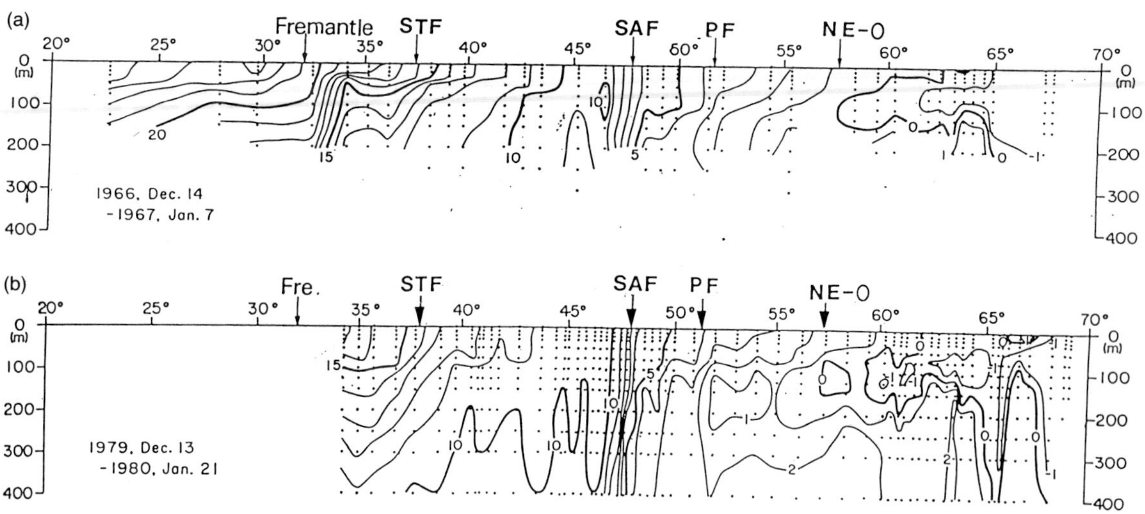

Fig. 2. Examples of temperature cross-section along the line from Fremantle to Syowa Station in 1966 (a) and 1979 (b). The sections are projected to the longitudinal plane. The latitude of Fremantle and the estimated positions of the fronts are shown in the upper margin of each figure.

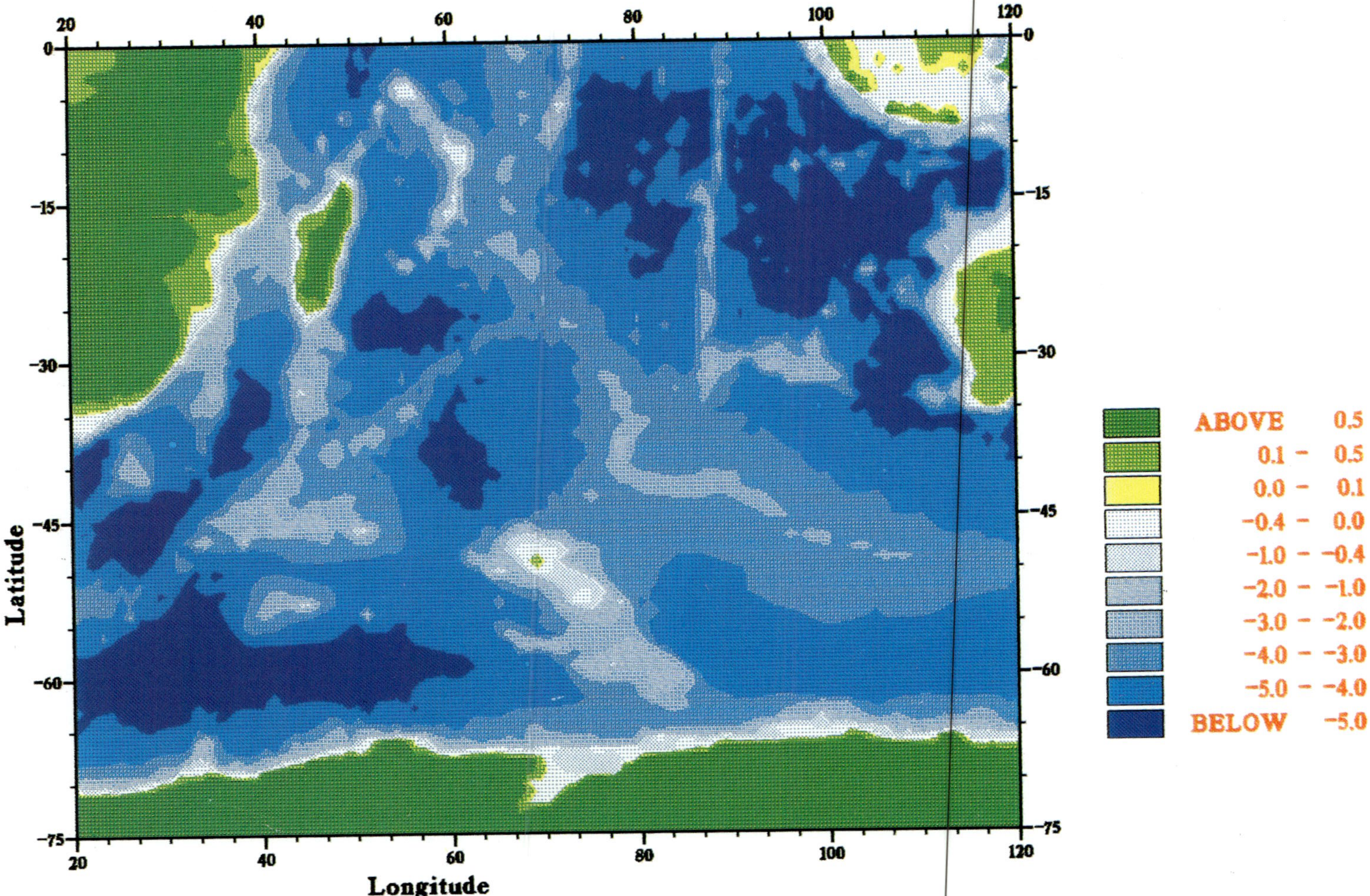

Fig. 3. Bathymetry and topography of the Indian sector of the Southern Ocean. The depth scale is in kilometres. Data: 5′ × 5′, National Geophysical Data Center, NOAA. From Cazenave, CNES-GRGS, Toulouse.

Table 2. *Mean latitudinal positions and ranges of different frontal zones in the Indian sector of the Southern Ocean*

	Sub-Antarctic Front (SAF)	Polar Front (PF)	Polar Front (NE-0)
100–110°E	47°05′S	51°04′S	56°16′S
	44°30′S–48°20′S	49°50′S–52°40′S	53°30′S–58°44′S
0–20°E	43°10′S	51°09′S	56°04′S
	42°06′S–45°25′S	49°22′S–52°56′S	54°44′S–59°16′S
40–55°E	45°33′S	49°45′S	53°56′S
	43°35′S–46°44′S	49°06′S–51°06′S	51°28′S–58°50′S

Source: After Nagata *et al.* (1988).

the Indian sector. This frontal zone is at a depth of 100 m and is limited by the Polar Front (PF), following the 3 °C isotherm and by the northernmost edge of the 0 °C isotherm contour. Some authors place the Polar Front at the latitude where sinking Antarctic Surface Water produces a belt of minimum salinity at 200 m depth. This convergence zone occurs near the latitude of strongest west winds and the maximum northward drift. These oceanic fronts show rather small temporal and spatial fluctuations if we define their position by the overall position of the ‘frontal zone’. Some of the variation of the front position may be regarded as a change in the fine structure of the frontal zone.

3. The *Sub-Antarctic Front*, defined by the position of the 7 °C isotherm at a depth of 100 m. It is more accurate to speak of a ‘frontal zone’ occurring between the 5 °C and 10 °C isotherms, with a depth between 100 m and 300 m.

Bottom topography and wind regime influence not only the position of these fronts but also the general circulation pattern and the generation and evolution of eddies. The bottom topography (Fig. 3) causes a heterogeneity in the Indian sector. In the west, the area is formed by a part of the Atlantic–Indian basin at 4000 m depth and, in the east by part of the Antarctic–Indian basin at 6500 m depth. These two entities are separated by a north–south partition, the Kerguelen Plateau, which is indicated, at the surface, by the islands of Kerguelen and Heard–MacDonald.

All of the oceanic sector to the north of 55°S is exposed to a regular regime of strong winds, principally W–E, which are linked to the permanent tropical anticyclone and the low-pressure circum-Antarctic area. These westerly winds, in terms of annual mean velocity, reach their maximum in the Indian sector ($\tau \geqslant 0.20$ to 0.25 N m^{-2}) even though they are minimal in the Pacific ($\tau \leqslant 0.102$ N m^{-2}). The maximum speed of the westerlies is found near 49°S in the Indian and Atlantic Oceans and near 55°S in the Pacific. This zonal irregularity is also present in the Antarctic Circumpolar Current, which is concentrated in the zone bounded by the Subantarctic Convergence and the Subtropical Convergence. Owing to the coalescence of these two fronts along the northern slope of the Kerguelen shelf (Gambcroni *et al.*, 1982), swift (50 cm s^{-1} at the surface) and deep-reaching geostrophic baroclinic currents are encountered; they carry 87% of the cross-sectional transport (Park *et al.*, 1989).

Coastal and Continental Shelf Zone (CCSZ) and Seasonal Ice Zone (SIZ) as exemplified by Prydz Bay

Well-documented studies of the Continental Shelf waters and of the Seasonal Ice Zone (SIZ) were made in the Weddell Sea and the Ross Sea. In these areas, the Marginal Ice Zone (MIZ), which can extend 50–250 km north of the ice-edge, produces more phytoplankton than the open ocean by a factor of at least two. This led Smith *et al.* (1988) and also Sullivan *et al.* (1988) to suggest a model of the conditions necessary for the MIZ phytoplankton bloom by identifying forcing functions for phytoplankton dynamics which were mostly related to meteorolo-

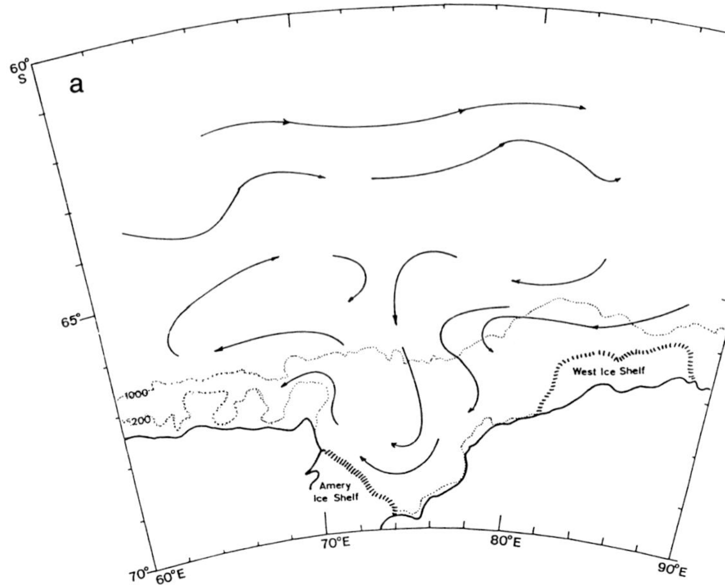

15 -28 February

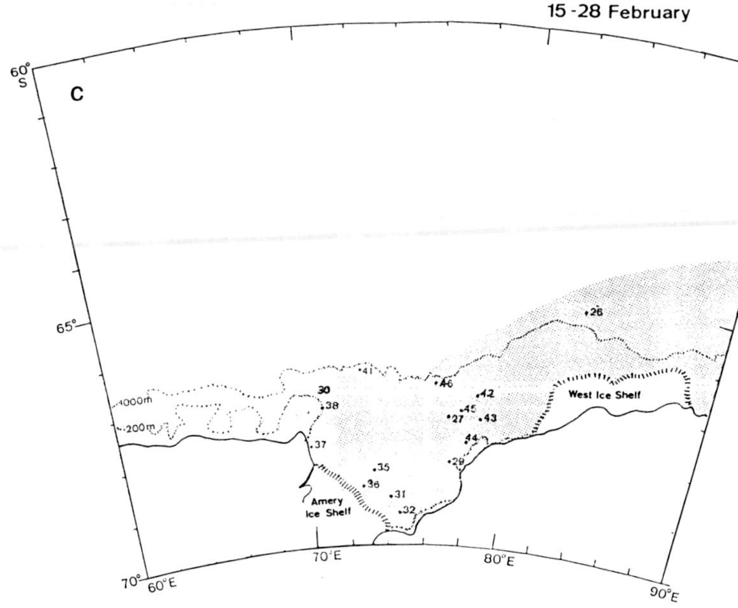

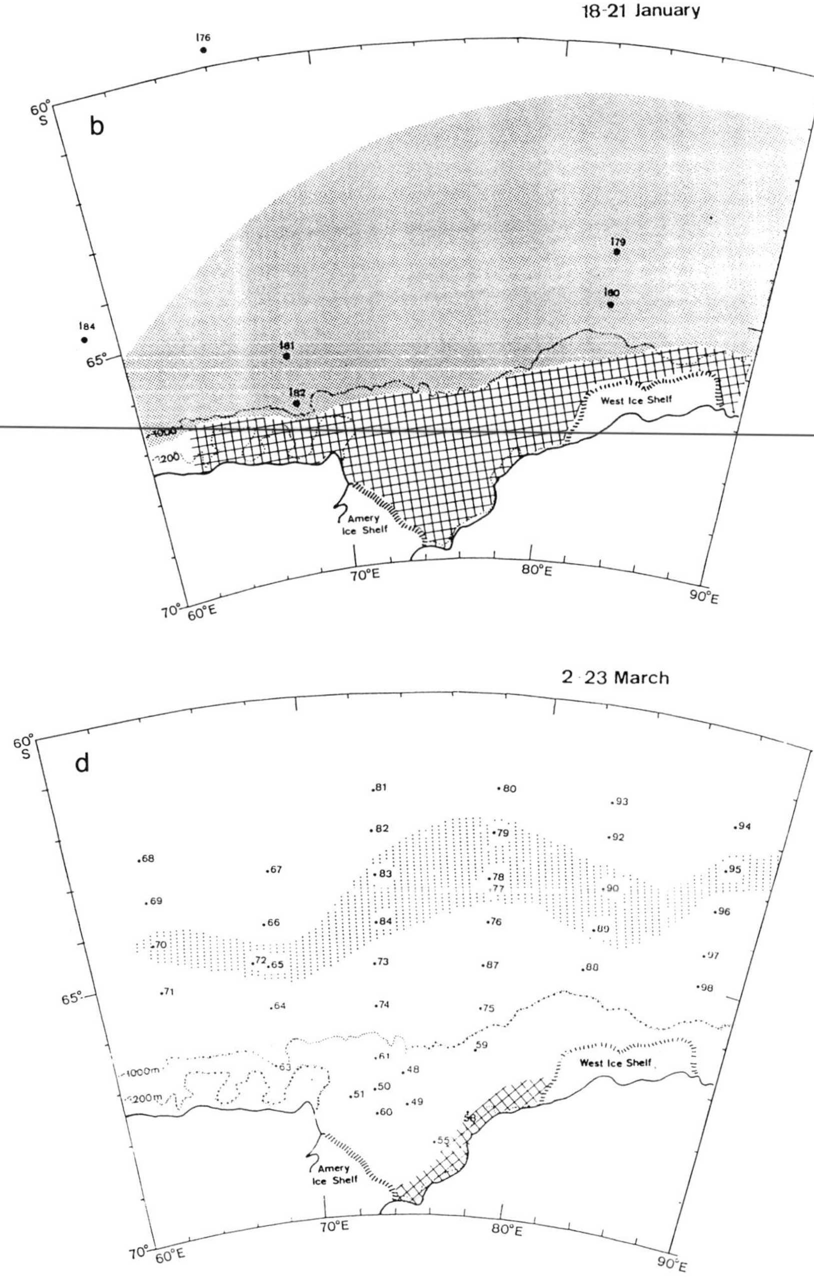

Fig. 4. Circulation of general surface water masses in the Prydz Bay area (a) and geographical distribution of the three major phytoplanktonic habitats at three periods of observation during the austral summer 1987: (b) early summer situation (18–21 January 1987); (c) end summer situation (15–28 February 1987); (d) early winter situation (2–23 March 1987). (a) After Smith *et al.* (1984); (b–d) after Lancelot *et al.* (1990).

Table 3. *Mean characteristics of the pelagic system in the main habitats of Prydz Bay*

Units: K_{max}, mg C mg Chl a^{-1} h^{-1}; α, mg C mg Chl a^{-1} h^{-1} (µmol m^{-2} s^{-1})$^{-1}$; I_k, µmol m^{-2} s^{-1}. Pr.Pr. is primary production.

	NO_3 (µM)	Si (µM)	Chl*a* (mg m^{-3})	Phytoplankton Pr.Pr. (mg C $m^{-2}d^{-1}$)	K_{max}	α	I_k
Open Ocean	21	38	0.14	170	4.17	0.070	60
Divergence Zone	17	27	0.25	240			
MIZ	13	21	2.10	460	1.50	0.057	26

Source: After Lancelot *et al.* (1990).

gical conditions. Smith *et al.* (1988) have used this conceptual model to calculate estimates of the primary production for the pack-ice zone. During the period 1973–88, the annual primary production was 28 ± 3.8 Tmol C y^{-1}, compared with about 103 Tmol C y^{-1} for the complete Antarctic Ocean. The blooms which commence in the zone of the pack-ice activity are generally dominated by ice-diatoms with a diameter of over 10 µm, among which *Nitzschia curta* predominates. Consequently, the *f* ratio exceeds 0.4 and even 0.6 (Nelson & Smith, 1986). A large fraction of the algal biomass is exported to the highest trophic levels and/or to the sediment.

Phytoplankton dynamics in Prydz Bay

Although much smaller, the Prydz Bay is, to a certain extent, comparable with the Ross Sea and the Weddell Sea as it also has a continental shelf (off Princess Elizabeth Land) and permanent ice-shelf areas. The shelf area of the Bay is free of ice only by the end of January. Dense pack-ice areas remain west of West Ice Shelf until March (Fig. 4). In Prydz Bay itself, a closed cyclonic gyre occurs, with a depression magnitude close to 0.02 dynamic metres in the centre of the gyre.

Three major phytoplankton habitats can be distinguished: the ice-covered zone, the marginal ice zone and open sea areas. On basis of data from about 70 stations, Lancelot *et al.* (1990) reported the mean characteristics of the pelagic system (Table 3) as follows.

> Although complete nutrient depletion was never observed, a significant reduction of all nutrients appeared in each zone when compared with the maximal values.
>
> The mean phytoplankton biomass of the MIZ is > 2 mg Chl *a* m^{-3}, one order of magnitude higher than that which is characteristic for the open sea. Notice that the maximum chlorophyll *a* concentrations do not reflect the large decrease observed in nutrient concentrations, indicating significant phytoplankton losses. High levels of biomass and production are restricted to the edge of the retreating sea-ice.
>
> Phytoplankton biomasses in the divergence band are higher but not significantly so relative to the extremely low values found for the open sea (0.14 mg Chl *a* m^{-3} as a mean).
>
> The maximum potential photosynthetic capacity (K_{max}) varied over a wide range depending on the habitat and the temperature, whereas the photosynthetic efficiency (α) did not significantly differ from one habitat to another (Table 3).
>
> The open-sea phytoplankton is dominated by small species (< 8 µm diameter), suggesting the predominance of flagellates over diatoms. In the MIZ, in contrast, more than 80% of the phytoplankton species are larger than 8 µm in diameter.

CCSZ data

The other available data for the CCSZ do not alter this scheme. In the coastal polynya, blooms have been observed with a long-term moored system (Fukuchi *et al.*, 1988). The bloom in Breid Bay (70°12′S, 24°19′E) continued for more than 2–3 months (December–February) and chlorophyll concentrations in the surface layer were 0.69–5.60 mg Chl *a* m^{-3}.

In contrast to Prydz Bay, Lützow–Holm Bay is not free of ice, and the fast-ice remains in the area south of the 1000 m isobath throughout summer (Fig. 5). In the fast-ice near Syowa Station (69°00′S,

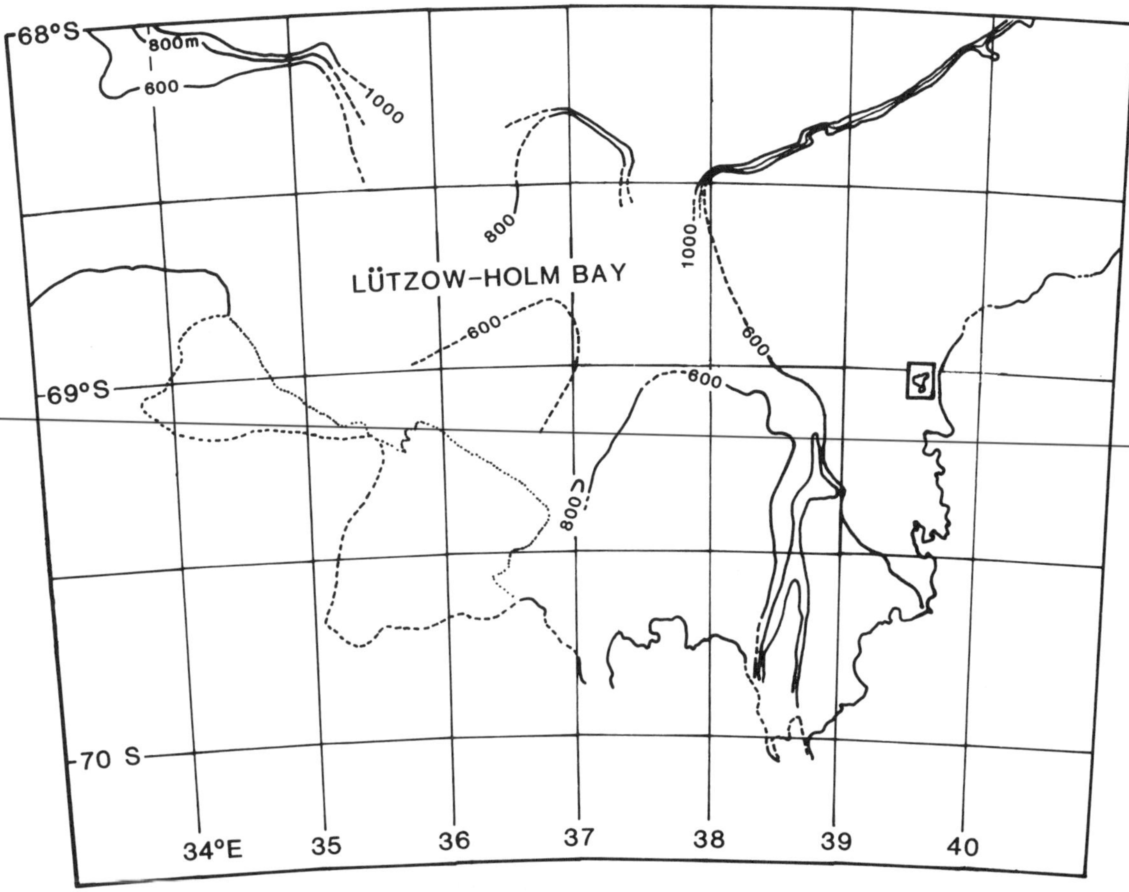

Fig. 5. Bottom topography of Lützow–Holm Bay, Antarctica. A square shows the Syowa Station area.

39°35′E), the chlorophyll *a* concentrations under the ice in late January increased to 11.30 mg m^{-3} and the integrated chlorophyll *a* stock (0–150 m) was 466.5 mg m^{-2}. Phytoplankton seems to grow for the four months (December–March) under the coastal fast-ice (Fukuchi *et al.*, 1984). In the open coastal zone, even during a presumed bloom period, the biomass is relatively low. Uno (1982) indicated a maximum of 1 mg Chl *a* m^{-3} at 100–120°E. The chlorophyll *a* concentrations were in the same range as those in the waters of the Gunnerus Bank (34°E) and off Lützow–Holm Bay (38°E) (Ohno *et al.*, 1987). In the same area, relatively high chlorophyll concentrations were recorded only in the pack-ice and the fast-ice environments. These high values (0.5 to 6 mg Chl *a* m^{-3}) do not represent a 'classic' pelagic bloom; they seem to be correlated with the release of ice-algae, which proliferate at the bottom of the sea-ice (Fukuda *et al.*, 1986). With the disappearance of the sea-ice, chlorophyll *a* concentrations decreased abruptly.

Permanently Open Ocean Zone (POOZ) and Polar Frontal Zone (PFZ)

General distribution of the biomass

To examine the trophic level of the POOZ, we have used the routine JARE surface measurements (Table 4). These observations, taken mainly in the Indian sector (Fig. 6), complement previous observations from the other sectors of the Antarctic Ocean.

Table 4. *Mean value and standard deviation of surface chlorophyll a* (mg m^{-3}) *in four different subsystems observed during nine JARE cruises of the Fuji in 1965–77*

Numerals in parentheses indicate number of data.

	CCSZ	POOZ	PFZ	Sub-Antarctic
Eastern Indian sector (middle–late December)	0.39±0.46 (125)	0.46±0.25 (108)	0.57±0.70 (5)	0.34±0.20 (54)
Western Indian sector (late February–early March)	0.58±0.74 (70)	0.25±0.16 (50)	0.28±0.16 (2)	0.19±0.12 (13)
Eastern Atlantic sector (late February–early March)	0.33±0.25 (48)	0.21±0.14 (49)	0.41±0.29 (3)	0.31±0.12 (29)

Source: After Fukuchi (1980).

Examination of some 460 values south of the Polar Front clearly confirm the oligotrophic nature of this ocean in terms of phytoplanktonic biomass: about half of the stations had chlorophyll *a* concentrations of less than 0.25 mg m^{-3}, with only 5% surpassing 1 mg Chl *a* m^{-3}. The average chlorophyll concentration appears to be only slightly higher than in the neighbouring subtropical system. These values are in the same range as those indicated by Ichimura & Fukushima in their pioneering work of 1963 (0.1–0.6 mg Chl *a* m^{-3}) and those of Yamaguchi & Shibata (1982) (0.1–0.4 mg Chl *a* m^{-3}).

It is not possible to separate the different effects of the geographic areas (Eastern Indian sector, Western Indian sector, Eastern Atlantic sector) from those of time. Regarding the POOZ, where there are few reasons to expect permanent geographic differences, we favour the hypothesis that these values indicate a higher level of primary production in December with average biomass equivalent to 0.46 mg Chl *a* m^{-3}. Subsequent campaigns after 1976 confirmed this tendency (Kanda & Fukuchi, 1979; Taniguchi *et al.*, 1986; Fukuchi *et al.*, 1985), the differences sometimes being accentuated and higher average December values reported. Thus, in December 1977, the average concentration was close to 0.8 mg Chl *a* m^{-3} and the maximum around 1 mg m^{-3}. Chlorophyll *a* concentrations of the order of 1 mg m^{-3} in surface waters and 50 mg m^{-2} in the 0–100 m layer were the maximum values encountered in the POOZ. These figures include the limited areas generally associated with the zone of eddies and vortices, frontal zones between different water masses and the zones above bottom topographic singularities. The positive effect of such singularities was suggested by Mordasova (1989) for the western part of the Atlantic zone of the Antarctic Ocean for the period January–March and by Jacques (1989) for the Indian sector.

Vertical distribution of the biomass

As expected, where winds are strong, the vertical distribution of phytoplankton is rather homogeneous in the POOZ in contrast to the MIZ. In the Indian sector, Watanabe & Nakajima (1982) showed an integrated chlorophyll *a* (mg m^{-2}): surface chlorophyll *a* (mg m^{-3}) ratio of 70 in the MIZ compared with 160 in the POOZ.

In the Indian sector (and very probably in the whole POOZ of the Antarctic Ocean), the vertical profiles of biomass and production generally show a maximum between the 50% and 10% light depths (Fukuchi & Tamura, 1982; Jacques & Minas, 1981). In the open-sea stations of the Weddell Sea, the fraction represented by production below 1% light depth relative to the total production reaches more than 30% (El-Sayed & Taguchi, 1981). In such cases, the subsurface chlorophyll *a* concentration may be a good indicator of integrated chlorophyll *a*. This may encourage routine observation of surface chlorophyll *a*, either from supply ships or from satellites (Kuroda & Fukuchi, 1982) (Fig. 7a). Subsurface production also is a good indicator of integrated production (Jacques & Minas, 1981, Fig. 7b).

Community structure

The classical concept of the Antarctic marine food chain gives large diatoms a predominant position in the pelagic system (Hasle, 1969). Since the early

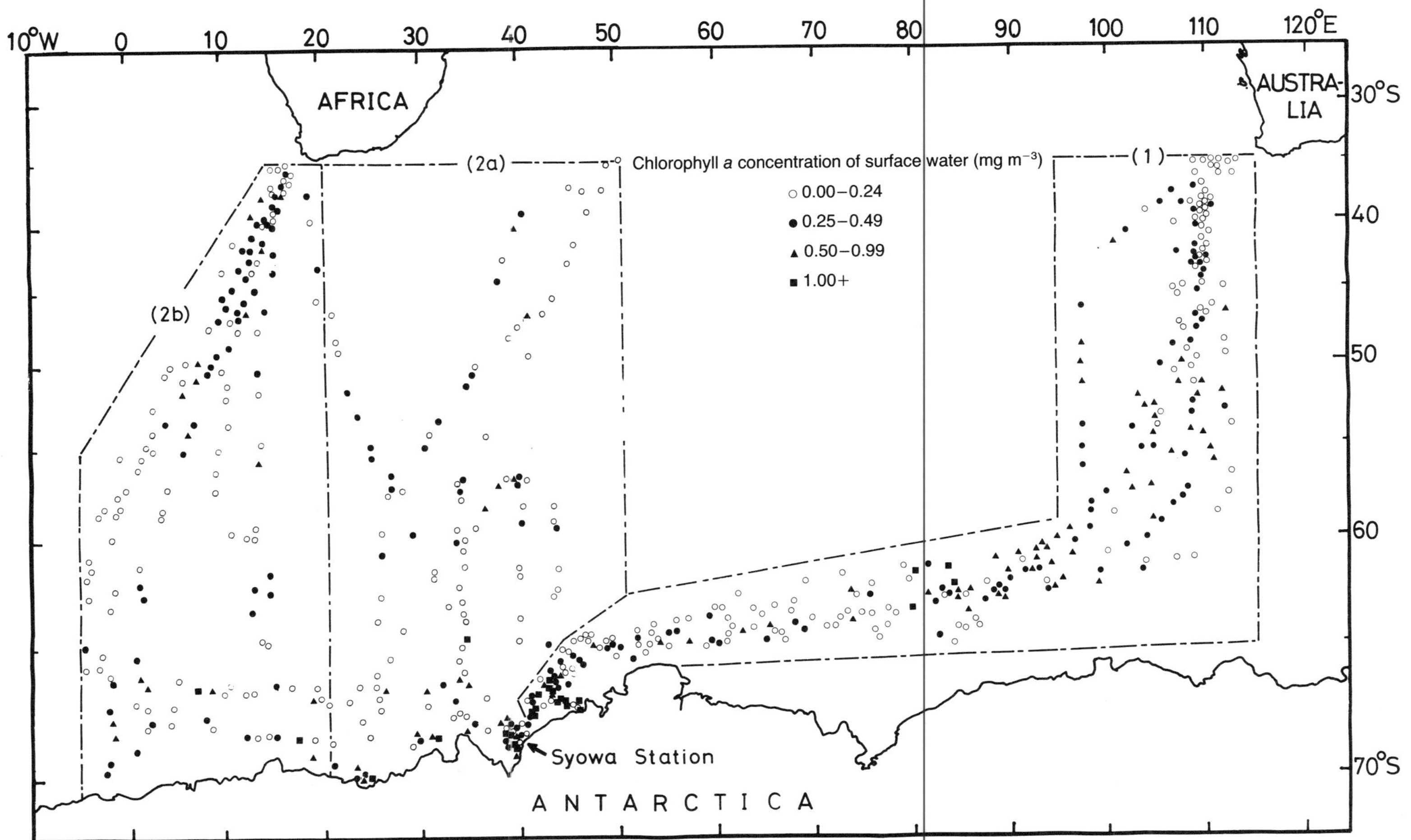

Fig. 6. Stations and surface chlorophyll concentration observed during the 9 JARE cruises in 1965–77. (1), Southward leg through the eastern Indian sector; (2a), northward leg through the western Indian sector; (2b), northward leg through the eastern Atlantic sector.

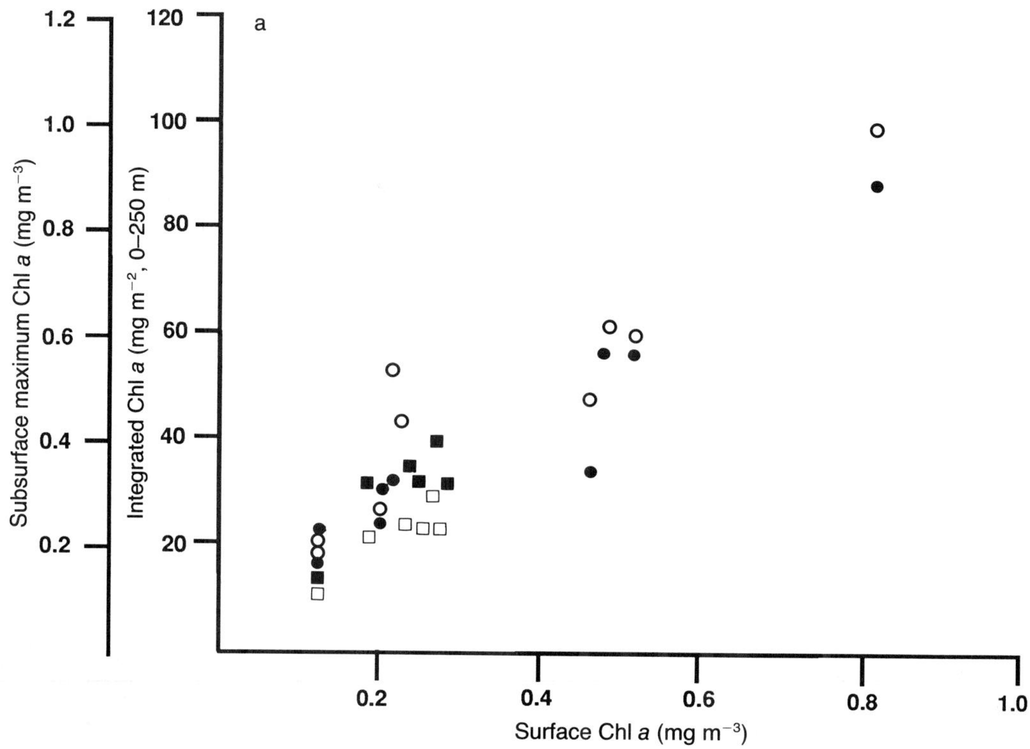

Fig. 7. Relations between integrated and surface concentration of phytoplankton. (a) Integrated chlorophyll *a* (closed symbols) or subsurface maximum (open symbols) against surface concentration. Data collected in the JARE 20 (square) and JARE 14 (circle) cruises. (b) Relations between integrated production and production at the 7% light level during the ANTIPROD 1 cruise in March 1977 ($r = 0.98^{**}$). (a) After Kuroda & Fukuchi (1982); (b) after Jacques & Minas (1981).

1930s, however, it has been recognized that algal taxa belonging to the 'less than 50 μm' communities may also be present in the Southern Ocean (Jacques, 1989; Jacques & Panouse, 1991; Weber & El-Sayed, 1987). These include small diatoms, prymnesiophytes and other flagellates (Gieskes & Elbrächter, 1986) and cyanobacteria (Marchant *et al.*, 1987).

Even if not well documented, the relative abundance of the picoplankton (diameter < 3 μm) seems quite low in the POOZ compared with the subtropical zone (Kosaki *et al.*, 1985). In the same water masses of the Indian sector, Hamada *et al.* (1985) showed that nanoplankton (3–10 μm) were the major contributors to the total phytoplankton stocks, but Sasaki (1984) demonstrated that the mean contribution of microplankton (> 20 μm) was about 65%. Using other JARE data, Hattori & Fukuchi (1989) also showed that species of diameter > 20 μm account for about 70% of the total phytoplankton stocks. They suggested that the regional variation of total phytoplankton chlorophyll biomass approximately reflects the change in stock of species of diameter > 20 μm.

In any case, diatoms dominate the phytoplankton community south of the Polar Front in the Indian

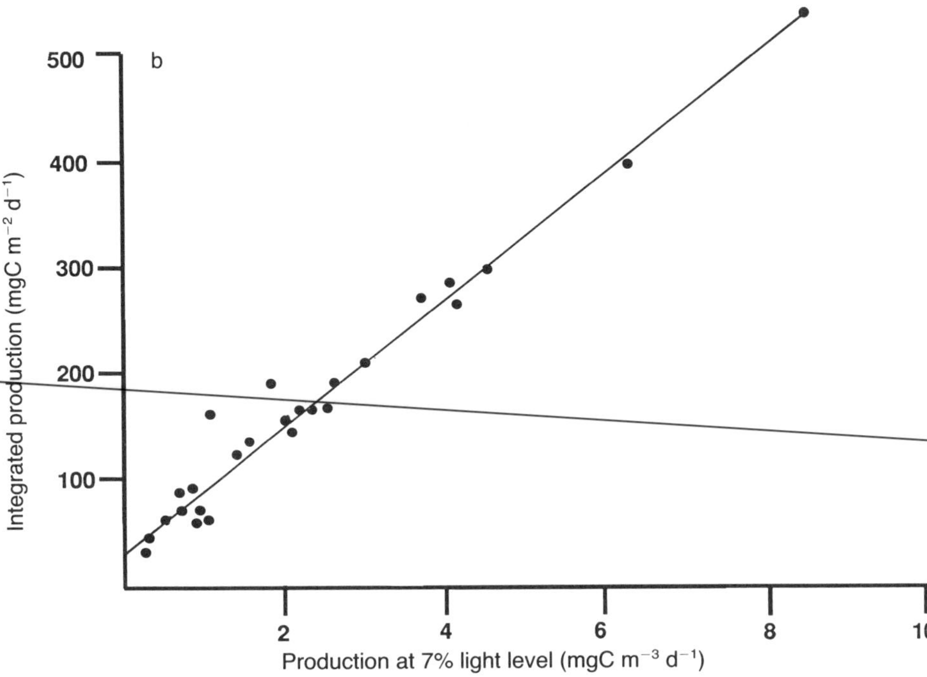

sector by contributing more than 70% to the total cell counts in that area (Jacques *et al.*, 1979; Kopcynska *et al.*, 1986) (Table 5) and, very probably, in the whole POOZ of the Southern Ocean. Depending on local or temporal variations, the diatom assemblages consist mainly of species either > 10 μm or < 10 μm in diameter, which may explain the mentioned discrepancies in the size fraction data. These size classes are characterized by:

1. dominant 'nanoplanktonic–weakly silicified' diatoms: *Nitzschia 'nana' (N. cylindrus, N. pseudonana), Chaetoceros dichaeta, C. atlanticum* var. *gracilis, C. simplex, Dactyliosolen tenuijunctus, N. prolongatoides, N. subcurvata*.

2. dominant 'microplanktonic–heavily silicified' diatoms: *Nitzschia kerguelensis, N. turgiduloides, Dactyliosolen antarcticus, Asteromphalus hookeri, A. parvulus, A. hyalinus, Chaetoceros criophilum, Thalassiosira antarctica*. At times, *N. kerguelensis* is by far the most abundant species (Jacques *et al.*, 1979); notice also the clear dominance of this diatom in the sediment of the antarctic abysses (Bareille *et al.*, 1991).

Is there a spring bloom in the POOZ?

The problem to be addressed is whether or not a definitive spring bloom, comparable to that occurring in the same latitudes in the northern hemisphere, occurs in the POOZ following the rapid increase in solar radiation. All the reviews on the Southern Ocean or on the Polar Oceans (Holm-Hansen *et al.*, 1977; Nemoto & Harrison, 1981; El-Sayed, 1984; Sakshaug & Holm-Hansen, 1984; Priddle *et al.*, 1986; Smith & Sakshaug, 1990) consider that the annual cycle is unimodal, with a delayed spring bloom with increasing latitude due to the light effect. The 'northern' region (PFZ) exhibits a peak in early December; the 'intermediate' region (POOZ), in January. If blooms are indicated by chlorophyll *a* values of over 1.5 mg m^{-3}, then the POOZ appears not to have a definitive

Table 5. *Percentage composition of phytoplankton south of the Antarctic Divergence (AD), between the AD and the Polar Front (PF) and north of the PF*

See also Fig. 1.

	South AD	AD–PF	North PF
Diatoms	71%	73%	27%
Dinoflagellates	3%	2.5%	7%
Monads and flagellates	26%	24.5%	65%

Source: After Kopczynska *et al.* (1986).

bloom during the period December–March. As far as we know, all data covering this period show values in the range 0.2–0.6, with some values reaching 1 mg Chl *a* m^{-3}.

It remains to be seen whether a bloom occurs in November. At equivalent latitudes in the northern hemisphere (the Antarctic POOZ is located approximately between 50°S and 60°S in the Indian and Atlantic sectors and about five degrees further south in the Pacific sector), the usual period for the spring bloom to develop is during May. Direct measurements during the same period (in November for the Antarctic Ocean) are practically non-existent. It is therefore necessary to adopt an indirect approach and look for an indicator of blooms (mooring of sediment traps will solve this problem). We have already touched upon this (Jacques, 1991) in examining the water characteristics of the superficial thermal minimum layer. In summer, a meridian transect extending across the POOZ clearly shows that this layer carries a record of the characteristics of the previous winter. In the middle of summer, nutrient concentrations are close to those at the surface during the mixing period; this observation demonstrates that a bloom does not occur.

During the first international BIOMASS experiment in 1980–1, three Japanese research/relief ships made several cruises into the Antarctic Ocean from early December to early March. Surface chlorophyll was observed on each cruise. Changes in the chlorophyll concentrations from December to March at some chosen latitude or water mass, such as the Polar Frontal Zone, illustrate seasonality in the chlorophyll concentrations. Overall means of chlorophyll concentrations in Table 4 are thought to indicate points of intersection of lines and curves in Fig. 8.

Limiting factors

The absence of a bloom in the POOZ seems to be certain. New production, as estimated from the actual consumption of nitrate, is around 0.3 g C m^{-2} d^{-1}, i.e. 27 g C m^{-2} over the three summer months (Jacques, 1991). These values are small considering

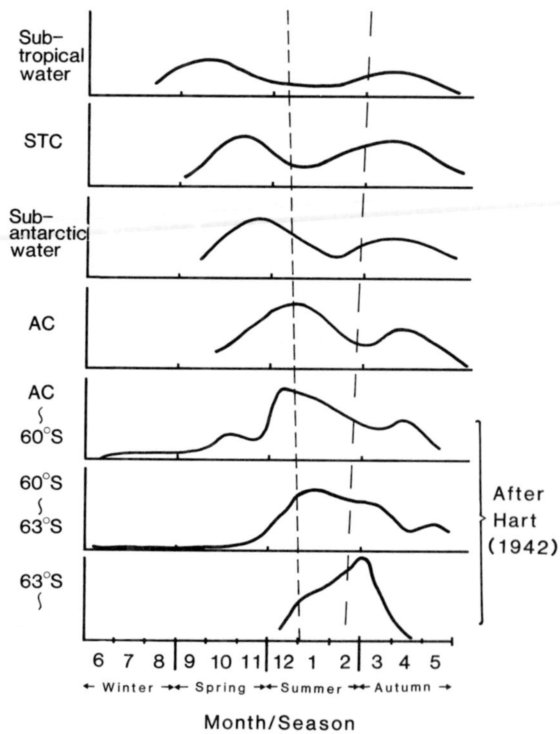

Fig. 8. Seasonal periodicity of phytoplankton chlorophyll in 7 different water masses. Lower three figures are redrawn from Hart (1942). Dashed lines indicate the annual relief cruise of Japanese icebreakers.

that the potential new production resulting from a complete exhaustion of the nutrient stock would be of the order of 200 g C m^{-2} during the same period (Fig. 9). The low silicate values in the northern part of the POOZ do not constitute a contradiction. The general pattern of nutrient distribution in the southern Indian Ocean agrees with reported observations in other sectors (Le Jehan & Tréguer, 1985). The N and P ratio anomalies are close to Redfield's model in near surface waters. The anomalies of the Si cycle (high $\Delta Si/\Delta N$) are related to the high demand of Antarctic diatoms for silicate (Jacques, 1983) and a high transport of biogenic silica toward the deep layer rather than to a diatom bloom.

The question, then, is not whether nutrients are limiting, but why nutrients are never reduced to limiting conditions. Two factors could explain the differences found between the southern and northern hemisphere: the turbulence and thus the available light, and trace elements. Typically, in summer, the mixing zone has the same depth range as the euphotic zone, about 60–100 m. Conditions are more favourable towards the Antarctic Divergence (with a shallower mixing zone and greater transparency) than towards the Polar Front (Jacques & Minas, 1981). Despite its conceptual utility, the Sverdrup approach in describing the onset of the spring bloom cannot be used explicitly in the Southern Ocean (Jacques & Minas, 1981; Smith & Sakshaug, 1990). As suggested by Smetacek & Passow (1990), the predictive ability of their approach in the context of spring bloom growth has been challenged by a few workers. It has been observed in the polar regions that at all bloom stations where chlorophyll *a* concentrations exceeded 2 mg m^{-3}, the pycnocline was situated at 20–40 m; 50 m may be the maximum depth for bloom development (Sakshaug & Holm-Hansen, 1984). The conditions normally encountered in the POOZ of the Indian Ocean impose a mixing zone > 50 m; therefore the residence time of phytoplankton in the upper layers is insufficient for blooming. Owing to the frequency of strong winds, loss processes surpass production processes; consequently a sustained bloom does not occur.

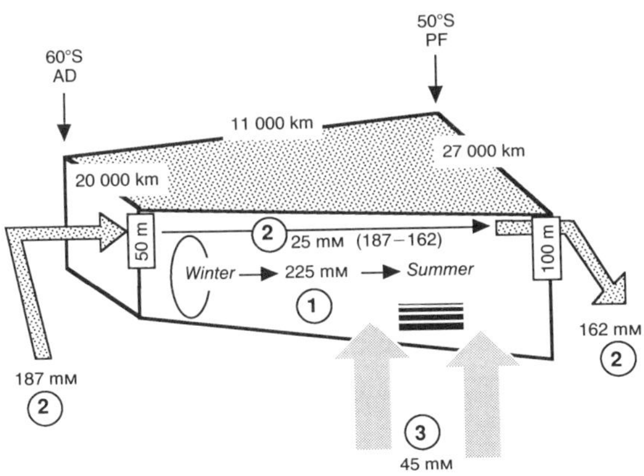

Fig. 9. Block diagram indicating the different nitrate ($N\text{-}NO_3\text{-}N$ in mM) reserves used by the phytoplankton in the Indian sector between the Antarctic Divergence (AD) and the Polar Front (PF). This nitrate depletion would lead to a new production over the 3-month summer period of: (1) winter vertical mixing, 225 mg C m^{-2} d^{-1}; (2) advection at the AD, 25 mg C m^{-2} d^{-1}; (3) diffusion through the pycnocline, 45 mg C m^{-2} d^{-1}.

Some other factors may control cell growth, among them silicates, at least in the northern part of the Antarctic Ocean (Jacques, 1983), and trace metals in nutrient-rich areas (Martin & Fitzwater, 1988; Martin *et al.*, 1989; Martin, 1990). The consequences of this hypothesis in terms of oceanic storage of CO_2 have been strongly contested (Davies, 1990; Keir, 1991; Peng & Broecker, 1991). However, the limitation of primary production by iron receives support from the following experimental results: reduced growth rate of phytoplankton in the absence of iron was observed in the Weddell and Scotia Seas (De Baar *et al.*, 1990). More remote sectors of the Southern Ocean, i.e. the POOZ, might in future be demonstrated to be more likely sites for trace metal limitation, as originally suggested by Hart (1942). In the Indian sector, the control of algal biomass by micro-and macrozooplankton grazing is poorly documented. It is not possible to apply to this sector results coming from either the SIZ or the Weddell and Scotia Seas (Lancelot *et al.*, 1991).

While it is now generally accepted that the relatively unproductive open ocean generates but a small flux of carbon onto the sea floor, the silica flux does not seem to fit into this scheme (Jones *et al.*, 1990). It is becoming more and more obvious that reported diatom accumulation in sediments does not necessarily imply a high level of production in the surface layer (Tréguer & Bennekom, 1991) as dissolution of silica material is clearly lowered in cold waters (Tréguer *et al.*, 1989). As a matter of fact, areas with highest accumulation rates of biogenic silica, which approximately correspond to the POOZ, coincide with that of minimum dissolution (estimated at *ca.* 20 ± 15%) of biogenic silica material (Bareille, 1991).

Vertical flux: the role of the POOZ

What is the importance of the POOZ compared with the other subsystems? It appears clear that superficial sediments are extremely rich in diatoms with a dominance of *Nitzschia kerguelensis* (reaching 80% of the total in species assemblages) (Bareille *et al.*, 1991). This is similar to the situation observed in surface waters (Sournia *et al.*, 1979). The antarctic abysses are therefore carpeted with silica to a much greater extent than in other systems (> 50% in mass). By contrast, the rate of accumulation appears to be low. Average sediment accumulation rates for siliceous sedimentation in the open Antarctic Ocean are between 0.5 and 2.0 cm kyr^{-1} (DeMaster, 1981). This is to be compared with 5–30 cm kyr^{-1} in some Antarctic continental shelves, where the primary productivity is of the same order as in high-productivity upwelling environments at low-altitude continental margins (Ledford-Hoffman *et al.*, 1986), and 5–17 cm kyr^{-1} in a band of siliceous sediments beneath the Polar Front (Cooke & Hays, 1982) based on the relative abundance of a radiolarian species as a function of depth. The sampling strategy concerns a part of the Atlantic sector where the maximum values were recorded, the Pacific sector, and the south-east and south-west Indian sectors. By taking into account the non-sampled areas where the rates are known to be less, the average rate of accumulation is close to 6 cm kyr^{-1}.

The POOZ therefore appears to be a system where the large amount of intact frustules in the sediment indicate a predominance of large diatoms in the pelagic system and their subsequent rapid sedimentation. It is now evident that during certain periods small plankton (the so-called 'microbial loop' regenerated community: bacteria, pico- and nanoplankton, ciliates, flagellates) dominate the pelagic system, but this pathway is probably not as important in the POOZ as in the other systems.

The ratio of sediment accumulation to primary production is certainly high, but because of the weak level of primary productivity and the scarcity of episodic events such as the spring bloom, the rate of accumulation is low. The contribution of this system to the total accumulation of silica in Antarctic sediments is therefore either very weak for the minimum observed rates (0.5 cm kyr^{-1}) or moderate if the higher values are confirmed (2.0 cm kyr^{-1}).

References

Abelmann A, Gersonde R (1991) Biosiliceous particle flux in the Southern Ocean. In: Tréguer P, Quéguiner B (eds) Biochemistry and circulation of water masses in the Southern Ocean. *Mar Chem* **35**: 503–36.

Bareille G (1991) Flux sédimentaires, silice biogène et détritique dans le secteur Indien de l'Océan Austral: paléoproductivité et évolution hydrologique au cours du dernier cycle climatique. PhD, Université de Bordeaux 1: 259 pp.

Bareille G, Labracherie M, Pichon J J, Turon J L, Labeyrie L (1991) Biogenic silica accumulation rate during the Holocene in the southeastern Indian Ocean. In: Tréguer P, Quéguiner B (eds) Biochemistry and circulation of water masses in the Southern Ocean. *Mar Chem* **35**: 537–51.

Cooke D W, Hays J D (1982) Estimates of Antarctic Ocean seasonal sea-ice cover during glacial intervals. In: Craddock C (ed) *Third International Symposium on Antarctic Geology and Geophysics*. University of Wisconsin Press, pp. 1017–25.

Davies A G (1990) Taking a cool look at iron. *Nature* **345**: 114–15.

Deacon G E R (1982) Physical and biological zonation in the Southern Ocean. *Deep-Sea Res* **29**: 1–15.

Deacon G E R (1983) Kerguelen, Antarctic and subantarctic. *Deep-Sea Res* **30**: 77–81.

De Baar H J W, Buma A G J, Nolting R F, Cadee G C, Jacques G, Tréguer P J (1990) On iron limitation of the Southern Ocean: experimental observations in the Weddell and Scotia Seas. *Mar Ecol Prog Ser* **65**: 105–22.

DeMaster D J (1981) The supply and accumulation of silica in the marine environment. *Geochim Cosmochim Acta* **45**: 1715–32.

El-Sayed S Z (1984) Productivity of the Antarctic waters – A reappraisal. In: Holm-Hansen O, Bolis L, Gilles R (eds) *Marine Phytoplankton and Productivity*. Springer-Verlag, Berlin, pp. 19–34.

El-Sayed S Z, Taguchi S (1981) Primary production and standing crop of phytoplankton along the ice-edge in the Weddell Sea. *Deep-Sea Res* **28**: 1017–32.

Fukuda Y, Ohno M, Fukuchi M (1986) Surface chlorophyll *a* distribution in marginal ice zone in Antarctica, 1984/85. *Mem Natl Inst Polar Res Spec Issue* **44**: 24–33.

Fukuchi M (1980) Phytoplankton chlorophyll stocks in the Antarctic Ocean. *J Oceanogr Soc Japan* **36**: 73–84.

Fukuchi M, Fukuda Y, Ohno M, Hattori H (1985) Surface phytoplankton chlorophyll distribution continuously observed in the JARE-26 cruise (1984/85) to Syowa station, Antarctica (Sibex II). *Mem Natl Inst Polar Res Spec Issue* **44**: 15–23.

Fukuchi M, Hattori H, Sasaki H, Hoshiai T (1988) A phytoplankton bloom and associated processes observed with a long term moored system in antarctic waters. *Mar Ecol Prog Ser* **45**: 279–88.

Fukuchi M, Tamura S (1982) Chlorophyll *a* distribution in the Indian sector of the Antarctic Ocean in 1977–1979. *Ant Rec* **74**: 143–62.

Fukuchi M, Tanimura A, Ohtsuka H (1984) Seasonal change of chlorophyll *a* under fast ice in Lützow-Holm Bay, Antarctica. *Mem Natl Inst Polar Res Spec Issue* **32**: 51–9.

Gamberoni L, Geronimi J, Jeannin P F, Murail J F (1982) Study of frontal zones in the Crozet-Kerguelen region. *Oceanol Acta* **5**: 289–99.

Gieskes W W C, Elbrächter M (1986) Abundance of nanoplankton-size chlorophyll-containing particles caused by diatom disruption in surface waters of the Southern Ocean (Antarctic Peninsula region). *Neth J Sea Res* **20**: 291–303.

Hamada E, Taniguchi A, Okazaki M, Naito Y (1985) Report on the phytoplankton pigments measured during the JARE-25 cruise to Syowa Station, Antarctica, November 1983 to April 1984. *JARE Data Rep* **103** (Mar Biol): 1–89.

Hart T J (1942) Phytoplankton periodicity in Antarctic waters. *Discovery Rep* **21**: 261–356.

Hasle G R (1969) An analysis of the phytoplankton of the Pacific Southern Ocean: abundance, composition, and distribution during the Brategg Expedition, 1947–1948. *Hvalrad Skr Norske Vidensk Akad* **52**: 1–168.

Hattori H, Fukuchi M (1989) Distribution of nano-, micro- and netplankton chlorophyll in the surface water of the Indian sector of the Southern Ocean, 1985/86. *Proc NIPR Symp Polar Biol* **2**: 26–33.

Holm-Hansen O, El-Sayed S Z, Franceschini G A, Cuhel R L (1977) Primary production and the factors controlling phytoplankton growth in the Southern Ocean. In: Llano G A (ed) *Adaptations within Antarctic Ecosystems*. Smithsonian Institution, Washington, D.C., pp. 11–50.

Ichimura S, Fukushima H (1963) On the chlorophyll content in the surface water of the Indian and the Antarctic Oceans. *Bot Soc Japan* **76**: 395–9.

Jacques G (1983) Some ecophysiological aspects of the Antarctic phytoplankton. *Polar Biol* **2**: 27–33.

Jacques G (1989) Primary production in the open Antarctic Ocean during the austral summer. A review. *Vie Milieu* **39**: 1–17.

Jacques G (1991) Is the concept of new production-regenerated production valid for the Southern Ocean? In: Tréguer P, Quéguiner B (eds) Biochemistry and circulation of water masses in the Southern Ocean. *Mar Chem* **35**: 273–86.

Jacques G, Descolas-Gros C, Grall J R, Sournia A (1979) Distribution du phytoplancton dans la partie Antarctique de l'Océan Indien en fin d'été. *Int Rev Ges Hydrobiol* **64**: 609–28.

Jacques G, Minas M (1981) Production primaire dans le secteur indien de l'océan Antarctique en fin d'été. *Oceanol Acta* **4**: 33–41.

Jacques G, Panouse M (1991) Biomass and composition of size fractionated phytoplankton in the Weddell–Scotia Confluence area. *Polar Biol* **11**: 315–28.

Jacques G, Tréguer P (1986) L'Océan Antarctique. In: Jacques G, Tréguer P (eds) *Ecosystèmes Pélagiques Marins*. Masson (Coll. Ecologie 19), Paris, pp. 101–62.

Jones E P, Nelson D M, Tréguer P (1990) Chemical oceanography. In: Smith W O Jr (ed) *Polar Oceanography*, part B: *Chemistry, Biology and Geology*. Academic Press, New York, pp. 407–76.

Kanda H, Fukuchi M (1979) Surface chlorophyll *a* concentration along the course of the Fuji to and from Antarctica in 1977–1978. *Ant Rec* **66**: 37–49.

Keir R S (1991) Ironing out greenhouse effects. *Nature* **349**: 198.

Kopczynska E E, Weber L H, El-Sayed S Z (1986) Phytoplankton species composition and abundance in the Indian sector of the Antarctic Ocean. *Polar Biol* **6**: 161–9.

Kosaki S, Takahashi M, Yamagushi Y, Aruga Y (1985) Size characteristics of chlorophyll particles in the Southern Ocean. *Trans Tokyo Univ Fish* **6**: 85–97.

Kuroda K, Fukuchi M (1982) Vertical distribution of chlorophyll *a* in the Indian sector of the Antarctic Ocean in 1972–1973. *Ant Rec* **74**: 127–42.

Lancelot C, Billen G, Mathot S (1990) Ecophysiology of phytoplankton and bacterioplankton growth in the Southern Ocean. Science Policy Office. ANTAR/**05**: 1–97.

Lancelot C, Billen G, Veth C, Becquevort S, Mathot S (1991) Modelling carbon cycling through phytoplankton and microbes in the Scotia–Weddell Sea area during sea ice retreat. In: Tréguer P, Quéguiner B (eds) Biochemistry and circulation of water masses in the southern ocean. *Mar Chem* **35**: 305–24.

Ledford-Hoffman P A, DeMaster D J, Nittrouer C A (1986) Biogenic-silica in the Ross Sea and the importance of Antarctic continental-shelf deposits in the marine silica budget. *Geochim Cosmochim Acta* **50**: 2099–110.

Le Jehan S, Tréguer P (1985) The distribution of inorganic nitrogen, phosphorus, silicon and dissolved organic matter in surface and deep waters of the Southern Ocean. In: Siegfried W R, Condy P R, Laws R M (eds) *Antarctic Nutrient Cycles and Food Webs (Proc 4th SCAR Symp Antarct Biol)*. Springer-Verlag, Berlin, pp. 22–9.

Lutjeharms J R E, Walters N M, Allanson B R (1985) Oceanic frontal systems and biological enhancement. In: Siegfried W R, Condy P R, Laws R M (eds) *Antarctic Nutrient Cycles and Food Webs (Proc 4th SCAR Symp Antarct Biol)*. Springer-Verlag, Berlin, pp. 11–21.

Marchant H J, Davidson A T, Wright S W (1987) The distribution and abundance of chroococcoid Cyanobacteria in the Southern Ocean. *Proc NIPR Symp Polar Biol* **1**: 1–9.

Martin J H (1990) Glacial-interglacial CO_2 change: the

iron hypothesis. *Paleoceanography* **5**: 1–13.

Martin J H, Fitzwater S E (1988) Iron deficiency limits phytoplankton growth in the north-east Pacific subarctic. *Nature* **331**: 341–43.

Martin J H, Gordon R M, Fitzwater S E (1989) Vertex: phytoplankton/iron studies in the Gulf of Alaska. *Deep-Sea Res* **35**: 177–96.

Mordasova N V (1989) Chlorophyll distribution in the Antarctic zone of the Atlantic Ocean. *Oceanology* **29**: 368–74.

Nagata Y, Michida Y, Unimura Y (1988) Variation of positions and structures of the oceanic fronts in the Indian Ocean sector of the Southern Ocean in the period from 1965 to 1987. In: Sahrhage D (ed) *Antarctic Ocean and Resources Variability*. Springer-Verlag, Berlin, pp. 92–8.

Nelson D M, Smith W O Jr (1986) Phytoplankton bloom dynamics of the Western Ross Sea ice edge. II. Mesoscale cycling of nitrogen and silicon. *Deep-Sea Res* **33**: 1389–412.

Nemoto T, Harrison G (1981) High latitude ecosystems. In: Longhurst A R (ed) *Analysis of Marine Ecosystems*. Academic Press, London, pp. 95–107.

Ohno M, Fukuda Y, Fukuchi M (1987) Vertical distribution and standing stocks of chlorophyll *a* in the coastal waters of the Antarctic Ocean. *Ant Rec* **31**: 93–108.

Park Y H, Charriaud E, Gambéroni L, Lamy A, Saint-Guily B (1989) Structure et variabilité du courant circumpolaire antarctique dans la région Kerguelen-Amsterdam. *CR Acad Sci Paris* **308** (Série II): 177–83.

Peng T H, Broecker W S (1991) Dynamical limitations on the Antarctic iron fertilization strategy. *Nature* **349**: 227–9.

Priddle J, Hawes I, Ellis-Evans J C, Smith T J (1986) Antarctic aquatic ecosystems as habitats for phytoplankton. *Biol Rev* **61**: 199–238.

Saijo Y (1964) Size distribution of photosynthesizing phytoplankton in the Indian Ocean. *J Oceanogr Soc Jap* **19**: 187–9.

Sakshaug E, Holm-Hansen O (1984) Factors governing pelagic production in polar oceans. In: Holm-Hansen O, Bolis L, Gilles R (eds) *Marine Phytoplankton and Productivity*. Springer-Verlag, Berlin, pp. 1–18.

Sasaki H (1984) Distribution of nano- and microplankton in the Indian sector of the Southern Ocean. *Mem Natl Inst Polar Res Spec Issue* **32**: 38–50.

Smetacek V, Passow U (1990) Spring bloom initiation and Sverdrup's critical-depth model. *Limnol Oceanogr* **35**: 228–34.

Smith N R, Zhaoquian D, Kerry K R, Wright S (1984) Water masses and circulation in the region of Prydz Bay, Antarctica. *Deep-Sea Res* **31**: 1125–47.

Smith W O Jr, Keene N K, Comiso J C (1988) Interannual variability in estimated primary productivity of the Antarctic Marginal Ice Zone. In: Sahrhage D (ed) *Antarctic Ocean and Resources Variability*. Springer-Verlag, Berlin, pp. 131–9.

Smith W O Jr, Sakshaug E (1990) Polar phytoplankton. In: Smith W O Jr (ed) *Polar Oceanography*, part B: *Chemistry, Biology and Geology*. Academic Press, London, pp. 477–525.

Sommer U (1986) Nitrate- and silicate-competition among Antarctic phytoplankton. *Mar Biol* **91**: 345–51.

Sournia A, Grall J R, Jacques G (1979) Diatomées et dinoflagellés planctoniques d'une coupe méridienne dans le sud de l'océan Indien (campagne "Antiprod 1" du Marion-Dufresne, mars 1977). *Bot Mar* **22**: 183–98.

Sullivan C W, McClain C R, Comiso J C, Smith W O Jr (1988) Phytoplankton standing crops within an Antarctic ice edge assessed by satellite remote sensing. *J Geophys Res* **93**: 12487–98.

Taniguchi A, Hamada E, Okazaki M, Naito Y (1986) Distribution of phytoplankton chlorophyll continuously recorded in the JARE-25 cruise to Syowa station, Antarctica (Sibex 1). *Mem Natl Inst Polar Res Spec Issue* **44**: 3–14.

Tréguer P, Bennekom A J van (1991) The annual production of biogenic silica in the Antarctic Ocean. *Mar Chem* **35**: 477–88.

Tréguer P, Gueneley S, Kamatani A, Quéguiner B (1989) Kinetics of dissolution of Antarctic diatoms and the biogeochemical cycle of silicon in the Southern Ocean. *Polar Biol* **9**: 397–403.

Tréguer P, Jacques G (1992) Dynamics of nutrients and phytoplankton and fluxes of carbon, nitrogen and silicon in the Antarctic Ocean. *Polar Biol* **12**: 149–62.

Uno S (1982) Distribution and standing stock of chlorophyll *a* in the Antarctic Ocean, from December 1980 to January 1981. *Mem Natl Inst Polar Res Spec Issue* **23**: 20–7.

Watanabe K, Nakajima Y (1982) Vertical distribution of chlorophyll *a* along 45°E in the Southern Ocean, 1981. *Mem Natl Inst Polar Res Spec Issue* **27**: 73–86.

Weber L H, El-Sayed S Z (1987) Contribution of the net-, nano- and picoplankton to the phytoplankton standing crop and primary productivity in the Southern Ocean. *J Plankt Res* **9**: 973–94.

Yamaguchi Y, Shibata Y (1982) Standing stock and distribution of phytoplankton chlorophyll in the Southern Ocean south of Australia. *Trans Tokyo Univ Fish* **5**: 111–28.

The zooplankton of the Antarctic Peninsula region

S. B. SCHNACK-SCHIEL & A. MUJICA

Introduction

THE SOUTHERN Ocean is well known for its latitudinal zonation. Three different large-scale subsystems coincide with the distribution of water masses and the ice cover. The ice-free oceanic West Wind Drift is dominated by herbivorous copepods, relatively small euphausiid species and salps. The seasonal pack-ice zone includes most of the East Wind Drift with krill (*Euphausia superba*) as the main component. The permanent pack-ice zone includes cold water along the continental shelf with the ice krill (*Euphausia crystallorophias*) and copepods as the major plankton organisms (G. Hempel, 1985).

In the southern Scotia Sea, cold water emerging from the Weddell Sea mixes with warmer eastward-flowing water reaching the Scotia Sea through the Drake Passage. This frontal zone, the Weddell–Scotia Confluence (WSC) separates these two water masses and forms a complex field of mesoscale eddies and meanders. The WSC is recognizable within the Atlantic sector of the Southern Ocean (Stein, 1986), and has a major effect on the temperature distribution and ice cover and, thus, on the distribution of zooplankton species. The WSC acts as a southern biological border for sub-Antarctic species that are typical inhabitants of the permanently ice-free West Wind Drift, e.g. *Calanus simillimis* (Vladimirskaya, 1978; Marin, 1987). Zooplankton communities north of the WSC, in the Antarctic Circumpolar Current, consist of more advanced developmental stages than in the Weddell Sea at the same time (Kanaeva, 1968; Vladimirskaya, 1978; Makarov, 1979; Marin, 1987). The composition of the developmental stages of the calanoid copepod *Calanoides acutus* changed from north to south along a transect from the open Scotia Sea to the ice-covered Weddell Sea (Fig. 1a). At the northern station (143), the population showed a bimodal stage structure with copepodite stage V from the previous year's generation as well as offspring of the current year. At the southern stations (149–53) adults from the previous year are dominant. The population occurs in greater numbers in the upper water layers at the stations in the Scotia Sea (143) and in the Confluence zone (147+149), whereas the bulk is found in deeper overwintering waters in the Weddell Sea (stations 151+153) (Fig. 1b).

The southern part of the Scotia Sea, i.e. west of the Antarctic Peninsula and the Bransfield Strait, belongs to the seasonal pack-ice zone (G. Hempel, 1985). The seasonal ice cover and the moving ice edge play an important role in development and vertical distribution pattern of zooplankton because the marginal ice zone acts as a frontal system with enhanced primary productivity (Hopkins & Torres, 1988). Thus, the zooplankton is subjected to a distinct seasonality of production, which according to

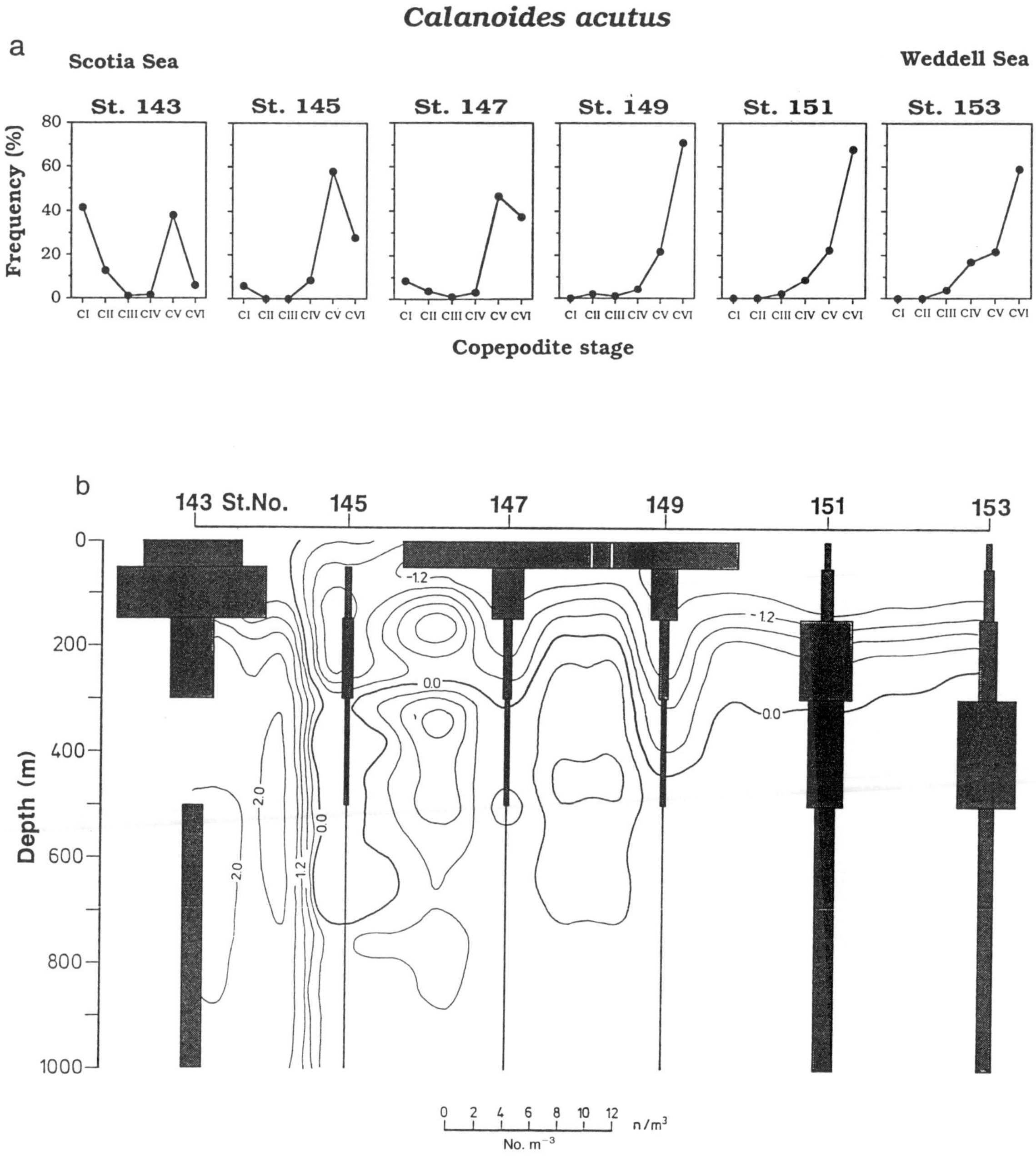

Fig. 1. Relative composition of developmental stages (a) and vertical distribution pattern (b) of *Calanoides acutus* in the top 1000 m along a transect from open Scotia Sea water (St. 143) across the Weddell–Scotia Confluence (St. 145–9) into the Weddell Sea pack-ice (St. 151–3), 26–30 November 1988 (S. Schnack-Schiel, unpublished data).

Clarke (1983) is the main adaptive regulator. While zooplankton standing stocks and composition in the upper 1000 m in the open sea and within the pack-ice in the western Weddell Sea in March 1986 are roughly equivalent, there are significant differences in the vertical distribution (Hopkins & Torres, 1988). Zooplankton biomass and respiratory activity (ETS) were relatively high in the ice edge region in November–December 1988, while low biomass and activity were encountered under the pack-ice (Fig. 2) (Schalk, 1990). The high ETS activities coincided with maximum concentrations of chlorophyll *a* (Jacques & Panouse, 1991).

Our knowledge of the complex geographical distribution of zooplankton in the Antarctic Peninsula region has been greatly improved since the BIOMASS Programme started (e.g. Jazdzewski *et al.*, 1982; Hopkins, 1985; Mujica & Asencio, 1985; Piatkowski, 1985*a*; Schnack *et al.*, 1985*a*; Witek *et al.*, 1985; Atkinson & Peck, 1988; Siegel & Piatkowski, 1990). In recent years, intensive studies on zooplankton have included distributional patterns, population structure and trophodynamics (e.g. Hagen, 1985; Schnack, 1985; Schnack *et al.*, 1985*b*; Marin, 1988; Huntley *et al.*, 1989; Atkinson, 1991; Huntley & Escritor, 1991, 1992).

This chapter summarizes the results of zooplankton research carried out in the vicinity of the Antarctic Peninsula (Atlantic Section) during BIOMASS. Most of the research reported herein was conducted by scientists from Argentina, Brazil, Chile, Germany, Great Britain, Poland and the U.S.A.

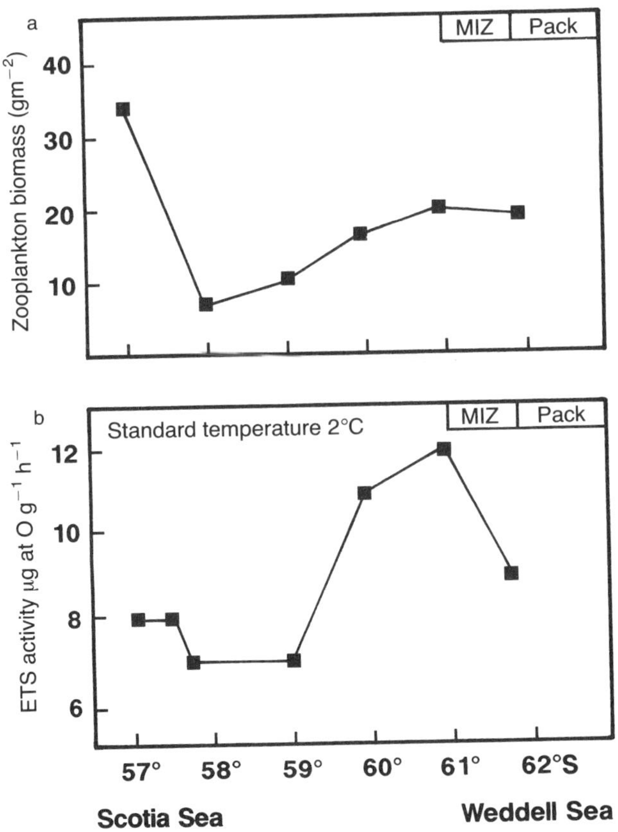

Fig. 2. Fresh biomass (a) and mass-specific ETS activities (b) of zooplankton along a transect running from open Scotia Sea water across the Weddell–Scotia Confluence into Weddell Sea pack-ice, 27–30 December 1988 (after Schalk, 1990).

Composition, abundance and biomass of zooplankton in the study area

Copepods and euphausiids by far dominate the zooplankton assemblages (Hardy & Gunther, 1935; Mujica & Torres, 1982; Chojnacki & Węglénska, 1984; Hopkins, 1985; Mujica & Asencio, 1985; Montú & Oliveira, 1986; Kittel *et al.*, 1988). In some years, however, they can be outnumbered by salps, mainly *Salpa thompsoni* (Huntley *et al.*, 1989; Mujica & Asencio, 1985; Witek *et al.*, 1985; Montú & Oliveira, 1986; Nast, 1986; Piatkowski, 1985*b*). Groups other than copepods, euphausiids and salps (such as polychaetes, molluscs, appendicularians, ostracods, amphipods, chaetognaths and coelenterates) contribute only moderately (<10%) to zooplankton biomass in the Antarctic Peninsula region (Mujica & Torres, 1982; Hopkins, 1985; Hopkins & Torres, 1988; Boysen-Ennen *et al.*, 1991).

The literature on Antarctic zooplankton is dominated by studies on a single species, *Euphausia superba*, the Antarctic krill, which as well as being of enormous ecological interest is also a resource of considerable commercial potential. Krill is often considered the most important species of the Southern Ocean; other zooplankton have therefore received less attention. For instance, Hopkins (1985) estimated the biomass of *E. superba* in Croker Passage west of the Antarctic Peninsula to be more than 15-fold higher than that of all other zooplankton, i.e. contributing 55 of 59 g dry mass per square metre. This, however, does not reflect the overall share of krill in the Antarctic. During FIBEX, January–March 1981, Brinton & Antezana (1984) estimated krill and zooplankton biomass. Non-krill zooplankton constituted 75% of the total zooplankton biomass including krill swarms in the Scotia Sea. Excluding the krill swarms, non-krill zooplankton accounted for 94%. Around Elephant Island, a region of usually high krill density, and in Bransfield Strait, non-krill zooplankton made up about 26% and 31%, respectively, of the total zooplankton biomass within krill swarms. Excluding the krill swarms, non-krill zooplankton contributed 60% in both areas. Non-krill zooplankton consisted mainly of copepods, chaetognaths, the amphipod *Themisto gaudichaudii* and the euphausiid *Thysanoessa macrura* (Brinton & Antezana, 1984). In Admiralty Bay, King George Island, copepods constituted 60–87% of the total summer biomass of the zooplankton, including euphausiids (Chojnacki & Węglénska, 1984). Boysen-Ennen *et al.* (1991) summarized biomass values of zooplankton from different regions in the Southern Ocean: mesoplankton (length <14.5 mm) biomass generally ranged from 0.8 to 3.6 g dry mass m^{-2} and macroplankton (length 14.5–>39.6 mm) from 0.2 to 2.4 g dry mass m^{-2}. Thus, mesoplankton biomass is higher than macroplankton biomass and is in the same order of magnitude as krill biomass. Because of the higher production : biomass ratio of copepods (4.5) compared with krill (1.0), Boysen-Ennen *et al.* (1991) concluded that copepods, which constitute the main component of the mesoplankton biomass, contribute most of the zooplankton production in the Antarctic.

Copepods

Copepods are the most diverse group, accounting for more than 70 species in the upper 1000 m of the water column in the western Weddell Sea (Hopkins & Torres, 1988). In the upper 200 m in Bransfield Strait, approximately 30 species were reported (Zmijewska, 1985). Most of these, however, were uncommon. The copepod assemblages are often dominated by cyclopoid copepods: *Oithona* spp. and *Oncaea* spp. account for 40–80% of the total copepod population by number (Jazdzewski *et al.*, 1982; Chojnacki & Węglénska, 1984; Hopkins, 1985; Schnack *et al.*, 1985*a*; Montú & Oliveira, 1986; Almeida Prado-Por, 1989). *Oithona similis* prefers the upper 200 m whereas *Oncaea* spp. are bathypelagic (Schnack *et al.*, 1985*a*; Zmijewska, 1988). Among the calanoid species, *Metridia gerlachei*, *Microcalanus pygmaeus* and *Ctenocalanus citer* are the most numerous ones and occur throughout the area (Jazdzewski *et al.*, 1982; Chojnacki & Węglénska, 1984; Hopkins, 1985; Schnack *et al.*, 1985*a*; Zmijewska, 1985; Montú & Oliveira, 1986; Dinofrio, 1987*a*). *Microcalanus pygmaeus* and *Metridia gerlachei* together made up 70% of all calanoids in Croker Passage between Gerlache Strait and Bransfield Strait (Hopkins, 1985) and up to 90% in the northern Bransfield Strait and South of Elephant Island (Schnack *et al.*, 1985*a*). Small copepod species (calanoid and cyclopoid) predominate by number, but large species such as *Calanoides acutus*, *Calanus propinquus*, *Rhincalanus gigas* and *Metridia gerlachei* generally make up a larger portion of the total biomass, i.e. about 50–90% (Chojnacki & Węglénska, 1984; Hopkins, 1985; Hopkins & Torres, 1988; Boysen-Ennen *et al.*, 1991).

The biology of two dominant species, *Calanoides acutus* and *Metridia gerlachei*, will now be discussed in more detail. *Calanoides acutus* undergoes a seasonal ontogenetic vertical migration; the abundance in the upper 1000 m in winter is only one third of the summer concentration around South Georgia and in the south-eastern Weddell Sea (Atkinson & Peck, 1988; Schnack-Schiel *et al.*, 1991). When comparing only the upper 200 m in Bransfield Strait, the concentration is over 20 times higher in January than in May (Fig. 3). The highest densities occur from December to March, when females ascend to the surface to spawn (Andrews, 1966), and in summer, when the offspring of the current year (copepodite stages I–III) are highly concentrated in the upper 100 m of the water column (Andrews, 1966; Huntley & Escritor, 1991). *Calanoides acutus* appears to spawn intensely over a short period and develops as a single identifiable cohort (Huntley & Escritor, 1991). Differences in abundance and vertical distribution are found within the Antarctic Peninsula region: the abundance of *C. acutus* is consistently higher in Gerlache Strait waters than in Bransfield Strait and southern Drake Passage. The population attains the overwintering copepodite stages CIV and CV by March. Ontogenetic descent

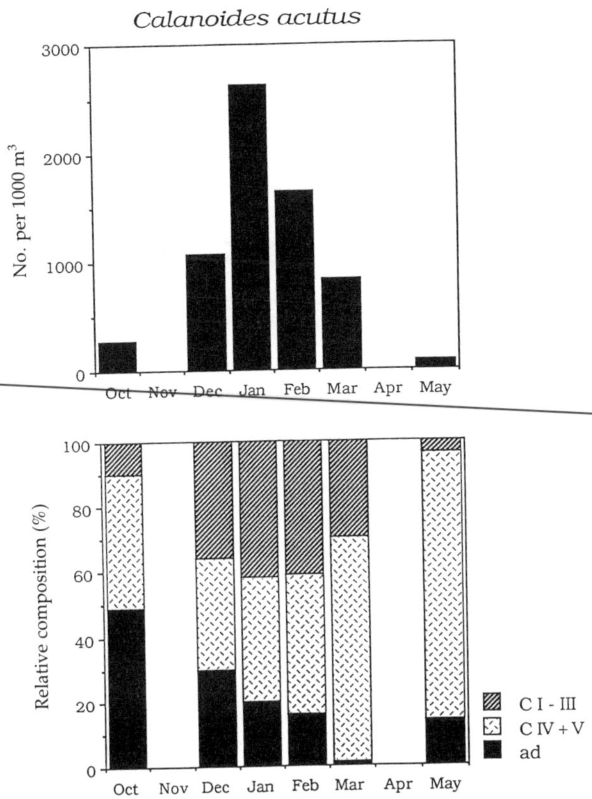

Fig. 3. Abundance and development structure of *Calanoides acutus* in the upper 200 m in the Bransfield Strait in different months (after Schnack *et al.*, 1985*a*; Zmijewska, 1985; Montú & Oliveira, 1986; Huntley & Escritor, 1991; S. Schnack-Schiel, unpublished data).

migration of the overwintering stages occurs in the Drake Passage by late March, but is delayed in the relatively food-rich waters of Gerlache Strait. In the Drake Passage in January the population is dominated by CI + II and is concentrated in the upper 200 m; by March, CIV + V predominate in the population and are most abundant at depths below 1000 m.

Unlike *C. acutus*, *Metridia gerlachei* does not seem to undergo seasonal ontogenetic vertical migration during winter (Atkinson & Peck, 1988). The highest abundance of *M. gerlachei* in the upper 200 m of the water column occurred in the Bransfield Strait in February and young copepodite stages (I–III) make up more than 60% of the total population (Fig. 4). Like *C. acutus*, higher numbers are consistently found in the Gerlache Strait (in January up to about 500 m^{-3}). However, a large proportion of the population occurs below 200 m (Huntley & Escritor, 1992). Spawning of *M. gerlachei* takes place before December and appears to be continuous during summer. The length of copepodite stages IV and V was consistently greater in phytoplankton-rich nearshore regions than in phytoplankton-poor offshore areas (Huntley & Escritor, 1992).

Euphausiids

The second dominant zooplankton group are euphausiids. Five species occur in the Antarctic Peninsula region: *Euphausia superba*, *E. crystallorophias*, *E. frigida*, *E. triacantha* and *Thysanoessa macrura*. All are known to have a circumpolar distribution pattern south of the Antarctic Convergence

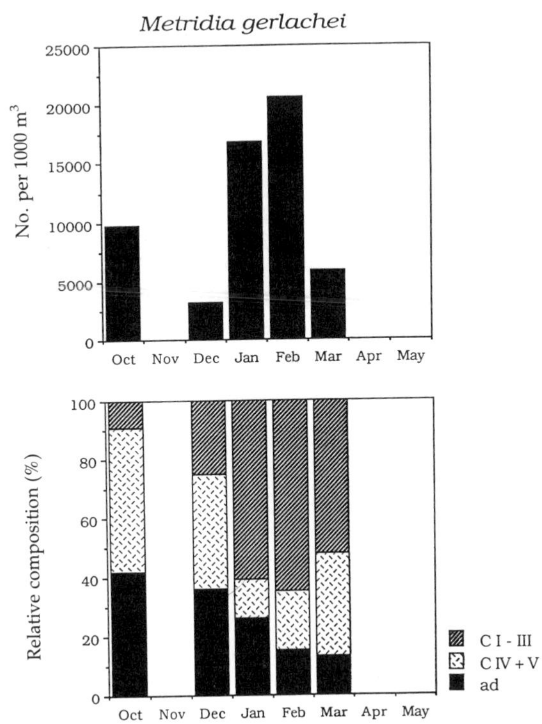

Fig. 4. Abundance and developmental structure of *Metridia gerlachei* in the upper 200 m in the Bransfield Strait in different months (after Schnack *et al.* 1985*a*; Zmijewska, 1985; Montú & Oliveira, 1986; Huntley & Escritor, 1992; S. Schnack-Schiel, unpublished data).

(Baker, 1954). *Euphausia superba* is dominant (Weigmann-Haass & Haass, 1980; Stepnik, 1982; Piatkowski, 1985*a*; Pires, 1989) and is discussed in separate chapters (Siegel & Kalinowski, this volume; Everson & Miller, this volume). *Thysanoessa macrura* is the second dominant euphausiid species (Kittel & Stepnik, 1983; Kittel *et al.*, 1985; Piatkowski, 1985*a*; Mujica & Asencio, 1990), and can sometimes outnumber *E. superba* (Nast, 1986; Piatkowski, 1987; Atkinson & Peck, 1988; Kittel, 1988; Mujica, 1989). The distribution pattern of *Thysanoessa macrura* is similar to that of *E. superba* (Weigmann-Haass & Haass, 1980; Piatkowski, 1985*a*). It is the most broadly distributed species, and there is no indication that it prefers certain water masses (Brinton, 1985; Piatkowski, 1985*a*; Makarov *et al.*, 1990). The ice-krill, *Euphausia crystallorophias*, is neritic and occurs over the shelf of the Bransfield Strait and west of the Antarctic Peninsula (John, 1936; Weigmann-Haass & Haass, 1980; Kittel & Stepnik, 1983; Piatkowski, 1987). Swarms of *E. crystallorophias* occur in Admiralty Bay, King George Island (Jackowska, 1980; Rakusa-Suszczewski & Stepnik, 1980; Stepnik, 1982), in coastal bays of the Gerlache Strait (Brinton & Townsend, 1991) and in the caldera of Deception Island (Piatkowski, 1985*a*; Everson, 1987). *Euphausia frigida* and *E. triacantha* are mainly confined to the midwater layer of the West Wind Drift and occur only at stations not influenced by Weddell Sea waters (John, 1936; Baker, 1959; Hempel & Marschoff, 1980; Weigmann-Haass & Haass, 1980; Kittel *et al.*, 1985; Witek & Kittel, 1985; Witek *et al.*, 1985; Piatkowski, 1985*a*; Pires, 1986). None of the two species tend to swarm (Kittel *et al.*, 1985; Piatkowski, 1985*a*).

The euphausiid species have different periods of reproduction. Spawning of *Thysanoessa macrura* commences as early as September; it starts to spawn earlier in the West Wind Drift than in the Weddell Sea (Makarov, 1979). At the same time *Euphausia frigida* and *E. triacantha* begin to spawn (Baker, 1959; Makarov, 1977), followed by *E. crystallorophias* (Stepnik, 1982). Spawning in *E. superba* takes place as late as between November and April (Hempel & Marschoff, 1980; Makarov *et al.*, 1990). The developmental ascent, as described by Marr (1962) for *E. superba*, is also true for the other Antarctic euphausiid species, except for the neritic *E. crystallorophias* (Makarov, 1979).

Salps

Salpa thompsoni was a dominant member of the zooplankton community in waters off the Antarctic Peninsula during SIBEX I in the austral summer of 1983–4, but also in other years (Mujica & Asencio, 1985; Piatkowski, 1985*b*). Nast (1986) counted a maximum of 889 individuals per 1000 m^3 north of Elephant Island in March 1985. On the other hand, salps were rarely found at stations influenced by Weddell Sea water (Mackintosh, 1934; Foxton, 1966; Mujica & Asencio, 1985; Piatkowski, 1985*a*; Witek *et al.*, 1985; Nast, 1986). In 1983–4 less salp biomass was found in December and January than in early March (Huntley *et al.*, 1989; Witek *et al.*, 1985). However, the distribution pattern was similar. The highest biomass was found in Bransfield Strait and the lowest near the South Shetlands. In December–January, biomass in the upper 200 m varied from 0.04 to 208 mg C m^{-2}; in mid-March, it varied from 49 to 671 mg C m^{-2}. In mid-March salps were distributed in three regions according to size, each region being uniform with respect to the size–frequency distribution: (1) the Bransfield Strait; (2) north of the South Shetlands; and (3) west of Elephant Island. Lengths were on average 20, 30 and 40 mm, respectively (Fig. 5). In mid-February to mid-March 1981 and in the austral summer of 1986–7, however, salps were only found at a few stations in the same region (Jazdzewski *et al.*, 1982; Huntley *et al.*, 1989). Coincidentally with the widespread occurrence of *S. thompsoni* in the austral summer 1983–4 there was a strikingly low abundance of other zooplankton, including the larvae of Antarctic krill (Montú & Oliveira, 1986). In the upper 200 m, larval *E. superba* averaged fewer than 2 m^{-2} near the South Shetland Islands in March 1984, compared with more than 50 m^{-2} in January–February 1981 (Brinton *et al.*, 1986).

Under favourable conditions, salps can rapidly respond to rises in the phytoplankton stock, with rapid multiplication that can give rise to enormous concentrations (Foxton, 1964). It has been suggested by Fraser (1962) that salps remove all particles from the water and thus deprive other grazers of their food. Grazing by *S. thompsoni* may have removed a major part of the daily primary production in March 1984, in contrast to <1% in January (Huntley *et al.*, 1989), the time at which krill larvae begin to appear in the plankton (Marr, 1962). Apart

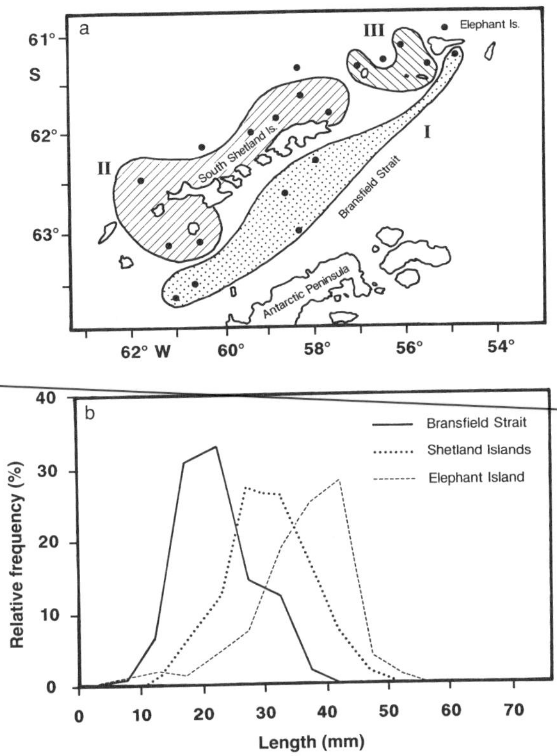

Fig. 5. (a) Geographical distribution of stations grouped by cluster analysis. Group I, Bransfield Strait; Group II, South Shetland Islands; Group III, Elephant Island. (b) Mean size frequency distribution of each group of *Salpa thompsoni* identified by cluster analysis (after Huntley *et al.*, 1989).

from the ability to remove the food resources of larval krill in March, *S. thompsoni* may represent a significant predation pressure directly on krill larvae. The comparatively low biomass of krill larvae in 1983–4 may be attributed partly to these two features (Huntley *et al.*, 1989). Over a 60 d period, the salps removed 10–30% of the krill larvae in the upper 200 m: this would have serious effects on the larval stock and could account for the one order of magnitude reduction in larval krill abundance in 1984 relative to 1981 (Brinton *et al.*, 1986).

Remaining zooplankton groups

Polychaetes are permanent components of the zooplankton (Mujica & Torres, 1982; Mujica & Asencio, 1985). The primary species is *Pelagobia longicirrata* (Jazdzewski *et al.*, 1982; Hopkins, 1985; Dinofrio, 1987*b*; Lana & Blankensteyn, 1987, 1989; Sicinski, 1988). The distribution of *P. longicirrata* along with that of *Rhynchonereella bongraini* is related to the penetration of colder Weddell Sea water (Witek *et al.*, 1985; Lana & Blankensteyn, 1987, 1989). *Tomopteris carpenteri* is common in the warmer surface water of the Bellingshausen Sea and Drake Passage. This species avoids Weddell Sea water (Piatkowski, 1985*a*; Witek *et al.*, 1985; Lana & Blankensteyn, 1987).

Among pteropods, *Limacina helicina* is predominant. The highest concentrations occur in the Drake Passage and in the waters of the Palmer Archipelago (Jazdzewski *et al.*, 1982; Witek *et al.*, 1985). *Clione limacina*, a predator on Limacinidae, is equally common but less abundant (Witek *et al.*, 1985). *Clio pyramidata* represents a characteristic member of oceanic communities (Piatkowski, 1987).

Appendicularians are mainly concentrated in the upper 100 m at stations on the shelf and near the northern slope of the South Shetlands and the Palmer Archipelago (Jazdzewski *et al.*, 1982).

Ostracods have significantly higher abundances in oceanic than in neritic waters (Hardy & Gunther, 1935; Boysen-Ennen & Piatkowski, 1988). In the Antarctic Peninsula region this group occurs regularly, with its highest concentrations in the Palmer Archipelago area (Witek *et al.*, 1985) and in the western entrance of the Bransfield Strait (Mujica & Torres, 1982). In Croker Passage *Conchoecia* (= *Alacia) belgica* is the most dominant species followed by *C. isocheira* and *Conchoecia* (=*Alacia) hettacra* (Hopkins, 1985).

Among true carnivorous zooplankton groups, hyperiid amphipods, chaetognaths and siphonophores are most predominant and exhibit a wide distribution. *Themisto gaudichaudii*, *Hyperiella dilatata* and *Vibilia antarctica* are the most frequent amphipods in the study area (Piatkowski, 1985*a*; Witek *et al.*, 1985; Montú & Oliveira, 1986; Dinofrio, 1987*b*; Atkinson & Peck, 1988). *Themisto gaudichaudii* and *Vibilia antarctica*, however, are absent at stations characterized by Weddell Sea water (Piatkowski, 1985*a*,*b*, 1987; Witek *et al.*, 1985; Weigmann-Haass, 1990). The latter species co-occurred with salps in December 1983 (Witek *et al.*, 1985).

Among chaetognaths, *Eukrohnia hamata* is the

most abundant (Hagen, 1985; Hopkins, 1985; Witek *et al.*, 1985; Dinofrio, 1987*b*). There are pronounced differences in the vertical distribution of the different chaetognath species. *Eukrohnia hamata* occurs in highest densities between 200 m and 750 m depth, but large numbers can be found throughout the water column. *Sagitta gazellae* occurs above 400 m depth, *S. marri* is mesopelagic and most found in waters between 200 and 750–1000 m depth (Hagen, 1985; Hopkins, 1985). *Sagitta maxima* appears to be the subantarctic counterpart of *S. marri* with respect to vertical distribution (David, 1958; Hagen, 1985). *Eukrohnia bathypelagica* inhabits the lower mesopelagic and upper bathypelagic layers. The rarely caught *E. bathyantarctica* and the four *Heterokrohnia* species are bathypelagic. No evidence was obtained that hydrographic factors have had an effect on chaetognath distribution in the vicinity of the Antarctic Peninsula (Hagen, 1985).

Diphyes antarctica and *Dimophyes arctica* are the prevailing siphonophore species and occur within the entire Antarctic Peninsula region, but with higher concentrations in oceanic surface waters (Piatkowski, 1985*a*; Witek *et al.*, 1985).

Zooplankton composition and distribution during FIBEX and SIBEX

During FIBEX, maximum zooplankton biomass occurred in the SW of the Bransfield Strait, on the northern shelf of the Antarctic Peninsula, and off King George Island, as well as between King George Island and Elephant Island. The poorest zooplankton zones were encountered in the central part of Bransfield Strait (Fig. 6) where the highest krill biomass was found (Mujica & Torres, 1982; Jazdzewski *et al.*, 1982). In February–March 1981 the total zooplankton in the upper 100 m layer of the Antarctic Zone area (southern Drake Passage) was 3–4 times higher than that of the Continental Zone waters (Bransfield Strait). The West Wind Drift waters of the Drake Passage were characterized by high abundances of radiolarians and *Limacina* sp. and the regular occurrence of the copepods *Rhincalanus gigas*, *Clausocalanus* sp. and *Calanoides acutus*. On the other hand, these species were scarce in Bransfield Strait waters where the copepods *Metridia gerlachei* and *Oncaea curvata* dominated (Jazdzewski *et al.*, 1982). Zooplankton in the 300–100 m water layer seems to be less diversified. In this layer, copepods dominate all other zooplankton groups in terms of number. However, the differences between the southern Drake Passage and Bransfield Strait are also reflected in that layer. *Metridia gerlachei* is a characteristic form of cold, continental waters of the Bransfield Strait and the South Shetlands shelf. It was also regularly caught in the Drake Passage and Scotia Sea, but only in the deeper and colder 300–100 m layer (Jazdzewski *et al.*, 1982; I. Hempel, 1985). In summer 1981, very high concentrations of krill and its larvae were found in the survey area (Brinton, 1985; Brinton *et al.*, 1986).

During SIBEX I in October 1983, krill abundance and biomass were relatively high in the Bransfield Strait, and consisted almost only of late furcilia larvae, juvenile and immature animals (Siegel, 1988). During this period, the pack-ice had already retreated from the survey area (Siegel, 1988). In summer 1983, however, following this early onset, a very greatly reduced krill biomass and a predominance of salps were encountered in the Bransfield Strait (Witek *et al.*, 1985). There was a consistently low krill biomass in the Bransfield Strait for a period of approximately 9 months, which coincided with unusually high water temperatures (Priddle *et al.*, 1991). Salps predominated in the study area from November 1983 until March 1984 (Nast, 1986; Witek *et al.*, 1985; Mujica & Asencio, 1985; Montú & Oliveira, 1986; Huntley *et al.*, 1989) and made up over 75% of the total zooplankton biomass (Witek *et al.*, 1985). Those stations where salps were not found were characterized by Weddell Sea water. The diversity and abundance of zooplankton other than salps were low (Witek *et al.*, 1985; Montú & Oliveira, 1986). Low zooplankton concentrations co-occur with highest concentrations of *Salpa thompsoni*. In areas of salp dominance, the widely distributed *Thysanoessa macrura* was the prevalent euphausiid (Kittel *et al.*, 1985; Mujica & Asencio, 1985). The 1983–4 season could be regarded as most unusual (Siegel, 1988). During SIBEX II, one year later in 1984–5, salps occurred only in small numbers (Montú & Oliveira, 1986). Copepods and euphausiids dominated the zooplankton assemblages. Large abundances of appendicularians were found in the south-western Bransfield Strait coinciding with high phytoplankton biomass (Mujica & Asencio, 1986).

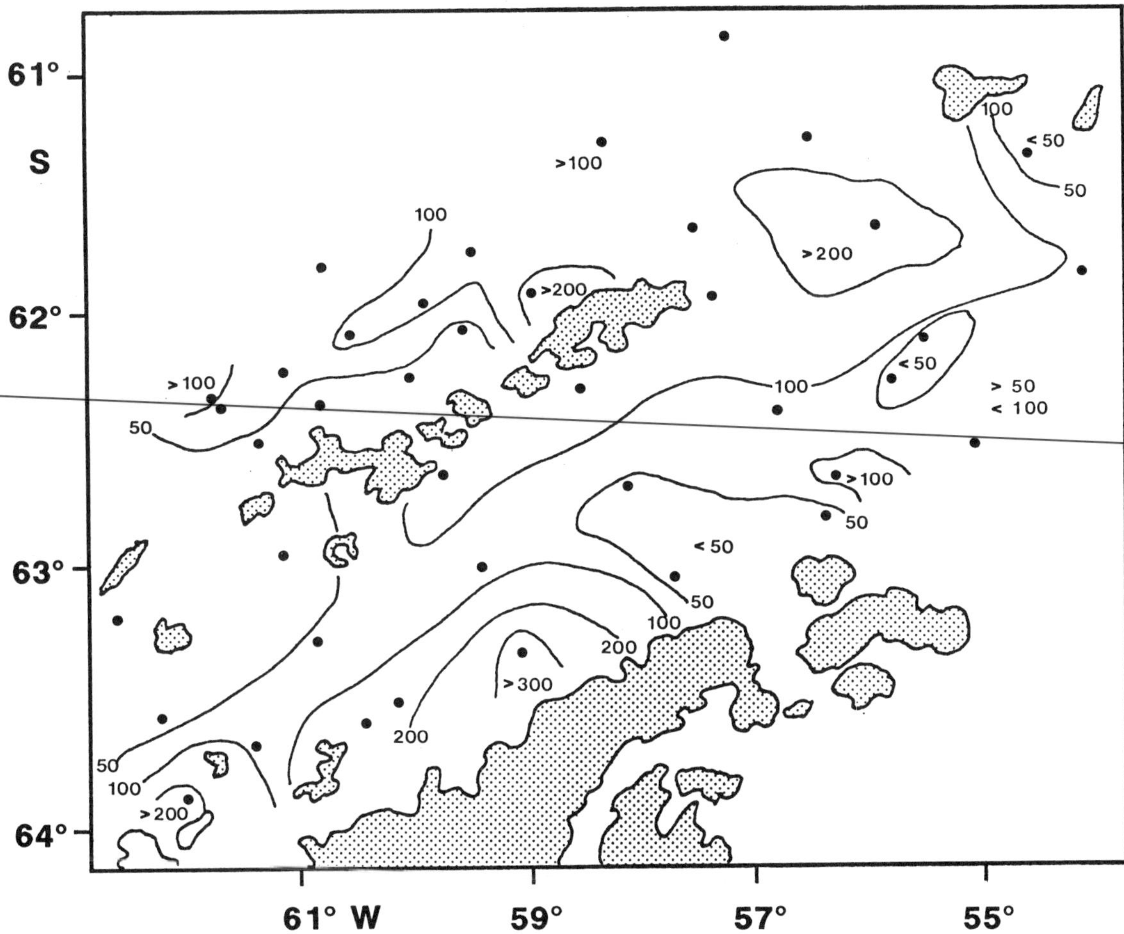

Fig. 6. Biomass distribution, expressed in millilitres per thousand cubic metres (displacement of wet volumes) during FIBEX (after Mujica & Torres, 1982).

Zooplankton assemblages

Zooplankton assemblages can be distinguished by differences in total concentration as well as species composition and abundance. Such differences are widely attributed to different water masses (Jazdzewski *et al.*, 1982; Witek *et al.*, 1985; Piatkowski, 1989; Siegel & Piatkowski, 1990).

The summer plankton fauna in waters off the Antarctic Peninsula was distinguished in communities by differences in species composition and abundance widely attributed to different water masses (Piatkowski, 1989). However, indicator species were only found for the oceanic community which comprises stations influenced by the West Wind Drift. Oceanic species like the hyperiid amphipod *Vibilia antarctica* and the pelagic polychaete *Tomopteris carpenteri* are typical components (Piatkowski, 1989). The macrozooplankton communities off the Antarctic Peninsula are largely identical to the macrozooplankton communities off the Weddell Sea (Boysen-Ennen & Piatkowski, 1988; Piatkowski, 1987). Based on four expeditions to the Antarctic Peninsula region at different seasons,

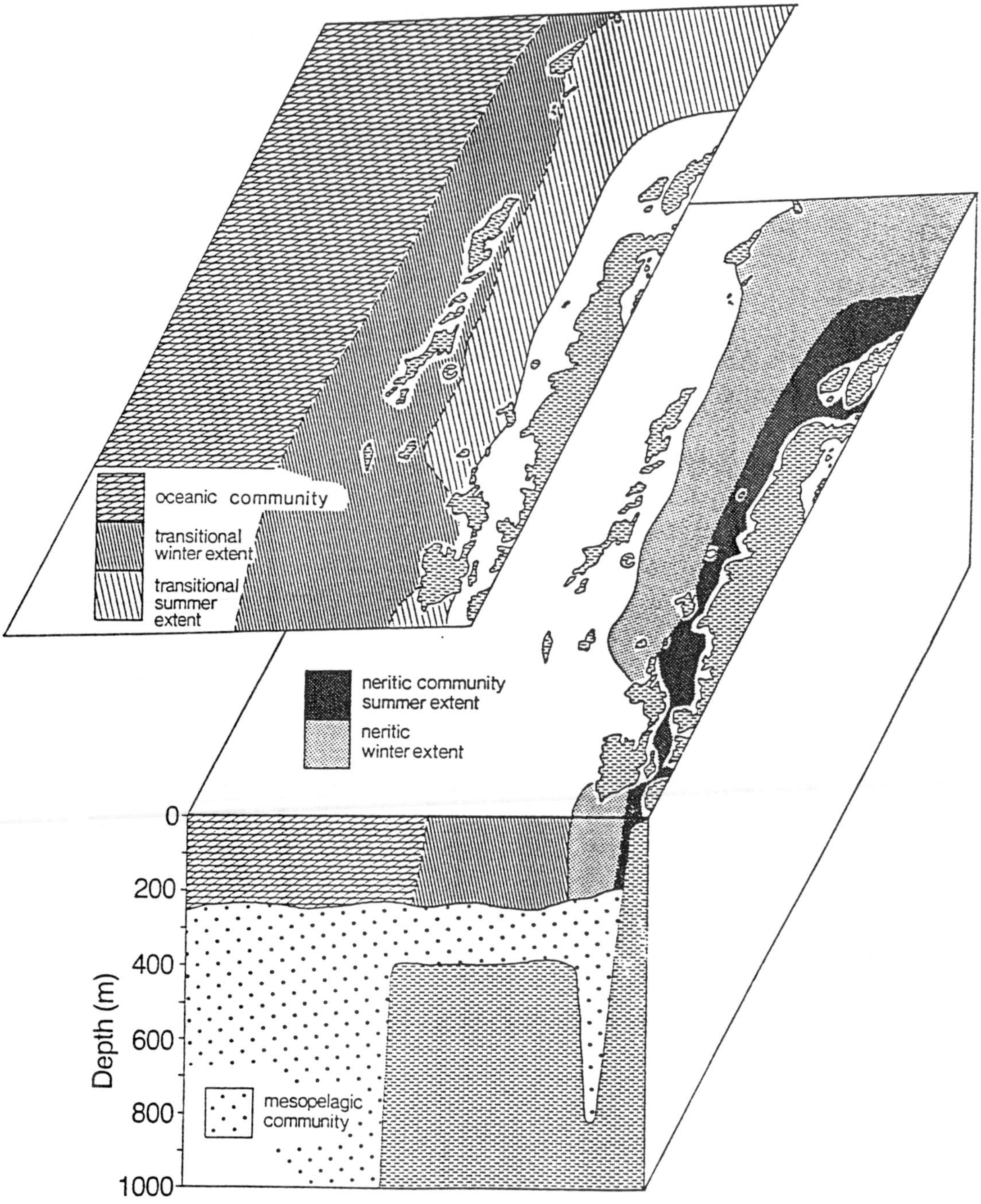

Fig. 7. Generalized picture for the seasonal distribution of macrozooplankton communities along the Antarctic Peninsula (after Siegel & Piatkowski, 1990).

Siegel & Piatkowski (1990) could define statistically three distinct macrozooplankton communities: an oceanic, a neritic and a mesopelagic one (Fig. 7). A so-called 'transitional cluster' represents a mixing zone created by frequent occurrence of species from both the oceanic and the neritic community. The differences in the composition of the zooplankton between site clusters were mainly due to changes in abundance rather than to presence or absence of particular species. The locations of the oceanic and the neritic community sites seemed to be relatively stable during seasons, while occurrence and abundance of most species changes with time of the year. It can roughly be summarized that the oceanic community is influenced by the West Wind Drift and the neritic one by Weddell Sea water in the Bransfield Strait, and by Bellingshausen Sea water off the Palmer Archipelago (Siegel & Piatkowski, 1990).

Conclusions

Research on zooplankton increased rapidly during the BIOMASS decade. Our knowledge of the biology and distribution patterns of different zooplankton groups has improved considerably. In the past few years, emphasis has been placed on the investigation at the community level. However, most studies were carried out in the open water zone during relatively short periods of the year, particularly during summer. Thus, there are still major gaps in our understanding of the life cycles of dominant zooplankton species, in particular the question of overwintering strategies and physiological adaptation. Microzooplankton, though common in Antarctic waters (e.g. Bröckel, 1981), has been neglected until recently. Microheterotrophs, for instance, can play an important role in Antarctic food webs, representing an efficient transfer path of organic matter from pico- and nanoplankton to higher trophic levels (Marchant & Murphy, this volume). Work is therefore required to quantify their role as grazers as well as prey in Antarctic pelagic ecosystems.

Trophic relations are strongly connected to planktonic life cycles. Thus, in order to understand trophodynamic processes, studies have to be done on all phases of the seasonal cycles, in open waters, in the marginal ice zone as well as in the pack-ice zone. The Antarctic sea-ice zone needs to be investigated more thoroughly to obtain a better understanding of how physical processes influence the dynamics of zooplankton populations. This aspect is especially important at high latitudes where seasonal activities fluctuate widely.

Acknowledgements

We thank Drs R Haass, W. Hagen and M. Huntley for critically reading the manuscript. Thanks are also due to Dr G. Dieckmann for linguistic improvements and to Mrs M. Puls for preparing some of the figures. This is contribution no. 545 of the Alfred-Wegener-Institut für Polar- und Meeresforschung, Bremerhaven, Germany.

References

Almeida Prado-Por M S (1989) Diel vertical migration of zooplankton at a fixed station in the Antarctic (Bransfield Strait, summer 1985). *Pesq Antárt Bras* **1**: 89–93.

Andrews K J H (1966) The distribution and life history of *Calanoides acutus* (Giesbrecht). *Discovery Rep* **34**: 117–62.

Atkinson A (1991) Life cycles of *Calanoides acutus*, *Calanus simillimus* and *Rhincalanus gigas* (Copepoda:Calanoida) within the Scotia Sea. *Mar Biol* **109**: 79–91.

Atkinson A, Peck J M (1988) Summer–winter differences in copepod distribution around South Georgia. *Polar Biol* **8**: 463–73.

Baker A de C (1954) The circumpolar continuity of Antarctic plankton species. *Discovery Rep* **27**: 201–18.

Baker A de C (1959) The distribution and life history of *Euphausia triacantha* Holt and Tattersall. *Discovery Rep* **29**: 309–40.

Boysen-Ennen E, Piatkowski U (1988) Meso- and macrozooplankton communities in the Weddell Sea, Antarctica. *Polar Biol* **9**: 17–35.

Boysen-Ennen E, Hagen W, Hubold G, Piatkowski U (1991) Zooplankton biomass in the ice-covered Weddell Sea, Antarctica. *Mar Biol* **111**: 227–35.

Brinton E (1985) The oceanographic structure of the eastern Scotia Sea – III. Distributions of euphausiid species and their developmental stages in 1981 in relation to hydrography. *Deep-Sea Res* **32**: 1153–80.

Brinton E, Antezana T (1984) Structures of swarming and dispersed populations of krill (*Euphausia superba*) in Scotia Sea and South Shetland waters during January–March 1981, determined by bongo nets. *J Crust Biol* **4** (Spec No 1): 45–66.

Brinton E, Huntley M, Townsend A W (1986) Larvae of *Euphausia superba* in the Scotia Sea and Bransfield Strait in March 1984–Development and abundance compared with 1981 larvae. *Polar Biol* **5**: 221–34.

Brinton E, Townsend A W (1991) Development rates and habitat shifts in the Antarctic neritic euphausiid *Euphausia crystallorophias*, 1986–87. *Deep-Sea Res* **38**: 1195–211.

Bröckel K v (1981) The importance of nanoplankton within the pelagic Antarctic ecosystem. *Kieler Meeresforsch* **5**: 61–7.
Chojnacki J, Węglénska T (1984) Periodicity of composition, abundance, and vertical distribution of summer zooplankton (1977/1978) in Ezcurra Inlet, Admiralty Bay (King George Island, South Shetland). *J Plankt Res* **6**: 997–1017.
Clarke A (1983) Life in cold water: the physiological ecology of polar marine ectotherms. *Oceanogr Mar Biol* **21**: 341–453.
David P M (1958) The distribution of the chaetognatha of the Southern Ocean. *Discovery Rep* **29**: 200–29.
Dinofrio E O (1987*a*) Planktonic copepods of waters near the South Orkney Islands. In: El-Sayed S Z, Tomo A P (eds) *Regional Symposium on Recent Advances in Antarctic Aquatic Biology with Special References to the Antarctic Peninsula. BIOMASS Sci Ser* **7**: 53–65.
Dinofrio E O (1987*b*) Quetognatos, Poliquetos y Anfipodos planctónicos de aquas cercanas a las islas Orcadas del Sul. *Contr Inst Ant Arg* **341**: 1–17.
Everson I (1987) Some aspects of the small scale distribution of *Euphausia crystallorophias*. *Polar Biol* **8**: 9–15.
Foxton P (1964) Seasonal variations in the plankton of Antarctic waters. In: Carrick R, Holgate M W, Prevost J (eds) *Biologíe Antarctique*. Hermann, Paris, pp. 311–18.
Foxton P (1966) The distribution and life history of *Salpa thompsoni*-Foxton with observations on a related species *Salpa gerlachei*-Foxton. *Discovery Rep* **34**: 1–116.
Fraser J H (1962) The role of ctenophores and salps in zooplankton production and standing crop. *Rapp P-V Reun Cons Perm Int Explor Mer* **153**: 121–3.
Hagen W (1985) On the distribution and population structure of Antarctic Chaetognatha. *Meeresforschung* **30**: 280–91.
Hardy A D, Gunther E R (1935) The plankton of the South Georgia whaling grounds and adjacent waters, 1926–1927. *Discovery Rep* **11**: 1–456.
Hempel G (1985) On the biology of polar seas, particularly the Southern Ocean. In: Gray J S, Christiansen M E (eds) *Marine Biology of Polar Regions and Effects of Stress on Marine Organisms*. John Wiley & Sons, Chichester, pp. 3–33.
Hempel I (1985) Vertical distribution of larvae of Antarctic krill, *Euphausia superba*. In: Siegfried W R, Condy P R, Laws R M (eds) *Antarctic Nutrient Cycles and Food Webs*. Springer-Verlag, Berlin, pp. 308–10.
Hempel I, Marschoff E (1980) Euphausiid larvae in the Atlantic sector of the Southern Ocean. *Meeresforschung* **28**: 32–47.
Hopkins T L (1985) The zooplankton community of Croker Passage, Antarctic Peninsula. *Polar Biol* **4**: 161–70.
Hopkins T L, Torres J J (1988) The zooplankton community in the vicinity of the ice edge, western Weddell Sea, March 1986. *Polar Biol* **9**: 79–87.
Huntley M, Escritor F (1991) Dynamics of *Calanoides acutus* (Copepoda:Calanoida) in Antarctic coastal waters. *Deep-Sea Res* **38**: 1145–67.
Huntley M, Escritor F (1992) Ecology of *Metridia gerlachei* Giesbrecht in the western Bransfield Strait, Antarctica. *Deep-Sea Res* **39**: 1027–55.
Huntley M E, Sykes P F, Marin V (1989) Biometry and trophodynamics of *Salpa thompsoni* Foxton (Tunicata:Thaliacea) near the Antarctic Peninsula in austral summer. *Polar Biol* **10**: 59–70.
Jackowska H (1980) Krill monitoring in Admiralty Bay (King George Island, South Shetland Islands) in summer 1979/1980. *Pol Polar Res* **1**: 117–25.
Jacques G, Panouse M (1991) Biomass and composition of size fractionated phytoplankton in the Weddell–Scotia confluence area. *Polar Biol* **11**: 315–28.
Jazdzewski K, Kittel W, Lotocki K (1982) Zooplankton studies in the southern Drake Passage and in the Bransfield Strait during austral summer (BIOMASS-FIBEX, February–March 1981). *Pol Polar Res* **3**: 203–42.
John D D (1936) The southern species of the genus *Euphausia*. *Discovery Rep* **14**: 195–325.
Kanaeva I P (1968) Distribution of mass copepods in the western part of the Scotia Sea. *Oceanology* **8**: 562–9.
Kittel W (1988) Euphausiid larvae in the Scotia Front west of Elephant Island (BIOMASS III, October–November 1986). *Pol Polar Res* **9**: 305–14.
Kittel W, Sicinski J, Luczak C (1988) Structure and biomass of zooplankton between King George Island and Elephant Island (BIOMASS III, October–November 1986). *Pol Polar Res* **9**: 265–76.
Kittel W, Stepnik R (1983) Distribution of *Euphausia crystallorophias*, *E. frigida*, *E. triacantha* and *Thysanoessa macrura* (Crustacea, Euphausiacea) in the southern Drake Passage and Bransfield Strait in February and March 1981. *Pol Polar Res* **4**: 7–19.
Kittel W, Witek Z, Czykieta H (1985) Distribution of *Euphausia frigida*, *Euphausia crystallorophias*, *Euphausia triacantha* and *Thysanoessa macrura* in the southern part of Drake Passage and in the Bransfield Strait during the 1983–1984 austral summer (BIOMASS-SIBEX). *Pol Polar Res* **6**: 133–49.
Lana P C, Blankensteyn A (1987) Distribution patterns of pelagic polychaetes in the southern Drake Passage and Bransfield Strait (January–February 1984). *Nerítica* **2**: 37–64.
Lana P C, Blankensteyn A (1989) Distribution patterns of pelagic polychaetes in the Bransfield Strait (January–February 1983). *Pesq Antárt Bras* **1**: 63–79.
Mackintosh N A (1934) Distribution of macrozooplankton in the Atlantic sector of the Antarctic. *Discovery Rep* **9**: 65–160.
Makarov R R (1977) Distribution of the larvae and some questions of the ecology of reproduction of the euphausiid *Euphausia frigida* Hansen, 1911 (Crustacea:Euphausiacea) in the southern part of the Scotia Sea. *Oceanology* **17**: 208–13.

Makarov R R (1979) Larval distribution and reproductive ecology of *Thysanoessa macrura* (Crustacea:Euphausiacea) in the Scotia Sea. *Mar Biol* **52**: 377–86.

Makarov R R, Menshenina L, Spiridonov V (1990) Distributional ecology of euphausiid larvae in the Antarctic Peninsula region and adjacent waters. *Proc NIPR Symp Polar Biol* **3**: 23–35.

Marin V (1987) The oceanographic structure of the eastern Scotia Sea–IV. Distribution of copepod species in relation to hydrography in 1981. *Deep-Sea Res* **34**: 105–21.

Marin V (1988) Qualitative models of the life cycles of *Calanoides acutus*, *Calanus propinquus* and *Rhincalanus gigas*. *Polar Biol* **8**: 439–46.

Marr J W S (1962) The natural history and geography of the Antarctic krill (*Euphausia superba* Dana). *Discovery Rep* **32**: 33–464.

Montú M, Oliveira I R (1986) Zooplanktonic associations, trophic relations and standing stock of krill and other groups of the community near Elephant Island (February–March 84/85). *Nerítica* **1**: 111–29.

Mujica A R (1989) Distribución del orden Euphausiacea en el estrecho Bransfield y zonas adyacentes (SIBEX-INACH, Enero-Febrero, 1984). *Pesq Antárt Bras* **1**: 35–44.

Mujica A R, Asencio V V (1985) Fish larvae, euphausiids and community structure of zooplankton in the Bransfield Strait (SIBEX-Phase I) 1984. *Ser Cient INACH* **33**: 131–54.

Mujica A R, Asencio V V (1986) Composisión y abundancia relativa del zooplancton antártico en el estrecho Bransfield (SIBEX-Fase II, Chile). *Ser Cient INACH* **35**: 95–104.

Mujica A R, Asencio V V (1990) Composisión y estados de desarollo de Euphausiacea en el strecho Bransfield (SIBEX-Fase II). *Ser Cient INACH* **40**: 55–69.

Mujica A R, Torres A G (1982) Qualitative and quantitative analysis of the Antarctic zooplankton. *Ser Cient INACH* **28**: 165–74.

Nast F (1986) Changes in krill abundance and in other zooplankton relative to the Weddell–Scotia Confluence around Elephant Island in November 1983, November 1984 and March 1985. *Arch FischWiss* **37**: 73–94.

Piatkowski U (1985*a*) Distribution, abundance and diurnal migration of macrozooplankton in Antarctic surface waters. *Meeresforschung* **30**: 264–79.

Piatkowski U (1985*b*) Maps of the geographical distribution of macrozooplankton in the Atlantic Sector of the Southern Ocean. *Ber Polarforsch* **22**: 55 pp.

Piatkowski U (1987) Zoogeographische Untersuchungen und Gemeinschaftsanalysen an antarktischen Makroplankton. *Ber Polarforsch* **34**: 150 pp.

Piatkowski U (1989) Macroplankton communities in Antarctic surface waters: spatial changes related to hydrography. *Mar Ecol Prog Ser* **55**: 251–9.

Pires A M S (1986) Vertical distribution of euphausiid larvae (Crustacea) in the Bransfield Strait during the First Brazilian Antarctic Expedition (Summer 1982/83). *An Acad Bras Cienc* **58**: 44–51.

Pires A M S (1989) Density and distribution of larval stages of Euphausiacea (Crustacea) in the Bransfield Strait and south of the Drake Passage in summer 1984. *Pesq Antárt Bras* **1**: 122–32.

Priddle J, Croxall J P, Everson I, Heywood R B, Murphy E J, Prince P A, Sear C B (1991) Large-scale fluctuations in distribution and abundance of krill – a discussion of possible causes. In: Sahrhage D (ed) *Antarctic Ocean and Resources Variability*. Springer-Verlag, Berlin, pp. 169–82.

Rakusa-Suszczewski S, Stepnik R (1980) Three species of krill from Admiralty Bay (King George, South Shetlands), in summer 1978/79. *Polsk Arch Hydrobiol* **27**: 273–84.

Schalk P H (1990) Biological activity in the Antarctic zooplankton community. *Polar Biol* **10**: 405–11.

Schnack S B (1985) Feeding by *Euphausia superba* and copepod species in response to varying concentrations of phytoplankton. In: Siegfried W R, Condy P R, Laws R M (eds) *Antarctic Nutrient Cycles and Food Webs*. Springer-Verlag, Berlin, pp. 311–23.

Schnack S B, Marschall S, Mizdalski E (1985*a*) On the distribution of copepods and larvae of *Euphausia superba* in Antarctic waters during February 1982. *Meeresforschung* **30**: 251–63.

Schnack S B, Smetacek V, Bodungen B v, Stegmann P (1985*b*) Utilization of phytoplankton by copepods in Antarctic waters during spring. In: Gray J S, Christiansen M E (eds) *Marine Biology of Polar Regions and Effects of Stress on Marine Organisms*. John Wiley & Sons, Chichester, pp. 65–81.

Schnack-Schiel S B, Hagen W, Mizdalski E (1991) A seasonal comparison of *Calanoides acutus* and *Calanus propinquus* (Copepoda:Calanoida) in the southeastern Weddell Sea, Antarctica. *Mar Ecol Prog Ser* **70**: 17–27.

Sicinski J (1988) Pelagic Polychaeta in the Scotia Front west of Elephant Island (BIOMASS III, October–November 1986). *Pol Polar Res* **9**: 277–82.

Siegel V (1988) A concept of seasonal variation of krill (*Euphausia superba*) distribution and abundance west of the Antarctic Peninsula. In: Sahrhage D (ed) *Antarctic Ocean and Resources Variability*. Springer-Verlag, Berlin, pp. 219–30.

Siegel V, Piatkowski U (1990) Variability in the macrozooplankton community off the Antarctic Peninsula. *Polar Biol* **10**: 373–86.

Stein M (1986) Variability of water masses and currents off the Antarctic Peninsula during SIBEX. *Arch FischWiss* **37**: 25–50.

Stepnik R (1982) All-year populational studies of Euphausiacea (Crustacea) in the Admiralty Bay (King George Island, South Shetland Islands, Antarctic). *Pol Polar Res* **3**: 49–68.

Vladimirskaya Ye V (1978) Age composition of winter population of abundant copepod species in the southern part of the Scotia Sea. *Oceanology* **18**: 202–4.

Weigmann-Haass R (1990) Taxonomie und Verbreitung von *Vibilia antarctica* STEBBING 1888 im antarktischen Teil des Atlantik. *Senckenbergiana Biol* **70**: 419–28.

Weigmann-Haass R, Haass G (1980) Geographische Verbreitung und vertikale Verteilung der Euphausiacea (Crustacea) während der Antarktis-Expedition 1975/76. *Meeresforschung* **28**: 19–31.

Witek Z, Kittel W (1985) Larvae of the species of the genus *Euphausia* (Euphausiacea, Crustacea) in the southern part of the Drake Passage and Bransfield Strait during the BIOMASS-SIBEX (December 1983–January 1984). *Pol Polar Res* **6**: 117–32.

Witek Z, Kittel W, Czykieta H, Zmijewska M I, Presler E (1985) Macrozooplankton in the southern Drake Passage and in the Bransfield Strait during BIOMASS-SIBEX (December 1983–January 1984). *Pol Polar Res* **6**: 95–115.

Zmijewska M I (1985) Copepoda in the southern part of Drake Passage and in Bransfield Strait during early summer 1983–1984 (BIOMASS-SIBEX, December–January). *Pol Polar Res* **6**: 79–93.

Zmijewska M I (1988) Vertical distribution and population structure of copepoda in a water column between King George Island and Elephant Island (BIOMASS III, October–November 1986). *Pol Polar Res* **9**: 283–304.

The macrozooplankton communities in the Prydz Bay region, Antarctica

G. W. HOSIE

Introduction

THE MAIN emphasis of the BIOMASS Programme was the assessment of stocks of the Antarctic krill, *Euphausia superba*. Krill were perceived as the key component of the Antarctic marine ecosystem, and also as a major source of protein for human consumption. During the course of BIOMASS, krill fishing developed rapidly into a major world fishery with a sustained yield in recent years of approximately 400 000 tonnes per year (FAO, 1991; Nicol, 1989).

In addition to studying the biology of krill, the BIOMASS Programme also provided the opportunity to improve our knowledge of the ecology of the other zooplankton, necessary for understanding all aspects of the Antarctic marine ecosystem. The degree of interaction between krill and other zooplankton species, e.g. herbivorous copepods and salps, is of particular interest. For example, Kawamura (1986, 1987) tentatively suggested there had been a 10- to 100-fold decrease in the abundances of herbivorous copepods in the areas west and east of Prydz Bay. This decrease was considered a likely consequence of competition for food between copepods and a surplus 150 million tonnes of unconsumed krill, a result of declined whale stocks. The salp *Salpa thompsoni* is another species capable of competing with krill for food but may also consume large numbers of krill larvae (Huntley *et al.*, 1989). Dietary studies of vertebrate predators have shown that zooplankton form important alternative pathways in the Prydz Bay food web (Williams, 1985, 1989) and also in the Weddell Sea (Boysen-Ennen *et al.*, 1991), further highlighting the need for research into zooplankton ecology in addition to krill studies.

Four of the eleven participating nations carried out official BIOMASS Programmes in the Prydz Bay region, i.e. Australia, Japan, South Africa and France. South Africa carried out a limited zooplankton study during SIBEX I, with results presented for broad taxonomic groups only, i.e. not as species (Boden, 1985). During the same season, Japan carried out a more detailed survey, with results again grouped for major taxa, although abundance data were produced for some copepod species from six sites between 62° and 64°S (Yamada & Kawamura, 1986). The Soviet Union also carried out similar BIOMASS-relevant research on the distribution and abundances of zooplankton in the Prydz Bay area in the mid-1980s (Budnichenko & Khromov, 1988; Pakhomov, 1989). During BIOMASS, Japan conducted extensive studies to the east of the Prydz Bay region south of Australia and also west of the region in the inshore margins around Syowa station (e.g. Fukuchi *et al.*, 1985; Kawamura, 1986; Tanimura *et al.*, 1986; Terazaki & Wada, 1986; Terazaki, 1989). However, there have been fewer zooplankton studies from Prydz Bay published in the mainstream

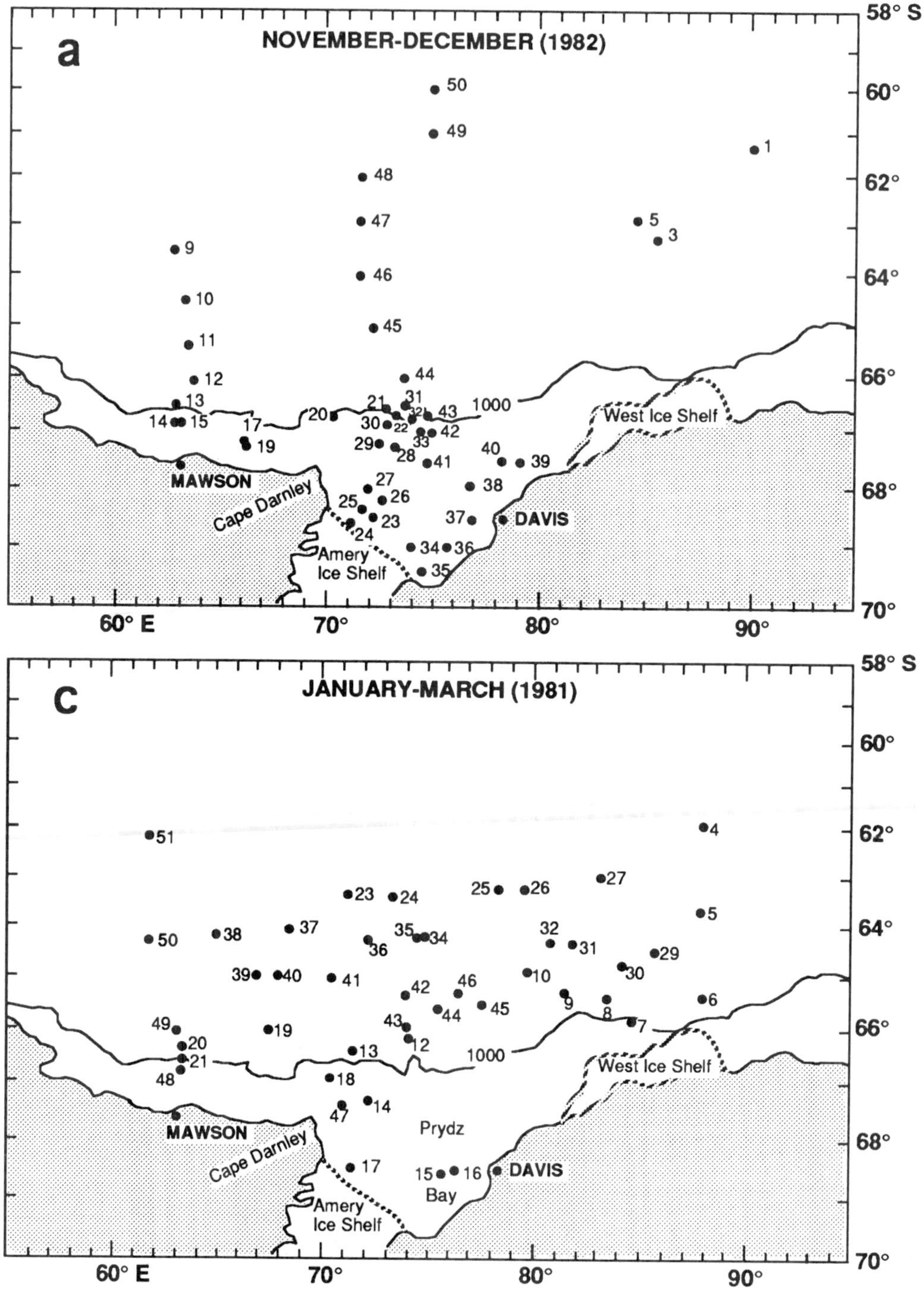

Fig. 1. Zooplankton sampling sites for the four surveys between 1981 and 1987. The 1000 m contour is shown.

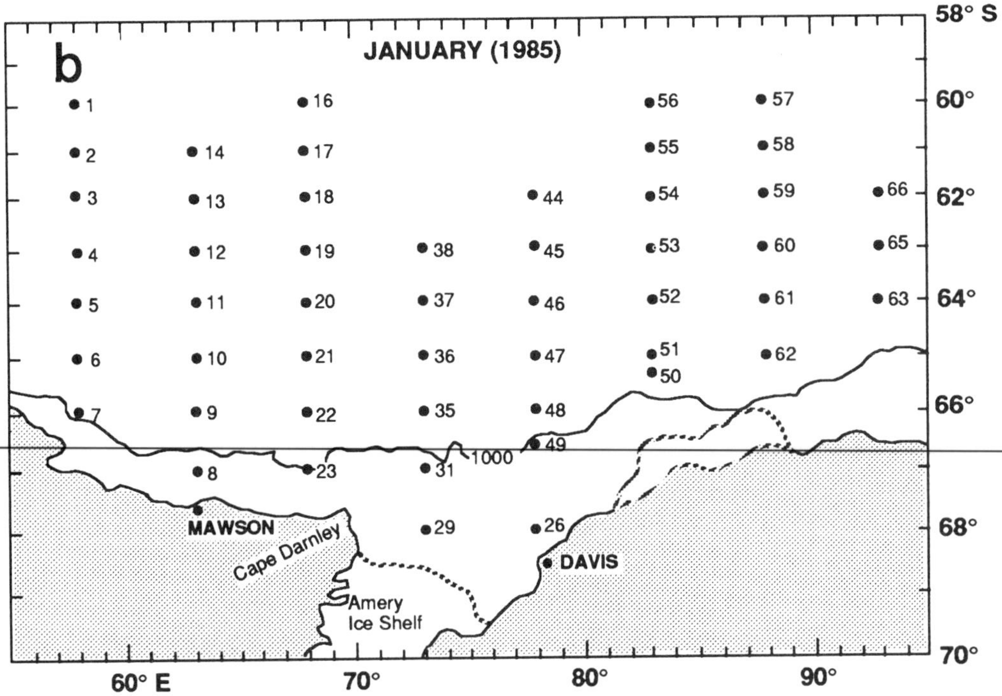
b
JANUARY (1985)
58° S
60°
62°
64°
66°
68°
70°
60° E
70°
80°
90°
MAWSON
Cape Darnley
Amery
Ice Shelf
DAVIS
1000

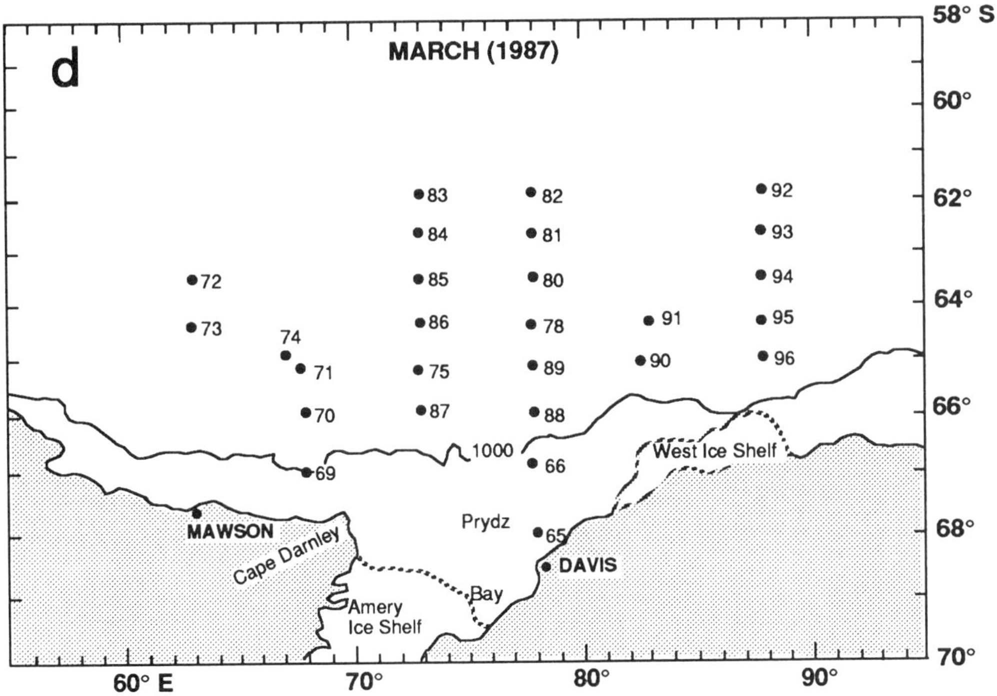
d
MARCH (1987)
58° S
60°
62°
64°
66°
68°
70°
60° E
70°
80°
90°
MAWSON
Cape Darnley
Amery
Ice Shelf
Prydz
Bay
DAVIS
West Ice Shelf
1000

Table 1. *Summary of sampling surveys*

ADBEX = Antarctic Division BIOMASS Experiment, AAMBER = Australian Antarctic Marine Biological Ecosystem Research.

Sampling period	Cruise name	No. sites	Haul	Sample depth (m)	Full sampling details (reference)
20 January to 10 March 1981	FIBEX	44	Horizontal	62+75[a]	Williams *et al.*, 1983
19 November to 19 December 1982	ADBEX 1	43	Horizontal Oblique[b]	0–180, various 0–120, various	Williams *et al.*, 1986
4–26 January 1985	SIBEX 2	51	Oblique	0–200	Ikeda *et al.*, 1986
7–23 March 1987	AAMBER 1	27	Oblique	0–200	Hosie *et al.*, 1991

[a] Principal sampling depth.
[b] Five hauls only.

literature in comparison with those on krill biology, which was treated as a higher priority. Information on community ecology is notably lacking for the Prydz Bay region, and only a few studies have been carried out in the Atlantic sector (Boysen-Ennen & Piatkowski, 1988; Hubold *et al.*, 1988; Piatkowski, 1989*a*,*b*). Siegel & Piatkowski (1990) made the only serious study of seasonal variability in community patterns in the Atlantic sector.

Rather than to review the few past zooplankton studies in Prydz Bay, the purpose of this chapter is to present new results that redress the lack of knowledge of zooplankton community ecology in that area. The Australian Antarctic Division carried out four geographically extensive surveys of the Prydz Bay region between 1981 and 1987, and covered the summer months of November–March. The resulting zooplankton data sets are the most comprehensive for the region produced under the auspices of BIOMASS. These data sets were analysed by a combination of multi- and univariate analytical techniques to define the zooplankton communities in the region, species composition and affinities, indicator species characteristic of each community, and variability in community patterns. These patterns were compared with various environmental parameters to generate hypotheses as to which factors govern zooplankton distributions. Two smaller surveys were also undertaken in January 1984 and September–November 1985. The former survey involved mainly aimed trawls at krill detected by hydroacoustics (Hosie *et al.*, 1988). During the 1985 survey, different net types and methods were employed to cope with the extensive ice cover (Hosie & Stolp, 1989). The zooplankton data obtained in both surveys were thus dependent on the sampling methodology, e.g. trawls targeted at krill, or on the net aperture and particularly mesh size. The two surveys were also limited in the number of samples, and were therefore not included in the present analyses.

Methods

The Prydz Bay region study area was defined for the present purpose as being from 58° to 93°E and south from 60°S to the Antarctic coast. Fig. 1 shows the station positions for the four sampling surveys in January–March 1981 (FIBEX), November–December 1982, January 1985 (SIBEX 2) and March 1987. Sampling methods on each cruise, together with sampling dates and number of sites sampled have been summarized in Table 1. Zooplankton were collected using a Rectangular Midwater Trawl (RMT8) with a nominal mouth area of 8 m^2 and mesh size of 4.5 mm (Baker *et al.*, 1973). During the 1981 and 1982 surveys, the volume filtered by the net was determined by multiplying the effective mouth area of the net by the distance travelled by the net, calculated as a function of ship's speed by the time the net was open. On subsequent voyages the net was equipped with a flowmeter. In calculating the volume filtered, the effects of towing speed and trajectory were taken into account (Roe *et al.*, 1980; Pommeranz *et al.*, 1982).

Towing speed varied between 1.5 and 4 knots. The routine sampling method employed on the first two surveys was a horizontal tow. In 1981 this was primarily to nominal depths of 62–75 m, and in 1982 sampling depth varied between 0 and 180 m. During the latter survey there was also five oblique hauls between 0 and 120 m. On the remaining surveys the net was towed obliquely between 0 and 200 m. A combined electromechanical net release and real-time depth sensor was usually mounted above the net. In oblique tows the net was kept open throughout. Complete details of the sampling programmes, e.g. sampling position, time, depth, conditions, etc., as well as details of catch composition, are provided by Williams *et al.* (1983, 1986), Ikeda *et al.* (1986) and Hosie *et al.* (1991).

Basic sorting of taxa was carried out on board ship. In particular, specimens of *E. superba* were removed, as were large and fragile zooplankton (jellyfish, salps, etc.). All specimens were preserved in Steedman's solution (Steedman, 1976) for later examination in the Antarctic Division laboratories where specimens were identified to the species level where possible and counted. For the purposes of this study, zooplankton were defined as all species collected by the RMT8. This included *E. superba* and ichthyoplankton, but not euphausiid larvae, which were collected by other more efficient means and were the subject of other studies (Hosie & Kirkwood, 1986; Hosie, 1991).

At each sampling site water samples were collected for phytoplankton pigment analysis, at depths of 0, 10, 25, 35, 50, 75, 100 and 200 m, using a General Oceanics rosette sampler with twelve 5 l Niskin bottles (Wright, 1987; Wright & Shearer, 1984). A Neil Brown Mark 3 CTD was mounted on the rosette, which provided continuous profiles of conductivity/salinity and temperature (Kerry *et al.*, 1987*a*,*b*; Kerry & Woehler, 1987; Woehler & Williams, 1988).

Data analyses

Species by sampling site matrices for each cruise, expressed as density values of number of individuals per thousand cubic metres, were analysed using both cluster analysis and non-metric multidimensional (NMDS) ordination. Sampling sites were first compared to define areas with similar species composition (q-type analysis). Multivariate and univariate statistical tests were subsequently applied to test the validity of observed patterns and to relate these patterns to environmental parameters. The inverse comparison analysis of species data (r-type analysis) was also undertaken to define species associations or affinities. A flow chart summarizing the numerical analyses is shown in Fig. 2 and is primarily a modification of the methodology recommended by Field *et al.* (1982) and Kruskal & Wish (1978). Multivariate analyses were carried out using BIOΣTAT II (Pimental, R. A. & Smith, J. D., 1985; Sigma Soft, Placentia, California).

Prior to the comparison of sampling sites, data were transformed using the $\log_{10}(x+1)$ function to reduce the bias of very high-abundance species and to give more weighting to species likely to have been undersampled, e.g. copepods. Cluster analysis was carried out first using the Bray–Curtis dissimilarity index (Bray & Curtis, 1957) coupled with unweighted pair group average linkage (UPGMA). The Bray–Curtis index was chosen because of its ability to deal with matrices with a high component of zero data entries, i.e. it will not be influenced by joint absences (Field *et al.*, 1982). In the January 1985 data set, nearly 70% of the matrix comprised zero values.

Sampling site cluster groups identified by the cluster analysis were further analysed to define indicator species which characterized each site group. Two types of indicator species were defined. The first type consisted of species which distinguished cluster groups by their frequency of occurrence, i.e. were more frequent in one cluster group than another. These 'frequency' type indicators were defined by using Field's information statistic ($2\Delta I_i$) (Field *et al.*, 1982). Every taxon (species), i, was compared between two cluster groups by the following formula:

$$2\Delta I_i = 2(I_{ti}-I_{1i}-I_{2i}), \qquad (1)$$

where I_{ti} = total information content of both cluster groups combined, $I_{ti} = N_t \log N_t - A_{ti} \log A_{ti} - (N_t - A_{ti}) \log(N_t - A_{ti})$; N_t = number of samples in both cluster groups together ('potential presences'); A_{ti} = number of samples in which species i is actually present; $(N_t - A_{ti})$ = number of samples from which species i is absent. Similarly, the information content I_{1i} and I_{2i} are obtained for clusters 1 and 2, respectively. Field's information statistic basically

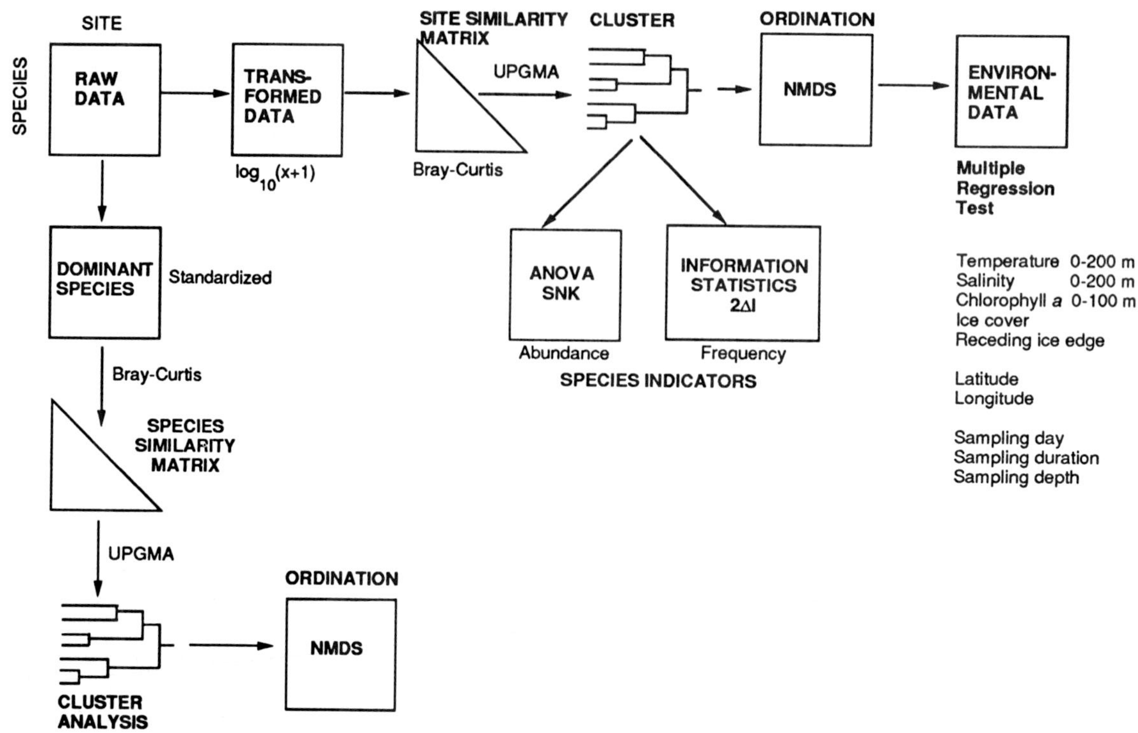

Fig. 2. Diagrammatic summary of the steps used in the multivariate analyses to define species relationships (comparison of species data), areas of common zooplankton composition (comparison of sampling sites), indicator species and possible environmental parameters affecting zooplankton distribution patterns. UPGMA, unweighted pair group average linkage; NMDS, non-metric multidimensional scaling; ANOVA/SNK, analysis of variance: Student Newman–Keuls multiple range test; $2\Delta I$, Field's information statistic (after Field *et al.*, 1982).

provides an index of the degree of difference in the frequency of occurrence of a species between two cluster groups, in relation to the number of sites. The resulting $2\Delta I_i$ values were compared with χ^2 values at 5% and 1% probability levels for 1 degree of freedom, on the premise that $2\Delta I_i$ approximates a χ^2 distribution (Field *et al.*, 1982).

Quite often a species was common to two or more site cluster groups, but had a significantly higher abundance in one particular group. Therefore, the second type of species indicator comprised species that distinguish a group by their higher abundance. These 'abundance' indicator species were defined by analysis of variance and further tested for significance using the Student Newman–Keuls (SNK) multiple range test (Zar, 1984). To avoid spurious low-abundance rare indicator species, and also to reduce computing time, only numerically dominant species were used in the ANOVA. These dominant species were the same as those defined and used in the comparison of species (see below). Reliable indicator species should ideally be abundant.

Following the cluster analyses the data sets were further subjected to non-metric multidimensional scaling (NMDS) ordination. This analysis provided a view or 'map' of the sampling site similarities in two-, three- or four-dimensional space, in addition to the cluster analysis dendrogram, which is fundamentally one-dimensional. NMDS is considered to be one of the most robust ordination techniques available, especially in coping with a large number of zero counts (Field *et al.*, 1982; Gray *et al.*, 1988). Another advantage of NMDS is that a 'stress' value is produced for each ordination (4,3,2 axes), indicating the goodness-of-fit between the original data and the derived 'map'. The lower the value, often expressed between 0 and 1, the better the fit. The stress value was then used, via the graphical-inflection method of Kruskal & Wish (1978) to select the least number of axes which adequately summarizes the data set. In this study a 2-axis ordination was found to suffice in all cases.

Prior to ordination, 'outlier' sample sites identified by cluster analysis were removed from the data set to obviate problems associated with such outliers, e.g. biasing or dominating the ordination, often compressing the distribution of the remaining sites (Gauch, 1982). Outliers are sites with a peculiar species composition and thus have a very low similarity with other sites.

New ordination scores derived from NMDS were compared by multiple regression analysis with various environmental parameters to determine which of these parameters may best explain the zooplankton distributions. The environmental parameters compared were: integrated temperature and salinity between 0 and 200 m using 2 m intervals; chlorophyll *a* integrated for 0–100 m in 1982 and 1985, and surface values in 1981 and 1987; pack-ice cover in tenths; and finally receding ice edge. There are distinct thermoclines and occasionally slight haloclines in the region, especially in the waters north of the continental shelf (Smith *et al.*, 1984; Kerry *et al.*, 1987*a*; Middleton & Humphries, 1989). However, there was no knowledge of where zooplankton were distributed in the water column in relation to these structures. Zooplankton abundances were integrated for the upper 200 m in the oblique trawls, whereas the horizontal trawls sampled only one depth layer at each site. Instead, integrated salinity and temperature values were used as indicative of scores of general oceanographic patterns, e.g. horizontal thermal gradients and fronts (Brandt & Wadley, 1981), for the purpose of comparison with zooplankton NMDS scores. The receding ice parameter was defined as the time in days from when the pack-ice was at its most northerly extent, taken as mid-October, to when the ice had receded or dissipated, exposing the sampling site. Pack-ice cover was estimated at the time of net sampling, and ice recession was measured for each season from the Northern Ice Limit maps produced weekly by the U.S. Navy–NOAA Joint Ice Center (Navy Polar Oceanography Center, Suitland), as well as consulting the Antarctic Division Antarctic Sea Ice Extent database (Jacka, 1983). Sampling methodology may cause variation in the zooplankton data. Hence, sampling day, the time since the start of a survey, duration of the haul, and sampling depth were included in the regression analyses. Time of day was not included in the analysis because of the limitations of tests to satisfactorily resolve angular–linear correlations (Zar, 1984). Latitude and longitude of sampling sites were also compared with the ordination scores to ascertain the type and degree of zonation in the distributions. The actual multiple regression analysis involved treating each parameter as the dependent variable and the ordination scores for each axes, which are a collective summary of the ecological data, as the independent variables. The direction of maximum correlation of the regression line is at an angle θ_r with the *r*th axis. The direction cosine, or regression weight, c_r of that angle is derived from the formula of Kruskal & Wish (1978):

$$c_r = b_r / \sqrt{(b_1^2 + b_2^2 + \ldots b_r^2)}, \qquad (2)$$

where b_1, b_2, etc. are the regression coefficients from the multiple regression $a + b_1x_1 + b_2x_2 + \ldots b_rx_r$, and x_r are the ordination scores of the axes.

Water circulation is one parameter which is expected to have a substantial effect on the distribution of zooplankton, but cannot be included in the regression analysis. Instead, the geographical distribution patterns of station groups, defined by multivariate analysis, were compared visually with known water currents shown in Fig. 3. Fig. 3a shows the geostrophic water flow in the Prydz Bay region derived from CTD data by R. A. Nunes Vaz and G. W. Lennon (personal communication, 1990). Fig. 3b shows a synthesis of other currents, known from sea-ice buoy and iceberg trajectories, current meters, and geostrophic flow (Fig. 3a), which are believed to influence the distribution of euphausiid larvae (Hosie, 1991) but may also affect zooplankton distributions.

Table 2. *Dominant species used in inverse cluster analyses and ordination of species, and also in the ANOVA/SNK analyses to define species indicators*

Dominant species were defined as those with a >4% numerical dominance for any given sampling site, for any sampling cruise. Species abbreviations used in the NMDS plots (Fig. 8) are shown.

Taxa	Abbreviation
Appendicularia	Ap
Calanoides acutus	Ca
Calanus propinquus	Cp
Clio pyramidata	Clp
Cyllopus lucasii	Cl
Electrona antarctica	Ela
Euchaeta antarctica	Ea
Eukrohnia hamata	Eh
Euphausia crystallorophias	Ec
Euphausia frigida	Ef
Euphausia superba	Es
Euphausia triacantha	Et
Gammaridae	Ga
Haloptilus ocellatus	Ho
Hyperiella dilatata	Hd
Ihlea racovitzai	Ir
Limacina helicina	Lh
Metridia gerlachei	Mg
Pagetopsis macropterus	Pm
Pleuragramma antarcticum	Pa
Rhincalanus gigas	Rg
Sagitta gazellae	Sg
Sagitta marri	Sm
Salpa thompsoni	St
Siphonophore (nectophores)	SN
Themisto gaudichaudii	Tg
Thysanoessa macrura	Tm
Tomopteris carpenteri	Tc

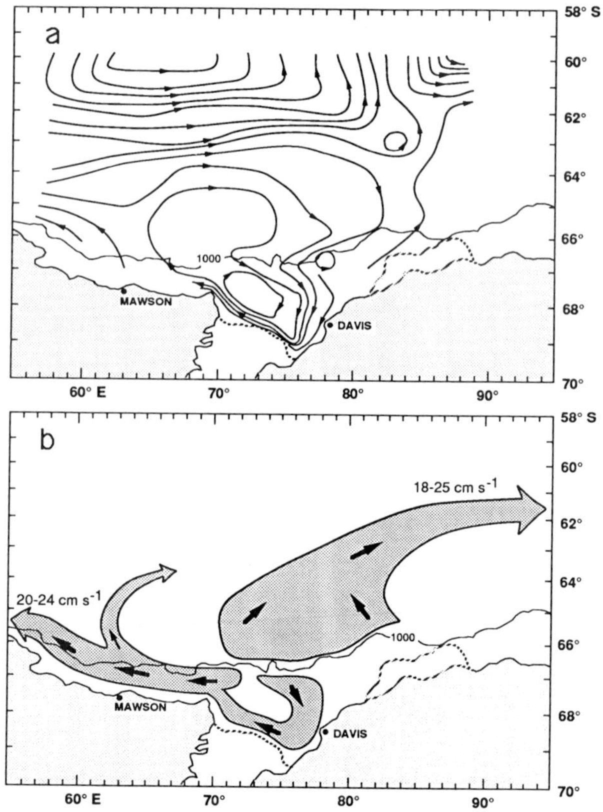

Fig. 3. Horizontal water circulation patterns in the Prydz Bay region. (a) The geostrophic water flow redrawn from the geopotential anomaly contours of R. A. Nunes Vaz and G. W. Lennon (personal communication, 1990). (b) Apparent dispersal routes of euphausiid larvae, likely to affect zooplankton distributions, determined from sea-ice buoy and iceberg trajectories, current meters and geostrophic flow. Current speeds shown were determined from iceberg trajectories. From Hosie (1991).

The inverse analysis of the data sets to define species affinities also involved cluster analysis as the first step, with Bray–Curtis index and UPGMA linkage followed by NMDS to map species associations in two-dimensional space. However, prior to analysis a data set was reduced to a subset of common dominant species to avoid spurious associations among very low-abundance rare species, caused by their chance occurrence at one site. Following the example of Field *et al.* (1982), dominant species were defined as those with a numerical dominance greater than 4% for any given sampling site, for any sampling cruise, providing that a species was regularly observed in all surveys. Table 2 lists these species along with their abbreviations used in the NMDS plots. There were 26 main species or taxa so defined, plus two additional species: *Ihlea racovitzai* and larvae of *Pagetopsis macropterus*. These additional species were only found in 1981 and 1982, but both were defined as indicator species of the inshore area and were therefore included in the analyses. There was also a higher frequency of sampling on the shelf during those cruises.

The species data were standardized according to Field *et al.* (1982), i.e.:

$$Y_{ij} = 100\ X_{ij} / \sum_{j=1}^{n} X_{ij}, \qquad (3)$$

where X_{ij} = abundance of the ith species in the jth sample; $\sum_{j=1}^{n} X_{ij}$ = summed abundance of the ith species over all samples; Y_{ij} = corresponding standardized score.

Results

Analyses for each sampling period are presented in the order of the months sampled, rather than in historical order. Throughout these descriptions, principal station cluster groups have been annotated with a number, and where necessary subgroups have a letter suffix, e.g. station Group 1, Group 2a. Capital letters have been used for species cluster groups, e.g. Group A, with numeric suffixes for subgroups.

November–December 1982

Three distinct cluster groups were defined at 70% dissimilarity, showing a clear zonation between inshore and offshore station groups (Figs 4a and 5a). The inshore, continental group of sites was further

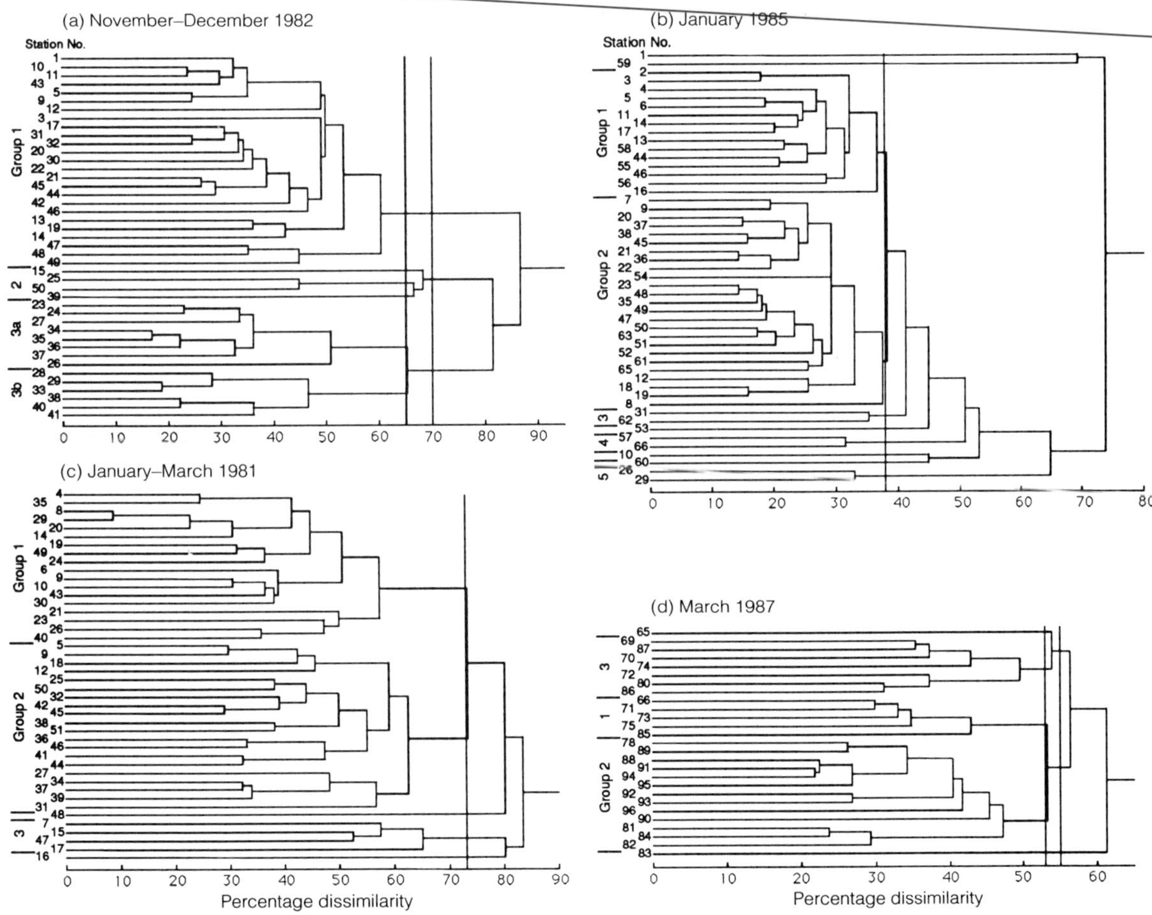

Fig. 4. Dendrograms of cluster analyses comparing sampling sites for each survey. The Bray–Curtis dissimilarity index was used for the comparison with UPGMA linkage, after $\log_{10}(X+1)$ transformation of species abundance data. Selection of station groups was arbitrary for each survey.

Table 3. *November–December 1982: frequencies of occurrence of indicator species distinguishing cluster groups defined in Fig. 4a*

Species in each sub-table are ranked according to information statistic. Species above the dotted line in each sub-table have a $2\Delta I>6.63$ ($p=0.01$), and those below the line have $2\Delta I>3.84$ ($p=0.05$). Maximum possible occurrences are; Group 1 = 25, Group 2+3 = 20, Group 3a = 8 and Group 3b = 6.

(a) Group 1 (oceanic) indicator species

Species	Group 1	(Group 2+3)
Calanus propinquus	25	5
Calanoides acutus	25	6
**Vanadis antarctica*	20	1
**Primno macropa*	21	2
**Clione limacina*	20	3
Tomopteris carpenteri	16	1
Rhincalanus gigas	18	2
**Tomopteris septentrionalis*	12	0
Sagitta gazellae	24	8
....		
Eukrohnia hamata	18	3
**Spongiobranchaea australis*	18	3
**Calycopsis borchgrevinki*	9	0
Haloptilus ocellatus	9	0
Thysanoessa macrura	24	10
Sagitta marri	8	0
Clio pyramidata	21	7
Appendicularia	11	1

(b) Group 2+3 (inshore) indicator species

Species	Group 2+3	Group 1
Euphausia crystallorophias	15	6
....		
**Hyperiella macronyx*	13	4
**Cryodraco antarcticus* larvae	7	0
*Channichthyidae larvae	6	0
Pleuragramma antarcticum	12	5
**Chionodraco* sp. larvae	10	3

(c) Group 3a indicator species from Group 3b

Species	Group 3a	Group 3b
None		
....		
**Hyperiella macronyx*	8	2
Salpa thompsoni	8	2

(d) Group 3b indicator species from Group 3a

Species	Group 3b	Group 3a
None		
....		
Euphausia superba	6	2

*Not dominant species; not used in SNK or inverse species analyses.

Table 4. *November–December 1982: Mean abundances, analysis of variance (F) and SNK multiple range tests of dominant species in cluster groups defined in Fig. 4a*

Analyses were carried out on $\log_{10}(x+1)$ transformed abundances (Zar, 1984); values shown are number of individuals 1000 m^{-3}. Species with significant differences in mean abundance are shown in bold text, while those with significantly higher abundances in a cluster group according to SNK analysis are underlined. For ANOVA p values; *, <0.05; **, <0.005; ***, <0.0005; –, not significant; d.f. = 3,39.

Species	Group 1 mean	Group 2 mean	Group 3a mean	Group 3b mean	*F*	*p*
Appendicularia	0.07	0.00	0.00	0.001	1.41	–
Calanoides acutus	339.04	1.09	0.68	2.01	21.30	***
Calanus propinquus	43.47	0.70	0.02	0.07	25.03	***
Clio pyramidata	2.37	0.04	0.08	0.02	5.81	**
Cyllopus lucasii	0.49	0.08	0.03	0.04	2.88	*
Electrona antarctica	0.02	0.00	0.00	0.00	0.43	–
Euchaeta antarctica	1.44	0.04	0.00	0.001	1.60	–
Eukrohnia hamata	1.52	0.29	0.01	0.37	2.82	–
Euphausia crystallorophias	4.92	0.70	938.02	407.79	33.54	***
Euphausia frigida	0.001	0.00	0.00	0.00	0.23	–
Euphausia superba	82.23	0.07	1.07	859.44	24.97	***
Gammaridae	0.03	0.10	0.02	0.10	1.71	–
Haloptilus ocellatus	0.65	0.00	0.00	0.00	2.19	–
Hyperiella dilatata	0.19	0.15	0.30	0.07	1.43	–
Ihlea racovitzai	0.21	0.00	0.00	0.00	0.45	–
Limacina helicina	1.16	0.08	1.59	0.07	1.25	–
Metridia gerlachei	17.41	0.12	0.08	0.10	2.41	–
Pagetopsis macropterus	0.00	0.01	0.02	0.00	2.91	*
Pleuragramma antarcticum	0.16	0.01	1.63	1.11	4.74	*
Rhincalanus gigas	39.53	0.45	0.00	0.07	4.05	*
Sagitta gazellae	4.74	0.33	0.06	0.12	10.17	***
Sagitta marri	0.07	0.00	0.00	0.00	1.12	–
Salpa thompsoni	3.30	103.92	3.86	1.68	8.03	***
Siphonophore (nectophore)	1.95	0.22	0.01	0.11	2.68	–
Themisto gaudichaudii	0.01	0.00	0.00	0.00	0.40	–
Thysanoessa macrura	65.49	0.25	0.11	1.56	10.07	***
Tomopteris carpenteri	0.44	0.00	0.01	0.00	2.42	–

divided at 65% dissimilarity into stations south of 68°S (Group 3a), which were characterized by *Euphausia crystallorophias*, *Pleuragramma antarcticum* and the larvae of channichthyid fish. Between this group and the continental shelf edge were stations (Group 3b) also characterized by high abundances of *E. crystallorophias* but also by the adult krill *E. superba*, in terms of both frequency and great abundance (Tables 3 and 4).

Cluster Group 1 was primarily an oceanic group of stations, with a few sites just on the continental shelf. Although there are 25 sites in this group, the spatial coverage is not sufficient to determine if there is further off-shore zonation. There were a number of frequency ($2\Delta I$) indicator species for the oceanic group, many with significantly high abundances (Tables 3 and 4).

Group 2 represented a very loose grouping of four sites. One site, station 50, is located in the far north of the Prydz Bay region, while three of the sites are located on the continental shelf. The cluster was fragmented at 65% dissimilarity, but at 81.3% was grouped with the other continental shelf stations. The NMDS plots also show the four sites as a distinct group (Fig. 6a) more closely aligned with Group 3. There was no unique species distinguishing this group from the others in terms of frequency indicators. Group 2 did have a significantly higher abundance of *Salpa thompsoni* (Table 4), which would be the prime reason why the northern station

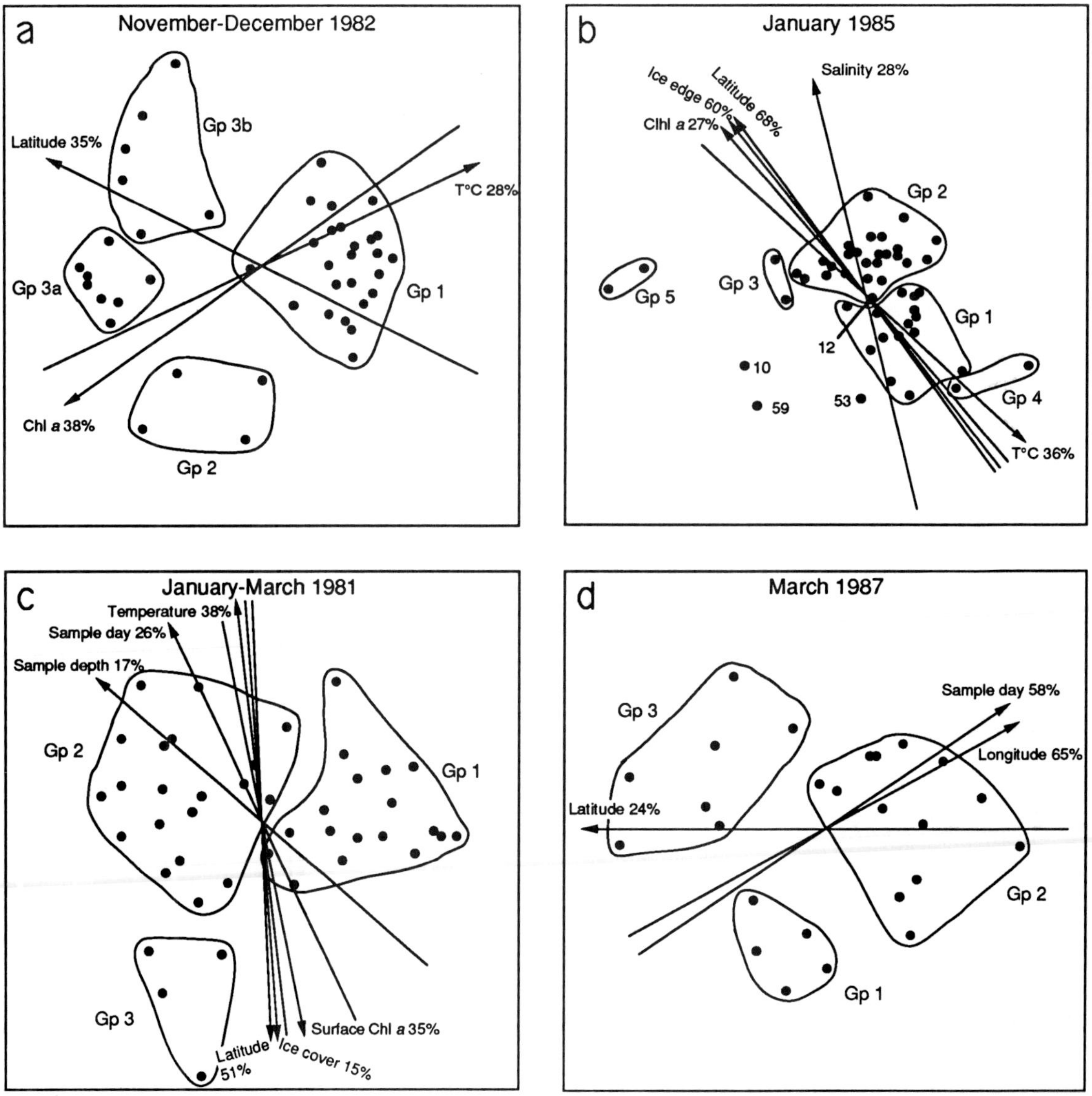

Fig. 6. Ordination plots of sampling sites for each survey using non-metric multidimensional scaling and Bray–Curtis dissimilarity index. Respective cluster groups identified in Fig. 4 are superimposed. Significant multiple regressions between ordination scores and environmental parameters are shown, as well as the fraction (%) of variance in the zooplankton data explained by the parameter (see Table 5). The direction of the regression line was determined from Equation 2 (Kruskal & Wish, 1978). Axis scales are relative in NMDS, based on non-metric ranking of dissimilarity, and are therefore not shown. Stress values are (a) 0.15, (b) 0.14, (c) 0.18 and (d) 0.18.

Table 5. *Multiple regression analyses between environmental parameters and NMDS scores for two-axis ordination of comparison of sampling sites*

Regression weights are derived from Equation 2. Adj. R^2 = adjusted coefficient of determination which gives the fraction of the variance accounted for by the explanatory variable (Jongman *et al.*, 1987). For ANOVA p values: *, <0.05, **, <0.005, ***, <0.0005; ns, not significant.

(a) November–December 1982

Variable	Direction cosines (regression weights) *X*	*Y*	Adj. R^2	*F*	d.f.	*p*
Chlorophyll *a*	−0.819	−0.573	0.378	6.785	2,17	*
Latitude	−0.897	0.442	0.347	12.151	2,40	***
Temperature	0.906	0.424	0.284	7.136	2,29	**
Salinity	—	—	0.040	1.623	2,28	ns
Sampling day	—	—	0.014	1.294	2,40	ns
Longitude	—	—	−0.007	0.848	2,40	ns
Sampling depth	—	—	−0.010	0.798	2,40	ns
Duration haul	—	—	−0.028	0.435	2,40	ns
Pack-ice cover	—	—	−0.040	0.198	2,40	ns

(b) January 1985

Variable	Direction cosines (regression weights) *X*	*Y*	Adj. R^2	*F*	d.f.	*p*
Latitude	−0.595	0.804	0.684	52.924	2,46	***
Ice recession	−0.617	0.787	0.604	37.653	2,46	***
Temperature	0.735	−0.678	0.355	11.725	2,37	***
Salinity	−0.234	0.972	0.279	8.559	2,37	**
Chlorophyll *a*	−0.652	0.759	0.271	8.805	2,40	**
Duration haul	—	—	0.065	2.660	2,46	ns
Sampling depth	—	—	0.062	2.590	2,46	ns
Longitude	—	—	0.002	1.044	2,46	ns
Sampling day	—	—	−0.015	0.644	2,46	ns

(c) January–March 1981

Variable	Direction cosines (regression weights) *X*	*Y*	Adj. R^2	*F*	d.f.	*p*
Latitude	0.047	−0.999	0.509	22.257	2,39	***
Temperature	−0.113	0.994	0.381	11.788	2,33	***
Chl *a* at surface	−0.199	−0.980	0.349	9.582	2,30	**
Sampling day	−0.431	0.902	0.263	8.334	2,39	**
Sampling depth	−0.760	0.650	0.173	5.282	2,39	*
Ice cover	0.077	−0.997	0.146	4.514	2,39	*
Ice recession	—	—	0.034	1.719	2,39	ns
Longitude	—	—	−0.028	0.437	2,39	ns
Duration haul	—	—	−0.031	0.382	2,39	ns
Salinity	—	—	−0.035	0.402	2,33	ns

(d) March 1987

Variable	Direction cosines (regression weights) *X*	*Y*	Adj. R^2	*F*	d.f.	*p*
Longitude	0.877	0.480	0.649	23.161	2,22	***
Sampling day	0.828	0.560	0.584	17.873	2,22	***
Latitude	−1.000	−0.016	0.237	4.737	2,22	*
Duration haul	—	—	0.132	2.824	2,22	ns
Surface Chl *a*	—	—	0.070	1.756	2,18	ns
Sampling depth	—	—	−0.021	0.750	2,22	ns
Temperature	—	—	−0.037	0.609	2,20	ns
Pack-ice cover	—	—	−0.053	0.397	2,22	ns
Ice recession	—	—	−0.085	0.062	2,22	ns

is linked with the inshore sites. *Antarctomysis maxima* was also in high abundance in this group but only at the one site close to the Amery Ice Shelf.

Much of the region was covered by pack-ice, but the degree of pack-ice cover did not explain any of the geographic variation (Table 5a). Integrated chlorophyll *a* accounted for most of the variation in the data at 38%, while 28% of the variation could be explained by mean temperature (Table 5a, Fig. 6a). Fig. 5a shows a quite distinct north–south pattern, but latitude explained only 35% of the data. In the three-axis ordination, latitude accounted for 65% of the variation (F = 27.0, d.f. = 3,39, p < 0.0005). Latitude was mainly associated with the third axis, and consequently was less evident when the ordination was reduced to two axes. There was no correlation between sampling parameters and the ordination scores.

In the inverse analysis, comparison of species, 27 of the taxa listed in Table 2 were analysed. The euphausiid *E. triacantha* was not collected in this particular year, but was regularly observed in other

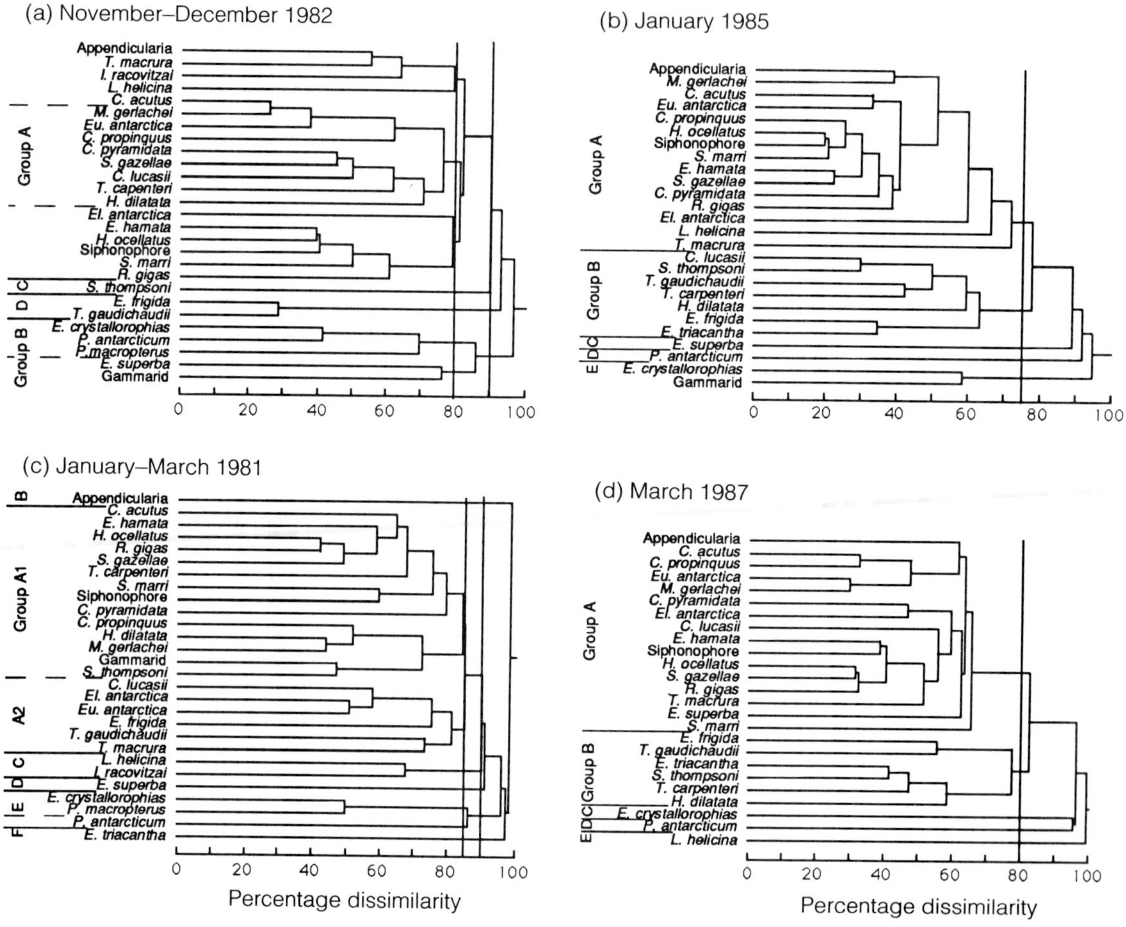

Fig. 7. Dendrograms of inverse cluster analyses comparing dominant species, defined as those with a >4% numerical dominance for any given sampling site, for any sampling cruise (Table 2). The Bray–Curtis dissimilarity index was used for the comparison with UPGMA linkage, after standardizing species abundance (Field *et al.* 1982). Selection of species groups were arbitrary for each survey.

years. Four groups of species were defined at 90% dissimilarity (Fig. 7a). The major group (A) comprised 19 species, with most being indicator species of the oceanic station Group 1 (Tables 3a and 4). This assemblage contained all of the dominant copepods, chaetognaths, siphonophores, and appendicularians, and the euphausiid *Thysanoessa macrura*. Three subgroups could be further defined at 80% dissimilarity but the NMDS analysis showed that these subgroups overlapped considerably (Fig. 8a).

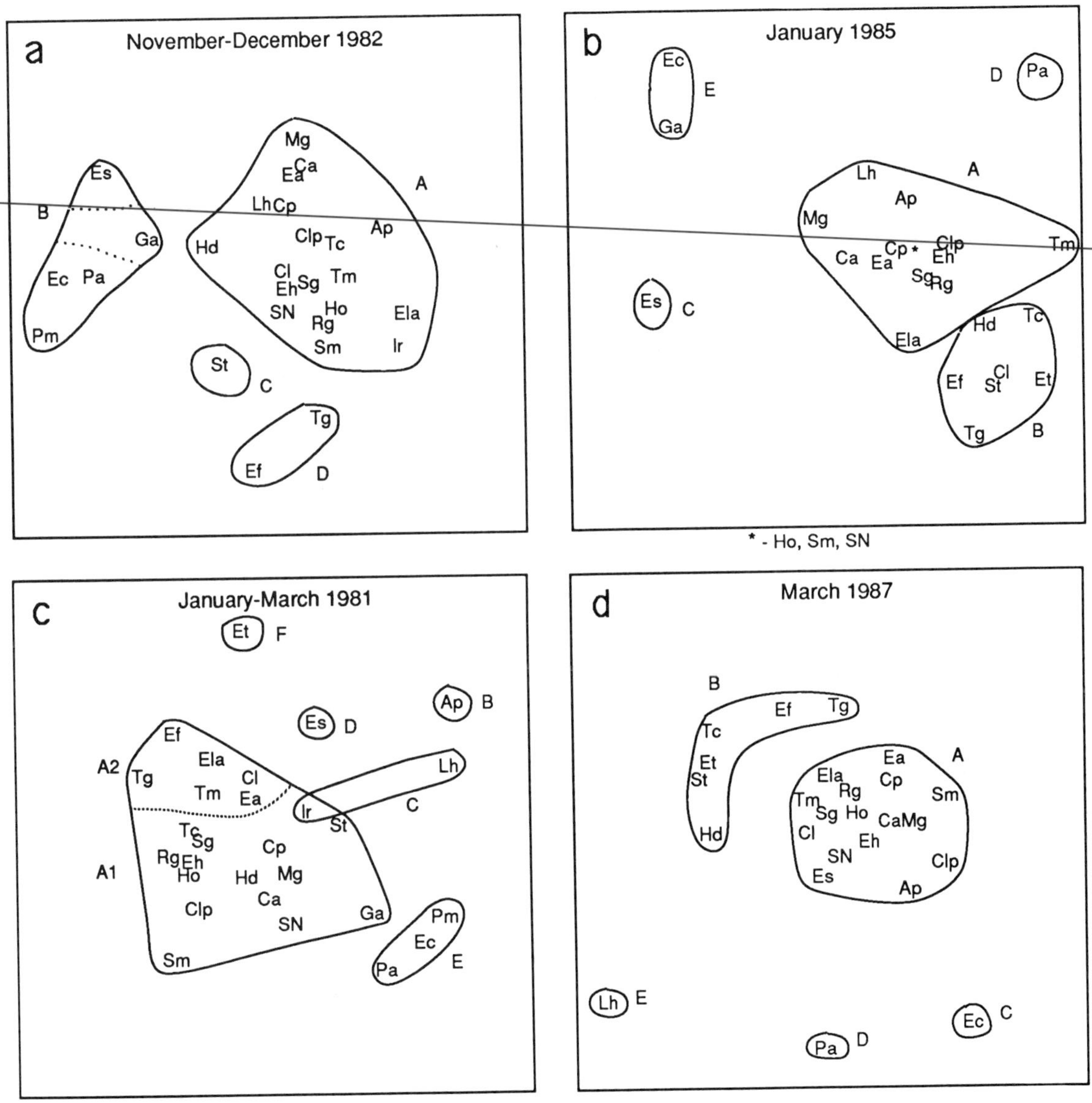

Fig. 8. NMDS inverse ordination plots comparing dominant species. Respective cluster groups identified in Fig. 7 are superimposed. Species abbreviations used in the plots are shown in Table 2. Axis scales are relative in NMDS and are therefore not shown. Stress values are (a) 0.17, (b) 0.14, (c) 0.22 and (d) 0.16.

The split at 90% would therefore appear justified. Group B mainly comprised species associated with the continental shelf. *E. crystallorophias*, *Pleuragramma antarcticum* and *Pagetopsis macropterus* were identified as indicator species of the southern part of Prydz Bay (Fig. 5a), while *E. superba* was an indicator of the northern part of the shelf. Gammarids, which were also part of Group B, were identified as indicator species of the shelf region in 1981 (see below), but not in 1982. The cluster analyses and NMDS plot both show that *E. superba* formed a subgroup (86% dissimilarity) from the three inner-shelf species, which reflects the difference between the shelf subgroups defined previously (Fig. 5a). The NMDS plot also indicates that the hyperiid *Hyperiella dilatata* has perhaps more affinity with the inshore species group (Fig. 8a). The salp *S. thompsoni* was clearly separate from other species. As mentioned above, this species characterized the station Group 2, particularly station 50 in the north where the catch consisted almost entirely of salps. The fourth group (D) comprised just two species, *Euphausia frigida* and *Themisto gaudichaudii*. The latter species was an indicator of northern waters in January 1985 and March 1987, while the larvae and adults of *E. frigida* are usually restricted to waters north of the divergence (John, 1936; Hosie, 1991).

January 1985

Cluster analysis showed two distinct outlier stations, 1 and 59. Station 1 (Fig. 1b) was characterized by a very high abundance of *S. thompsoni*, which represented more than 98% of the biomass of the catch. Station 59 returned the lowest abundance for any sampling site with very few taxa caught compared with the other stations. Subsequent re-analysis, after removing these stations, produced exactly the same groupings as the first cluster analysis for the rest of the stations.

At 38% dissimilarity I defined two main groups with 15 and 25 stations each, three smaller groups of two stations each, and three sites which were ungrouped (Figs 4b and 5b). Group 5, comprising the continental shelf sites 26 and 29, was the only group distinct from the rest of the sites at 64% dissimilarity. This group was characterized by *E. crystallorophias* and *Pleuragramma antarcticum*, but also by the complete absence of prominent species such as *Haloptilus ocellatus*, *Sagitta marri* and *Salpa thompsoni* (Tables 6 and 7). The lack of clear separation amongst the rest of the cluster groups, and the low dissimilarity, indicates some degree of homogeneity in the zooplankton composition in the offshore stations. None the less, the geographic plot of stations shows a further partitioning, mainly latitudinally, amongst the offshore stations (Fig. 5b). The boundary between Group 1, the northern stations, and Group 2, occupying the middle of the Prydz Bay region, does meander to some degree. Further, the NMDS plot (Fig. 6b) shows that Station 12 in Group 2 is perhaps more aligned with Group 1. These two groups, as well as Groups 3 and 4, share common abundance indicator species, e.g. *Eukrohnia hamata* and *Sagitta gazellae*, which distinguish these groups from the inshore stations (Table 6). These two species were also the only indicator species of Group 3, which was closely aligned with Group 2 (Fig. 6b). Group 2 was also characterized by the higher abundance of a number of other species compared with their abundance in Group 1, e.g. *Calanus propinquus* and siphonophores (Table 6.).

The salp *Salpa thompsoni* also had a wide geographic range, although abundances decreased southwards, and was not found in Groups 3 and 5. However, this species dominated the northern stations with its exceptionally high abundances and is thus the most notable indicator species in Groups 1 and 4. North of 63°S this species comprised 73% of the total zooplankton biomass by fresh mass, and 90% north of 62°S. The outlier station 1 was also noted for a high proportion of salps in the catch and should thus belong to station Group 1. The high abundance of salps would undoubtedly have masked the contribution of other species as indicators for Group 1.

Apart from the abundance indicators, there were a few frequency indicator species that discriminated Groups 1 and 2, these being *Vibilia* sp. and *Electrona antarctica* larvae for Group 1, and appendicularians for Group 2 (Table 7). The latter taxon was also an abundance indicator. The myctophid *E. antarctica* occurred at all Group 1 stations, but also at 15 of the 25 sites in Group 2. Consequently this species had a $2\Delta I$ score that was only significant at the 5% level, while appendicularians and *Vibilia* sp. were both significant at 1%.

The north-eastern Group 4 was characterized by

Table 6. *January 1985: mean abundances, analysis of variance (F) and SNK multiple range tests of dominant species in cluster groups defined in Fig. 4b*

Analyses were carried out on $\log_{10}(x+1)$ transformed abundances (Zar, 1984); values shown are number of individuals 1000 m^{-3}. Species with significant differences in mean abundance are shown in bold text, while those with significantly higher abundances in a cluster group according to SNK analysis are underlined. For ANOVA p values:*, <0.05; **,<0.005; ***,<0.0005; –, not significant. DF =4,41.

Species	Group 1 mean	Group 2 mean	Group 3 mean	Group 4 mean	Group 5 mean	F	p
Appendicularia	0.19	0.83	1.64	0.00	0.00	4.84	**
Calanoides acutus	5.51	15.97	9.93	2.28	3.57	9.01	***
Calanus propinquus	5.38	24.13	10.16	6.60	2.78	13.70	***
Clio pyramidata	5.49	18.90	2.15	7.09	1.91	6.23	**
Cyllopus lucasii	1.21	0.22	0.00	11.48	0.00	19.52	***
Electrona antarctica	0.69	0.45	0.00	1.10	0.00	2.22	–
Euchaeta antarctica	0.85	3.47	1.16	6.27	1.05	10.89	***
Eukrohnia hamata	26.75	41.11	15.41	1.07	2.04	11.41	***
Euphausia crystallorophias	0.00	0.88	0.31	0.00	14.48	13.32	***
Euphausia frigida	2.25	1.91	0.00	15.75	0.00	2.66	*
Euphausia superba	3.61	8.20	14.49	1.72	6.06	1.20	–
Euphausia triacantha	0.22	0.53	0.00	3.09	0.00	2.38	–
Gammaridae	0.00	0.01	0.05	0.00	0.62	8.17	***
Haloptilus ocellatus	0.97	3.53	0.71	0.60	0.00	11.77	***
Hyperiella dilatata	0.20	0.08	0.06	0.61	0.00	2.89	*
Limacina helicina	0.19	1.03	0.06	0.10	0.00	1.30	–
Metridia gerlachei	0.17	4.54	4.53	0.06	2.29	6.71	***
Pleuragramma antarcticum	0.01	0.17	0.00	0.00	0.12	0.19	–
Rhincalanus gigas	40.87	38.54	7.51	2.35	1.17	7.02	***
Sagitta gazellae	17.84	17.38	4.65	6.21	0.56	9.35	***
Sagitta marri	1.25	3.04	0.89	0.00	0.00	12.98	***
Salpa thompsoni	113.46	6.86	0.00	489.68	0.00	25.64	***
Siphonophore (nectophore)	9.42	19.27	5.68	2.41	1.60	17.04	***
Themisto gaudichaudii	0.42	0.07	0.00	9.99	0.00	25.22	***
Thysanoessa macrura	4.70	8.16	4.88	11.75	0.68	1.83	–
Tomopteris carpenteri	0.47	0.38	0.00	5.61	0.00	2.80	*

the high abundance of the two hyperiids *Cyllopus lucasii* and *T. gaudichaudii*, the copepod *Euchaeta antarctica*, and *Sagitta gazellae*, but also by the absence of the chaetognath *S. marri*, which occurred in all Group 1 and 2 stations.

The geographic plot of station groups displays a distinct latitudinal zonation. Not surprisingly then, latitude produced the highest correlation value with ordination scores explaining 68% of the variation in the 2-axis ordination (Table 5b). The receding ice edge was the best of the environmental variables, explaining 60% of the variation. Integrated values of temperature, salinity and chlorophyll *a* explained 36%, 28% and 27% of the variation, respectively. Fig. 6b shows the ice, temperature and chlorophyll *a* regressions nearly paralleling the latitude variable. None of the sampling parameters was correlated with the ordination scores.

In the inverse analysis of species, all 26 of the major dominant species were analysed. The two additional species *Ihlea racovitzai* and *Pagetopsis macropterus* were not observed in this year. Five groups were defined at 75% dissimilarity: two main groups, a minor group of two species, and two single-species groups (Fig. 7b). In the larger group of 15 species (Group A), all three chaetognath species, two of the six copepod species, the siphonophores, and *Clio pyramidata* were indicator species of the main oceanic station Group 2 in the centre of the region (Fig. 4b). Group A also contained *Metridia gerlachei*, *Calanoides acutus*, *Rhincalanus gigas*, and appendicularians which occurred in

Table 7. *January 1985: frequencies of occurence of indicator species distinguishing cluster groups defined in Fig. 4b*

Species in each sub-table are ranked according to information statistic. Species above the dotted line in each sub-table have a $2\Delta I > 6.63$ ($p=0.01$), and those below the line have $2\Delta I > 3.84$ ($p=0.05$). Maximum possible occurrences are: Group 1 = 15, Group 2 = 25, Group 5 = 2; total number of sites = 49.

(a) Group 1 (northern oceanic) indicator species

Species	Group 1	Remaining sites
**Vibilia* sp.	13	7
Electrona antarctica	15	17
....		
None		

(b) Group 2 (southern oceanic) indicator species

Species	Group 2	Remaining sites
Appendicularia	21	7
....		
**Tomopteris septentrionalis*	18	7

(c) Group 5 (neritic) indicator species

Species	Group 5	Remaining sites
None		
....		
Euphausia crystallorophias	2	2
Pleuragramma antarcticum	2	3

*Not dominant species; not used in SNK or inverse species analyses.

higher abundances in the station Group 2, though this was not considered significant in the SNK multiple range test (Table 6). *S. gazellae* and *Eukrohnia hamata* were also abundance indicator species of the northern station group (Group 1). The next species group (B) of seven species contained three species, *Cyllopus lucasii*, *Salpa thompsoni* and *T. gaudichaudii*, which were defined as indicator species for the northern station group. In addition, the two euphausiids have distributions known to be confined primarily to waters of the Antarctic Circumpolar Current (ACC) (John, 1936; Baker, 1966; Kittel *et al.*, 1985; Smith & Schnack-Schiel, 1990). The NMDS analysis shows a close association between the two species Groups A and B (Fig. 8b), but also suggests that the larvae of *Electrona antarctica*, a frequency indicator species of the northern stations, have as much affinity with Group B as with Group A. Group E comprised the inshore species assemblage, i.e. *Euphausia crystallorophias* and gammarids. *Pleuragramma antarcticum* was also a shelf indicator species, but was isolated from the other species (Figs 7b and 8b). *E. superba* was also classified as a single-species group at the 75% dissimilarity level. The NMDS analysis more clearly shows the dissociation of *E. superba* from the other species (Fig. 8b).

January–March 1981

Three main cluster groups were defined at 73% dissimilarity (Figs 4c and 5c). Group 3 comprised four continental shelf stations. In addition to *E. crystallorophias*, gammarids and the larvae of *P. antarcticum*, a number of other species were defined as indicator species for shelf stations, including siphonophores and the copepod *Calanoides acutus*

Table 8. *January–March 1981: frequencies of occurrence of indicator species distinguishing cluster groups defined in Fig. 4c*

Species in each sub-table are ranked according to information statistic. Species above the dotted line in each sub-table have a $2\Delta I > 6.63$ ($p=0.01$), and those below the line have $2\Delta I > 3.84$ ($p=0.05$). Maximum possible occurrences are: Group 1 = 18, Group 2 = 20, Group 3 = 4.

(a) Group 2 indicator species from Group 1

Species	Group 2	Group 1
Eukrohnia hamata	19	3
Haloptilus ocellatus	12	1
Calanoides acutus	19	8
**Vanadis antarctica*	16	5
Rhincalanus gigas	20	12
Tomopteris carpenteri	20	12

(b) Group 3 (inshore) indicator species from Group 1

Species	Group 3	Group 1
Pagetopsis macropterus	4	1
Gammaridae	4	2
*Hippolytidae larvae	3	0

*Not dominant species; not used in SNK or inverse species analyses.

(Tables 8 and 9). The latter two taxa were previously described as indicators of the oceanic stations. Station 16 (Fig. 1c) west of Davis was defined as an outlier, but distantly grouped with the inshore sites at 79.9% dissimilarity. This site comprised only 8 taxa and 71 specimens in total, with a density of 2.4 individuals 1000 m^{-3}, which is considerably less than the mean density of Group 3 zooplankton of 63.0 ± 78.6 individuals 1000 m^{-3}. The most abundant species at station 16 were *E. crystallorophias*, and the larvae of *P. antarcticum* and *Pagetopsis macropterus*. This similarity in composition explains the link with Group 3 sites.

Group 2 represents the main group of stations lying east–west in the middle of the region (Fig. 5c). This group was distinguished by much the same species, copepod and chaetognaths (Tables 8 and 9), that characterized the oceanic groups described above, i.e. Group 1 in Fig. 5a and Group 2 in Fig. 5b. Group 1 separates Group 2 oceanic stations from the inshore stations but also borders Group 2 to the north. The single indicator species of this group was *E. superba*, which represented nearly 80% of Group 1 catches in terms of density (Table 9). Station 48 was an outlier, distantly linked with Groups 1 and 2 at 80.3% dissimilarity (Fig. 4c). This station, just north of Mawson (Fig. 1c), had the highest abundance of *S. thompsoni* of any site, as well as a very high abundance of *Clio pyramidata*. At other stations with a high proportion of *C. pyramidata* there were no salps. This coupled with the low abundance of other species would explain the lack of similarity with other sites.

In the NMDS ordination, latitude explained 51% of the variation in the data, while mean temperature and surface chlorophyll *a* values accounted for 38% and 35% of the variation, respectively (Table 5c; Fig. 6c). The only other environmental parameter correlated with the ordination scores was pack-ice cover, which explained less than 15% of the variation. The time in days since the start of the sampling period explained 26% of the observed pattern.

Table 9. *January–March 1981: mean abundances, analysis of variance (F) and SNK multiple range tests of dominant species in cluster groups defined in Fig. 4c*

Analyses were carried out on $\log_{10}(x+1)$ transformed abundances (Zar, 1984); values shown are number of individuals 1000 m^{-3}. Species with significant differences in mean abundance are shown in bold text, while those with significantly higher abundances in a cluster group according to SNK analysis are underlined. For ANOVA p values: *,<0.05; **,<0.005; ***,<0.0005; –, not significant; d.f. =2,39.

Species	Group 1 mean	Group 2 mean	Group 3 mean	*F*	*p*
Appendicularia	0.03	0.00	0.00	2.15	—
Calanoides acutus	2.08	9.24	21.35	7.88	**
Calanus propinquus	4.40	12.61	1.84	2.03	—
Clio pyramidata	0.97	6.36	0.05	6.49	**
Cyllopus lucasii	0.36	0.66	0.03	0.43	—
Electrona antarctica	0.33	0.29	0.00	0.43	—
Euchaeta antarctica	0.13	1.69	0.66	2.28	—
Eukrohnia hamata	0.02	1.12	0.16	4.54	*
Euphausia crystallorophias	0.002	0.004	7.92	14.62	***
Euphausia frigida	0.82	0.32	0.05	0.32	—
Euphausia superba	168.46	3.20	0.86	62.44	***
Euphausia triacantha	0.00	0.01	0.00	0.54	—
Gammaridae	0.01	0.03	0.10	3.48	*
Haloptilus ocellatus	0.002	0.13	0.01	5.99	*
Hyperiella dilatata	0.01	0.09	0.02	3.04	—
Ihlea racovitzai	0.01	0.37	1.00	4.03	*
Limacina helicina	0.05	0.02	0.69	4.05	*
Metridia gerlachei	0.06	0.84	0.74	2.53	—
Pagetopsis macropterus	0.002	0.001	0.91	23.19	***
Pleuragramma antarcticum	0.03	0.02	22.82	35.64	***
Rhincalanus gigas	0.20	13.86	0.31	20.29	***
Sagitta gazellae	0.21	2.68	0.07	10.22	***
Sagitta marri	0.001	0.03	0.00	0.97	—
Salpa thompsoni	0.62	0.55	0.53	0.01	—
Siphonophore (nectophore)	0.05	0.33	0.61	3.29	*
Themisto gaudichaudii	2.69	4.31	0.02	0.63	—
Thysanoessa macrura	28.30	15.84	0.84	1.31	—
Tomopteris carpenteri	0.23	0.58	0.01	2.54	—

In addition, the choice of sampling depth, this being a cruise when horizontal trawls were employed, also accounted for some 17% of the variation in the data.

Six groups were defined in the comparison of species cluster analysis at the 90% dissimilarity level, comprising a major group of 20 species and a number of single- and double-species groups. The main group (Group A) could be split into two subgroups at the 85% level. The NMDS analysis shows that a split at this level is valid. Group A1 is similar in composition to the main species assemblages discussed above (Groups A in Figs 7a,b) and for 1987 (see below). This group also contained the indicator species for the main oceanic group of stations (Group 2 in Fig. 5c). Gammarids were also found in this group and yet were defined as frequency and abundance indicators of the shelf group of stations (Fig. 5c). The NMDS analysis (Fig. 8c) does show the gammarids more closely aligned with the Group E shelf species of *E. crystallorophias*, *Pleuragramma antarcticum* and *Pagetopsis macropterus*. Group A2 contained species described as being associated with northern stations in 1985, but also contained three species, *Euchaeta antarctica*, *Electrona antarctica* and *Thysanoessa macrura* described in the previous 1982 and 1985 surveys as belonging to the main species group of copepod–chaetognath–siphonophores (Group A in Fig. 8a,b). In turn, *S. thompsoni* was included in Group A1, whereas in other years (Fig. 7a,b) it was either isolated or associated with the *Euphausia frigida*, *E. triacantha*, *Cyllopus lucasii* and *Themisto gaudichaudii* species group (Group B in Fig. 8b). Nor was *E. triacantha* associated with these species in this particular sampling season (Fig. 8c). This may well be an artefact of analysis caused by this species' low abundance in 1981. Appendicularians were also in very low abundance and were also a single-species cluster group. However, the NMDS analysis does group appendicularians with *Limacina helicina*. This latter species was grouped with the salp *I. racovitzai* in the cluster analysis (Fig. 7c), but in the ordination *I. racovitzai* was separate from *L. helicina* and grouped with Group A1. *E. superba* was clearly dissociated from the other species at 91% dissimilarity. As noted above, this species occurred in high abundance in station Group 1 (Fig. 5c) and was the only indicator species for that group of stations.

Overall, there is some disagreement between the cluster and ordination results, and some interchange of species between species groups, as well as different and additional indicator species for the shelf area. These differences may be attributable in part to some degree of geographical mixing of species groups detected by the extended sampling, with sampling period accounting for 26% of the variation in the comparison of sampling sites. The interchange of species between Groups A1 and A2 could also be due to the lack of sampling north of 62°S, resulting in the poor representation of northern species and therefore not being discerned as a distinct group as observed in 1985.

March 1987

The cluster analysis produced two main groups and one outlier, station 83 in the north (Fig. 1d) at the 55% dissimilarity level (Fig. 4d). Station 83 was dominated by the hyperiid *T. gaudichaudii*, which represented 55% of the catch in terms of fresh mass. At 53% dissimilarity, three main groups were defined, plus an additional outlier, station 65 near Davis (Fig. 1d). This station was dominated by *E. crystallorophias*, representing 85% of the catch by fresh mass, and yet this site was linked to Group 3, which had no *E. crystallorophias* representation (Table 10). The NMDS plot (Fig. 6d) shows three distinct main station groups, and validates the selection at 53%.

The major group (Group 2) comprised most of the stations in the eastern part of the region and was distinguished from the western sites by very high abundance of the same copepod and chaetognath species that characterized the main oceanic groups in other years. Group 3 stations were characterized by low abundances: the mean total zooplankton density for the dominant species was nearly 10 times lower than that of Group 2 (Table 10). There were no indicator species for Group 3. The predominant species in Group 1 sites was *E. superba*, which represented 76–95% of the catch fresh mass. *Thysanoessa macrura* was an indicator species of both Group 1 and Group 2, i.e. Group 3 was distinguished by the low abundance of this species (Table 10).

There were few frequency indicators. Group 2 was distinguished from the remaining sites by the chaetognath *Sagitta marri*, which occurred in 11 of the 13 Group 2 sites and three of the remaining 12. Group 2 was specifically separated from Group 1 by the hyperiid *Primno macropa* (not a dominant species), which was found in nine of the Group 2 sites and not in any of Group 1. The $2\Delta I$ information scores for *S. marri* and *P. macropa* were only significant at 5%.

Latitudinal zonation was a feature of the previously discussed data sets. The March 1987 data set, however, exhibited a distinct longitudinal zonation. In the NMDS ordination, longitude explained 65% of the variation, while latitude accounted for 24% (Table 5d; Fig. 6d). The time in days since the start of the sampling period also explained a significant amount (58%) of the observed pattern. In the 3-axis ordination, integrated temperature and ice recession explained 75% ($F = 22.9$, d.f. $= 3,19$, $p < 0.0005$) and 27% ($F = 4.0$, d.f. $= 3,21$, $p = 0.02$) of the variation, respectively. However, these parameters were mainly correlated with the third axis (regression weight = −0.994) and were not significant variables when the ordination was reduced to two axes (Table 5d).

Gammarids, *I. racovitzai* and *Pagetopsis macropterus* were not observed in this season and therefore only 25 species were used in the inverse species analysis. Two major groups were defined at 82%. Group A comprised the same main species group

Table 10. *March 1987: mean abundances, analysis of variance (F) and SNK multiple range tests of dominant species in cluster groups defined in Fig. 4d*

Analyses were carried out on $\log_{10}(x+1)$ transformed abundances (Zar, 1984); values shown are number of individuals 1000 m^{-3}. Species with significant differences in mean abundance are shown in bold text, while those with significantly higher abundances in a cluster group according to SNK analysis are underlined. For ANOVA *p* values:*, <0.05; **, <0.005; ***, <0.0005; –, not significant. d.f. =2,22.

Species	Group 1 mean	Group 2 mean	Group 3 mean	*F*	*p*
Appendicularia	0.00	0.55	0.04	2.10	–
Calanoides acutus	4.75	141.45	12.30	19.22	***
Calanus propinquus	1.54	67.61	3.80	19.93	***
Clio pyramidata	2.35	30.44	0.89	1.74	–
Cyllopus lucasii	0.87	2.07	0.11	1.30	–
Electrona antarctica	0.03	0.40	0.00	2.49	–
Euchaeta antarctica	0.45	6.11	0.18	4.87	*
Eukrohnia hamata	1.79	13.28	4.39	4.50	*
Euphausia crystallorophias	0.15	0.05	0.00	0.90	–
Euphausia frigida	0.43	5.11	0.00	2.13	–
Euphausia superba	44.95	11.44	0.23	12.84	***
Euphausia triacantha	0.55	0.95	0.00	1.09	–
Haloptilus ocellatus	0.56	2.30	0.61	5.34	*
Hyperiella dilatata	0.09	0.09	0.10	0.00	–
Limacina helicina	0.00	0.00	0.01	1.32	–
Metridia gerlachei	1.12	7.40	0.06	5.05	*
Pleuragramma antarcticum	0.50	0.00	0.03	2.00	–
Rhincalanus gigas	3.34	40.05	8.84	8.73	**
Sagitta gazellae	1.88	5.14	1.61	5.35	*
Sagitta marri	0.04	1.81	0.08	4.96	*
Salpa thompsoni	2.23	27.09	0.13	1.47	–
Siphonophore (nectophore)	2.60	4.21	3.36	0.25	–
Themisto gaudichaudii	3.01	10.88	1.62	0.72	–
Thysanoessa macrura	5.40	7.17	0.78	11.41	***
Tomopteris carpenteri	0.02	0.05	0.01	0.61	–

identified in the other three years, i.e. all the copepod and chaetognath species, siphonophores, appendicularians and *T. macrura*. Nine of the species were abundance indicators of the main station Group 2 in the eastern half of the Prydz Bay region (Fig. 5d). Unlike previous years, *E. superba* was clearly linked with this group of species (Figs 7d and 8d). Group B comprised *Themisto gaudichaudii* (dominant at station 83), *Salpa thompsoni*, and the two euphausiids *E. frigida* and *E. triacantha*. *Tomopteris carpenteri* and *Hyperiella dilatata* were also associated with these species. This species grouping was also defined in the 1985 data set (compare Groups B in Fig. 7b,d). *C. lucasii*, previously associated with *E. frigida* and *Themisto gaudichaudii* in 1981 and 1985, was linked with the main species Group A. The remaining three species, *E. crystallorophias*, *Pleuragramma antarcticum* and *L. helicina*, formed single-species groups. The continental shelf group of species was not a distinct group in the comparison of species. This is mainly because sampling was restricted to just one site west of Davis where *E. crystallorophias* was the dominant species. Although *P. antarcticum* was also collected at this site, the two species did have different distributions. *P. antarcticum* was also found at stations 70 and 71, while *E. crystallorophias* was found at stations 66 and 96 (Fig. 1d).

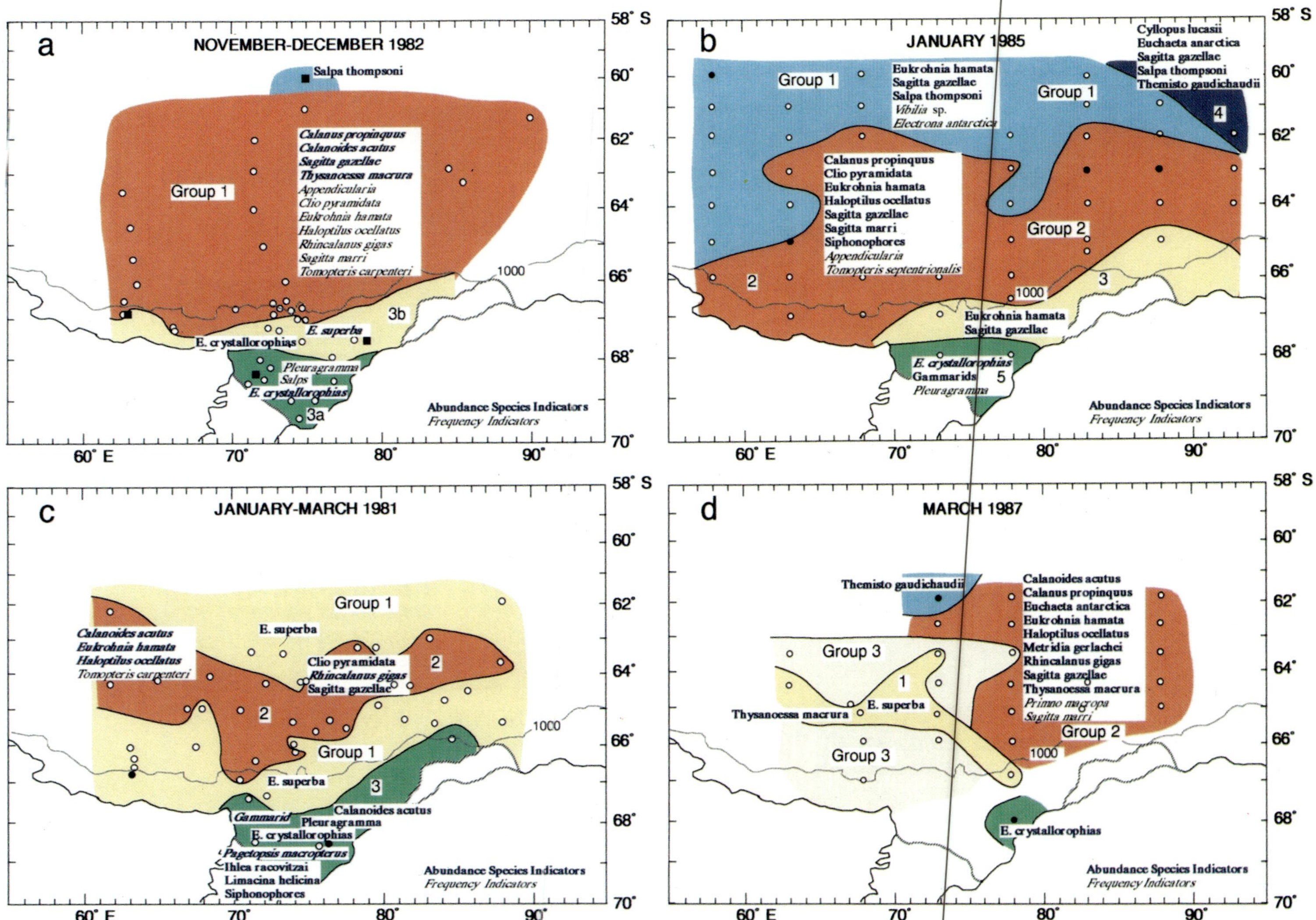

Fig. 5. Geographical distribution of station groups defined by cluster analyses shown in Fig. 4. Indicator species listed in blue bold type are those characterizing an area (station group) by their unique higher abundance identified by SNK analysis (Tables 4, 6, 9 and 10). Indicator species in black italic type are those with a higher frequency of occurrence in that station group as identified by Field's information statistic (Tables 3, 7 and 8). Note: in many cases species were both abundance and frequency indicator species. Station groups with common species composition between years have the same shading colour (see Discussion in text, and Fig. 9). Open circles, sampling sites; filled circles, 'outlier' sampling sites not grouped in the cluster analyses; squares in (a), Station Group 2 in November–December 1982.

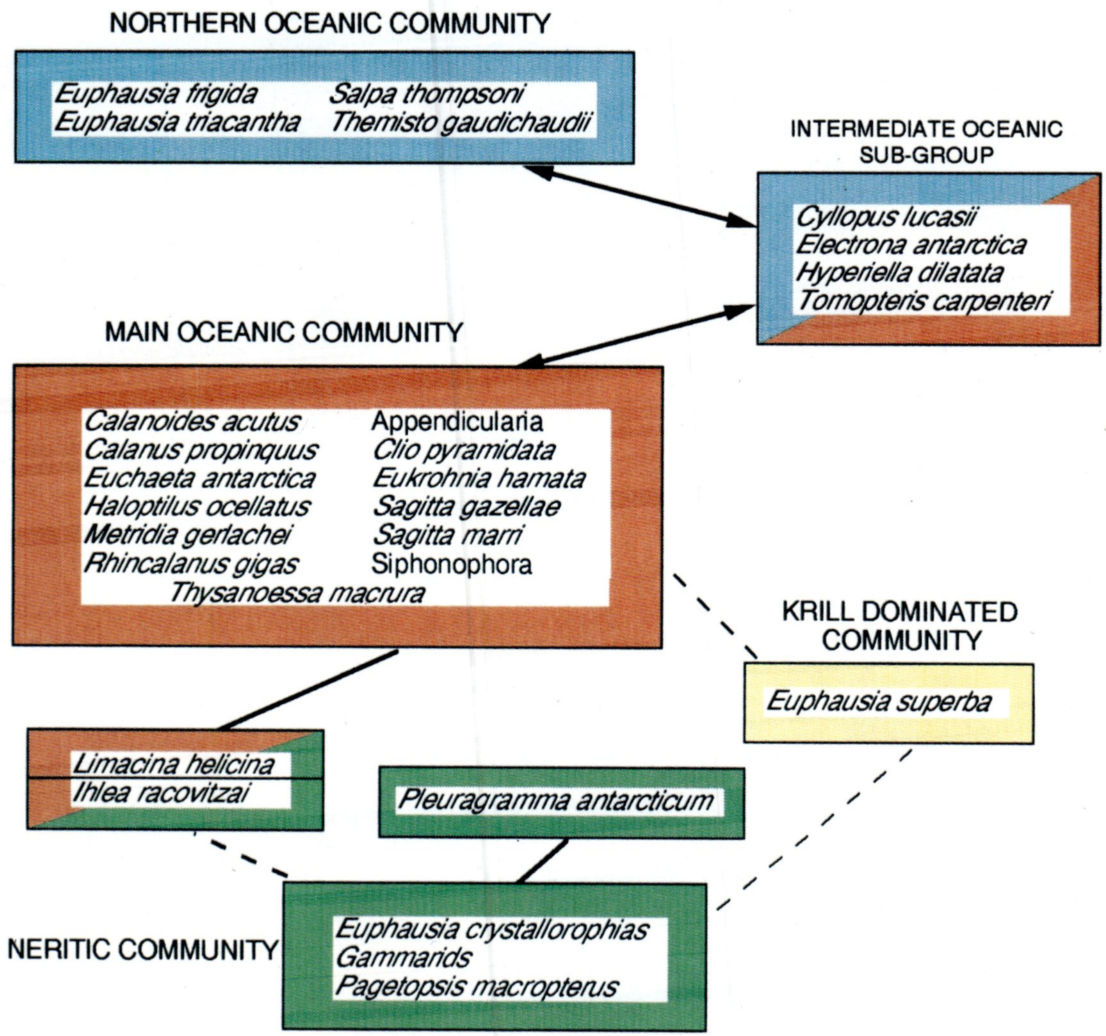

Fig. 9. Zooplankton communities derived from the comparison of species analyses (Figs 7 and 8) and indicator species defined for the station groups. Each assemblage is coded with a colour corresponding to the areas in Fig. 5 where that assemblage predominated.

Discussion

Species assemblages

There was considerable consistency in the species associations defined for each year, despite different sampling methodology and sampling at different times of the year. Further, there was good correspondence between the species groups and the indicator species defined for the station groups. Consequently, the various species associations shown in Figs 7 and 8 can be combined into four principal species assemblages or communities that can be related back to specific geographical areas where these assemblages are dominant. These communities are the neritic, the main oceanic community, a northern oceanic community and an *E. superba*-dominated community (Fig. 9). Each assemblage is coded with a colour corresponding to the areas in Fig. 5 where that assemblage predominated.

Neritic community

The neritic or shelf group was characterized by *E. crystallorophias*. When present, gammarids and the larvae of *Pagetopsis macropterus* formed a close association with this euphausiid. This assemblage dominated the inshore region of Prydz Bay south of 67°30′S, although these species were found elsewhere on the shelf. *Pleuragramma antarcticum* is also a neritic species but only formed a partial association with the other three species. Differences in the respective distributions of *P. antarcticum* and *E. crystallorophias* are probably the main reason why there is a slight dissociation. The larvae of *P. antarcticum*, especially those less than 25 mm in length, are readily carried offshore. Their distribution closely follows the western boundary of the Prydz Bay gyre (Fig. 3) (see Williams & Duhamel, this volume, for a more detailed discussion). The larvae of *E. crystallorophias* are also carried offshore by the gyre, whereas the adult *E. crystallorophias* are confined to the shelf (Hosie, 1991). The larvae of both *P. antarcticum* and *E. crystallorophias* are useful indicators of water of Prydz Bay or shelf origin moving offshore.

Main oceanic community

The main oceanic assemblage comprised the bulk of the numerically dominant species found in the region. This included all the abundant copepod species, particularly the herbivores *Calanus propinquus*, *Calanoides acutus* and *R. gigas*, as well as the associated carnivores such as chaetognaths. Chaetognaths are one of the principal predators of copepods, with *Eukrohnia hamata* alone responsible for reducing the copepod biomass by as much as 12% (Øresland, 1990). Included in this assemblage are the euphausiid *Thysanoessa macrura*, the pteropod *Clio pyramidata*, and appendicularians and siphonophores collectively. All species in this assemblage are notable for having wide circumpolar distributions. For example, the copepods (Vervoort, 1951), the three chaetognaths (O'Sullivan, 1982) and *T. macrura* (Mauchline & Fisher, 1969) have distributions that extend south from the Antarctic Convergence onto or near the continental shelf. The copepod *Haloptilus ocellatus* is usually found south of 60°S (Vervoort, 1951). During this study, members of this assemblage were sometimes collected at shelf stations, and hence overlapped the distribution of the neritic assemblage. In 1981 (Fig. 5c), siphonophores and *Calanoides acutus* were also abundance indicator species of the shelf region. A Polish survey in 1969, where a much finer plankton mesh (50 μm) was used than that of the present study, found copepodites of *C. acutus*, *Calanus propinquus* and *M. gerlachei* widely distributed over the shelf region of Prydz Bay (Zmijewska, 1983). While the species of the main oceanic community have extensive geographic distributions, the present study has shown they were generally more abundant in, and thus dominated, the region between the continental shelf edge and 62–63°S. The northern limit approximates the southern boundary of the Antarctic Circumpolar Current (ACC). Pakhomov (1989) also found pteropods and siphonophores to be major abundant groups in the area south of the Antarctic Divergence (AD) in the Prydz Bay region, but not north of the divergence, nor in shelf waters. The oceanic and neritic zones were distinguished both by the differences in species composition as revealed by the information statistic, and also by the differences in abundances of species common in both areas.

The salp *I. racovitzai* and the pteropod *L. helicina* were two species that were not distinctly associated with either the neritic or oceanic communities. Both species were associated primarily with the main oceanic community but were also abundance indicators of the neritic zone in 1981. *L. helicina* has been described as a typical inhabitant of the shelf regions in the Weddell and Ross Seas (Smith & Schnack-Schiel, 1990), but also as being a member of both the neritic and oceanic communities around the Antarctic Peninsula (Piatkowski, 1989*a*; Siegel & Piatkowski, 1990). Little is known of the detailed distribution and ecology of *I. racovitzai* other than that it has a wide circumpolar distribution south of the convergence but may prefer waters colder than 0 °C (Foxton, 1971). Neither *I. racovitzai* nor *L. helicina* can be considered as reliable indicator species.

Northern oceanic community

Differences in species abundances was perhaps the main feature distinguishing the northern oceanic and main oceanic communities, rather than differences in species composition. The northern assemblage comprised four main species, the euphausiids *Euphausia frigida* and *E. triacantha*, the hyperiid *Themisto gaudichaudii* and the salp *S. thompsoni*. These species have been described previously as being confined to or more abundant in the waters of the West Wind Drift (Baker, 1966; Foxton, 1966; Kane, 1966; Smith & Schnack-Schiel, 1990). Both *S. thompsoni* and *T. gaudichaudii* were collected at most stations north of the shelf, but had significantly higher abundances in the northern waters, generally north of 62°S, where they were abundance indicator species of that region. The exceptionally high abundance of salps in January 1985 would have concealed the importance of some of the species of the main oceanic assemblage which were also found in northern waters. Pakhomov (1989) also observed higher abundances of *S. thompsoni* and *T. gaudichaudii* in waters north of the Antarctic Divergence in Prydz Bay in both February–March 1985 and 1986. Pakhomov reported that *E. frigida* was also more abundant north of the divergence. In the present study, neither *E. frigida* nor *E. triacantha* was found to be an indicator of northern waters. Both species were sparsely distributed through most of the oceanic area. *E. frigida* was more abundant in the two north-eastern stations in January 1985 (Group 4 in Table 6 and Fig. 5b), but this was not significant in terms of SNK, nor did the two species have significantly high abundances in any other station group. None the less, these species did form associations with *S. thompsoni* and *T. gaudichaudii* (Fig. 8b,d), although this is less evident in January–March 1981 when northern waters were not sampled in detail. The larvae of *E. frigida* are confined to waters of the ACC (Hosie, 1991) and are perhaps better indicators than the adults.

Electrona antarctica, *Cyllopus lucasii*, *Hyperiella dilatata* and *Tomopteris carpenteri* formed an intermediate oceanic subgroup of species that were, at different times, associated with either the northern assemblage or the main oceanic assemblage. Species of *Cyllopus* and *Vibilia* are obligate symbionts of salps (Madin & Harrison, 1977), and Ainley *et al.* (1988) reported the close association of these taxa with *S. thompsoni* in the Weddell Sea. Similarly, in January 1985, *C. lucasii* and *Vibilia* sp. were commonly distributed in the northern waters where *S. thompsoni* was most abundant. All three species were indicators of the northern sites (Fig. 5b). Moreover, both the cluster (Fig. 7b) and NMDS analyses (Fig. 8b) show a very close affinity between *C. lucasii* and *S. thompsoni* in January 1985. The NMDS analyses also showed that in November 1982 the hyperiid *H. dilatata* had a closer affinity to the neritic species than with the main oceanic assemblage (Fig. 8a). In January (1985) *H. dilatata* had a near-equidistant affinity with the main oceanic and northern species communities (Fig. 8b), while in March 1987 this species was clearly associated with the northern species group (Fig. 8d). Siegel & Piatkowski (1990) described the same situation in the Antarctic Peninsula region, with *H. dilitata* moving from the neritic community in spring (November) to the oceanic in summer (February). They concluded that this species was not a reliable indicator species. Although the polychaete *T. carpenteri* and the larvae of the myctophid *Electrona antarctica* are members of the intermediate subgroup, their oceanic distributions are not in dispute. The myctophid was never found on the shelf during this study (see Williams & Duhamel, this volume, for a more detailed discussion), while *T. carpenteri* was only found at two shelf sites, once each in 1981 and 1982.

Krill-dominated community

Euphausia superba was, for most of the time, dissociated from all other species. Krill did form a distant association with the neritic species in 1982 and a more substantial association with the main oceanic assemblage in 1987. However, in those surveys and in 1981, krill were important abundance indicators of geographic areas distinct from the main oceanic and neritic distributions. In 1985, krill were not indicators of any station group, but were most abundant in station Group 3 near the shelf edge between the oceanic and neritic zones (Table 6). Hydroacoustic studies in that season also indicated a higher biomass of krill in the shelf edge area (Higginbottom *et al.*, 1988). Studies by Pakhomov (1989), Bibik & Yakolev (1991) and Ichii (1990) have also shown that krill are more abundant near the continental shelf edge between 60° and 80°E in Prydz Bay. The shelf region was proposed as a major spawning region for krill (Hosie, 1991). The results from 1981, 1982 and 1985 of the present study have also shown that krill are primarily localized and are dominant near the continental shelf edge in a transition zone between the neritic and oceanic assemblages. In the krill-dominated areas the non-krill zooplankton abundance is consistently lower (Fig. 10). In turn, the areas of high zooplankton abundance, generally dominated by the oceanic

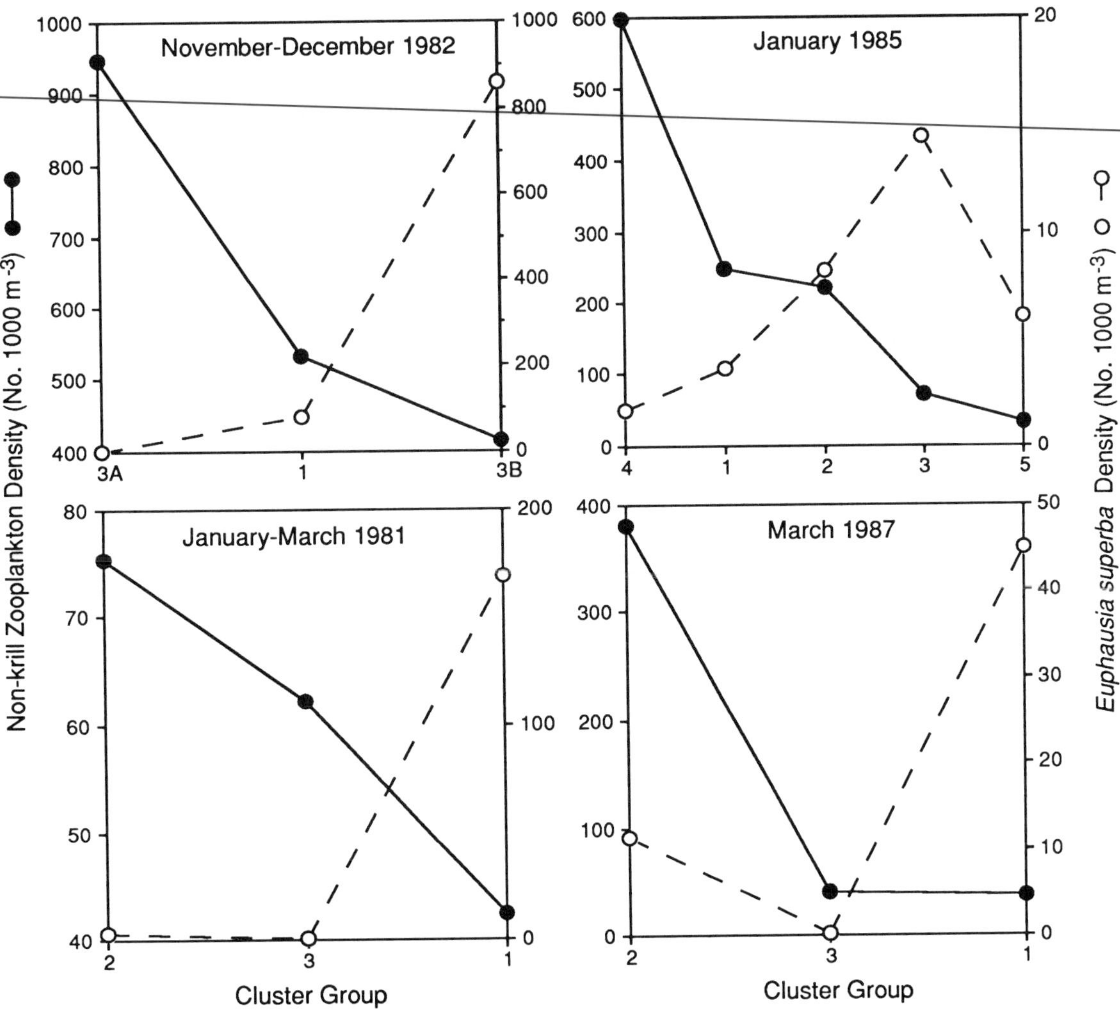

Fig. 10. Comparison of mean *E. superba* density with remaining mean-non-krill zooplankton density for each station group for the four surveys (Tables 4, 6, 9 and 10). Station groups are ordered according to increasing krill abundances. Note: station group order for January 1985 (b) is near to north–south.

assemblages, are areas where krill were in very low abundance.

The low zooplankton abundances in the krill-dominated areas may be a result of *E. superba* excluding other species from these areas via competition. Fig. 11 shows that the combined mean abundance of the herbivorous copepods *Calanoides acutus*, *Calanus propinquus* and *R. gigas* in the krill-dominated areas decreases in relation to increases in krill abundance for the four surveys. Decreases in copepod numbers would also result in associated decreases in carnivores such as chaetognaths, which depend on copepods. The reverse situation of herbivorous zooplankton excluding krill from the main oceanic community is possible, although at a localized level krill would be expected to exclude other species from their living space as krill formed dense aggregations. A superswarm sampled in Prydz Bay was devoid of other species of zooplankton (Higginbottom & Hosie, 1989).

An alternative to the 'exclusion' theory is that krill may simply dominate a region which is unsuitable for either the neritic or the oceanic species. Neritic species would presumably avoid an oceanic existence, whereas the shelf edge is either the southern limit of oceanic species distributions or their abundance decreases towards and onto the shelf. In this study a number of species were consistently in very low abundance or more often absent from the shelf or near the shelf area, i.e. appendicularians, *Electrona antarctica*, *Haloptilus ocellatus*, *Sagitta marri*, *Cyllopus lucasii* and *T. carpenteri*, as well as all members of the northern oceanic community. In addition, a number of widely distributed species had abundances that diminished toward the continent, i.e. *R. gigas*, *S. gazellae*, *Eukrohnia hamata*, *Clio pyramidata* and siphonophores. Budnichenko & Khromov (1988) observed a similar north–south decrease in abundance of *R. gigas* in Prydz Bay in 1984. Fig. 10 shows a clear trend of decreasing non-krill zooplankton abundance from the northern station groups towards and onto the shelf in January 1985. Fig. 10 is also consistent with the theory of exclusion through competition. The exclusion theory could account for the distribution of the krill-dominated community to the north of the main oceanic community in 1981, as well as the western distribution of the krill community away from the shelf edge in 1987, whereas the 'unfavourable conditions' theory could not. On the other hand, the creation of large discrete areas of few zooplankton through exclusion alone is difficult to imagine and would perhaps be better explained by the 'unfavourable conditions' alternative. Both theories have merit and there is no reason why both situations could not be combined insofar as the exclusion of the zooplankton from an area is facilitated by their initial diminution in numbers. Regardless of which theory prevails, it would appear that competition for resources is minimised through geographical separation of the main herbivorous groups: *Euphausia crystallorophias* in the inner shelf area, copepods further offshore, and krill primarily in between with almost no overlap with salps in the far north.

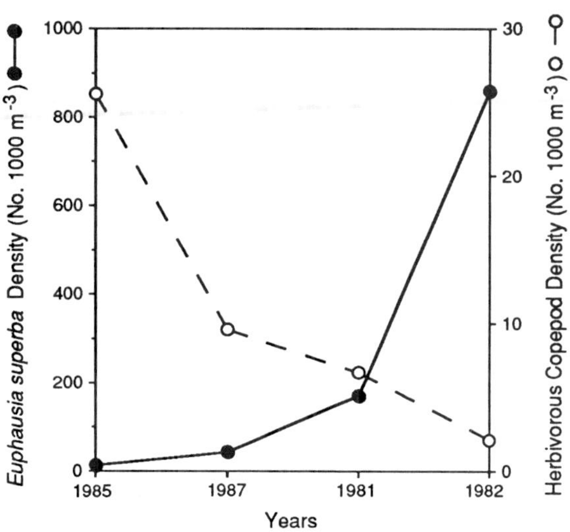

Fig. 11. Comparison of mean *E. superba* density with combined mean abundance of the herbivorous copepods *C. acutus*, *C. propinquus* and *R. gigas* in the krill-dominated community areas for the four surveys. Survey years are ordered by increasing krill abundance.

There was much speculation and discussion during the BIOMASS Colloquium as to whether *E. superba* is really as important in Prydz Bay, and in other areas, as originally perceived. Certainly, a greater part of the oceanic region is dominated by copepods and salps, while the inner shelf is dominated by *E. crystallorophias*, an important dietary component of various vertebrates (Puddicombe & Johnstone, 1988; Williams, 1985, 1989). This study has shown, without doubt, that *E. superba* is the principal dominant species of large areas of the southern Prydz Bay region. These areas overlap the summer distribution of minke whales (Ichii, 1990) and also of various fish and birds that at times feed extensively on krill (Montague, 1984; Williams, 1985, 1989; Green, 1986; Green & Johnstone, 1988; Ichii, 1990; Armstrong & Siegfried, 1991). Krill should still be viewed as the most important single species in the Prydz Bay area.

Geographical variability and controlling factors

Sampling resolution of the shelf region was perhaps not ideal for an area of rapid transition between two communities. The amount of sampling that was carried out suggests little variability in the northern boundary of the neritic group, with this community consistently dominating the region south of 67°S. It is more difficult to comment on the geographical variability in the northern oceanic community because this area was only covered in detail in 1985. There is no evidence to suggest much variability in the geography of the West Wind Drift, especially in relation to the southern boundary. Indeed, there is evidence of consistency in the northern zooplankton, with Pakhomov (1989) describing near-identical distributions and relative abundances of salps and hyperiids in the West Wind Drift in Prydz Bay for the January–February periods of both 1985 and 1986.

The greatest amount of variability was displayed by the main oceanic and krill-dominated communities. For three of the years there was strong latitudinal zonation, although the north–south width varied somewhat, while there was a distinct longitudinal zonation between the two communities in March 1987. There are species that will favour or live in certain temperature and salinity regimes: for example, *E. crystallorophias* lives in waters of predominantly <-1 °C, while *Salpa thompsoni* is restricted to the warmer waters of the West Wind Drift. Temperature did account for much of the zooplankton data variation in three of the four years (Table 5). In January 1985 all four community zones were well represented and, along with temperature, salinity did explain some of the variation in the zooplankton distributions. The main oceanic and krill communities dominated the regression analyses in the other three years, and it would seem that horizontal salinity gradients have less or no effect on the collective zooplankton distributions between and within these communities. It remains to be determined what effect vertical stratification of the water column has on community distributions.

The presence and duration of pack-ice cover is an important factor affecting the structure of the pelagic food web (Smith & Schnack-Schiel, 1990). This influence comes firstly from the rotting early summer ice, which provides a favourable substrate facilitating high production of ice-algae and thus an important and rich food source for krill and zooplankton (Smetacek *et al.*, 1990). Secondly, as the ice recedes the meltwater stabilizes the surface water layers where phytoplankton blooms rapidly develop (Hart, 1942; Smetacek *et al.*, 1990). The effects of ice recession could not be determined in November–December 1982, as much of the region was still covered by ice. Neither did the degree of pack-ice cover explain any of the data variation. Ice usually dissipates by the beginning of January in Prydz Bay (Jacka, 1983). In January 1985, ice recession did account for a substantial amount of the variation, more so than chlorophyll *a*. Ice recession was also proposed as a factor influencing the distribution pattern of *Thysanoessa macrura* developmental stages during the same period (Hosie, 1991). The zooplankton and ice recession correlation may also be coincidental, since the two parameters both have a latitudinal pattern. Fig. 6b shows the close alignment in ice and latitude regressions. There was no similar correlation between ice recession and zooplankton observed in the later months of 1981, 1987, but then this may not be expected 2–3 months after the ice has gone.

Smith *et al.* (1988) cautioned that the 'snapshot' approach of comparing biological and environmental events from the same point in time, as described above, may not adequately reveal true relationships because different rates of environmental processes and community responses are involved. Despite such differences between the phyto- and zooplankton, chlorophyll *a* was another variable, along with temperature, that accounted for much of the variation in three of the four years. Presumably, at the start of summer, the plankton interactions are driven by the receding ice. However, the exact relationship or nature by which chlorophyll affects zooplankton distributions, particularly in later months, is not clear and still needs to be established. For example, in this study, no significant relationship was found between chlorophyll *a* and any zooplankton community parameter, e.g. abundance or number of taxa at a site. None the less, by the end of summer, as shown by the March 1987 patterns, chlorophyll *a* no longer has an influence on the distribution patterns of the zooplankton and instead the north-easterly current became the single important controlling factor (Fig. 3b).

Various studies have demonstrated consistency in the horizontal water circulation patterns in the Prydz Bay region between years, with little intra-annual variation (Khimitsa, 1976; Tchernia & Jeannin, 1983; Smith *et al.*, 1984; Allison, 1989; Hodgkinson *et al.*, 1988, 1991*a*,*b*; Middleton & Humphries, 1989). Currents were expected to have a major influence on the distribution of the zooplankton throughout the summer sampling period. The currents near the shelf, as shown in Fig. 3b, are believed to be responsible for the dispersion of euphausiid larvae (Hosie, 1991). Apart from the West Wind Drift, the north-easterly current out of Prydz Bay (Fig. 3b) was the only other major current that would appear to have had an effect on zooplankton distributions at the community level. There was some indication in January 1985 that the main oceanic community was carried in a north-easterly direction, following the dispersal route of euphausiid larvae (Hosie, 1991). This pattern may also be an artefact of the analyses, as the boundary between the main and northern oceanic communities was based on a relatively low level of dissimilarity (38%). The likely effects of the north-easterly current were more pronounced in March 1987, when the main oceanic community was concentrated in the eastern part of the Prydz Bay region. No effect of this current was evident in 1981 or 1982. Neither would it seem that the Prydz Bay gyre or the East Wind Drift had any appreciable effect on community distributions, although both currents have been identified as transporting euphausiid larvae and larvae of neritic fish species offshore (Hosie, 1991; Williams & Duhamel, this volume).

Sampling methods

The correlation between ordination scores and sampling day in the January–March 1981 data indicates that a significant time component was introduced into the data set by the protracted sampling period (Table 5c). More than likely, during the course of the 48-day sampling period, the very communities being sampled and defined had undergone a substantial change, possibly in distribution or composition. This complicated the analyses and interpretation of the data, especially in relation to defining possible causal effects. A sampling period of this duration is therefore not recommended. Instead, sampling surveys like the present study, should not extend beyond approximately 30 days, which was the length of the 1982 survey where no time component was detected.

The length of the sampling period was also correlated with ordination scores in March 1987 (Table 5d). This relationship, however, was probably not a valid result, but a statistical artefact caused by the sampling progressing from west to east coupled with a longitudinal zonation in the zooplankton. Sampling day was strongly correlated with longitude (r= 0.78, F=35.83, p<0.001, d.f.=24); Fig. 6d shows the close alignment of the sampling day and longitude regressions.

Analysis of the 1981 data was further complicated by the fact that the horizontal tows at fixed depths explained some of the variation in the data. Different vertical migration patterns exhibited by zooplankton passing through a fixed sampling layer will probably produce a bias in the data. Some species may not be adequately sampled at all. Oblique sampling over a greater depth range, e.g. 200 m, will mask the vertical migration bias, and is thus a better method in order to determine large-scale community structures and patterns.

Conclusions

Kawamura's (1986, 1987) suggestion that the zooplankton composition of Antarctic waters may have changed over a 50–60 year period was based on a few samplings over a wide geographical range, west and east of the Prydz Bay region. Given the degree of variability in the distribution of the krill and especially main oceanic/copepod dominated communities displayed in this study, it is quite possible that Kawamura may not have detected a change in composition but change in community distributions. None the less, any indication of changes in zooplankton composition warrants further investigation. While there is some geographical variation in community distributions, the species assemblages themselves appear to be reasonably constant between years. Similarly, there is a remarkable consistency in the oceanography of the Prydz Bay region in terms of temperature and salinity profiles (Kerry & Woehler, 1987; Kerry *et al.*, 1987*a*,*b*; Woehler & Williams, 1988), description of the gyre for various years (Khimitsa, 1976; Smith *et al.*, 1984; Middleton and Humphries, 1989; R. A. Nunes Vaz & G. W. Lennon, personal communication, 1990), and other currents identified by current meter moorings, sea-ice buoy and iceberg drifts (Tchernia & Jeannin, 1983; Hodgkinson *et al.*, 1988, 1991*a*,*b*; Allison, 1989). Therefore, it is quite possible that the various community distribution patterns observed may in fact be repeatable phenomena. At the very least, discrete areas can be identified where each of the four communities are persistent, thus facilitating future research into the dynamics of each community.

Perhaps the most important message of this study is that, for the greater part of the Prydz Bay region, *E. superba* was not widely distributed in great abundance. Instead, there were a number of assemblages dominated by zooplankters other than krill. *E. superba* would appear to be associated more with the shelf edge, rather than with any other feature such as ice, but within that area where various predators feed, *E. superba* is the predominant species, almost to the exclusion of other zooplankton. Despite the dominance of other species in other areas, *E. superba* is still the most important single species in Prydz Bay.

Acknowledgements

I am extremely grateful to Dr S. Wright for providing unpublished chlorophyll *a* data and to Drs R. Nunes Vaz and G. Lennon for providing their unpublished results on horizontal water circulation. I am also grateful to Dr S. Nicol for advice and comments on the manuscript. Messrs R. Williams, J. M. Kirkwood and D. B. O'Sullivan collected and processed specimens on FIBEX and ADBEX I. I thank Dick Williams for identifying ichthyoplankton in subsequent years. A special acknowledgement goes to T. G. Cochran, J. Kitchener and M. Stolp for their assistance, both on ship and later in the laboratory, in the processing of samples.

References

Ainley D G, Fraser W R, Daly K L (1988) Effects of pack ice on the composition of micronekton communities in the Weddell Sea. In: Sahrhage D (ed) *Antarctic Ocean Resources and Variability*. Springer-Verlag, Berlin, pp. 140–6.

Allison I (1989) Pack-ice drift off East Antarctica and some implications. *Ann Glaciol* **12**: 1–8.

Armstrong A J, Siegfried W R (1991) Consumption of Antarctic krill by minke whales. *Antarct Sci* **3**: 13–18.

Baker A de C (1966) The latitudinal distribution of *Euphausia* species in the surface waters of the Indian Ocean. *Discovery Rep* **33**: 309–34.

Baker A de C, Clarke M R, Harris M J (1973) The NIO combination net (RMT1+8) and further developments of rectangular midwater trawls. *J Mar Biol Ass UK* **53**: 167–84.

Bibik V A, Yakolev V N (1991) The status of krill (*Euphausia superba* Dana) resources in CCAMLR statistical divisions 58.4.2 and 58.4.3 from 1986 to 1989. Results of acoustic surveys. In: *Selected Scientific Papers Presented to the Scientific Committee of CCAMLR 1990*, SC-CAMLR-SSP/7. CCAMLR, Hobart, pp. 163–73.

Boden B P (1985) The South African SIBEX I cruise to the Prydz Bay region, 1984: VIII. The plankton encountered during the survey. *S Afr J Antarct Res* **15**: 28–32.

Boysen-Ennen E, Hagen W, Hubold G, Piatkowski U (1991) Zooplankton biomass in the ice-covered Weddell Sea, Antarctica. *Mar Biol* **111**: 227–35.

Boysen-Ennen E, Piatkowski U (1988) Meso- and macrozooplankton communities in the Weddell Sea, Antarctica. *Polar Biol* **9**: 17–35.

Brandt S B, Wadley V A (1981) Thermal fronts as ecotones and zoogeographic barriers in marine and freshwater systems. *Proc Ecol Soc Aust* **11**: 13–26.

Bray J R, Curtis J T (1957) An ordination of the upland forest communities of southern Wisconsin. *Ecol Monogr* **27**: 325–49.

Budnichenko E V, Khromov N C (1988) (Biomass of mesoplankton, age composition and the distribution of dominant species in relation to hydrological structure of the Sudruzhestva and Kosmonavtov Seas.) In: *Kompleksniye Issledovaniya Ekosistemi Pelagiali v Moryach Sodruzhestva i Kosmonabtov. Sbornik Nauchnich Trudov. (Multi-disciplinary Investigations of Pelagic Ecosystems in the Sodruzhestva and Kosmonavtov Seas. Collection of Scientific Papers.)* VNIRO Publishers, Moscow, pp. 83–107. (In Russian.)

FAO (1991) *FAO Year Book 1989. Fisheries Statistics. Catches and Landings*. Vol. 68. Food and Agriculture Organization of the United Nations, Rome, 518 pp.

Field J G, Clarke K R, Warwick R M (1982) A practical strategy for analysing multispecies distribution patterns. *Mar Ecol Prog Ser* **8**: 37–52.

Foxton P (1966) The distribution and life history of *Salpa thompsoni* Foxton with observations on a related species *Salpa gerlachei*. *Discovery Rep* **34**: 1–116.

Foxton P (1971) On *Ihlea magellanica* (Apstein) (Tunicata: Salpidae) and *Ihlea racovitzai* (van Beneden). *Discovery Rep* **35**: 179–98.

Fukuchi M, Tanimura A, Ohtsuka H (1985) Zooplankton community conditions under sea ice near Syowa Station, Antarctica. *Bull Mar Sci* **37**: 518–28.

Gauch H G Jr (1982) *Multivariate Analysis in Community Ecology*. (Cambridge Studies in Ecology). Cambridge University Press, 298 pp.

Gray J S, Aschan M, Carr M R, Clarke K R, Green R H, Pearson T H, Rosenberg R, Warwick R M (1988) Analysis of community attributes of the benthic macrofauna of Frierfjord/Langesundfjord and in a mesocosm experiment. *Mar Ecol Prog Ser* **46**: 151–65.

Green K (1986) Food of the Cape Pigeon (*Daption capense*) from Princess Elizabeth Land, East Antarctica. *Notornis* **33**: 151–4.

Green K, Johnstone G W (1988) Changes in the diet of Adelie penguins breeding in East Antarctica. *Aust Wildl Res* **15**: 103–10.

Hart T J (1942) Phytoplankton periodicity in Antarctic surface waters. *Discovery Rep* **21**: 261–356.

Higginbottom I, Hosie G W (1989) A super swarm of krill in Prydz Bay, Antarctica. *Mar Ecol Prog Ser* **58**: 197–203.

Higginbottom I R, Kerry W R, Wayte S E (1988) The distribution and abundance of krill in the region of Prydz Bay, Antarctica. A report of hydroacoustic surveys carried out on the research cruises FIBEX, ADBEX-II, and SIBEX-II. *ANARE Res Notes* **62**: 1–46.

Hodgkinson R P, Colman R S, Kerry K R, Robb M S (1988) Water currents in Prydz Bay, Antarctica in 1985. *ANARE Res Notes* **59**: 1–127.

Hodgkinson R P, Colman R S, Robb M, Williams R (1991*a*) Current meter moorings in the region of Prydz Bay, Antarctica, 1986. *ANARE Res Notes* **81**: 1–130.

Hodgkinson R P, Colman R S, Robb M, Williams R (1991*b*) Current meter moorings in the region of Prydz Bay, Antarctica, 1987. *ANARE Res Notes* **82**: 1–68.

Hosie G W (1991) Distribution and abundance of euphausiid larvae in the Prydz Bay region, Antarctica. *Antarct Sci* **3**: 167–80.

Hosie G W, Ikeda T, Stolp M (1988) Distribution, abundance and population structure of the Antarctic krill (*Euphausia superba* Dana) in the Prydz Bay region, Antarctica. *Polar Biol* **8**: 213–24.

Hosie G W, Kirkwood J M (1986) Euphausiid larvae collected from the Prydz Bay region during the *Nella Dan* cruise (SIBEX I). *Mem Natl Inst Polar Res, Spec Issue* **40**: 110–16.

Hosie G, Kitchener J, Stolp M, Cochran T G, Williams R (1991) AAMBER 86/87 Cruise Krill/Zooplankton Sampling Data. *ANARE Res Notes* **79**: 1–54.

Hosie G W, Stolp M (1989) Krill and zooplankton in the western Prydz Bay region, September–November 1985. *Proc NIPR Symp Polar Biol* **2**: 34–45.

Hubold G, Hempel I, Meyer M (1988) Zooplankton communities in the southern Weddell Sea. *Polar Biol* **8**: 225–33.

Huntley M E, Sykes P F, Marin V (1989) Biometry and trophodynamics of *Salpa thompsoni* Foxton (Tunicata: Thaliacea) near the Antarctic Peninsula in Austral summer, 1983–1984. *Polar Biol* **10**: 59–70.

Ichii T (1990) Distribution of Antarctic krill concentrations exploited by Japanese krill trawlers and minke whales. *Proc NIPR Symp Polar Biol* **3**: 36–56.

Ikeda T, Hosie G W, Stolp M (1986) SIBEX II cruise krill/zooplankton data. *ANARE Res Notes* **32**: 1–70.

Jacka T H (1983) A computer data base for Antarctic sea ice extent. *ANARE Res Notes* **13**: 1–54.

John D D (1936) The southern species of the genus *Euphausia*. *Discovery Rep* **14**: 193–324.

Jongman R H G, Braak C J F ter, Tongeren O F R van (1987) *Data Analysis in Community and Landscape Ecology*. Pudoc, Wageningen, 299 pp.

Kane J E (1966) The distribution of *Parathemisto gaudichaudii* (Geur) with observations on its life history in the 0° to 20°E sector of the Southern Ocean. *Discovery Rep* **34**: 163–98.

Kawamura A (1986) Have marine Antarctic ecosystems changed? – A tentative comparison of present and past macrozooplankton abundances. *Mem Natl Inst Polar Res, Spec Issue* **40**: 197–211.

Kawamura A (1987) Two series of macrozooplankton catches with the N70V net in the Indian Ocean sector of the Antarctic Ocean. *Proc NIPR Symp Polar Biol* **1**: 84–9.

Kerry K R, Woehler E J (1987) Oceanographic data: Prydz Bay region – Antarctic Division BIOMASS Experiment I, MV *Nella Dan*, November–December 1982. *ANARE Res Notes* **52**: 1–166.

Kerry K R, Woehler E J, Robb M S (1987*a*) Oceanographic data: Prydz Bay region – SIBEX II, MV *Nella Dan*, January 1985. *ANARE Res Notes* **53**: 1–122.

Kerry K R, Woehler E J, Wright S, Dong Z (1987*b*) Oceanographic data: Prydz Bay region – FIBEX, MV *Nella Dan*, January–March 1981. *ANARE Res Notes* **49**: 1–114.

Khimitsa V A (1976) Investigations of geostrophic currents in the Antarctic Zone of the Indian Ocean. *Oceanology* **16**: 131–3.

Kittel W, Witek Z, Czykieta H (1985) Distribution of *Euphausia frigida*, *Euphausia crystallorophias*, *Euphausia triacantha*, and *Thysanoessa macrura* in the southern part of Drake Passage and in the Bransfield Strait during the 1983–1984 austral summer (BIOMASS-SIBEX). *Pol Polar Res* **6**: 133–49.

Kruskal J B, Wish M (1978) *Multidimensional Scaling*. Sage Publications, Beverly Hills, 93 pp.

Madin L P, Harrison G R (1977) The association of Amphipoda Hyperiidae with gelatinous zooplankton – I. Association with Salpidae. *Deep-Sea Res* **24**: 449–63.

Mauchline J, Fisher L R (1969) The biology of euphausiids. *Adv Mar Biol* **7**: 1–454.

Middleton J H, Humphries S E (1989) Thermohaline structure and mixing in the region of Prydz Bay, Antarctica. *Deep-Sea Res* **36**: 1255–66.

Montague T L (1984) The food of Antarctic Petrels (*Thalassoica antarctica*). *Emu* **84**: 244–5.

Nicol S (1989) Who's counting on krill? *New Scient* **124**: 20–3.

Oresland V (1990) Feeding and predation impact of the chaetognath *Eukrohnia hamata* in Gerlache Strait, Antarctica Peninsula. *Mar Ecol Prog Ser* **63**: 201–9.

O'Sullivan D (1982) A guide to the chaetognaths of the Southern Ocean and adjacent waters. *ANARE Res Notes* **2**: 1–57.

Pakhomov E A (1989) (Macroplankton distributions in the central part of the Indian Southern Ocean during the summer seasons of 1984–1986.) In: *Antarktika. Dokladi Komissii. Mezhduvedomstvennaya Komissiya po Izucheniyoo Antarktiki. Vipusk 28.* (The Antarctic. The Committee Reports. Soviet Committee of Antarctic Research No. 28.) Nauka Publishers, Moscow, pp. 145–57. (In Russian.)

Piatkowski U (1989*a*) Macroplankton communities in Antarctic surface waters: spatial changes related to hydrography. *Mar Ecol Prog Ser* **55**: 251–9.

Piatkowski U (1989*b*) Macrozooplankton communities from Weddell Sea surface waters, Antarctica. *Pesq Antárt Bras* **1**: 1–10.

Pommeranz T, Hermann C, Kühn A (1982) Mouth angle of the Rectangular Midwater Trawl (RMT1+8) during paying out and hauling. *Meeresforschung* **29**: 267–74.

Puddicombe R A, Johnstone G W (1988) The breeding season diet of Adelie penguins at the Vestfold Hills, East Antarctica. *Hydrobiologia* **165**: 239–53.

Roe H S J, Baker A de C, Carson R M, Wild R, Shale D M (1980) Behaviour of the Institute of Oceanographic Science's rectangular midwater trawls: theoretical aspects and experimental observations. *Mar Biol* **56**: 247–59.

Siegel V, Piatkowski U (1990) Variability in the macrozooplankton community off the Antarctic Peninsula. *Polar Biol* **10**: 373–86.

Smetacek V, Scharek R, Nöthig E-M (1990) Seasonal and regional variation in the pelagial and its relationship to the life history cycle of krill. In: Kerry K R, Hempel G (eds) *Antarctic Ecosystems. Ecological Change and Conservation.* Springer-Verlag, Berlin, pp. 103–14.

Smith S L, Schnack-Schiel S B (1990) Polar Zooplankton. In: Smith W O Jr (ed) *Polar Oceanography*, Part B: *Chemistry, Biology, and Geology*. Academic Press, San Diego, pp. 527–98.

Smith R W, Bernstein B B, Cimberg R L (1988) Community-environmental relationships in the benthos: Applications of multivariate analytical techniques. In: Soule D F, Kleppel G S (eds) *Marine Organisms as Indicators*. Springer-Verlag, New York, pp. 247–326.

Smith N R, Dong Z, Kerry K R, Wright S (1984) Water masses and circulation in the region of Prydz Bay, Antarctica. *Deep-Sea Res* **31**: 1121–47.

Steedman H F (1976) Zooplankton fixation and preservation. *Monogr Oceanogr Methodol* **4**: 1–359.

Tanimura A, Fukuchi M, Hoshiai T (1986) Seasonal change in the abundance of zooplankton and species composition of copepods in the ice-covered sea near Syowa station, Antarctica. *Mem Natl Inst Polar Res, Spec Issue* **40**: 212–20.

Tchernia P, Jeannin P F (1983) *Quelques Aspects de la Circulation Océanique Antarctique Révélés par l'Observation de la Dérivé d'Icebergs (1972–1983).* Muséum National d'Histoire Naturelle, CNRS, Paris, 474 pp.

Terazaki M (1989) Distribution of chaetognaths in the Australian sector of the Southern Ocean during the BIOMASS SIBEX cruise (KH-83–4). *Proc NIPR Symp Polar Biol* **2**: 51–60.

Terazaki M, Wada M (1986) Euphausiids collected from the Australian sector of the Southern Ocean during the BIOMASS SIBEX cruise (KH-83–4). *Mem Natl Inst Polar Res, Spec Issue* **40**: 97–109.

Vervoort W (1951) Plankton copepods from the Atlantic sector of the Antarctic. *K Ned Akad Wet (Versl Gewone Vergad Afd Natuurkd Dl)* **48**: 1–156.

Williams R (1985) Trophic relationships between pelagic fish and euphausiids in Antarctic waters. In: Siegfried W R, Condy P R, Laws R M (eds) *Antarctic Nutrient Cycles and Food Webs*. Springer-Verlag, Berlin, pp. 452–9.

Williams R (1989) Australian research on Antarctic bird and seal diets. *Selected Papers Presented to the Scientific Committee of CCAMLR 1988*, SC-CAMLR-SSP/5. CCAMLR, Hobart, Part II, pp. 231–49.

Williams R, Kirkwood J M, O'Sullivan D B (1983) FIBEX cruise zooplankton data. *ANARE Res Notes* **7**: 1–163.

Williams R, Kirkwood J M, O'Sullivan D B (1986) ADBEX I cruise zooplankton data. *ANARE Res Notes* **31**: 1–108.

Woehler E J, Williams R (1988) Oceanographic data: Prydz Bay region – Australian Antarctic Marine Biological Ecosystem Research 86/87 MV *Nella Dan*, February–April 1987. *ANARE Res Notes* **63**: 1–136.

Wright S W (1987) Phytoplankton pigment data: Prydz Bay region – SIBEX II, MV *Nella Dan*, January 1985. *ANARE Res Notes* **58**: 1–102.

Wright S W, Shearer J D (1984) Rapid extraction and high-performance liquid chromatography of chlorophylls and carotenoids from marine phytoplankton. *J Chromatogr* **294**: 281–95.

Yamada S, Kawamura A (1986) Some characteristics of the zooplankton distribution in the Prydz Bay region of the Indian Ocean sector of the Antarctic Ocean in the summer of 1983/84. *Mem Natl Inst Polar Res, Spec Issue* **44**: 86–95.

Zar J H (1984) *Biostatistical Analysis*, 2nd edn. Prentice-Hall, Englewood Cliffs, New Jersey, 718 pp.

Zmijewska M I (1983) Copepoda (Calanoida) from Prydz Bay (Antarctica, Indian Ocean Sector). *Pol Polar Res* **4**: 33–47.

Discussant's report: primary production in the Antarctic pelagial – a view from the North

E. SAKSHAUG

IN THE PAST TWO DECADES knowledge of the Antarctic ecosystems has increased considerably. The BIOMASS Programme has been instrumental in shaping the present picture of the distribution of phytoplankton and krill biomass in the Antarctic Ocean. Prior to the initiation of BIOMASS the picture of the Antarctic Ocean was one of bountiful harvestable marine resources, i.e. krill, seals and whales. Today's picture is one which is characterized by low overall productivity, with high productivity being patchy and confined to coastal regions (El-Sayed 1984; Smith & Sakshaug, 1990; Smith & Schnack-Schiel, 1990). The three chapters in this section verify this picture.

When BIOMASS was concluded, the ecological research programmes ISHTAR (Bering Sea; Walsh *et al.*, 1989) and Pro Mare (Barents Sea; Sakshaug *et al.*, 1991) were also winding up. Here we will discuss the Antarctic data in the light of recent knowledge about the Barents Sea.

In the Antarctic Ocean, defined as the region south of the oceanic Polar Front (about 35 million km^2), the ice-free zone constitutes 40–85%, whereas the seasonally ice-covered zone occupies 10–45% of the area, depending on season. Thus, during most of the growth season, considerably more than half of the Antarctic Ocean is ice-free. This zone, except on shelves, is represented by HNLC (High-Nutrient, Low-Chlorophyll) waters. Such waters are not found in the Barents Sea.

The Antarctic HNLC waters may be among the least productive ice-free waters anywhere: figures for primary productivity are of the order of 0.14 g C d^{-1} or 16 g C m^{-2} yr^{-1} (Holm-Hansen *et al.*, 1977). This looks paradoxical considering that macronutrient (nitrate, phosphate, silicate) concentrations are the highest known (El-Sayed, 1987; Sakshaug, 1989). It follows from the chapter by Jacques & Fukuchi (this volume) that virtually all of the production in HNLC waters is regenerative. In contrast, where sea-ice is melting, daily productivity may be as high as 0.5–1 g C d^{-1} (Smith & Nelson, 1986), or 60–120 g C m^{-2} yr^{-1}. Thus the zone of melting ice and shelf waters may be responsible for about 40% (i.e. about 300 Mt C) of the total primary productivity of the Antarctic Ocean. This may include virtually all the new production.

In the Barents Sea, which is divided by the oceanic Polar Front into a northern half with polar surface waters and seasonal ice, and a southern, permanently ice-free half with Atlantic surface waters, the overall primary productivity is about five times higher than in the Antarctic Ocean, i.e. about 110 g C m^{-2} yr^{-1} compared with 20–30 g C m^{-2} yr^{-1}. This difference is explained by the large area of HNLC waters in the Antarctic Ocean. In the Barents Sea, in contrast, the permanently ice-free waters are more productive, about 160 g C m^{-2} yr^{-1} (of which about 55% is 'new'), compared with 60–90 mg C m^{-2} yr^{-1} in the northern half (Sakshaug & Slagstad, 1992). The main reason is that wind (regular passages of atmospheric low pressures) causes a depth of mixing of 40–50 m in the southern half, in contrast to a persistent pycnocline at 20–35 m in the northern half. The melting of seasonal sea-ice in the northern half in the main concentrates a lower overall productivity into a narrow zone along the ice edge.

The vexing question with respect to the Antarctic Ocean thus is: which factor is controlling phytoplankton biomass and production in HNLC waters? Three main hypotheses have been advanced: iron-limitation, light-limitation (deep vertical mixing) and high grazing pressure, none of which should be considered as entirely independent. The iron limitation hypothesis, which has generated much controversy (Martin *et al.*, 1990), was put forward at about the time when BIOMASS field exercises were concluded and was therefore not tackled by BIOMASS investigators. This question, however, has not yet been solved, and it will not be discussed here; the interested reader is referred to the review by Cullen (1991).

Thanks to BIOMASS, Antarctic krill stock estimates are fairly good. If we assume a stock of 100 Mt (fresh mass) and a distribution confined to seasonally ice-covered waters (set at 20 million km^2), the average krill biomass, in terms of carbon, is about 0.15 g m^{-2}. For the whole Barents Sea, however, the figure is twice as high (Pedersen, 1992). Moreover, the average biomass of *Calanus* populations in the Barents Sea is about five times the average krill biomass, i.e. about 1.5 g C m^{-2} (Tande, 1991). Considering that the average primary productivity in Antarctic seasonally ice-covered waters may not be very different from the average for the Barents Sea, one might expect about the same biomass of herbivores in both areas, assuming that the grazing efficiency is not very different. If so, the biomass of herbivorous zooplankton, other than krill, in the Antarctic seasonally ice-covered waters should be at least five times larger than the overall average krill stock. The two zooplankton reviews presented in this section here, which succinctly sum up the state of knowledge at the conclusion of BIOMASS, clearly demonstrate the importance of copepods in Antarctic waters. However, studies of zooplankton, other than krill, to a large extent were not given their rightful place in the BIOMASS Programme because of the heavy emphasis of the BIOMASS planners on krill investigations.

While it is praiseworthy that a synthesis has been made of the plankton in the little-known Prydz Bay region, an effort to synthesize phytoplankton–zooplankton–krill interactions in the Atlantic and Indian sectors is sorely missed. Considering the wealth of data available at the BIOMASS Data Centre, it is to be hoped that someone will soon undertake such a synthesis.

References

Cullen J J (1991) Hypotheses to explain high-nutrient concentrations in the open sea. *Limnol Oceanogr* **36**: 1578–99.

El-Sayed S Z (1984) Productivity of Antarctic waters. A reappraisal. In: Holm-Hansen O, Bolis L, Gilles R (eds) *Marine Phytoplankton and Productivity*. Springer-Verlag, Berlin, pp. 19–34.

El-Sayed S Z (1987) Biological productivity of the Antarctic waters. In: El-Sayed S Z, Tomo A P (eds) *Regional Symposium on Recent Advances in Antarctic Aquatic Biology with Special References to the Antarctic Peninsula* (BIOMASS Sci Ser 7). SCAR Publications, Cambridge, pp. 1–21.

Holm-Hansen O, El-Sayed S Z, Franceschini G A, Cuhel R L (1977) Primary production and the factors controlling phytoplankton growth in the Southern Ocean. In: Llano G A (ed) *Adaptations Within Antarctic Ecosystems*. Gulf, Houston, pp. 11–50.

Martin J H, Gordon R M, Fitzwater S E (1990) Iron in Antarctic waters. *Nature* **345**: 156–8.

Pedersen O P (1992) Modeling and simulation of vertical carbon flux in the Barents Sea. M.Eng. thesis, University of Trondheim. (In Norwegian.)

Sakshaug E (1989) The physiological ecology of polar phytoplankton. In: Alexander V, Rey L (eds) *Proc 6th Conf Comité Arctique International*. Brill, Leiden, pp. 61–89.

Sakshaug E, Hopkins C C E, Øritsland N A (eds) (1991) *Proc Pro Mare Symp Polar Marine Ecology. Polar Res* **10**: 1–662.

Sakshaug E, Slagstad D (1992) Sea ice and wind: effects on primary productivity in the Barents Sea. *Atmos Ocean* **30**: 579–91.

Smith W O Jr, Nelson D M (1986) Importance of ice edge phytoplankton production in the Southern Ocean. *BioScience* **36**: 251–7.

Smith W O Jr, Sakshaug E (1990) Polar phytoplankton. In: Smith W O Jr (ed) *Polar Oceanography*. Academic Press, New York, pp. 477–525.

Smith S L, Schnack-Schiel S B (1990) Polar zooplankton. In: Smith W O Jr (ed) *Polar Oceanography*. Academic Press, New York, pp. 527–98.

Tande K S (1991) *Calanus* in North Norwegian fjords and in the Barents Sea. *Polar Res* **10**: 389–408.

Walsh J J, McRoy C P, Blackburn T H, Coachman L K, Goering J J, Henriksen K, Andersen P, Nihoul J, Parker P L, Springer A M, Tripp R B, Whitledge T E, Wirick C D (1989) The role of Bering Strait in the carbon/nitrogen fluxes of polar marine ecosystems. In: Alexander V, Rey L (eds) *Proc 6th Conf Comité Arctique International*. Brill, Leiden, pp. 90–120.

Antarctic marine stocks: krill

Krill mesoscale distribution and abundance: results and implications of research during the BIOMASS Programme

I. EVERSON & D. G. M. MILLER

Introduction

IN 1976, when the BIOMASS Programme was founded, it was recognized that there was a pressing need for intensive study of the Southern Ocean, paying particular attention to krill, *Euphausia superba*. A long list of research programmes was identified with the overall aims of gaining a better understanding of the fundamental biology of krill, the extent of the interdependence of higher trophic levels on krill stocks, and the interrelations between krill swarming and the hydrographic features of Antarctic waters (Anon, 1977).

This paper considers the questions of mesoscale distribution of krill and the applications of acoustics to determining distribution and abundance. These topics were highlighted as warranting specific research effort when the BIOMASS Programme began (Anon, 1977). We discuss the progress that has been made in these areas in the light of the aims of BIOMASS (El-Sayed, this volume) and their relevance and application to the Commission for the Conservation of Antarctic Marine Living Resources (CCAMLR), an organization that can be fairly judged to have its roots in BIOMASS.

In this paper we begin at the largest spatial scale and discuss the question of stock separation; we then consider progress that has been made in estimating standing stock, and finally consider the commercial fishery for krill in the Southern Ocean. For discussion of the smaller-scale distribution of krill and particularly its aggregation dynamics, the reader is referred to Siegel & Kalinowski (this volume).

Stock separation

From observed distributional features, various inferences have been drawn as to whether local krill concentrations are essentially isolated from one another or whether substantial intermixing occurs. Such information, if definitive, has profound implications for management since it will ultimately facilitate the determination of the extent to which local and/or regional krill concentrations may be treated as separate 'stocks' for management purposes (i.e. as 'management units'; Butterworth, 1984).

Prior to the implementation of BIOMASS, there was considerable speculation, based on distributional information, that supported the idea that certain high-density regions might contain self-maintaining krill stocks (Mackintosh, 1973). Makarov (1973) even suggested the existence of distinct races of krill. More recently, Lubimova *et al.* (1982) considered that krill in the Indian and Pacific sectors are concentrated within closed circulatory systems. From this they postulated the existence of isolated, self-maintaining populations in most, if not all, such systems. In these terms, krill in each of the high-density regions identified by Lubimova *et al.*

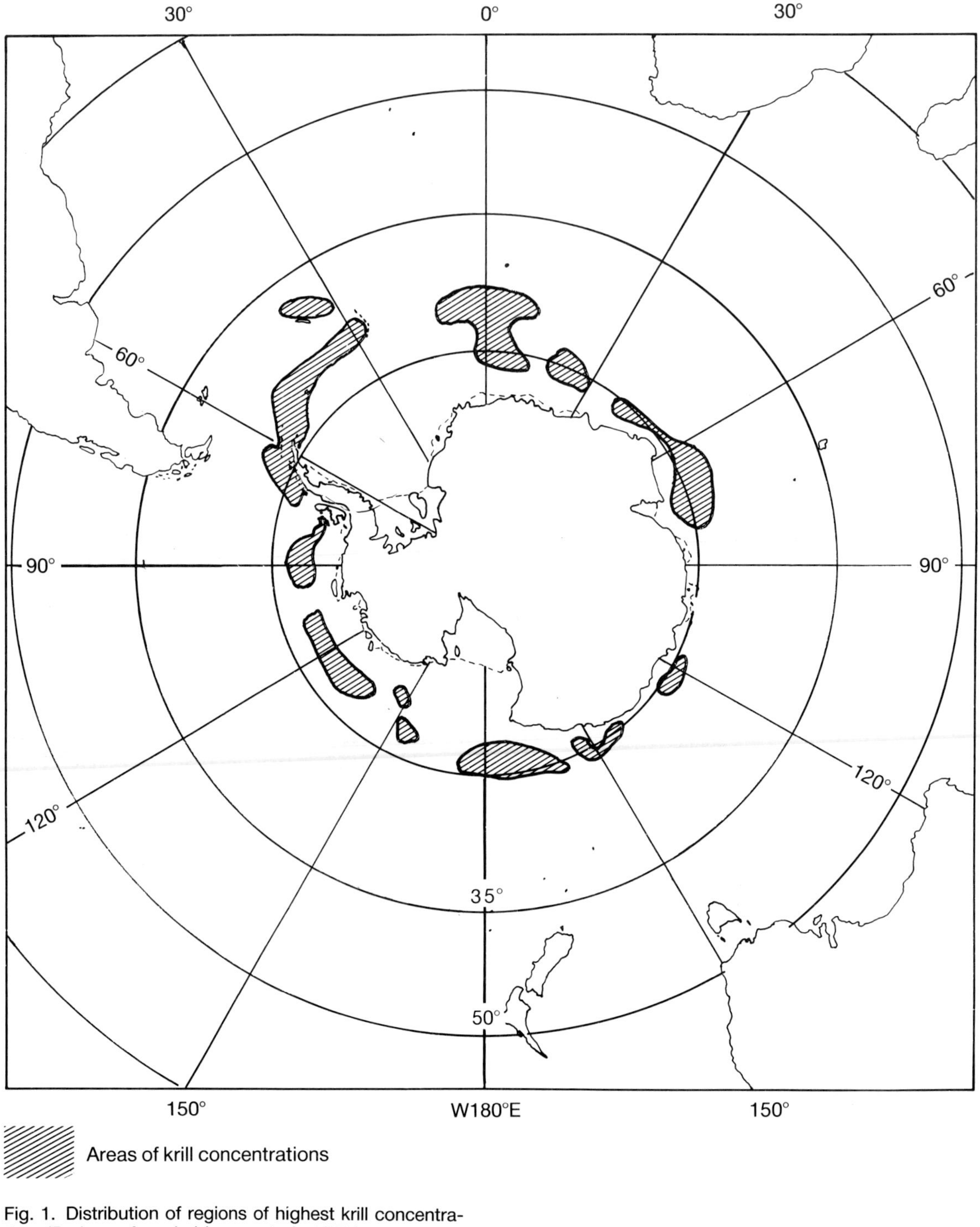

Fig. 1. Distribution of regions of highest krill concentration. (Redrawn from Lubimova *et al.*, 1982.)

(Fig. 1) may be considered as belonging to separate stocks. In support of their argument, Lubimova *et al.* (1982) noted that certain biological characteristics (size, age structure, sex ratio, etc.) were homogeneous within the regions and heterogeneous between them. In contrast, biological characteristics of krill in both the West Atlantic and West Indian sectors were on the whole very variable; from this, Lubimova *et al.* (1982) concluded that one self-supporting population (which they referred to as the 'Weddell Sea population') exists in these two regions.

The hypothesis of Lubimova *et al.* (1982) rests heavily on the strength, scale and persistence of the proposed closed circulatory systems. The general lack of common zonal distribution features in long-term records from the Indian and Pacific Sectors, such as is shown by the Japanese commercial krill catches in Fig. 2, and the fact that water circulation is probably largely influenced by atmospheric conditions (Miller & Hampton, 1989*a*) raises doubts that any such circulatory features are sufficiently persistent, either temporally or spatially, to result in permanent stock separation. General mobility and possible passive mixing of krill under the ice during winter would be expected to break down any stock isolation further.

Siegel (1986) attempted to separate krill populations found around the Antarctic Peninsula through an analysis of morphometric variability. On the basis of seven morphometric measurements per individual animal on 995 specimens, he was unable to distinguish any differences between krill from the north Bellingshausen Sea, the Drake Passage, the Bransfield Strait, and the northern and southern Weddell Sea. Similarly, Fevolden & George (1984) were unable to track the derivations of krill stocks west of the Peninsula using length distribution information alone.

Because of the many inherent difficulties associated with morphometric studies (particularly as related to differences in individual krill growth characteristics; Miller & Hampton 1989*a*), electrophoretic analysis of variations in the structure of enzymic protein has emerged as a favoured alternative to the problem of separating krill stocks. Early electrophoretic analyses (Ayala *et al.*, 1975; Fevolden & Ayala, 1981) suggested the existence of at least two discrete krill populations in the Antarctic Peninsula region. Although differences in average heterozygosity and allele frequencies were small, taken together with estimates of genetic distance they did indicate separation between animals in the Bellingshausen and Weddell–Scotia Seas. Unfortunately,

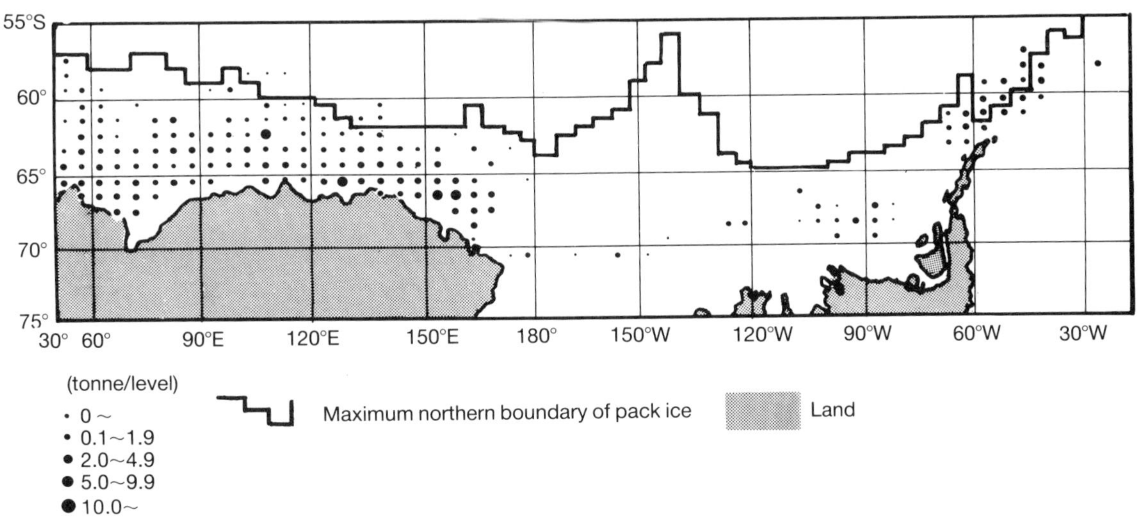

Fig. 2. Krill densities derived from catch per haul data from the Japanese commercial krill fishery during the period from 1973–4 to 1982–3.

sample sizes were small; this, together with the complex oceanographic conditions prevailing in the region (cf. Stein, 1982), casts some doubt on the validity of the final conclusions (Schneppenheim & MacDonald, 1984).

Subsequently, various workers associated with BIOMASS (MacDonald & Schneppenheim, 1983; Schneppenheim & MacDonald, 1984; Fevolden, 1986; Kuhl & Schneppenheim, 1986; MacDonald *et al.*, 1986; Fevolden & Schneppenheim, 1988, 1989) have expanded these original studies. Samples of krill were collected from locations in the Weddell Sea, the Scotia Sea, the Ross Sea around the Antarctic Peninsula, and Prydz Bay. Analyses of allele distributions indicate that such samples were all from a single breeding population. Similarly, W Grant & D Miller (unpublished data) found no detectable allelic-frequency differences over a wide geographical spread (90°W to 160°E) of samples collected during SIBEX.

On the other hand, Fevolden (1984) has demonstrated that there may be significant correlations between specific environmental variables and the heterozygosity of certain individual krill enzymes. In other words, physical barriers may give rise to particular enzymic adaptations, in turn inducing genetic subdivision of populations. Fevolden (1984) urges that future investigations attempt to determine whether environmentally induced differences can withstand pronounced gene flow between populations of krill (on both a macro- and a micro-scale) thereby allowing for the detection of allelic differences between such populations.

Grant (1983) has calculated that a very small number of migrant krill would be sufficient to account for the low levels of genetic divergence observed in early studies (e.g. Fevolden & Ayala, 1981). On this basis, it seems unlikely that either habitat or clinal divergence plays a significant role in separating krill populations. Grant (1983) thus concludes that if genetic subdivisions do indeed occur, they are probably the result of adaptive divergence (as implied by Fevolden (1984)) following the imposition of some physical barrier or barriers to migration. These barriers may no longer exist, and such genetic heterogeneity as has been observed to date could represent genetic convergence through reassociation rather than divergence from an original stock. Similarly, in a recent review of information on krill's biochemical genetics and population structure, Fevolden (1988) concludes that in the majority of studies Hardy–Weinberg expectations are induced by environmental selection rather than being the product of isolated self-maintaining stocks. The fact that significant deviations from Hardy–Weinberg expectations have been reported in only a few cases (e.g. Fevolden, 1986) is therefore very strong evidence that the krill so far analysed all belong to the same breeding unit and that substantial gene flow has prevented general genetic differentiation of the species.

It must be concluded, therefore, that despite the progress resulting either directly or indirectly from BIOMASS the identification of separate krill stocks (by genetic or any other means) has not been clearly demonstrated. More work is required; possible alternative procedures should be attempted before any of the approaches carried out to date can be rejected or accepted with confidence. For the time being, therefore, it seems best to treat stock separation as related to krill as an open question and one which BIOMASS has not really answered. For this reason, as has been suggested by Butterworth (1984), a practical set of arbitrary stock boundaries will probably have to be accepted as an initial precautionary measure to spread the krill harvest, thereby reducing the possibility of overexploiting potentially localized stocks. Such an approach is totally congruent with recent thinking in the genetics of fisheries management, as indicated by Gauldie (1991), which argues in favour of migration-linked stocks exhibiting heterogeneous growth rates.

Distribution and standing stock estimation

Acoustic techniques

Background

For many years echosounders have been used to detect fish; at the time of the formation of BIOMASS, moves were well advanced to use acoustic data quantitatively to estimate abundance. At that time the main emphasis was towards pelagic fish species such as cod (*Gadus morhua*), herring (*Clupea harengus*) and anchovy (Engraulidae), all of which give strong echoes because they possess swimbladders (Forbes & Nakken, 1972). Zooplankton, which were frequently noted on echo charts, were generally treated as 'noise'. BIOMASS recognized that there was useful information in this 'noise' and provided a springboard which encouraged many groups to investigate it with a view to using acoustics as a tool for studying krill distribution and abundance.

There are two key factors to consider when assessing acoustics for quantitative work (Clay & Medwin, 1977): firstly, the size of the object under consideration; and secondly, the wavelength (λ) of the sound pulse (Urick, 1975). If the wavelength is much less than the radius of the equivalent average diameter (l) of the object then normal geometric principles apply. On the other hand if the wavelength is much larger than l then the scattering cross-section varies as the fourth power of the wavelength. Between these two types is an intermediate zone where scattering depends on the size, geometry and orientation of the scatterer.

At 38 kHz, a frequency commonly used for fish work, the wavelength is about 4 cm, equal to the mean length for many krill but, owing to the elongated shape of the animals, much larger than the radius of an equivalent surface. At 120 kHz the wavelength is 1.25 cm and at 200 kHz it is 0.75 cm, much nearer to what we might consider as an equivalent radius for a medium to large krill. At these frequencies krill are likely to fall into the category of 'geometric scatterers'. There are therefore some advantages in choosing a high frequency for studying krill. These advantages are offset, however, by the very much higher absorption of sound in water at higher frequencies. Choice of frequencies is therefore a trade-off between the operational range of the echosounder and the size of the scatterers that need to be quantified. The compromise that most workers have used for krill is 120 kHz.

Target strength

Knowledge of the target strength, the ratio of the intensities of reflected to incident sound at one metre from the target, is essential if absolute estimates of abundance are to be made using acoustic survey data. BIOMASS recognized this and highlighted it as a topic for study (Anon, 1977).

The topic may be investigated in several ways: theoretical modelling; measurements on encaged aggregations; and *in situ* measurements on swarms or on single individuals. For objects of uniform composition and regular shape the laws of physics can provide a good indication of the target strength of an object. The information that is required is the shape of the object and the difference in its acoustic properties when compared with seawater. It is for this reason that high-purity copper or tungsten carbide spheres have been used to provide calibration standards for acoustic systems (Foote *et al.*, 1987). These allow echosounders to be calibrated to an accuracy of 0.1 dB.

Unfortunately, krill are neither regular in shape nor constant in composition and hence pose problems in modelling their target strength. An early approach was to model the target strength of krill by an equivalent shape and volume of a series of small spheres and then determine the acoustic properties of such a composite structure. Johnson (1977) and Greenlaw (1977), using a fluid sphere model, estimated target strengths for euphausiids of 40 mm total length to be about −76 dB and of 27 mm length to be −82 dB. Recent work, using new measurements of velocity of sound in krill (Foote, 1990; Foote *et al.*, 1990), has provided similar values.

There are many problems in undertaking studies on encaged krill. The animals themselves, being holoplanktonic, are quite fragile; large numbers need to be used for individual experiments; and the experiments need to be conducted in localities where a good depth of high-quality oceanic water is available. These logistic difficulties have meant that few studies have been undertaken. An extensive study reported by Everson *et al.* (1990) and Foote *et al.* (1990) measured target strength at 38 and 120 kHz over a small size range of individuals. Results have also been reported by Kasatkina (1991) from studies at 136 kHz which, if transformed according to the method proposed by Greene *et al.* (1991) to provide estimated values at 120 kHz, are close to those of the other encaged experimental results. All these results are within the range of the theoretical models described above but are significantly lower than those derived during the First Post-FIBEX Workshop (Anon, 1981) from an examination of information available at that time.

Developments in echosounder technology and signal processing, particularly with the introduction of the dual beam, and, more recently, split beam systems, has made it possible to estimate target strength *in situ*. These systems rely on measuring signal strength of targets that are on the main beam axis (dual beam) or determining where an individual target is within the beam and computing its theoretical signal strength as if it were on the beam axis

Table 1. *Comparison of recent values of target strength at 120 kHz for individual krill of 40 mm total length with that used for the initial FIBEX analysis*

Target strength	Data source
−63.8	Anon, 1984*b*
−72.9	Everson *et al.*, 1988
−71.5	Anon, 1989
−71.6	Hewitt & Demer, 1991
−71.4	Kasatkina, 1991
−72.7	Watkins, 1991
−71.6	Greene *et al.*, 1991

(split beam). The method also requires that only a single target be present within a pulse length, a constraint which limits the effectiveness to densities of less than one krill per cubic metre. Using a split beam system, Hewitt & Demer (1991) describe results over a range of sizes of krill which are close to those of the encaged experiments described above.

Other direct measurements have involved partitioning the acoustic parameter mean volume backscattering strength (MVBS) by the density of krill in the sampled volume, either in a trawl net (Kasatkina, 1991), from stereophotography (Watkins, 1991) and free-swimming single krill in a large tank (T J Pauly & J D Penrose, in preparation). All these experiments yielded results close to those of the encaged studies.

These recent studies have been considered by the CCAMLR Krill Working Group and are presented in Fig. 3 and as a summary, standardized to a krill of length 40 mm measured at 120 kHz, set out in Table 1. Combining all these estimates, but pending

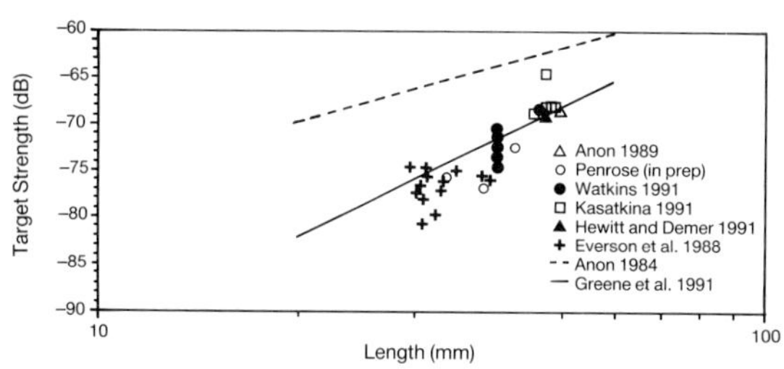

Fig. 3. Krill target strength estimates from various sources.

publication of further analysis, the Working Group recommended the following interim definition of krill target strength (TS):

$$\text{TS (dB)} = (34.85 \times \log_{10} (\text{total length in mm})) - 127.45.$$

Minimum detection level

Direct information from echosounders is derived from two sources: quantitative data on the MVBS derived from a specific location; and marks on echo charts. The latter give a signature to the type of scatterer from which an experienced analyst can derive much useful information. Taken to its extreme this would imply that clear paper on the echo chart indicates an absence of krill. We need to ask the question: is this reasonable?

Echosounder systems installed on ships have an inherent noise level; modern circuitry has served to reduce it but it is unlikely ever to be eliminated. Furthermore there is also noise generated within the sea by waves, the movement of the ship or other factors. It is normal practice to set the chart recorder of an echosounder so that, under most operating conditions, no darkening or colouring of the echo chart occurs due to these effects. While engaged on a survey this background level can be seen clearly on an echointegrator output as a consistent, but low, MVBS when the echo chart is clear. Such a background level might be, for example, −80 dB. It is impossible to state, based on the acoustic record, whether there are any krill within such a zone and yet, if there are small krill with a mean TS of −73 dB in the area and these contribute to half of the background level, this may indicate a density of one krill per ten cubic metres. Kalinowski (in Anon, 1984*b*) estimated that the minimum detection level for krill would be approximately 0.18 g m^{-2} of sea surface over the top 100 m. This value was estimated using the BIOMASS equation for target strength; hence the actual density, using the revised target-strength-to-size equation described above, is approximately ten times this value, a result close to that outlined above. As a density this may seem a trivially small amount, but averaged over the volume of ocean producing a 'no-krill' acoustic signal it could amount to many tonnes. Some consideration must therefore be given to this problem; currently, the only way that it can be addressed is by the use of nets.

Near-surface estimation

Krill are known to migrate to the surface under certain conditions. Generally, but not always, this occurs during the hours of darkness and is the period when the krill are most susceptible to predation from flying birds. In line with one of the aims of BIOMASS, to elucidate the contribution of krill to the dynamics of its main predators, one must consider its near-surface distribution.

Acoustics can be used to address the problem by two possible approaches. The first is to use conventional downward-directed echosounders with the transducer sited very close to the surface; the presence of continued swell and frequent waves makes this approach essentially impracticable. The second approach is to use an echosounder with the transducer directed up towards the surface. This technique has been used successfully (Everson & Bone, 1986) although owing to the pulse length, beam angle and sea surface irregularities there is still a 'dead zone' of at least one metre.

Application of acoustic techniques

Within the BIOMASS Programme there have been two major field studies, FIBEX and SIBEX, both of which have included large research components involving the study of the distribution and standing stock of krill.

The FIBEX study

It is no surprise, bearing in mind the wide-ranging aims of BIOMASS, that the First International BIOMASS Experiment (FIBEX) (Anon, 1980*a*,*b*) was conceived as a multidisciplinary study. This caused a few problems in attempting to schedule all the relevant sampling procedures into a limited period of time and utilizing all the available sampling vessels. This problem was at the forefront of the minds of the *ad hoc* Group on Krill Survey Design (Anon, 1980*a*,*b*). That group agreed that the primary objectives of the krill acoustic survey would be:

> to examine the methodology for future assessments of total krill abundance;
> to provide instantaneous pictures of the abundance of krill in the SW Atlantic and SW Indian sectors of the Southern Ocean.

At that time the geographical limits to krill distribution had been well established (Everson, 1977) but although krill were known to concentrate in some areas there was little information to predict where such areas might occur at the time of the study. The only information indicated a tendency for krill concentrations to occur on the shelf or shelf break.

The survey design had therefore to be flexible, to cover large areas, but to be able to sample intensively in regions of high concentration. Furthermore it had to provide some contingency for adverse weather. The smaller area coupled with the availability of a larger number of research vessels in the Atlantic sector meant that a very detailed design was possible. In the Indian Ocean sector the area actually covered on the survey was two and a half times that in the Atlantic sector with less than half the number of ships; consequently the survey design was much broader-scale and less sophisticated.

In the Atlantic sector the overall area for the study was divided into a series of blocks, small enough to be surveyed by single vessels. In some cases there was overlap between adjacent blocks but generally they were separated. Each block was to be surveyed in two phases by a series of randomly spaced parallel north–south transects. The first phase would cover the whole of the block and identify regions of high krill abundance; the second phase would be a further series of randomly spaced transects within the high krill density region found on the first phase. The number of transects allocated to each vessel was sufficient to occupy the time available fully, even if the whole region contained krill. The Scientist in Charge would regularly assess how many transects could be surveyed in the remaining time available and, in the event that the vessel lost time, transects would be omitted by random selection of those remaining unsurveyed at the time of assessment. This approach meant that, providing the vessel completed the first phase, a series of transects would have been surveyed that were independent and also representative of the whole region.

FIBEX results

In spite of the immense difficulties of bringing together ten research vessels from eleven countries and arranging a coordinated effort, the field sampling worked extremely well and is a credit to the planning group (Anon, 1980*a*) and all the scientific

Table 2. *Estimated standing stock values from the FIBEX survey*

Location or ship	Corrected standing stock (kt)	CCAMLR Subarea	Total subarea standing stock (kt)
Indian Ocean		58	25 357
Agulhas	3 428		
Marion Dufresne	45		
Kaiyo Maru	12 988		
Nella Dan	8 896		
Antarctic Peninsula		48.1	4 642
West Drake	214		
North Drake	112		
South Drake	34		
Central Bransfield	877		
East Bransfield	1 315		
Walther Herwig	2 091		
South Orkneys		48.2	8 615
Odissey (A)	6 654		
Odissey (B)	646		
Eduardo L Holmberg	1 315		
South Georgia		48.3	2 220
Odissey	2 220		

Source: Results from Anon (1984*b*), corrected to take account of revised estimates of acoustic target strength in SC-CAMLR (1991).

teams. Furthermore, within six months of the completion of the fieldwork a preliminary workshop was held at which much of the data checking, validation and preliminary analyses were undertaken (Anon, 1981). Plans for the analysis of the acoustic data were discussed and agreed (Anon, 1984*a*) leading to a further workshop meeting at Frankfurt in 1984 (Anon, 1984*b*).

Recent studies, referenced above, have demonstrated that the acoustic target strength to size relationship reported in Anon (1984*b*) is not correct. The results do provide valid indices of abundance and, if multiplied by an appropriate correction factor, will provide absolute values. The CCAMLR Krill Working Group considered the questions of krill target strength and the FIBEX standing stock estimates and used a multiplication factor of 5.62 to convert the Anon (1984*b*) results to total standing stock for the Atlantic sector. We have used the same factor here to estimate standing stock within specific survey blocks; the results are summarised in Table 2. The highest krill densities were, as expected, found in all three subareas of the Atlantic sector with the highest value at the Eastern end of the Bransfield Island. This was an area where a 'superswarm' was encountered during the study (Mathisen & Macaulay, 1983; Macaulay *et al.*, 1984); this formed the focus for much additional study. The much larger area covered in the Indian Ocean sector means that the mean density must be multiplied by a larger amount to achieve a large estimated standing stock.

In addition to these survey estimates of standing stock the FIBEX acoustic data set also provided much useful information on the distribution of krill swarms. The random spacing of the transects, while maintaining a statistical rigour for the estimates, did cause problems for other types of analyses where regular spacing of sampling effort is the normal requirement. FIBEX also gave a clear demonstration that large multiship surveys can be organized, and furthermore demonstrated that acoustic techniques can be used for the estimation of krill standing stock.

The SIBEX study

For the Second International BIOMASS Experiment (SIBEX), research was concentrated on studies of mesoscale processes rather than on synoptic estimates as had been the case with FIBEX (Anon, 1982). The core of the project was a series of cooperative field studies in three main areas. In the Atlantic sector this was the Bransfield Strait and adjacent waters, in the Indian Ocean the Eastwind Drift zone between longitudes 60°E and 80°E, while in the Pacific sector work would be concentrated in the area of the oceanic gyres around 160°E.

The programme was arranged so that a standard set of transects and sampling stations would be sampled in sequence from a series of ships over a period of two years. In practice eight ships participated in the first SIBEX season (1983–4) and ten in the second season. Coverage was concentrated in the summer months, the period when production rates were likely to undergo the greatest variation (Anon, 1990*a*). Standing stock estimates from the different surveys show considerable variation both within and between seasons although at the time of the preparation of this paper investigations relating the krill distribution to other factors have not been completed (Anon, 1990*a*).

Commercial krill fishery

Catch levels

When BIOMASS began the krill fishery was still in its infancy. Although exploratory fishing had been in progress from the late 1960s it was not until the 1977–8 season that the total catch reached 100 000 tonnes. At that time the catch was considered large in terms of a crustacean fishery but small when compared to the estimated total krill production. In those early years the Antarctic sectors of the Southern Ocean, FAO Areas 48, 58 and 88, only extended as far north as 60°S; this meant that much of the krill catch was reported to Fishery Commissions dealing with temperate or even tropical fisheries. The boundaries of these areas were changed in 1977 so as to extend as far north as the Polar Front (Everson, 1977), bringing all the catches within sensible ecological boundaries.

Coincident with the extension of the FAO Areas came the introduction of subareas to divide up the total catch by what were assumed to be the major fishing localities. Although some historical catches could be assigned to appropriate subareas, prior to the 1979–80 season the bulk of the catches could only be indicated on an area basis.

During the 1980s there was an expansion in the fishery with over half a million tonnes being reported in the 1981–2 season. This was followed by a decline two years later, probably as a result of the fleets transferring their attention to the finfish (Miller, 1989), and then an increase to around four hundred thousand tonnes for the last four seasons of the decade. The largest catches have come consistently from the Atlantic Ocean sector although large catches were reported from the Indian Ocean sector in the early 1980s. Total reported catches by area are set out in Table 3.

In the Atlantic Ocean sector the main fishing areas are in the Antarctic Peninsula (subarea 48.1), South Orkneys (48.2) and South Georgia (48.3) (Table 4). Significant catches have been reported from all three with the largest, over a quarter of a million tonnes, coming from subareas 48.2 and 48.3 (Anon, 1990*b*,*c*).

Locations of commercial fishing

The aims of BIOMASS and the requirements of CCAMLR, that krill exploitation should be controlled in such a way that the requirements of dependent species should be taken fully into account, has meant that it is not only necessary to know how much krill has been caught but also the location of capture. The provision of this more detailed information, although clearly necessary on scientific grounds, does involve releasing data that might be commercially sensitive and labour-intensive to obtain. Initially there was, therefore, some reluctance on the part of the commercial operators to make this information available (SC-CAMLR, 1984); thankfully, the situation has improved with the recent provision of some very useful data sets (e.g. Endo & Ichii, 1989; Marin *et al.*, 1991).

In the Atlantic sector data have been reported to CCAMLR by ten-day period and for rectangles half a degree of latitude by a degree of longitude (approximately 30 square miles). An analysis of these data by Everson & Goss (1991) demonstrated

Table 3. *Total reported krill catches (t) for Statistical Areas in the Southern Ocean*

Year	Statistical Area 48	58	88	Total
1973	559	0	0	59
1974	19 339	446	0	19 785
1975	41 352	2 677	0	44 029
1976	885	4 750	0	5 635
1977	75 359	12 801	3 356	91 516
1978	78 940	52 855	554	132 349
1979	266 534	63 732	2 862	333 128
1980	356 974	116 816	3 181	476 971
1981	288 868	155 030	4 234	448 132
1982	373 586	147 978	6 637	528 201
1983	138 351	79 645	10 637	228 633
1984	104 680	25 734	641	131 055
1985	180 807	5 932	4 721	191 460
1986	425 871	15 910	3 892	445 673
1987	346 505	29 557	394	376 456
1988	364 173	6 490	0	370 663
1989	394 314	217	0	394 531

Source: Data from Anon (1990*b*,*c*).

Table 4. *Annual krill catches* (t) *by subarea within the Atlantic Ocean sector*

Year	Subarea 48.1	Subarea 48.2	Subarea 48.3
1980	49 435	173 765	133 774
1981	92 859	173 765	135 252
1982	69 023	257 695	46 868
1983	531	125 605	11 480
1984	32 128	64 112	8 440
1985	11 592	123 830	45 385
1986	42 113	241 673	142 084
1987	70 515	19 779	256 206
1988	78 918	94 659	190 492
1989	105 554	82 406	206 354

Source: Data from Anon (1990*b*,*c*).

that during the winter the fishery was concentrated on the shelf around South Georgia but in the spring it moved south to the South Orkneys and Antarctic Peninsula to concentrate on the shelf areas there during the summer. This pattern appears to be consistent from year to year (Anon, 1990*c*) and indicates that the bulk of the krill catch is taken from the shelf or shelf break region.

The fine-scale data reported to CCAMLR, although providing useful information, are still the result of grouping data, a process which can easily conceal important spatial and temporal variation. Some studies have been undertaken which consider the data on a haul-by-haul basis. Ichii (1987) looked in detail at the fishing operations of a commercial trawler off Wilkes Land. The vessel found, and fished successfully, several krill patches near to the shelf break in February and March 1986. In each case the size of the concentrations was of the order of tens of kilometres and was well defined by the distribution of the fishing locations. More recently, Marin *et al.* (1991) have recorded the local movements of krill patches as identified by fishing activity to the north of Livingstone Island. Catches were made on the shelf; the patches that were being fished were moving along the coast at a speed consistent with observed local current speeds.

Catch per unit of effort as an index of abundance

Traditionally, in fisheries, catch per unit of effort (CPUE) is used to provide an index of abundance. The basic method relies on the resource being randomly distributed over a wide area, but when the target species is aggregated, the index is highly sensitive to the effort index used as well as the distribution of the targets. Expressed simply, the fishery becomes concentrated on small dense patches and the catch per haul remains high until most of the patch is caught. In spite of this, useful information has been obtained from consideration of simple catch rates.

Everson (1988) showed that catch per hour of tow was increasing from season to season even though the catch per tow had stabilized at around seven tonnes per haul. Within the data set there were several periods, particularly in the Atlantic sector, when average catch rates were more than twenty tonnes per hour. Some of these were known to be coincident with the presence of what have been described as 'superswarms'. These high catch rates may indicate that the superswarm phenomenon may have been present in most seasons during the early 1980s.

An effort index based on towing time has some merits but is of limited application, because of the tendency of krill to aggregate noted above, in providing an abundance index. Other indices of effort must therefore be considered. Everson (1977) considered the problem and concluded that searching time, without defining it, would probably provide a good index. In line with these considerations CCAMLR commissioned a study to determine the types of effort index that would be most sensitive to changes in krill abundance.

Two consultants undertook the study: Dr Butterworth considered the Japanese and Dr Mangel the USSR krill fishery. Thanks to a high degree of cooperation between these two consultants and the fishing companies in Japan and the USSR they were able to follow the activities within fleets and then to model the krill distribution in localized areas (Butterworth, 1989; Mangel, 1989). They concluded that, because each krill concentration was composed of a number of individual swarms, hauls were generally being completed within the geographical limits of a concentration. Under these circumstances catch per hour of fishing provides a reasonable index of the density within a concentration. Because fishing is focused within these concentrations, this simple index gives no information of the conditions outside the concentration.

Recent analyses of haul-by-haul data from the Chilean krill fishery by Marin *et al.* (1991) indicated that the fishing fleet followed patches of krill along the coast to the north of Livingstone Island. The rate of movement was similar to that of the coastal current in the area. Information, obtained independently of the fishery, on the distribution of patches is available from research vessel surveys (Miller & Hampton, 1989*b*) which could indicate how the aggregations that are actually fished differ from others in the region or elsewhere in the Southern Ocean. Additional information on the distribution of concentrations is likely to be available from 'scouting vessels' which, in the (former) USSR fishery, move away in search of new concentrations. Such information might be used, in conjunction with the density within the patches estimated by CPUE, to provide a composite index of abundance (SC-CAMLR, 1989). These studies, although focused on limited data sets, have indicated some potentially very profitable lines for further investigation which we hope will be followed up in the near future.

Krill fishery management

It is no coincidence that in 1977, the year following the formation of BIOMASS, the Antarctic Treaty Consultative Parties at their eighth meeting decided to seek agreement to a wide-ranging conservation convention which would deal not only with the direct effects of harvesting specific organisms, but also with the indirect effects of exploitation on other species. The outcome of this meeting and subsequent discussions was the Convention for the Conservation of Antarctic Marine Living Resources, which was completed on 20 May 1980. By 8 March 1982, eight of the original signatory states, two of whom were fishing in the Convention Area, had ratified the Convention, thus fulfilling the requirements of the Convention to enter into force. The Convention, eagerly anticipated by the BIOMASS community, entered into force on 7 April 1982, and the Commission commenced operation in July 1982. During the ten meetings of the Commission and Scientific Committee there has been much discussion over possible approaches to the management of the krill fishery.

The current annual krill catch, generally less than half a million tonnes a year, is very small when compared with even the most conservative estimates of krill annual production. However, uncertainty over the production estimates and the requirements of dependent predators means that even this modest catch may be excessive if concentrated into one small area. Against this background recent discussions within CCAMLR have raised the question of setting a precautionary limit to the krill catch, particularly in those areas where predators are likely to be affected by local krill depletion. Recognition that most of the commercial krill fishery is concentrated on the shelf or at the shelf break, in many cases close to predator breeding sites, has highlighted this need even further. At its recent meeting the CCAMLR Krill Working Group proposed a precautionary limit for the krill fishery of 1.5 million tonnes for the fishery in Area 48 (SC-CAMLR, 1991). Subsequently this recommendation has been incorporated into the first conservation measure that CCAMLR has enacted for krill (CCAMLR, 1991). While welcoming this new conservation measure we feel that this is only a first step in establishing control over the fishery. Within the Atlantic Ocean sector 1.5 million tonnes may be a safe interim level; however, if this catch were taken from waters adjacent to the major breeding grounds of land-based predators it could, even at this level, be too high. We therefore feel it is important to set limits within the overall figure but on a much smaller spatial scale. Currently the obvious spatial level to choose is the statistical subarea.

The figure of 1.5 million tonnes was determined using the results of the FIBEX survey, applying a correction for target strength and making an allowance for turnover rate (SC-CAMLR, 1991). The full report of the Post-FIBEX Workshop (Anon, 1984*b*) was not available to the Working Group, who were unable therefore to partition this total figure between subareas. Precise determination of the proportion to allocate to the subareas is a major task requiring rerunning of all the processing algorithms, but in the meantime we have made some calculations which are a reasonable reflection of the correct values.

The Krill Working Group applied a factor of 5.62, equivalent to a difference in mean target strength of 7.5 dB, to the standing stock estimate from the Atlantic sector in Anon (1984*b*) and took ten per cent of this value as the suggested precautionary limit. The Krill Working Group further recommended that division of this overall limit should be made on a subarea basis as a matter of priority. We have used exactly the same approach to determining a precautionary limit for each of the three major subareas as was used for the whole area. The estimated values, derived from Table 2, are as follows:

subarea 48.1: 464 000
subarea 48.2: 861 500
subarea 48.3: 222 000

At this stage, even though they must be considered preliminary, when viewed in conjunction with Table 4 these figures indicate clearly that, although the catch in subareas 48.1 and 48.2 is well within this proposed precautionary level, the catch in subarea 48.3 (South Georgia) is close to it and in one year at least has exceeded it. It is clear now that CCAMLR, having taken the momentous step of setting a precautionary limit to the krill fishery, must now look very closely at the level of fishing activity within each subarea.

Conclusions

From its inception the BIOMASS Programme identified many of the key topics for study with respect to broad-scale research. Many of these were the subject of intensive study and in the case of FIBEX have formed the basis for establishing a precautionary limit to the krill fishery. Current ideas on the main factors controlling the distribution of krill and its movement between adjacent areas (the so-called 'krill flux') have their origins in ideas developed as part of BIOMASS. In consequence BIOMASS can be judged a success both within the formal time frame for the programme and also in providing guidance for subsequent krill research.

Acknowledgements

The success that BIOMASS has undoubtedly achieved is due not only to a large number of scientists from many countries working together but also to a small number of scientists who have guided the programme. We would like to acknowledge the efforts put into the programme by the many scientists around the world in organizing the field work and processing and analysing the resultant data. We know a few personally; we know some by name; doubtless there are many whose names do not appear in the BIOMASS Reports; to all we say thank you for a job well done. Certain 'guiding lights' in the programme deserve special mention: Ole Mathisen, who set up and chaired the Krill Acoustic Working Group; he latterly handed over to Ian Hampton, who steered the group through the analysis of acoustic data; David Cram, whose firm and decisive management pulled together the first data analysis workshop from which subsequent workshops, organized by Ian Hampton, assisted by Holger Klindt, George Jolly and Chris Ricketts, followed. All their efforts are gratefully acknowledged. Finally we thank Cathy Goss and Christine Phillips for providing the comprehensive bibliography on which this review is based.

References

Anon (1977) *Biological Investigations of Marine Antarctic Systems and Stocks*. (ed El-Sayed S Z). BIOMASS Scientific Series, Vol 1, 79 pp.

Anon (1980*a*) BIOMASS Technical Group on Programme Implementation and Coordination. *BIOMASS Rep Ser* **13**: 11–22.

Anon (1980*b*) Report of ad hoc group on Krill Survey Design. *BIOMASS Rep Ser* **14**: 1–14.

Anon (1981) Post-FIBEX Data Interpretation Workshop. *BIOMASS Rep Ser* **20**: 1–38.

Anon (1982) Meeting of the Group of Specialists on Southern Ocean Ecosystems and their Living Resources. *BIOMASS Rep Ser* **24**: 1–63.

Anon (1984*a*) Meeting of Working Party on Acoustics to Prepare for Post-FIBEX Acoustic Workshop, 1984. *BIOMASS Rep Ser* **37**: 1–60.

Anon (1984*b*) Report on Post-FIBEX Acoustic Workshop. *BIOMASS Rep Ser* **40**: 1–106.

Anon (1989) Target strength estimation of Antarctic Krill, *Euphausia superba*, by cooperative experiments with commercial trawlers. Scientific Committee for the Conservation of Antarctic Marine Living Resources, SC-CAMLR-VIII/BG/30.

Anon (1990*a*) SIBEX Acoustic Data Validation and Analysis Workshop. *BIOMASS Rep Ser* **65**: 1–62.

Anon (1990*b*) *Statistical Bulletin*, Vol 1 (1970–1979). Commission for the Conservation of Antarctic Marine Living Resources, CCAMLR-SB/90/1.

Anon (1990*c*) *Statistical Bulletin*, Vol 2 (1980–1989). Commission for the Conservation of Antarctic Marine Living Resources, CCAMLR-SB/90/2.

Ayala F J, Valentine J W, Zumwalt G S (1975) An electrophoretic study of the Antarctic zooplankton *Euphausia superba*. *Limnol Oceanogr* **20**: 636–40.

Butterworth D S (1984) Antarctic ecosystem management. In: *Selected Papers Presented to the Scientific Committee of CCAMLR, 1982–1984*, Part II. CCAMLR, Hobart, pp. 14–41.

Butterworth D S (1989) A simulation study of krill fishing by an individual Japanese trawler. In: *Selected Papers Presented to the Scientific Committee of CCAMLR, 1988*. CCAMLR, Hobart, pp. 1–108.

CCAMLR (1991) *Report of the Tenth Meeting of the Commission for the Conservation of Antarctic Marine Living Resources*. CCAMLR-X.

Clay C S, Medwin H (1977) *Acoustical Oceanography: Principles and Applications*. John Wiley, New York. 413 pp.

Endo Y, Ichii T (1989) CPUE's body length and greenness of Antarctic krill during 1987/88 season in the fishing ground north of Livingstone Island. In: *Selected Papers Presented to the Scientific Committee of CCAMLR,1989*. CCAMLR, Hobart, pp. 229–82.

Everson I (1977) *The Living Resources of the Southern Ocean*. Southern Ocean Fisheries Survey Programme. GLO/SO/77/1.

Everson I (1988) Can we satisfactorily estimate variation in krill abundance? In: Sahrhage D (ed) *Antarctic Ocean and Resources Variability*. Springer-Verlag, Berlin, pp. 199–208.

Everson I, Bone D G (1986) Detection of krill (*Euphausia superba*) near the sea surface: Preliminary results using a towed upward-looking echo-sounder. *Br Antarct Surv Bull* **72**: 61–70.

Everson I, Bone D G, Watkins J L, Foote K G (1988) *Target Strengths of Antarctic Krill (*Euphausia superba*)*. Scientific Committee for the Conservation of Antarctic Marine Living Resources, SC-CAMLR-VII/BG/30.

Everson I, Goss C M (1991) Krill fishing activity in the Southwest Atlantic. *Antarct Sci* **3**: 351–8.

Everson I, Watkins J L, Bone D G, Foote K G (1990) Implications of a new acoustic target strength for abundance estimates of Antarctic krill. *Nature* **345**: 338–40.

Fevolden S E (1984) Biotic and physical environmental impact on genetic variation of krill. *J Crust Biol* **4** (Spec No 1): 206–23.

Fevolden S E (1986) Genetic variation of *Euphausia superba* Dana in the Antarctic Peninsula waters. *Sarsia* **71**: 169–75.

Fevolden S E (1988) Biochemical genetics and

population structure of *Euphausia superba*. *Comp Biochem Physiol* **90**B(3): 507–13.

Fevolden S E, Ayala F J (1981) Enzyme polymorphism in Antarctic krill (Euphausiacea); genetic variation between population and species. *Sarsia* **66**: 167–81.

Fevolden S E, George R Y (1984) Size frequency pattern of *Euphausia superba* in the Antarctic Peninsula waters in the austral summer of 1983. *J Crust Biol* **4** (Spec No 1): 107–22.

Fevolden S E, Schneppenheim R (1988) Genetic population structure of *Euphausia superba* Dana in the Atlantic Sector of the Southern Ocean as demonstrated by different electrophoretic techniques. *Polar Biol* **9**: 1–8.

Fevolden S E, Schneppenheim R (1989) Genetic homogeneity of krill (*Euphausia superba* Dana) in the Southern Ocean. *Polar Biol* **9**: 533–9.

Foote K G (1990) Speed of sound in *Euphausia superba*. *J Acoust Soc Am* **87**(4): 1405–8.

Foote K G, Everson I, Watkins J L, Bone D G (1990) Target strengths of Antarctic krill (*Euphausia superba*). *J Acoust Soc Am* **87**(1): 16–24.

Foote K G, Knudsen H P, Vestnes G, MacLennan D, Simmonds E J (1987) *Calibration of Acoustic Instruments for Fish Density Estimation: a Practical Guide*. International Council for the Exploration of the Sea, Copenhagen, 1987, VIII, 69 pp. (Cooperative research report No 144).

Forbes S T, Nakken O (eds) (1972) *Manual of Methods for Fisheries Resource Survey and Appraisal*, part 2, *The Use of Acoustic Instruments for Fish Detection and Abundance Estimation* (FAO Manual of Fisheries Science No. 5). 138 pp.

Gauldie R W (1991) Taking stock of genetic concepts in fisheries management. *Can J Fish Aquat Sci* **48**: 722–31.

George R Y (1984) Ontogenetic adaptations in the Antarctic krill *Euphausia superba* in relation to temperature and pressure. *J Crust Biol* **4** (Spec No 1): 252–61.

Grant W S (1983) Population genetics of krill and comparison with other marine organisms. *Ber Polarforsch* **4**: 246–66.

Greene C H, Stanton T K, Wiebe P H, McClatchie S (1991) Acoustic estimates of Antarctic krill. *Nature* **349**: 110.

Greenlaw C K (1977) Backscattering spectra of preserved zooplankton populations. *J Acoust Soc Am* **62**(1): 44–52.

Hewitt R, Demer D A (1991) Krill abundance. *Nature* **353**: 310.

Ichii T (1987) Observation of fishing operation and distributional behaviour of krill on a krill trawler off Wilkes Land during the 1985/86 season. In: *Selected Papers Presented to the Scientific Committee of CCAMLR, 1987*. CCAMBLR, Hobart, pp. 335–68.

Johnson R K (1977) Sound scattering from a fluid sphere revisited. *J Acoust Soc Am* **61**: 375–7.

Kasatkina S (1991) Target strengths of krill at 136 and 120 kHz. Commission for the Conservation of Antarctic Marine Living Resources, WG-KRILL-91/29.

Kuhl S, Schneppenheim R (1986) Electrophoretic investigations of genetic variation in two krill species, *Euphausia superba* and *E crystallorophias* (Euphausiidae). *Polar Biol* **6**: 17–23.

Lubimova T G, Makarov R R, Maslenikov V V, Shevtsow V V, Shust K V (1982) (The ecological peculiarities, stocks and role of *E superba* in the trophic structure of the Antarctic ecosystem.) (Translation of a report originally published in Russian. Trudy, VNIRO, Moscow.) In: *Selected Papers Presented to the Scientific Committee of CCAMLR, 1982–1984*. CCAMLR, Hobart, pp. 391–505.

Macaulay M C, English T S, Mathisen O A (1984) Acoustic characterization of swarms of Antarctic krill (*Euphausia superba*) from Elephant Island and Bransfield Strait. *J Crust Biol* **4** (Spec No 1): 16–44.

MacDonald C M, Schneppenheim R (1983) Breeding structure and stock identity in the Antarctic krill, *Euphausia superba*. *Ber Polarforsch* **4**: 240–5.

MacDonald C M, Williams R, Adams M (1986) Genetic variation and population structure of krill (*Euphausia superba* Dana) from the Prydz Bay region of Antarctic waters. *Polar Biol* **6**: 233–6.

Mackintosh N A (1973) Distribution of post-larval krill in the Antarctic. *Discovery Rep* **36**: 95–106.

Makarov R R (1973) Some peculiarities of reproduction of *Euphausia superba* Dana. In: *All Union Conference on Macroplankton* (Abstract of papers). VNIRO, Moscow, pp. 34–5.

Mangel M (1989) Analysis and modelling of the Soviet Southern ocean krill fleet, II: Estimating the number of concentrations and analytical justification for search data. In: *Selected Papers Presented to the Scientific Committee of CCAMLR, 1989*. CCAMLR, Hobart, pp. 283–322.

Marin V, Mujica A, Eberhard P (1991) Chilean Krill Fishery: Analysis of the 1991 Season. CCAMLR, WG-KRILL/91/39.

Mathisen O A, Macaulay M C (1983) The morphological features of a super-swarm of krill, *Euphausia superba*. *Mem Natl Inst Polar Res Spec Issue* **27**: 153–64.

Miller D G M (1989) Commercial krill fisheries in the Antarctic 1973 to 1988. In: *Selected Papers Presented to the Scientific Committee of CCAMLR, 1989*. CCAMLR, Hobart, pp. 229–82.

Miller D G M, Hampton I (1989*a*) Biology and ecology of the Antarctic krill (*Euphausia superba* Dana): a review. BIOMASS Scientific Series No 9, 166 pp.

Miller D G M, Hampton I (1989*b*) Krill aggregation characteristics: spatial distribution patterns from hydroacoustic observations. *Polar Biol* **10**(1): 125–34.

Pauly T J, Penrose J D Laboratory target strength measurements of free swimming krill. (In preparation.)

SC-CAMLR (1984) *Report of the Third Meeting of the Scientific Committee*. Commission for the

Conservation of Antarctic Marine Living Resources, SC-CAMLR-III.
SC-CAMLR (1988) *Report of the Seventh Meeting of the Scientific Committee*. Commission for the Conservation of Antarctic Marine Living Resources, SC-CAMLR-VII.
SC-CAMLR (1989) *Report of the Workshop on the Krill CPUE Simulation Study*. Commission for the Conservation of Antarctic Marine Living Resources, SC-CAMLR-VIII, Annex 4.
SC-CAMLR (1991) *Report of the Working Group on Krill*. Commission for the Conservation of Antarctic Marine Living Resources, SC-CAMLR-X, Annex 5.
Schneppenheim R, MacDonald C M (1984) Genetic variation and population structure of krill (*Euphausia superba*) in the Atlantic sector of Antarctic waters off the Antarctic Peninsula. *Polar Biol* **3**: 19–28.
Shimadzu Y (1984) A brief summary of Japanese fishing activity relating to Antarctic krill, 1972/73–1982/83. In: *Selected Papers Presented to the Scientific Committee of CCAMLR, 1982–1984*, Part 1. CCAMLR, Hobart, pp. 439–71.
Siegel V (1986) Structure and composition of the Antarctic krill stock in the Bransfield Strait (Antarctic Peninsula) during the Second International BIOMASS Experiment (SIBEX). *Arch FischWiss* **37**: 51–72.
Stein M (1982) Fischereiozeanographische Untersuchungen während FIBEX 1981. *Arch FischWiss* **33**: 35–51.
Urick R J (1975) *Principles of Underwater Sound*. McGraw Hill, New York, 384 pp.
Watkins J L (1991) Krill target strength estimated by underwater photography and acoustics. Commission for the Conservation of Antarctic Marine Living Resources, WG-KRILL-91/40.

Krill demography and small-scale processes: a review

V. SIEGEL & J. KALINOWSKI

Introduction

IN THE EARLY 1970s it became evident that there was an increasing interest in the potential exploitation of Antarctic marine living resources. SCAR (the Scientific Committee on Antarctic Research) saw the need to develop a comprehensive international research programme to expand the scientific knowledge of the open sea ecosystem, which was based mainly on the *Discovery* expeditions in the 1930s. The principal objective of the BIOMASS (Biological Investigations of Marine Antarctic Systems and Stocks) Programme was 'to gain a deeper understanding of the structure and dynamic functioning of the marine Antarctic ecosystem as a basis for the future management of potential living resources' (BIOMASS, 1977). Special emphasis was given to those elements which were believed to contribute most to the general energy flow through the system or were of priority interest for exploitation. In both cases, the krill (*Euphausia superba*) was obviously one of the major key elements.

A number of research programmes were developed by BIOMASS with the view that they should be modified with increasing knowledge over time. Some of the specific proposals we will consider in this contribution include:

to describe the spatial distribution of krill in order to determine ranges of biomass densities and the shape, extent and stability of swarms, especially diel changes;

to determine growth rate and longevity using laboratory and field experiments, and collection and analysis of length composition data for krill collected in all seasons and areas (BIOMASS, 1977).

Krill aggregations

The ability of krill (*Euphausia superba*) to aggregate in swarms is a feature which distinguishes this species from other Antarctic crustaceans, although *Euphausia crystallorophias* may also sometimes occur in swarms (Everson, 1987; O'Brien, 1987*a*). Based on the results of many national expeditions to the Antarctic in general and the results of BIOMASS in particular, we have attempted to answer the following questions:

What are the forms of aggregations?
What are the physical parameters of aggregation?
What are the causes of aggregation formation?
What is the characteristic of diurnal vertical distribution?

Forms of krill aggregations

To begin, we must define an 'aggregation', since 'aggregation' is a very general term for underdis-

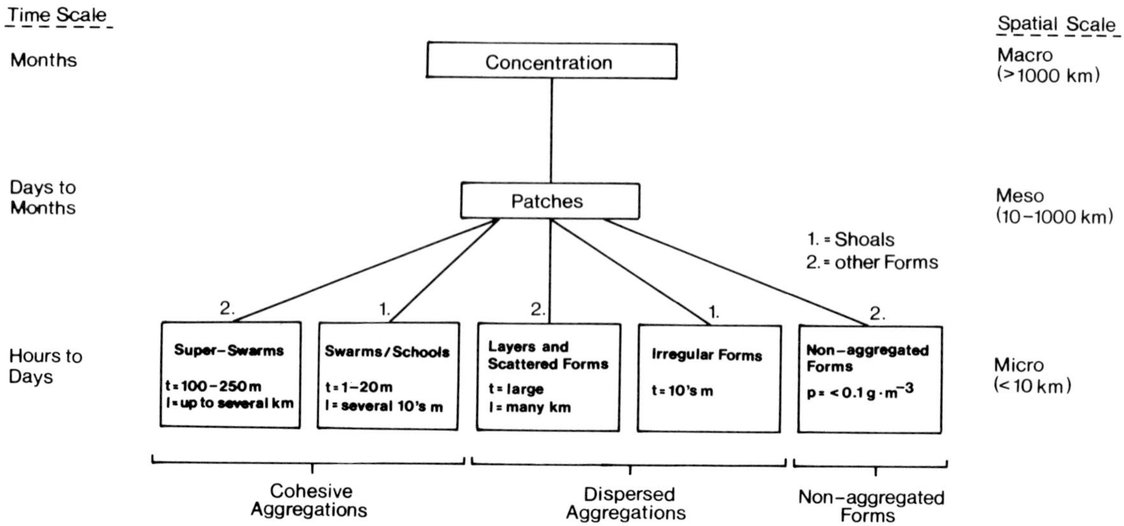

Fig. 1. Classification of krill aggregations (modified from Kalinowski & Witek (1985*a*) and Miller & Hampton (1989*a*), *t* = thickness; *l* = intercepted length; *p* = density.

persed non-random groupings. A concept of classification of euphausiid aggregations was developed by Mauchline (1980*a*) including the ranges of approximate densities. During BIOMASS this classification scheme was modified by Kalinowski & Witek (1985*a*) and Miller & Hampton (1989*a*) (Fig. 1) while Murphy *et al.* (1988) looked at scales over which interactions occur.

Aggregations are classified as 'concentrations' or 'patches' according to the scale of the underlying processes. Concentrations are macroscale features extending over distances of several 100 km, while patches are mesoscale features controlled by more local conditions. In other words, a concentration is the sum of patches distributed over a certain area; for example, during SIBEX (Second International BIOMASS Experiment) patches of krill were observed in the Bransfield Strait and the Peninsula region (Fig. 2), forming one big concentration.

Patches can be subdivided in several ways. One possibility is to follow a spatial and time scale classification which would lead to microscale aggregation (less than 10 km) superswarms, swarms, layers and irregular forms (Fig. 1). Time scale changes occur within hours to a few days. A second possible classification scheme considers the degree of internal organization. Shoals are usually constituent parts of patches and include swarms (no parallel orientation of krill) and schools (with parallel orientation). Behavioural mechanisms of the species are probably more important in maintaining shoals, rather than environmental physical parameters (Mauchline, 1980*a*). Other forms of microscale aggregations would then comprise superswarms, layers and irregular forms. Difficulties arise with the second scheme in distinguishing between swarms and schools when only indirect observation methods (e.g. hydroacoustic techniques) are available.

Physical parameters of microscale aggregations

To discuss the problems of the physical parameters of aggregations, it is necessary to take into consideration the methods used in aggregation determination. Until the 1970s, underwater observations or biological sampling were used. Since the mid-1970s hydroacoustic observations have dominated. By applying this technique it is possible to determine the aggregation's length, thickness and density (Table 1). There are considerable differences between the 'transect' and the 'aggregation' model.

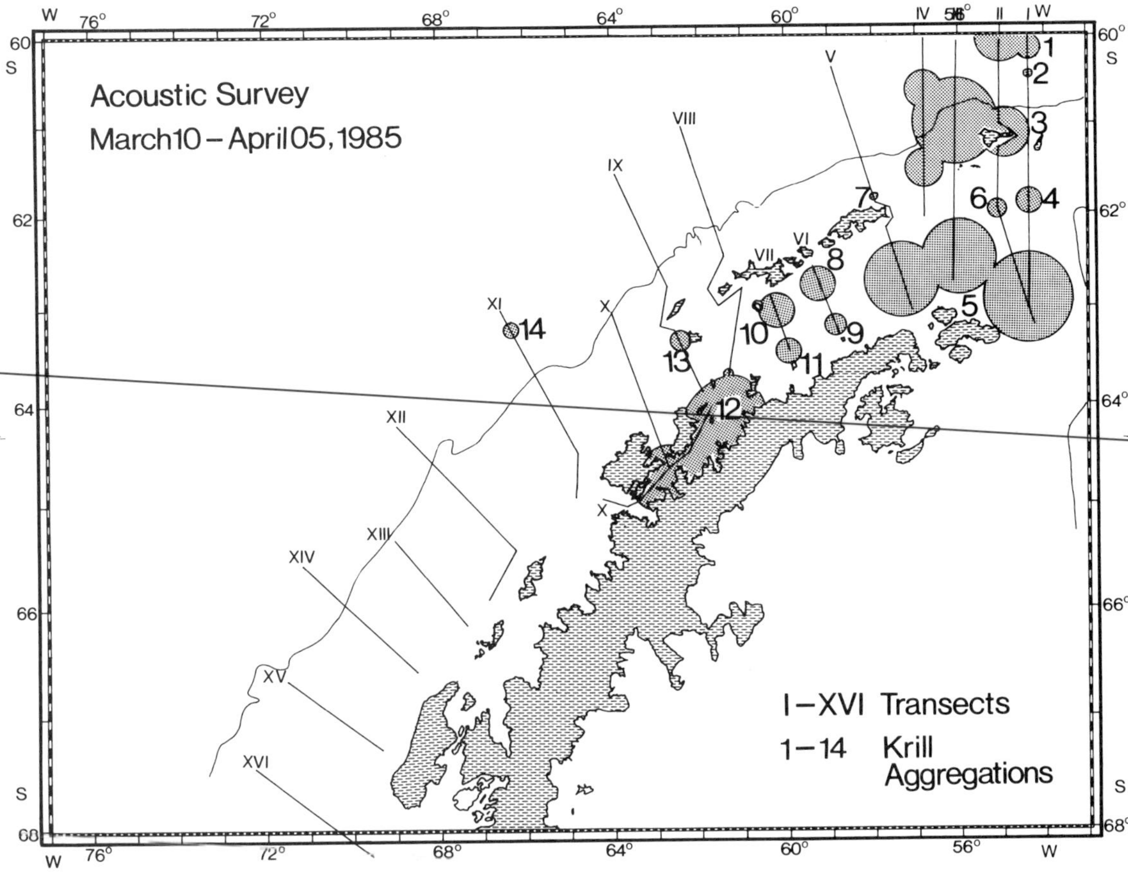

Fig. 2. Distribution of krill patches along hydroacoustic transects off the Antarctic Peninsula during SIBEX survey in March 1985.

In the first analysis the integrator intervals are defined as a transect. The aggregation model assumes that the shape of the aggregation is cylindrical and that the area of the surface equals the area of an ellipse. Although smaller swarms are generally round or oval in shape, larger swarms are often of irregular shape or ribbon-like (Marr, 1962; Nemoto *et al.*, 1981; Hamner *et al.*, 1983) depending probably on the wind direction. While measurements of linear dimensions of swarms can easily be obtained from echograms, there is still much controversy concerning the calculation of swarm size and density. During BIOMASS a formula for krill acoustic target strength was established, but recent measurements (e.g. Foote *et al.*, 1990) indicate that TS values may be much lower than this formula suggests and hence that the calculated density values should be increased.

Another problem concerns mean values given for many parameters. Kalinowski & Witek (1981) and Miller & Hampton (1989*b*) clearly demonstrate that the distribution of swarm parameters is strongly positively skewed. Parsons *et al.* (1977) have suggested that logarithmic transformation is the most convenient transformation to be applied to correct biological data of this kind. The transformed frequencies are then log-normally distributed with the geometric mean best describing average values. Both the devi-

Table 1. *Physical parameters of krill swarms*

Acoustic estimates are not always based on the BIOMASS target strength formula. Symbols: *, geometric mean; **, arithmetic mean; D, day; N, night.

Author	Length (m)	Density	Region	Method
Marr, 1962	sev. hundred	1 krill in^{-3}	Scotia Sea	biological sampling
Ragulin, 1969	25	several kg m^{-3}	South Georgia	underwater observations
Makarov *et al.*, 1970	–	up to 15 kg m^{-3}	South Georgia	biological sampling
Nemoto & Nasu, 1975	–	up to 5 kg m^{-3}		nets and acoustics
Doi & Kawakami, 1979	–	100 g m^{-3**}	60–66°S, 170–140°E	analysis of echogram, commercial trawls
Nemoto *et al.*, 1981	2–300	generally less than 1.5 kg m^{-3}	90°W to 90°E	echosurvey
Kalinowski & Witek, 1981	34, 40*	327, 223 g m^{-3**}	South Georgia	echosurvey
Witek *et al.*, 1988	27–46*	28–80 g m^{-3**}	Bellingshausen Sea Drake Passage	
Everson, 1982	–	D −74.69 dB m^{-3} N −82.53 dB m^{-3}	South Georgia	echosurvey
Everson, 1983	–	D −56.80 dB m^{-3} N −74.23 dB m^{-3}	South Georgia	echosurvey
Guzman & Marin, 1983	–	400 g m^{-3}	Bransfield Strait	echosurvey
Macaulay *et al.*, 1984	100–500	40–500 g m^{-3}	Elephant Island Bransfield Strait	echosurvey
Hamner, 1984	–	20000–30000 krill m^{-3}	South Shetlands Gerlache Strait	underwater observations
Hampton, 1985	–	mass 31 g m^{-3}	15–30°E	echosurvey
Anon, 1986 (FIBEX):				
RV *Agulhas*	13**	25 g m^{-3**}	15–30°E	echosurvey
RV *Holmberg*	539**	48 g m^{-3**}	South Orkneys	
RV *Prof. Siedlecki*	91**	94 g m^{-3**}	Drake Passage, Bransfd. Str.	
RV *Walther Herwig*	55**	–	Elephant Island	
RV *Itzumi*	460**	59 g m^{-3**}	Bransfield Strait	
Higginbottom & Hosie, 1989	–	44–182 g m^{-3**} max 1530 g m^{-3}	Prydz Bay	echosurvey
Miller & Hampton, 1989*b*	17.1* 34.0	18.8 g m^{-3}	Indian Sector Prydz Bay	echosurvey

ation from the mean and the variance of results can then best be explained by using the smaller geometric mean or median.

Krill normally have a tendency to form small, more compact swarms (Watkins *et al.*, 1986). This is well illustrated by data (Fig. 3) on the intercepted length of swarms along transects.

Results from the SIBEX cruise in March 1985 indicate that the diameter of swarms may range from 9 to 1800 m with a median value of 45 m (ln = 3.8). This result is consistent with measurements made by Kalinowski & Witek (1985*b*) for the Antarctic Peninsula region. Swarms were predominantly at depths of between 6 to 180 m, but were concentrated mostly at about 36 m with a mean thickness of 11 m (V. Siegel, unpublished data). Mean swarm thickness during FIBEX (First International BIOMASS Experiment) was about 5 m with no difference between the Atlantic and Indian Ocean sector (Anon, 1986). Miller & Hampton

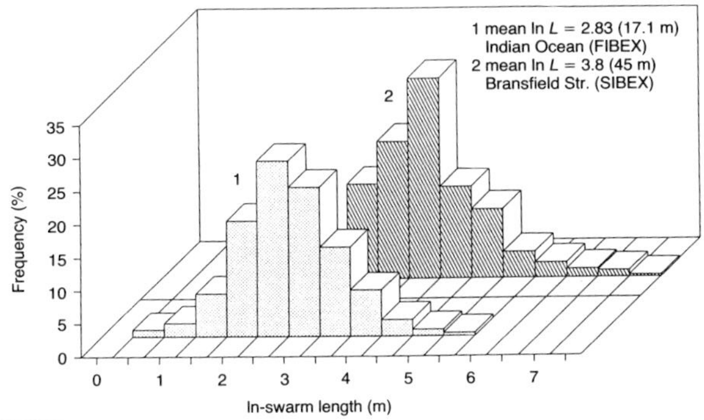

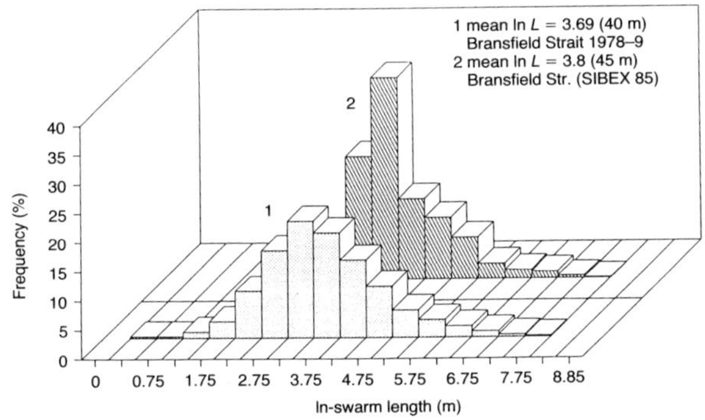

Fig. 3. Frequency distribution of intercepted swarm length obtained from hydroacoustic measurements (after Kalinowski & Witek (1981) from the western Atlantic sector; Miller & Hampton (1989b) from the Indian Ocean sector; V. Siegel (unpublished data) from the Antarctic Peninsula region). Data are log-transformed to allow direct comparison of published results.

(1989*a*) have stressed that swarms during FIBEX were larger and more dense on average in the Atlantic sector (40–45 m) than in the Indian Ocean sector (17 m, ln = 2.8 in Fig. 3). This would coincide with Hampton's (1985) observations that larger swarms are often found in regions with high current shear (the West Atlantic region is much more topographically structured than the oceanic Indian Ocean sector) while swarms tend to be smaller in regions of more laminar flow.

Swarms are the most frequent type of microscale aggregations, followed by irregular forms. The occurrence of superswarms is a rare event, although their existence has been described several times in great detail. Superswarms were observed in the Gerlache Strait (Cram *et al.*, 1979), north of Elephant Island (Mathisen & Macaulay, 1983; Macaulay *et al.*, 1984; Shulenberger *et al.*, 1984), in the Bransfield Strait (V. Siegel, unpublished data), off South Georgia (Kalinowski & Witek, 1982), off Enderby Land (Kanda *et al.*, 1982) and in the Prydz Bay region (Higginbottom & Hosie, 1989). The size of such swarms can be as large as 150 km^2 (Macaulay *et al.*, 1984) and extend over 11 km (Higginbottom & Hosie, 1989) or even 21.6 km (V. Siegel, unpublished data). Maximum density can reach 1530 g m^{-3} (Higginbottom & Hosie, 1989), although the average density is generally between 40 and 500 g m^{-3} (Macaulay *et al.*, 1984) (see also Table 1). Other estimates indicate a maximum density from 1.5 to 33 kg m^{-3} (Moiseev, 1970; Makarov *et al.*, 1970; Nemoto & Nasu, 1975; Nemoto *et al.*, 1981), but these estimates should be treated with caution given the differences in the acoustic methods employed by various investigators.

Superswarms and layers extend over a large area and may sustain their integrity for a number of days, but are best distinguished by the larger vertical dimension of superswarms (100–250 m) and the lower mean density (about 10 g m^{-3}) in layers. Kalinowski & Witek (1985*a*) noted that the presence of other species within layers may cause difficulties for the krill in coordinating their aggregation behaviour. The same holds true for irregular aggregation forms, which are often multispecific and have less well-defined boundaries.

The causes of swarm formation

It follows from analyses of uneven spatial distribution of marine organisms (cf. Stavn, 1971; Parsons *et al.*, 1977) that the following factors play an essential role:

- environmental boundaries;
- advective effects such as wind or water transport, including small-scale variations due to turbulence;
- social behaviour;
- intraspecific relations (coactive factors determined by interspecific competition).

It is obvious from Table 2 that most of these factors are applicable to krill and would be responsible for aggregation formation. The question that still arises

Table 2. *Causes for and environmental limits on krill swarm formation*

Author	Main causes and environmental limits	Region
Marr, 1962	inflow of the current branches from the Weddell Sea	South Georgia
Maslennikov, 1972	presence of system currents flowing in opposite direction	South Georgia
Wolnomiejski *et al.*, 1978	occurrence of eddies	South Georgia
Rakusa-Suszczewski, 1978	no elements of environmental limits directly influence distribution	West Atlantic
Mauchline, 1980*a*	environmental parameters and social behaviour	theoretical consideration
Kils, 1979	oxygen limits	laboratory experiments
Witek *et al.*, 1981	no elements of environmental parameters *T*, *S*, O_2, nutrients, phytoplankton	West Atlantic
Stein & Rakusa-Suszczewski, 1984	topography of bottom, which influences direction of water masses and hydrodynamic process	Bransfield Strait
Kalinowski & Witek, 1985*b*; Witek *et al.*, 1988	hydrodynamic processes and social behaviour	West Atlantic
Hampton, 1985	hydrodynamic forces, thermocline	Indian Sector
Weber & El-Sayed, 1985	no physical and chemical variability	Indian Sector
Weber *et al.*, 1986	phytoplankton (2–20 km spatial scale)	
Kalinowski & Witek, 1985*a*	presence of daylight for patch formation	West Atlantic
Loeb & Shulenberger, 1987	temperature (infusion of cold water) and wind direction	Elephant Island
Everson & Murphy, 1987	hydrodynamic processes (passive current-borne movement)	King George Island
El-Sayed, 1988	phytoplankton at scales 2–20 km	Southern Ocean
Murphy *et al.*, 1988	environmental phenomena and food availability	Southern Ocean
Priddle *et al.*, 1988	environmental phenomena and active reaction for disadvantageous environmental conditions	South Georgia Bransfield Strait
Maslennikov & Solyankin, 1988	variability of hydrological conditions	West Atlantic
Makarov *et al.*, 1988	stable eddies and influence of water masses	West Atlantic

is what are the dominant factors, since conflicting explanations are found in the literature. Summarizing the causes contained in Table 2, it seems that at the large and mesoscales, environmental factors are dominant; these factors may affect concentration with krill acting as a passive element subject to currents and water movements.

At the spatial and temporal microscale, krill may respond to various other environmental and biological factors, resulting in the formation of swarms. There is considerable confusion in the published literature concerning the causes of swarm formation. It therefore seems most wise to follow Nicol's proposal (cf. Miller & Hampton, 1989*a*) whereby a distinction is made between proximate causes (factors that serve to trigger aggregation) and ultimate effects (adaptive advantages of aggregation). Several proximate causes may be considered. The most important seem to be:

light;
phytoplankton;
physiological stimuli;
hydrographic mechanisms.

Such factors involve an active response by the species concerned and these in turn induce or maintain microscale aggregation patterns.

For cohesive formations the presence of day light seems to play an important role by allowing individuals to maintain visual contact (Kalinowski & Witek, 1985*a*). Barber (1961) concluded that chemo- and rheotaxis are major factors whereby contact is maintained between individuals of a number of crustacean species. Hamner *et al.* (1983) thought rheotaxis was particularly obvious for krill schooling supplied by the wake of preceeding animals, which would be most effective during the night.

Phytoplankton concentrations have been proposed as stimuli for krill swarming (Table 2), but krill swarms occur in clear as well as phytoplankton-rich water. Even inverse relationships between swarming and phytoplankton availability had been observed (e.g. Nast & Gieskes, 1986) and it is not clear from the data whether this effect is due to intense grazing or avoidance of phytoplankton-rich areas. In contrast Weber *et al.* (1986) used a detailed spectral analysis to indicate a positive correlation between *in vivo* fluorescence and krill biomass at scales of 2–20 km. Small-scale measurements made by Priddle *et al.* (1990) suggested that microscale patchiness in food supply seems likely to contribute most to swarm-to-swarm differences. These studies elucidated the critical need to resolve finer spatial scales for the interpretation of phytoplankton–zooplankton interactions.

Some authors have suggested that certain key environmental factors such as temperature (Hampton, 1985; Loeb & Shulenberger, 1987) or dissolved oxygen concentration (Kils, 1979) may play a role in determining krill aggregations; conversely, other studies have found no evidence for any direct association between these factors and aggregating behaviour (Witek *et al.*, 1981; Weber & El-Sayed, 1985).

The high biomass of swarms in regions of high current shear has been attributed to the presence of meanders and eddies (Witek *et al.*, 1982) and highly variable bottom topography (e.g. Stein & Rakusa-Suszczewski, 1984). Again there are also observations that eddies do not necessarily concentrate krill (Everson & Murphy, 1987).

These factors would cause a more passive concentration of krill, but this alone seems to be insufficient to explain the formation of such unusually cohesive aggregations as swarms (Miller & Hampton, 1989*a*). A simple passive aggregation factor can hardly explain the homogeneity of length, sex and developmental stages often observed within swarms and the often observed heterogeneity between neighbouring swarms (Marr, 1962; Watkins, 1986; Watkins *et al.*, 1986). Hamner *et al.* (1983) have suggested that differences in the swimming speeds of different-sized krill might cause segregation, although smaller and larger krill have on occasion been found to co-occur in the same swarm. Consequently Watkins *et al.* (1986) proposed that larger krill may actively reduce swimming speed to remain associated with smaller animals.

The ultimate effects of krill swarming seem to be more obvious. Several advantages had been discussed and summarized by Miller & Hampton (1989*a*) in detail. These include:

feeding;
reproduction;
moulting;
interaction with predators.

Feeding in summer occurs over a wide range of phytoplankton concentrations, suggesting that krill do not have to disperse to find food. Krill in swarms often have full stomachs; Hamner *et al.* (1983) observed that feeding was more intense when swarms were small and dense. They thus concluded that non-oriented swarms are feeding, while parallel orientation by swarming krill is an indication of active foraging. In their analysis on variation in krill feeding between swarms Priddle *et al.* (1990) could not link gut fullness to animal length, sex and maturity stage or to the properties of the swarms (swarm size, depth, time, density). Their conclusion was that the observed variability reflects the patchiness of the food source.

A possible advantage of aggregation is to improve foraging efficiency (to search for food resources?) as suggested by Kanda *et al.* (1982). In winter, the phytoplankton food supply in the water column is minimal and krill predominantly live directly beneath the ice where they are dispersed (Garrison *et al.*, 1986; O'Brien, 1987*a*; Marschall, 1988; Siegel *et al.*, 1990). Almost no swarming could be detected in winter in the open water zone of the entire Antarctic Peninsula area (Sahrhage, 1989; Siegel, 1989) until the onset of the pack-ice retreat in spring when krill switch from the cryopelagic to the pelagic habitat and recommence swarming (Cuzin-Roudy & Schalk, 1989; Siegel *et al.*, 1990). This seasonal change in behaviour intimates a strong association between krill feeding and aggregation behaviour.

Another seasonal aspect of swarming is probably due to the reproductive cycle. Aggregation may stimulate other animals in a swarm to mate (Naito *et al.*, 1986) and obviously facilitates location of gravid females for the transfer of spermatophores. This enhances overall reproductive success.

Krill from the same swarm are often found to be of the same moulting stage (Clarke, 1976; Buchholz, 1985). Morris (1985) developed an integrated model of krill swarming, feeding and moulting behaviour, with the soft, most endangered post-moult stages living in deeper water layers during the day and only swarming near the surface to feed at night. Hamner *et al.* (1983) even observed synchronized moulting within a swarm and concluded that discarded moults may act as a decoy for predators.

Further interactions with predators include evasion (escape response) (O'Brien, 1987*a*,*b*) and confusion of predators (although this is no protection against mass predators, such as whales) as well as it being difficult for predators to locate aggregations. In the case of a predator finding a swarm, a saturation effect will save at least some of the prey individuals.

Diurnal vertical migration and dispersal of swarms

Vertical migration of krill appears to be a common behaviour, although a recognizable pattern has only rarely been discernable. Hardy & Gunther (1935) reported greater quantities of krill collected at the sea surface at night than during the day. Marr (1962) observed krill rising near the surface at night and migrating to deeper waters in daytime. Everson (1984) indicated that in the open ocean there is likely to be a well-defined diurnal migration pattern between the surface and about 100 m depth during daytime. For a time station south of Elephant Island, Nast (1979) calculated a mean depth of 20 m for krill concentrations at night, while the species moved down to an average depth of 136 m during the day.

These data suggest a much stronger migration pattern than the one described by Mauchline & Fisher (1969) with krill at 10–100 m during the day, while mean depth at night was 0–70 m. Off South Georgia, Kalinowski (1978) found krill swarms during the day to about 50 m, whereas at night krill dispersed throughout the upper 150 m. Witek *et al.* (1981) observed krill swarms at depths from 10 to over 200 m. About 90% of these swarms were found in the upper 90 m during the day and 40 m during the night. The onset of dusk and dawn determined the beginning of the upward and downward movement of aggregations, respectively. However, Everson (1984) found a different situation off South Georgia and could not detect any clearly defined vertical migration, although swarms observed during daylight dispersed at night time.

Everson (1983) suggested that the thermocline had little effect on vertical migration of krill. However, Hampton (1985) reported that in the Indian Ocean krill concentrated near the top of the thermocline, and there was no significant difference between day and night depth distribution. Siegel (1986) investigated vertical distribution of krill in the Bransfield Strait and the northern Bellingshausen Sea during the summer of 1982, and found that 89% of the total krill occurred in the upper 65 m in an area where the thermocline was at about 60 m depth. In the Bransfield Strait the cold winter water layer is mostly absent and the thermocline is very weak, if present at all. In this area only 50% of the krill was restricted to the upper 65 m during the day. Further data from the summer of 1989–90 indicated that more than 96% of the total krill stock occurred in the 60 m surface layer above the thermocline (V. Siegel, unpublished data).

Several more studies have found no evidence for any detectable vertical migration pattern or dispersal during the night in various parts of the Southern Ocean such as the western Atlantic sector or the Prydz Bay region (Shulenberger *et al.*, 1984; Higginbottom & Hosie, 1989). Further analysis of the 1982 data revealed no day–night differences in the vertical distribution pattern. If a diurnal vertical migration existed at that time, the range of migration must have been smaller than the chosen depth strata of 0–65 m, 65–130 m, 130–200 m, etc. Daly & Macaulay (1988) found no diel trends for mean cumulative biomass with depth. Azzali & Kalinowski (1993) did not observe vertical movements of krill aggregations in the Ross Sea during summer, when daylight was almost continuous.

Diel changes in swarm parameters are also not very evident (see Table 1). Witek *et al.* (1981) demonstrated that in the Antarctic Peninsula region krill dispersed during the evening and there was a corresponding distinct decrease in the number of swarms detected acoustically, while swarms became larger and less dense. Everson (1982, 1983) reported that the density for day and night was markedly different in a krill patch, with daytime values being 6.3 times greater. Hampton (1985) also found lower krill density at night. Miller & Hampton (1989*b*), however, could not detect any significant day–night differences in the Indian Ocean sector for various swarm parameters (i.e. swarm depth, thickness, length, density and biomass within-swarm).

Given the above problem, it seems that often – but not always – density during the night is lower and swarm length greater than during the day. Some swarms probably disperse completely, while others exhibit reduced physical dimensions of swarms. The above changes may be attributed as representative of active response by krill to changes in light intensity. Therefore, diel changes may not be so evident in aggregations where the coordination of such activities is limited (e.g. superswarms) or in regions without distinct differences between day and night light intensity or food availability.

Several attempts have been made to generalize diel changes in krill vertical migration. The 12-hour model of feeding activity proposed by Pavlov (1969, 1974) as a function of vertical migration is too simple an approach to explain the observed complexity of aggregation dynamics. Godlewska & Klusek (1987) have therefore proposed a mathematical model for krill migration patterns and have suggested that the cycle depends on developmental stage (body length), light conditions and the region of occurrence. Everson & Ward (1980) suggested a periodicity linking feeding and swarming cycles which depends on food availability and the size of the animals concerned. In this case feeding to repletion would be variable for large and small animals, thereby resulting in a different amplitude of migration for different-sized krill. The stimulus for food requirement would then override the light stimulus. Morris & Ricketts (1984) determined a 'realistic model of krill feeding' in relation to both depth and time proposing that the link between feeding, swarming and vertical migration is not as strong as previous data have suggested. This confirmed the conclusion of Morris *et al.* (1984), who also found that the highest krill densities occurred above the thermocline.

Summarizing the previous discussion, the following can be said. First, the primary dependence of diurnal migration upon changes in light intensity seems to be an important event, although differences in latitudinal light regimes should be considered. Furthermore, feeding activity and food availability appear as primary factors which markedly affect both aggregation and vertical migration. Second, although krill tend to migrate towards the surface and disperse at night, this is by no means a consistent or regular feature of krill swarming. Therefore all the factors mentioned above do not appear to operate independently, but may be interlinked. The models presently available thus seem to be applicable only to certain situations. It must be concluded that a more integrated model still needs to be created to describe the multifactorial interrelationships which make diel aggregation and vertical migration patterns such complex and dynamic processes.

Effects of aggregation on population studies

Aggregation has important implications for the estimation of population trends. The marked spatial and temporal changes in aggregation affect catch rates and obscure underlying trends of abundance (Miller & Hampton, 1989*a*). Because of krill patchiness, net sampling will not provide realistic estimates of absolute abundance, while acoustic estimation may also face severe problems (day–night effects, and swarm type and shape). Both aggregation and some inhomogeneity between swarms tend to bias survey results or at least increase survey variance. This would also increase errors inherent in determining population age structure.

Differences in body length distribution occur between adjacent swarms (Watkins *et al.*, 1986). The study of Watkins *et al.* (1990) shows that samples from a single swarm are unlikely to be representative of the local population. When only a few net samples are used, the confidence intervals of the estimates become wider and the precision of the estimate is adversely affected. On the other hand, the differences between geographical sites and the separation into distinct areas of dominant size classes and developmental stages (Siegel, 1988; Anon, 1991) are much more pronounced than interswarm differences. Pooling samples from areas with similar krill composition will at least show a mesoscale effect, i.e. will smooth local interswarm variability in any subsequent population parameter analysis. Careful consideration of representative samples – probably replicate hauls as suggested by Watkins *et al.* (1990) – from all parts of the stock would thus be necessary in order to obtain less biased relative abundance estimates and to describe demographic parameters.

Krill demography

Longevity

It is well established that krill spawn during the summer season between mid-December and March (Fraser, 1936). Calyptopis I is the first larval stage to occur at the surface and begin feeding (Marr, 1962). It is then about 18–21 days after hatching (Witek *et al.*, 1980; Ross *et al.*, 1988) and is 1–2 mm in length (Fraser, 1936; Ikeda, 1984). At the end of summer, larvae moult to the early furcilia stages and in early winter they reach an average size of about 4–10 mm (Fraser, 1936; Witek *et al.*, 1980; Siegel, 1989) (Fig. 4a). Larval growth seems to continue during winter (Daly, 1990), but slows down by late March (Huntley & Brinton, 1991). In August F6 predominate; until the very beginning of the following spring season (October) mean lengths have been found within the range 12–17 mm. Before December, most larvae grow to an average size between 16–20 mm while they moult into the juvenile stage (Fraser, 1936; Siegel, 1986; Hosie *et al.*, 1988; Daly, 1990) (Fig. 4a). The variation in average length at a given time is thus primarily due to interannual and secondly to latitudinal variability in the onset of the spawning season (Witek *et al.*, 1980; Siegel, 1986). Defining 1 December as the earliest possible date for larvae occurrence (Siegel, 1987), while the onset can be as late as February, it becomes evident that larval development lasts 8–10 months.

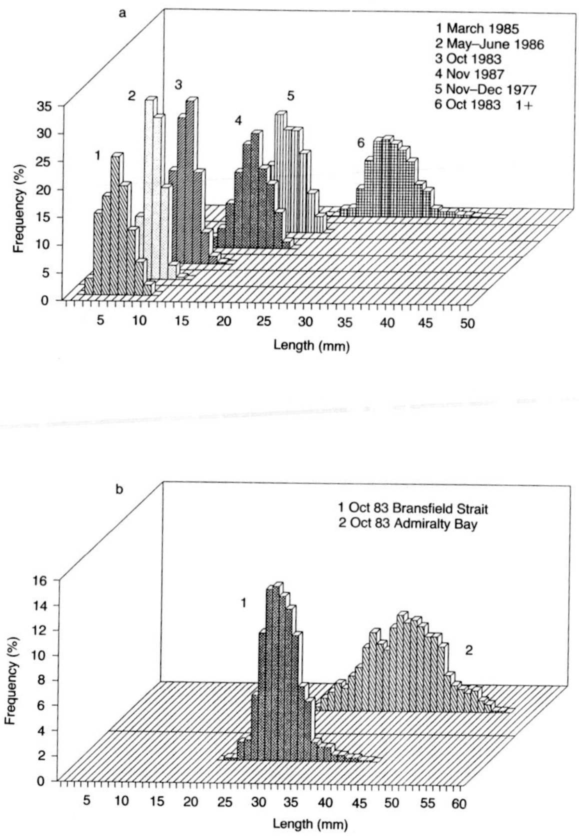

Fig. 4. Unimodal length frequency distributions of Krill: (a) age groups 0 (1–4 = larvae, 5–6 = juveniles) sampled in the Antarctic Peninsula region during March 1985, May–June 1986, October 1983, November 1987, November–December 1977; (b) 1+ juveniles and larger stages, from the October 1983 SIBEX cruise.

The relatively short spawning event during the summer season (Mauchline & Fisher, 1969) is reflected by the distinct normal distribution of the krill length frequency, e.g. in October 1983 during SIBEX or in November 1987 (Fig. 4b). The unimodal length–frequency distributions obtained for age group 0 can also be observed for the juvenile age group 1+. This result is a strong argument against Mackintosh's (1972) assumption of a second early winter spawning period. The presence of these two distinctly separated modal length groups can clearly be attributed to age classes 0 and 1+; the presence of further larger polymodal size groups at the same time in spring (Fig. 4) makes it impossible to agree with earlier assumptions that krill larvae grow to juveniles in one year, pass the juvenile stage and reach the adult stage during the next season, and then spawn after two years prior to dying soon after (Ruud, 1932; Bargmann, 1945; Nemoto, 1959; Marr, 1962). Apart from this, several other observations lead to the conclusion of a longer, multiyear life cycle in krill. These include that:

- adult krill still feed and moult after spawning;
- female ovaries recover after spawning (Makarov, 1975) and even rejuvenate (Denys & McWhinnie, 1982; Cuzin-Roudy, 1987);
- large adults are found in a resting stage during winter until early spring often in pack-ice-covered areas (Marschall, 1988; Siegel *et al.*, 1990; O'Brien, 1987*a*).

Ivanov (1970) and Aseev (1983) interpreted polymodal length frequencies as being indicative of a mixture of age-classes, and analysed modal sizes more carefully with graphical methods. These two authors concluded a lifespan of 4 and 5 years, respectively. Ikeda & Dixon (1982), however, observed shrinkage in krill body size subject to starvation over a long period under laboratory conditions. They suspected that this response might be interpreted as an overwintering strategy for times when the food supply is at a minimum. In this case, size structure and related age composition features would be obscured by shrinkage, rendering interpretation of age group information from length frequency distributions very difficult.

However, body shrinkage and the disturbance of length–frequency distributions have never been observed under natural conditions. At the end of winter, larval and juvenile length peaks are still clearly defined as narrow unimodal length frequencies and can easily be related to those of the preceding autumn size composition. Furthermore, observations of krill feeding under the ice during winter (Garrison *et al.*, 1986; Stretch *et al.*, 1988; O'Brien, 1987*a*; Marschall, 1988; Bergström *et al.*, 1990) suggest that food limitation is not so severe as to induce body shrinkage and obviously is not a problem for krill overwintering under natural conditions. McClatchie *et al.* (1991), who analysed year-round data from Admiralty Bay, addressed the question of body shrinkage; they suggested that the decrease in mean length of krill during winter results from predation reducing the larger, but less abundant, krill at a time when large numbers of juveniles are recruiting to the population.

Therefore, the analysis of length–frequency distributions still seems to be an adequate method for determining krill age. The analysis of modal length classes by Ivanov (1970) and Aseev (1983) faced the problem that older, overlapping groups were difficult to separate. This problem can partly be reduced by using the statistical distribution mixture analysis described by Macdonald & Pitcher (1979). A maximum of 5–6 age groups have been distinguished by Siegel (1986, 1987) and Hosie *et al.* (1988) using this procedure in the Atlantic sector and Prydz Bay, respectively. Observed differences between the two sets of results can be explained by the scarcity of 55–60 mm krill in samples from the Prydz Bay region.

Another approach to estimating krill longevity has been used by Ettershank (1983, 1984, 1985). He utilized the age pigment lipofuscin to determine metabolic age, a method originally described from studies on insects with a short lifespan and a terminal moult. Nicol (1987), however, has demonstrated that there are certain critical limitations in applying the chemical lipofuscin technique to ageing. Rosenberg *et al.* (1986) also pointed out the difficulties of relating metabolic age and real time as well as the unknown proportion of pigment that might be lost during ecdysis. Nicol *et al.* (1991) have reported the results of an experiment under laboratory conditions where the age of the krill was known, and showed that it was possible to separate at least the first two age groups by using the age-pigment (initially referred to as lipofuscin) fluorescence technique.

The interpretation of lipofuscin levels has resulted in identification of three age groups for gravid females (Ettershank, 1984). This is identical to the results obtained by distribution mixture analyses (Siegel, 1986, 1987). Further studies indicated a five-year lifespan for adult males – these include the so-called subadult stages of other investigators – and a single age group for the juvenile stage (Ettershank, 1985). Considering almost one year of larval development, this results in a total of seven age groups or a lifespan of six years. This is consistent with the findings of Berman *et al.* (1989), which were obtained by the fluorescence technique and image analysis of size, but is one year longer than the lifespan proposed by Siegel (1986, 1987). It seems possible, however, that the oldest, strongly overlapping length groups cannot be separated by the distribution mixture analysis and that Siegel's group 5+ consists of more than one age class. Rosenberg *et al.* (1986) estimated the growth rate by the step function and concluded that it would take a minimum of 6–7 years for krill to approach a maximum length of 60 mm.

Growth

In applying these results several attempts have been made to estimate krill growth parameters. Various growth curves calculated by Siegel (1986, 1987) and Rosenberg *et al.* (1986) best fit the earlier theoretical growth curves derived by Mauchline (1980*b*), if the annual growth period is reduced to 90 days. Reducing the annual growth period to this length increases the time taken to attain sexual maturity to three years – which is congruous with field data

Table 3. *Krill growth parameters of von Bertalanffy equation (VBGF)*

Yearly growth parameter, K	L_∞	Method	Reference
0.445/0.429	62.4/62.5	Ford–Walford plot	Siegel, 1986
0.445/0.4018	61.8/63.8	nonlinear regression VBGF	Siegel, 1986
0.4728	61.0	modified VBGF (Pauly & Gaschütz, 1979)	Siegel, 1986
(0.27) 0.43–0.47	60.0	VBGF (seasonal growth)	Rosenberg *et al.*, 1986
0.478/0.354	63.3/61.3	nonlinear regression VBGF	Siegel, 1987
0.8	—	ELEFAN	McClatchie, 1988

Table 4. *Daily growth rates of Antarctic krill*

Daily growth rate (mm d^{-1})	Comments	Reference
0.0354	laboratory experiments	Murano *et al.*, 1979
0.07	laboratory experiments 22–44 mm length	Ikeda *et al.*, 1985
0.024–0.044	laboratory experiments	Poleck & Denys, 1982
0.047	laboratory experiments juvenile krill	Ikeda & Thomas, 1987
0.13 =0.032	theoretical approach, in 90 d growth period per year, mean annual growth rate	Mauchline, 1980b
0.058	calculated from field samples in summer for 41 mm mean length	Kanda *et al.*, 1982
0.33	calculated from shift of 2 mm in modal length class of 30 mm sized krill over 6 d period	Clarke & Morris, 1983
0.14	in summer for 30 mm length class	Rosenberg *et al.*, 1986
0.083–0.156	laboratory experiments for 32 mm length class	Buchholz, 1988
0.033	mean annual growth rate for all age groups	Siegel, 1986
0.12	juvenile age group 1+ in summer	
0.07	subadult age group 2+ in summer	
0.13	0 age group in summer	McClatchie, 1988
0.025	adult ⩾3+ age group in summer	
0.01–0.048	in winter	
0.07	0 age group in winter	Daly, 1990

indicating age-at-first-spawning for the age group 3+ (Siegel, 1986) – and a lifespan of at least 5 years (Miller & Hampton, 1989*a*). It is also in agreement with the results of Rosenberg *et al.* (1986), who found that yearly growth performances (K values) are approximately 25% of those for the high growth season, so indicating that maximal growth occurs over only three months during early summer.

Some of the parameters of the von Bertalanffy growth curve are given in Table 3. A comparison shows that K values range mainly between 0.40 and 0.47 in the data presented by Rosenberg *et al.* (1986) and Siegel (1986, 1987). Only McClatchie (1988), who used Stepnik's (1982) data for December 1978 to November 1979 and then applied the ELEFAN method (Pauly & David, 1980), obtained a much higher growth rate with $K = 0.8$, which is at variance with other results.

Published data on daily growth rates are summarized in Table 4. Daily growth rates are often difficult to compare, since growth is a function of size, age, maturation and feeding conditions, as well as environmental conditions, and has a strong seasonal component. Although laboratory experiments mostly exhibit constant environmental conditions, they often suffer from lack of knowledge concerning the necessary requirements for development. This leads to artifacts during rearing experiments. These often

Table 5. *Estimates of krill natural mortality*

M:*Z*	Method	Area	Reference
5.5	Edmondson's method		Kawakami & Doi, 1979
	abundance data of length groups	Scotia Sea	Brinton & Townsend, 1984
2.31	larval to subadults, 1–2 years old		
0.51	2–3 years old		
0.78–1.17	linearized catch curve data	Ant. Peninsula	Siegel, 1986
0.88–0.96	linearized catch curve data		Siegel, 1992
0.94–0.99	Pauly (1980) formula on VBGF and *M* relationship		Siegel, 1986
0.8–1.35	$M = 2 \times K$ of VBGF	Scotia Sea	Priddle *et al.*, 1988
0.5	1-cumulative length–frequency	*Discovery* data	Basson & Beddington, 1989
0.45–0.65	length-dependent predation curve		

result in low or negative growth rates being reported (0.024–0.07 mm^{-1} in Table 4). Such data are similar to a mean annual growth rate integrated over the entire life span (Siegel, 1986). Daily growth rates calculated for the summer growth period, however, have been estimated to be as high as Mauchline's (1980*b*) theoretically expected value for a 90 day growth period per year (i.e. around 0.13 mm d^{-1}) (Table 4). From these results it becomes obvious that the growth potential of krill is comparable to that of other temperate euphausiid species (e.g. *Meganyctiphanes norvegica*) (Buchholz, 1988), although the annual growth period is certainly shorter in polar latitudes.

Mortality

Information on krill mortality rates is rare in the published literature. Estimates obtained by different methods have been summarized by Miller & Hampton (1989*a*) and discussed in some detail by Siegel (1992). Mortality rates, including recent estimates, are listed in Table 5.

Calculations based on length-at-age data are sensitive to the species' lifespan. Reducing the lifetime will also strongly affect growth rates while increasing the rate of mortality. Natural mortality (*M*) and the von Bertalanffy growth constant *K* are related. The conversion factor $M/K = 2$ used by Priddle *et al.* (1988) was taken from clupeoid fish, but may range between 1.5 and 2.5 for other species and should therefore only be regarded as a rough estimator of *M*.

In principle, longevity is more closely related to mortality than to the growth factor *K* and the asymptotic length L_∞ (Sparre *et al.*, 1989). Following Alagaraja (1984), the proposed lifespan of a species would be at the age at which 99% of a cohort has died when it has been exposed to natural mortality only. Then:

$$M_{1\%} = -\ln (0.01) / T_m,$$

where T_m is longevity in years.

Recently published results on the number of krill age groups range from 6, including the 0-age group (i.e. a lifespan of 5 years; Siegel, 1987) to a maximum of 7 or 8 (i.e. longevity of 6–7 years; Ettershank, 1984, 1985; Rosenberg *et al.*, 1986). Based on this maximum range, krill mortality would then be expected to be within the range $M = 0.66$–0.92. This is within the range of several recent estimates, and corresponds to a survival rate of 40–51% for postlarval age groups from one year to the next. It also indicates that some estimates of the mean level of *M* are unexpectedly high, at least for postlarval krill.

Table 6. *Estimates of P/B ratio for krill and other euphausiid species*

Species	*P/B*	Area	Reference
Euphausia pacifica	3.0	California Pacific Coast	Lasker, 1966
E. pacifica	8.7	Oregon Pacific Coast	Mauchline, 1977
E. krohni	6.0	North Atlantic	Lindley, 1982*c*
Thysanoessa raschi	4.0	Canada Atlantic Coast	Berkes, 1977
T. longicaudata	2–8	North Atlantic (1–2 generations/year)	Lindley, 1978
T. inermis and T. raschi	1.3–4.2	North Atlantic	Lindley, 1980
T. gregaria	6.0	North Atlantic	Lindley, 1982*c*
Meganyctiphanes norvegica	2.3	Scottish coast	Mauchline, 1977
M. norvegica	1.3–6.3	North Sea, waters west of British isles	Lindley, 1982b
Nyctiphanes couchi	4.0–5.5	waters off British isles	Lindley, 1982*b*
Nematoscelis megalops	5.4	North Atlantic	Lindley, 1982*c*
Nyctiphanes australis	13.3–14.5	Tasmania	Ritz & Hosie, 1982
Euphausia lucens	3.9–8.9	Benguela Current	Stuart & Pillar, 1988
Euphausiacea	3.3–4.8	oceanic North Atlantic	Lindley, 1982*a*
	>12	neritic NW and NE Atl.	
E. superba	1.8	Antarctic	Allen, 1971
E. superba	1.19		Yamanaka, 1983
E. superba	1.19–2.77	Indian Ocean sector	Miller *et al.*, 1985
E. superba	0.8–1.1	Antarctic Peninsula	Siegel, 1986
E. superba	0.83–1.1	Antarctic Peninsula	Siegel, 1992

Production index

Owing to the difficulties already mentioned in the determination of absolute krill abundance and biomass, the estimation of total production (*P*) is subject to similar problems. Data on relative abundance may, however, help to estimate production indices. Several attempts have been made to determine production to biomass (*P/B*) ratios for krill. These were comprehensively summarized by Miller & Hampton (1989*a*); Table 6 gives an overview of *P/B* estimates for krill and other euphausiid species.

Although some estimates of the krill production seem to be high, these can be explained by the underlying assumption of a short life cycle. Robertson (1979) has demonstrated a general relationship between the *P/B* ratio and maximum lifespan (*A*) for various Mollusca, Echinodermata and Crustacea. Inserting the suggested maximum age for krill of 5–7 years in Robertson's regression – the oldest age group still participates in spawning and is thus some months older than 5 to 7 years – would result in expected *P/B* ratios of 1.37 and 1.08, respectively, which is within the reported range of values for krill.

Euphausiid species from temperate waters show a much higher *P/B* ratio, which can be explained by a longer seasonal growth period, higher turnover rate, shorter lifespan and the production of several generations per year. *Thysanoessa longicaudata*, for example, produces two or three generations annually in different areas, and the dependent *P/B* ratio varies between 2 and 8 (Lindley, 1978). *Nyctiphanes australis* reproduces continuously over most of the year; for this short-lived, subtropical species the *P/B* ratio exceeds a value of 13 (Ritz & Hosie, 1982). Antarctic krill has a multiyear life cycle, a short annual growth period, a long larval and juvenile phase and a relatively low production rate.

Conclusions

This review of krill aggregating behaviour indicates that current explanations of the proximate causes for krill swarming and vertical migration remain contradictory and unclear. Further research is obviously required and should:

distinguish more clearly between proximate causes and ultimate effects;

consider the different scales of different types of aggregations (concentrations, patches and superswarms are probably influenced by different

factors than microscale swarms/schools);
pay attention to ontogenetic differences in swarming (state of congregation or dispersal);
consider possible geographical differences: swarming might differ in various regions or local areas (shelf or oceanic zones);
consider the seasonal aspect;
determine diurnal effects (vertical migration, dispersion).

Concerning krill population dynamics, much information has been gained during the past decade of BIOMASS-related research activities, on krill stock composition and seasonal mesoscale fluctuations, as well as longevity, growth, mortality and productivity, at least in some parts of the Southern Ocean. However, we do not know if the population dynamics of krill differ in various subareas (e.g. South Georgia), where krill live close to the species' northern distributional boundary, or in areas covered by ice. Information from pack-ice areas is especially sparse; this field will need much more attention during future research activities.

References

Alagaraja K (1984) Simple methods for estimation of parameters for assessing exploited fish stocks. *Ind J Fish* **31**: 177–208.

Allen K R (1971) Relation between production and biomass. *J Fish Res Bd Can* **28**: 1573–81.

Anon (1986) Post-FIBEX Acoustic Workshop. *BIOMASS Rep Ser* **40**: 1–106.

Anon (1991) Non-acoustic Krill data analysis workshop. *BIOMASS Rep Ser* **66**: 1–16.

Aseev Y P (1983) (Size structure and longevity of krill population from the Indian Ocean sector of the Antarctic). In: Lubimova T G (ed) (*Antarctic Krill. Distribution Pattern and Environment*). *Sb Nauch Tr VNIRO*, pp. 103–10. (In Russian.)

Azzali M, Kalinowski J (1993) Biomass of krill in the Ross Sea (Pacific sector) in relation to BIOMASS FIBEX/SIBEX results. (In press.)

Barber S B (1961) Chemoreception and thermoreception. In: Waterman T H (ed) *The Physiology of Crustacea*, Vol II. Academic Press, New York, pp. 109–28.

Bargmann H E (1945) The development and life-history of adolescent and adult krill *Euphausia superba*. *Discovery Rep* **23**: 103–76.

Basson M, Beddington J R (1989) Preliminary analysis of growth and mortality rates of Antarctic krill (*Euphausia superba*) from length frequency samples. *Antarct Spec Top* **1989**: 51–5.

Bergström B I, Hempel G, Marschall H P, North A W, Siegel V, Strömberg J O (1990) Spring distribution, size composition and behaviour of krill *Euphausia superba* in the western Weddell Sea. *Polar Rec* **26**: 85–9.

Berkes F (1977) Ecological significance of the neglecta form of *Thysanoessa inermis* (Kroyer) (Euphausiacea). *Crustaceana* **33**: 39–46.

Berman M S, McVey L, Ettershank G (1989) Age determination of Antarctic krill using fluorescence and image analysis of size. *Polar Biol* **9**: 267–71.

BIOMASS (1977) *Biological Investigations of Marine Antarctic Systems and Stocks (BIOMASS).* Volume 1: *Research Proposals*, 77 pp.

Brinton E, Townsend A W (1984) Regional relationships between development and growth in larvae of Antarctic krill, *Euphausia superba*, from field samples. *J Crust Biol* **4**: 224–46.

Buchholz F (1985) Moult and growth in euphausiids. In: Siegfried W R, Condy P R, Laws R M (eds) *Antarctic Nutrient Cycles and Food Webs*. Springer-Verlag, Berlin, pp. 339–45.

Buchholz F (1988) On the way of life of the Antarctic and the northern krill *Euphausia superba* and *Meganyctiphanes norvegica*. Comparative investigations of moult physiology and growth in the field and the laboratory. *Ber Inst Meereskd Kiel* **185**: 1–245. (In German with English summary.)

Clarke A (1976) Some observations of krill (*Euphausia superba* Dana) maintained alive in the laboratory. *Br Antarct Surv Bull* **43**: 111–18.

Clarke A, Morris D J (1983) Towards an energy budget for krill: The physiology and biochemistry of *Euphausia superba* Dana. *Polar Biol* **2**: 69–86.

Cram D L, Agenbag J J, Hampton I, Robertson A A (1979) SAS Protea cruise, 1978: The general results of the acoustics and remote sensing study, with recommendations for estimating the abundance of krill (*Euphausia superba* Dana). *S Afr J Antarct Res* **9**: 3–14.

Cuzin-Roudy J (1987) Gonad history of the Antarctic krill *Euphausia superba* Dana during its breeding season. *Polar Biol* **7**: 237–44.

Cuzin-Roudy J, Schalk P H (1989) Macrozooplankton – biomass, development and activity. In: Hempel I, Schalk P H, Smetacek V (eds) *The Expedition ANTARKTIS VII/3 (EPOS leg 2) of RV "Polarstern" in 1988/1989*. *Ber Polarforsch* **65**: 146–54.

Daly K L (1990) Overwintering development, growth and feeding of larval *Euphausia superba* in the Antarctic marginal ice zone. *Limnol Oceanogr* **35**: 1564–76.

Daly K L, Macaulay M C (1988) Abundance and distribution of krill in the ice edge zone of the Weddell Sea, austral spring 1983. *Deep-Sea Res* **35**(1): 21–41.

Denys C J, McWhinnie M A (1982) Fecundity and ovarian cycles of the Antarctic krill *Euphausia superba* (Crustacea, Euphausiacea). *Can J Zool* **60**: 2414–23.

Doi T, Kawakami T (1979) The estimation of krill abundance in the Antarctic by analysis of echogram. In: Doi T (ed) *Comprehensive Report on the Population of Krill* Euphausia superba *in the Antarctic*. Tokai Reg Fish Res Lab, pp. 23–33.

El-Sayed S Z (1988) Seasonal and interannual variabilities in Antarctic phytoplankton with reference to krill distribution. In: Sahrhage D (ed) *Antarctic Ocean and Resources Variability*. Springer-Verlag, Berlin, pp. 101–19.

Ettershank G (1983) Age structure and cyclical annual size change in the Antarctic krill, *Euphausia superba* Dana. *Polar Biol* **2**: 189–93.

Ettershank G (1984) A new approach to the assessment of longevity in Antarctic krill *Euphausia superba*. *J Crust Biol* **4**: 295–305.

Ettershank G (1985) Population age structure in males and juveniles of the Antarctic krill, *Euphausia superba* Dana. *Polar Biol* **4**: 199–201.

Everson I (1982) Diurnal variations in mean volume backscattering strength of an Antarctic krill (*Euphausia superba*) patch. *J Plank Res* **4**(1): 155–62.

Everson I (1983) Variations in vertical distribution and density of krill swarms in the vicinity of South Georgia. *Mem Natl Inst Polar Res Spec Issue* **27**: 84–92.

Everson I (1984) Zooplankton. In: Laws R M (ed) *Antarctic Ecology*, vol II. Academic Press, London, pp. 463–90.

Everson I (1987) Some aspects of the small scale distribution of *Euphausia crystallorophias*. *Polar Biol* **8**: 9–15.

Everson I, Murphy E (1987) Mesoscale variability in the distribution of krill *Euphausia superba*. *Mar Ecol Prog Ser* **40**(1–2): 53–60.

Everson I, Ward P (1980) Aspects of Scotia Sea zooplankton. *Biol J Linn Soc* **14**: 93–101.

Foote K G, Everson I, Watkins J L, Bone D G (1990) Target strengths of Antarctic krill (*Euphausia superba*) at 38 and 120 kHz. *J Acoust Soc Am* **87**(1): 16–24.

Fraser F C (1936) On the development and distribution of the young stages of krill (*Euphausia superba*). *Discovery Rep* **14**: 1–192.

Garrison D L, Sullivan C W, Ackley S F (1986) Sea ice microbial communities in Antarctica. *BioScience* **36**: 243–50.

Godlewska M, Klusek Z (1987) Vertical distribution and diurnal hydroacoustical observations, SIBEX, December 1983/January 1984. *Polar Biol* **8**: 17–22.

Guzman O, Marin V (1983) Hydroacoustic and photographic techniques applied to study the behaviour of krill (*Euphausia superba*). *Mem Natl Inst Polar Res* **27**: 129–52.

Hamner W M (1984) Aspects of schooling in *Euphausia superba*. *J Crust Biol* **4**(Spec No 1): 67–74.

Hamner W M, Hamner P P, Strand S W, Gilmer R W (1983) Behavior of Antarctic krill, *Euphausia superba*: Chemoreception, feeding, schooling and moulting. *Science* **220**: 433–5.

Hampton I (1985) Abundance, distribution and behaviour of *Euphausia superba* in the Southern Ocean between 15° and 30°E during FIBEX. In: Siegfried W R, Condy P R, Laws R M (eds) *Antarctic Nutrient Cycles and Food Webs*. Springer-Verlag, Berlin, pp. 294–303.

Hardy A C, Gunther E R (1935) The plankton of the South Georgia whaling grounds and adjacent waters, 1926–1927. *Discovery Rep* **11**: 1–456.

Higginbottom I R, Hosie G W (1989) Biomass and population structure of a large aggregation of krill near Prydz Bay, Antarctica. *Mar Ecol Prog Ser* **58**: 197–203.

Hosie G W, Ikeda T, Stolp M (1988) Distribution, abundance and population structure of the Antarctic krill (*Euphausia superba* Dana) in the Prydz Bay region, Antarctica. *Polar Biol* **8**: 213–24.

Huntley M, Brinton E (1991) Mesoscale variation in growth and early development of *Euphausia superba* Dana in the western Bransfield Strait region. *Deep-Sea Res* **38**: 1213–40.

Ikeda T (1984) Development of the larvae of the Antarctic krill (*Euphausia superba* Dana) observed in the laboratory. *J Exp Mar Biol Ecol* **75**: 107–17.

Ikeda T, Dixon P (1982) Body shrinkage as a possible overwintering mechanism of the Antarctic krill, *Euphausia superba* Dana. *J Exp Mar Biol Ecol* **62**: 143–51.

Ikeda T, Dixon P, Kirkwood J (1985) Laboratory observations on moulting, growth and maturation in Antarctic krill (*Euphausia superba* Dana). *Polar Biol* **4**: 1–18.

Ikeda T, Thomas P G (1987) Moulting interval and growth of juvenile Antarctic krill (*Euphausia superba*) fed different concentrations of the diatom *Phaeodactylum tricornutum* in the laboratory. *Polar Biol* **7**: 339–43.

Ivanov B G (1970) On the biology of the Antarctic krill (*Euphausia superba* Dana). *Mar Biol* **7**: 340–51.

Kalinowski J (1978) Vertical migration of krill in the region of South Georgia, February–March 1976. *Polsk Arch Hydrobiol* **25**(3): 573–83.

Kalinowski J, Witek Z (1981) The physical parameters of krill aggregations in the western Antarctic. *ICES Biol Oceanogr Cttee CM 1981/L* **19**: 1–4.

Kalinowski J, Witek Z (1982) Forms of Antarctic krill aggregations. *ICES Biol Oceanogr Cttee CM 1982/L* **60**: 1–8.

Kalinowski J, Witek Z (1985*a*) *Scheme for Classifying Aggregations of Antarctic Krill*. BIOMASS Handbook No 27, 12 pp.

Kalinowski J, Witek Z (1985*b*) Elementy biologii, formy grupowego wystepowania kryli. *Studia i Materialy MIR, Gdynia*, B52.

Kanda K, Takagi K, Seki Y (1982) Movement of the larger swarms of Antarctic krill *Euphausia superba* off Enderby Land during 1976–1977 season. *J Tokyo Univ Fish* **68**: 24–42.

Kawakami T, Doi T (1979) Natural mortality of krill and density in swarms. In: Doi T (ed) *Comprehensive Report on the Population of Krill,* Euphausia superba *in the Antarctic*. Tokai Reg Fish Res Lab, pp. 19–21.

Kils U (1979) Performance of Antarctic krill, *Euphausia superba*, at different levels of oxygen saturation. *Meeresforschung* **27**(1): 35–47.

Lasker R (1966) Feeding, growth, respiration, and carbon utilization of a euphausiid crustacean. *J Fish Res Bd Can* **23**: 1291–317.

Lindley J A (1978) Population dynamics and production of euphausiids. I. *Thysanoessa longicaudata* in the North Atlantic Ocean. *Mar Biol* **46**: 121–30.

Lindley J A (1980) Population dynamics and production of euphausiids. II. *Thysanoessa inermis* and *T. raschi* in the North Sea and American coastal waters. *Mar Biol* **59**: 225–33.

Lindley J A (1982*a*) Continuous plankton records: Geographical variations in numerical abundance, biomass and production of euphausiids in the North Atlantic Ocean and the North Sea. *Mar Biol* **71**: 7–10.

Lindley J A (1982*b*) Population dynamics and production of euphausiids. III. *Meganyctiphanes norvegica* and *Nyctiphanes couchi* in the North Atlantic Ocean and the North Sea. *Mar Biol* **66**: 37–46.

Lindley J A (1982*c*) Population dynamics and production of euphausiids. IV. *Euphausia krohni*, *Nematoscelis megalops* and *Thysanoessa gregaria*, and eight rare species in the North Atlantic Ocean. *Mar Biol* **77**: 1–6.

Loeb V J, Shulenberger E (1987) Vertical distributions and relations of euphausiid populations off Elephant Island, March 1984. *Polar Biol* **7**: 363–73.

Macaulay M C, English T S, Mathisen O A (1984) Acoustic characterization of swarms of Antarctic krill (*Euphausia superba*) from Elephant Island and Bransfield Strait. *J Crust Biol* **4**(Spec No 1): 16–44.

Macdonald P D M, Pitcher T I (1979) Age-groups from size-frequency data: A versatile and efficient method of analysing distribution mixtures. *J Fish Res Bd Can* **36**: 987–1001.

Mackintosh N A (1972) Life cycle of Antarctic krill in relation to ice and water conditions. *Discovery Rep* **36**: 1–94.

Makarov R R (1975) (A study of the second maturation of female euphausiids (Eucarida, Euphausiacea).) *Zool Zh* **54**: 670–81. (In Russian.)

Makarov R R, Maslennikov V V, Solyankin E V, Spiridonov V A, Yakovlev V N (1988) Variability in population density of Antarctic krill in the western Scotia Sea in relation to hydrological conditions. In: Sahrhage D (ed) *Antarctic Ocean and Resources Variability*. Springer-Verlag, Berlin, pp. 231–6.

Makarov R R, Naumov A G, Shevtsov V V (1970) The biology and the distribution of the Antarctic krill. In: Holdgate M W (ed) *Antarctic Ecology*, vol I. Academic Press, London, pp. 173–6.

Marr J W S (1962) The natural history and geography of the Antarctic krill (*Euphausia superba* Dana). *Discovery Rep* **32**: 33–464.

Marschall H P (1988) The overwintering strategy of Antarctic krill under the pack-ice of the Weddell Sea. *Polar Biol* **9**: 129–35.

Maslennikov V V (1972) (On the effects of water dynamics on the distribution of *Euphausia superba* Dana in the area of South Georgia.) *Trudy VNIRO* **75**: 107–17. (In Russian.)

Maslennikov V V, Solyankin E V (1988) Patterns of fluctuations in the hydrological conditions of the Antarctic and their effects on the distribution of Antarctic krill. In: Sahrhage D (ed) *Antarctic Ocean and Resources Variability*. Springer-Verlag, Berlin, pp. 209–13.

Mathisen O A, Macaulay M C (1983) The morphological features of a super-swarm of krill, *Euphausia superba*. *Mem Natl Inst Polar Res* **27**: 153–64.

Mauchline J (1977) Estimating production of mid-water organisms. In: Andersen N R, Zahuranec B J (eds) Oceanic sound scattering prediction. *Mar Sci* **5**: 361–400.

Mauchline J (1980*a*) *Studies on Patches of Krill* Euphausia superba *Dana*. BIOMASS Handbook No 6, 36 pp.

Mauchline J (1980*b*) The biology of mysids and euphausiids. *Adv Mar Biol* **18**: 1–681.

Mauchline J, Fisher L R (1969) The biology of euphausiids. *Adv Mar Biol* **7**: 1–454.

McClatchie S (1988) Food-limited growth of *Euphausia superba* in Admiralty Bay, South Shetland Islands, Antarctica. *Cont Shelf Res* **8**: 329–45.

McClatchie S, Rakusa-Suszczewski S, Filcek K (1991) Seasonal growth and mortality of *Euphausia superba* in Admiralty Bay, South Shetland Islands, Antarctica. *ICES J Mar Sci* **48**: 335–42.

Miller D G M, Hampton I (1989*a*) Biology and ecology of the Antarctic krill (*Euphausia superba* Dana): a review. *BIOMASS Sci Ser* **9**: 1–166.

Miller D G M, Hampton I (1989*b*) Krill aggregation characteristics: Spatial distribution patterns from hydroacoustic observations. *Polar Biol* **10**: 125–34.

Miller D G M, Hampton I, Henry J, Abrams R W, Cooper J (1985) The relationship between krill food requirements and phytoplankton production in a sector of the southern Indian Ocean. In: Siegfried W R, Condy P R, Laws R M (eds) *Antarctic Nutrient Cycles and Food Webs*. Springer-Verlag, Berlin, pp. 362–71.

Moiseev P A (1970) Some aspects of the commercial use of the krill resources of the Antarctic seas. In: Holdgate M W (ed) *Antarctic Ecology*, vol I. Academic Press, London, pp. 213–16.

Morris D J (1985) Integrated model of moulting and feeding of Antarctic krill *Euphausia superba* off South Georgia. *Mar Ecol Prog Ser* **22**: 207–17.

Morris D J, Everson I, Ricketts C, Ward P (1984) Feeding of krill around South Georgia. II. Relations between feeding activity, environment and vertical distribution. *Mar Ecol Prog Ser* **20**: 203–6.

Morris D J, Ricketts C (1984) Feeding of krill around South Georgia. I. A model of feeding activity in relation to depth and time of day. *Mar Ecol Prog Ser* **16**: 1–7.

Murano M, Segawa S, Kato M (1979) Moult and growth of Antarctic krill in laboratory. *Trans Tokyo Univ Fish* **3**: 99–106.

Murphy E J, Morris D J, Watkins J L, Priddle J (1988) Scales of interaction between Antarctic krill and the environment. In: Sahrhage D (ed) *Antarctic Ocean and Resources Variability*. Springer-Verlag, Berlin, pp. 120–30.

Naito Y, Taniguchi A, Hamada E (1986) Some observations on swarms and mating behavior of

Antarctic krill (*Euphausia superba* Dana). *Mem Natl Inst Polar Res* **40**: 178–82.

Nast F (1979) The vertical distribution of larval and adult krill (*Euphausia superba*) on a time station south of Elephant Island, South Shetlands. *Meeresforschung* **27**: 103–18.

Nast F, Gieskes W C (1986) Phytoplankton observations relative to krill abundance around Elephant Island in November 1983. *Arch FischWiss* **37**: 95–106.

Nemoto T (1959) Food of baleen whales with reference to whale movements. *Scient Rep Whales Res Inst* **14**: 149–290.

Nemoto T, Doi T, Nasu K (1981) Biological characteristics of krill caught in the Southern Ocean. *BIOMASS Sci Ser* **2**: 47–63.

Nemoto T, Nasu K (1975) Present status of exploitation and biology of krill in the Antarctic. *Oceanol Intern* **75**: 353–60.

Nicol S (1987) Some limitations on the use of the lipofuscin ageing technique. *Mar Biol* **93**: 609–14.

Nicol S, Stolp M, Hosie G W (1991) Accumulation of fluorescent age pigments in a laboratory population of Antarctic krill *Euphausia superba* Dana. *J Exp Mar Biol Ecol* **146**: 153–61.

O'Brien D P (1987*a*) Direct observations of the behaviour of *Euphausia superba* and *Euphausia crystallorophias* (Crustacea, Euphausiacea) under pack ice during the Antarctic spring 1985. *J Crust Biol* **7**: 437–48.

O'Brien D P (1987b) Description of escape responses of krill (Crustacea: Euphausiacea), with particular reference to swarming behavior and the size and proximity of the predator. *J Crust Biol* **7**: 449–57.

Parsons T R, Takahashi M, Hargrave B (1977) *Biological Oceanographic Processes*. 2nd edn. Pergamon Press, Oxford.

Pauly D (1980) On the interrelationships between natural mortality, growth parameters, and mean environmental temperature in 175 fish stocks. *J Cons Int Explor Mer* **39**: 175–92.

Pauly D, David N (1980) A BASIC program for the objective extraction of growth parameters from length frequency data. *ICES Statistics Cttee CM 1980/D* **7**: 1–14.

Pauly D, Gaschütz G (1979) A simple method for fitting oscillating length growth data, with a program for pocket calculators. *ICES Demers Fish Cttee:Pelagic Fish Cttee CM 1979/*G **24**, 1–26.

Pavlov V Y (1969) (The feeding of krill and some features of its behaviour.) *Trudy VNIRO* **66**: 207–22. (In Russian.)

Pavlov V Y (1974) O charaktere swjazi mezdu pitaniem i nekotorymi osobennostiami powedenija *Euphausia superba* Dana. (Über den Charakter einer Verbindung zwischen der Ernährung und einigen Besonderheiten des Verhaltens von *Euphausia superba* Dana.) *Trudy VNIRO* **99**(5): 104–16. (In Russian.)

Poleck T P, Denys C J (1982) Effect of temperature on the moulting, growth and maturation of the Antarctic krill *Euphausia superba* (Crustacea: Euphausiacea) under laboratory conditions. *Mar Biol* **70**: 255–65.

Priddle J, Croxall J P, Everson I, Heywood R B, Murphy E J, Prince P A, Sear C B (1988) Large-scale fluctuations in distribution and abundance of krill – a discussion of possible causes. In: Sahrhage D (ed) *Antarctic Ocean and Resources Variability*. Springer-Verlag, Berlin, pp. 169–82.

Priddle J, Watkins J, Morris D, Ricketts C, Buchholz F (1990) Variation of feeding by krill in swarms. *J Plankt Res* **12**(6): 1189–205.

Ragulin A G (1969) Underwater observations on krill. *Trudy VNIRO* **66**: 231–4. (In Russian. Published in English (1970) by Fish Lab Lowestoft 92, 3 pp.)

Rakusa-Suszczewski S (1978) Environmental conditions within krill swarms. *Polsk Arch Hydrobiol* **25**(3): 585–8.

Ritz D A, Hosie G W (1982) Production of the euphausiid *Nyctiphanes australis* in Storm Bay, South-Eastern Tasmania. *Mar Biol* **68**: 103–8.

Robertson A I (1979) The relationship between annual production: biomass ratio and lifespans for marine macrobenthos. *Oecologia* **38**: 193–202.

Rosenberg A A, Beddington J R, Basson M (1986) Growth and longevity of krill during the first decade of pelagic whaling. *Nature* **324**: 152–4.

Ross R M, Quetin L B, Kirsch E (1988) Effect of temperature on developmental times and survival of early larval stages of *Euphausia superba* Dana. *J Exp Mar Biol* **121**: 55–71.

Ruud J T (1932) On the biology of southern Euphausiidae. *Hvalrad Skr* **2**: 1–105.

Sahrhage D (1989) Hydroacoustic detection of krill during 'Polarstern' cruises ANT V/1 (1986) and ANT VI/2 (1987). *Arch FischWiss* **39**: 73–80.

Shulenberger E, Wormuth J H, Loeb V J (1984) A large swarm of *Euphausia superba*: Overview of patch structure and composition. *J Crust Biol* **4**: 75–95.

Siegel V (1986) Untersuchungen zur Biologie des antarktischen Krill, *Euphausia superba*, im Bereich der Bransfield Strasse und angrenzender Gebiete. *Mitt Inst Seefisch Hamburg* **38**: 1–244.

Siegel V (1987) Age and growth of Antarctic Euphausiacea (Crustacea) under natural conditions. *Mar Biol* **96**: 483–95.

Siegel V (1988) A concept of seasonal variation of krill (*Euphausia superba*) distribution and abundance west of the Antarctic Peninsula. In: Sahrhage D (ed) *Antarctic Ocean and Resources Variability*. Springer-Verlag, Berlin, pp. 219–30.

Siegel V (1989) Winter and spring distribution and status of the krill stock in Antarctic Peninsula waters. *Arch FischWiss* **39**: 45–72.

Siegel V (1992) Assessment of the krill (*Euphausia superba*) spawning stock off the Antarctic Peninsula. *Arch FischWiss* **41**: 101–30.

Siegel V, Bergström B, Strömberg J O, Schalk P H (1990) Distribution, size frequencies and maturity stages of krill, *Euphausia superba*, in relation to sea-ice in the northern Weddell Sea. *Polar Biol* **10**: 549–57.

Sparre P, Ursin E, Venema S C (1989) Introduction to tropical fish stock assessment. *FAO Fish Tech Pap* **306**: 1–337.

Stavn R H (1971) The horizontal-vertical distribution hypothesis: Langmuir circulation and *Daphnia* distribution. *Limnol Oceanogr* **16**(2): 453–66.

Stein M, Rakusa-Suszczewski S (1984) Meso-scale structure of water masses and bottom topography as the basis for krill distribution in the SE Bransfield Strait February–March 1981. *Meeresforschung* **30**: 73–81.

Stepnik R (1982) All year populational studies of Euphausiacea (Crustacea) in the Admiralty Bay (King George Island, South Shetland Islands Antarctic). *Pol Polar Res* **3**: 49–68.

Stretch J J, Hamner P P, Hamner W M, Michel W C, Cook J, Sullivan C W (1988) Foraging behavior of Antarctic krill *Euphausia superba* on sea ice micro-algae. *Mar Ecol Prog Ser* **44**: 131–9.

Stuart V, Pillar S C (1988) Growth and production of *Euphausia lucens* in the southern Benguela Current. *J Plank Res* **10**: 1099–112.

Watkins J L (1986) Variation in the size of Antarctic krill, *Euphausia superba* Dana, in small swarms. *Mar Ecol Prog Ser* **31**: 67–73.

Watkins J L, Morris D J, Ricketts C, Murray A W A (1990) Sampling biological characteristics of krill: effect of heterogeneous nature of swarms. *Mar Biol* **107**: 409–15.

Watkins J L, Morris D J, Ricketts C, Priddle J (1986) Differences between swarms of Antarctic krill and some implications for sampling krill populations. *Mar Biol* **93**: 137–46.

Weber L H, El-Sayed S Z (1985) Spatial variability of phytoplankton and the distribution and abundance of krill in the Indian sector of the Southern Ocean. In: Siegfried W R, Condy P R, Laws R M (eds) *Antarctic Nutrient Cycles and Food Webs*. Springer-Verlag, Berlin, pp. 284–93.

Weber L H, El-Sayed S Z, Hampton I (1986) The variance spectra of phytoplankton, krill and water temperature in the Antarctic Ocean south of Africa. *Deep-Sea Res* **33**: 1327–43.

Witek Z, Grelowski A, Kalinowski J (1982) Formation of Antarctic krill concentrations in relation to hydrodynamic processes and social behaviour. *ICES Biol Oceanogr Cttee CM 1982/L* **59**: 1–10.

Witek Z, Kalinowski J, Grelowski A (1988) Formation of Antarctic krill concentrations in relation to hydrodynamic processes and social behaviour. In: Sahrhage D (ed) *Antarctic Ocean and Resources Variability*. Springer-Verlag, Berlin, pp. 237–44.

Witek Z, Kalinowski J, Grelowski A, Wolnomiejski N (1981) Studies of aggregations of krill (*Euphausia superba*). *Meeresforschung* **28**: 228–43.

Witek Z, Koronkiewicz A, Soszka G J (1980) Certain aspects of the early life history of krill, *Euphausia superba* Dana (Crustacea). *Pol Polar Res* **1**: 97–115.

Wolnomiejski N, Chlapowski K, Porebski J, Garbacik-Wesolowska A (1978) Obserwacje nad ekologia antarktycznego kryla (*Euphausia superba* Dana). *Studia i Materialy MIR*, Gdynia, A23.

Yamanaka I (1983) Interaction among krill, whales and other animals in the Antarctic ecosystem. *Mem Natl Inst Polar Res* **27**: 220–32.

Krill energetics: seasonal and environmental aspects of the physiology of *Euphausia superba*

L. B. QUETIN, R. M. ROSS & A. CLARKE

Introduction

THE BIOMASS Programme (1976–91) has promoted the most intensive study of the ecology of the antarctic krill, *Euphausia superba* Dana, since the pioneering work of the *Discovery* Investigations from 1926 onwards for 14 years. The *Discovery* investigators learned much about the distribution of krill during its life cycle and in relation to the physical characteristics of its environment. For example, krill were found patchily distributed in the Southern Ocean, confined almost exclusively to regions south of the Antarctic Convergence, and seasonally associated with pack-ice. Based on catches of embryos and early larvae deep in the water column and later larval stages at the surface the developmental ascent of the larval stages of krill was described. Non-feeding larval stages were found to spend considerable time in Circumpolar Deep Water. Adult krill were found to be herbivores and to exist primarily in patches or swarms near the surface. Spawning was associated principally with shelf or slope waters with hatching occurring in the open sea over deeper water. The above are just a few examples of the abundant and valuable information gained during the *Discovery* Investigations. The BIOMASS era has greatly added to and expanded upon the earlier observations of the *Discovery* investigators. A significant advance during the BIOMASS period has been the development of physiological and biochemical studies of krill. These studies not only have an intrinsic value in attempting to understand the energetics of a pelagic organism living at low temperatures in high latitudes; they also provide necessary data for integration of krill dynamics within large-scale models of the pelagic system of the Southern Ocean.

In this chapter we review current knowledge of the energetics of *Euphausia superba*, emphasizing those topics where knowledge has developed since the previous review by Clarke & Morris (1983). We base our discussion around the energy budget, and progress from topics of uncertainty and debate to topics with more general agreement. We conclude with a discussion of the more ecological aspects of the interface between the energetics and the environment of krill. Our review is thus complementary to the recent major review of krill biology by Miller & Hampton (1989).

The energy budget

The traditional formulation of the energy budget for aquatic organisms is:

$$C = P + R + U + F,$$

where C is the total energy consumed (also often termed ingestion, I), P is the energy diverted to growth and gamete synthesis (together termed production), R is the energy used in respiration, U is the energy lost through excretion, and F is the energy egested as faeces.

This formulation of the energy budget equation is too broad to allow a detailed understanding of marine invertebrate energetics. In particular it does not reflect the separate elements of respiratory demand, and obscures the important relationships between production and respiration (Clarke, 1987). A better formulation is thus:

$$C \text{ (or } I) = (P_s + P_g) + (R_b + R_a + R_s + R_g) + U + F,$$

where P_s and P_g represent somatic growth (including exuviae) and gamete synthesis, respectively, and the respiration term is subdivided into R_b (basal or maintenance costs), R_a (the costs of all activity, including locomotion and feeding) and R_s and R_g (the respiratory costs of production for the synthesis of new somatic and reproductive tissue, respectively). R_s and R_g are the energy lost as heat during the synthesis of new tissue because according to current best estimates synthesis is only 70–80% efficient (Parry, 1983). R_s and R_g are now believed to account for most of the so-called Specific Dynamic Action (SDA, the increase in metabolic rate due to the ingestion of some foods) observed following feeding (Jobling, 1983; Vahl, 1983).

The various components of the energy budget will vary with age, season and factors such as food availability and temperature. A complete understanding of the energetics of krill will therefore necessarily involve a knowledge of the physiology of all life stages, and how this responds to environmental variables.

Filtration and egestion rates

In the past decade we have changed our view of filtration rate in krill, from an individual filtering only millilitres per hour to one capable of filtering litres of seawater per hour. Clarke & Morris (1983) and Morris (1984) reviewed previous studies of filtration rate in *Euphausia superba* up to that time, and suggested that many of the published values were too low because in traditional closed-volume filtration experiments most of the particles are initially removed more rapidly than the sampling interval can follow. Thereafter the krill are refiltering water from which most of the particles have already been removed, and filtration activity itself decreases. A simple scaling up of techniques used successfully for copepods does not necessarily allow for the much greater filtration activity of large euphausiids. However, Price *et al.* (1988) found filtration rates of krill feeding in 50 l volumes were 2 times higher than in 5 l volumes. Since krill densities were kept equal in the two volumes, lower filtration rates in the smaller volume were probably primarily caused by more frequent encounters with the walls of the smaller 5 l container compared with the larger 50 l container, resulting in a greater disruption of feeding behaviour.

The effect of chamber volume is clear, but even the highest rates (those of Price *et al.*, 1988, based on a chamber volume of 50 l) are lower than the filtration rate calculated from field ingestion rates by Quetin *et al.* (1988) for adult krill in summer (Fig. 1). Ikeda & Bruce (1986) have estimated filtration rates of up to 26 l h^{-1} from carbon requirements of mature females. With the typical range of 0.1 to 1 μg chlorophyll *a* per litre for non-bloom and oceanic chlorophyll concentrations found in the Southern Ocean (Smith & Sakshaug, 1990), filtration rates much higher than those measured in small chambers are necessary to meet the known energy requirements of krill (Morris, 1984). However, the extremely high filtration rates reported by Morris (1984) are statistical artefacts based on particle sizes that were present at only low concentrations, and so only the biologically meaningful data have been plotted in Fig. 1.

The only published data available for faecal egestion (Clarke *et al.*, 1988) compared laboratory and field estimates, which were similar, and showed that faecal egestion rates were high, a result consistent with high filtration rates in the summer. Egestion rates indicated a total energy intake of 17–28% body mass per day in adults, much higher than previous estimates for *Euphausia superba* but similar to estimates for other euphausiids. This represents a fundamental change in our view of krill energetics to an organism capable of a high throughput of energy, and hence of either fast growth rates or high reproductive output.

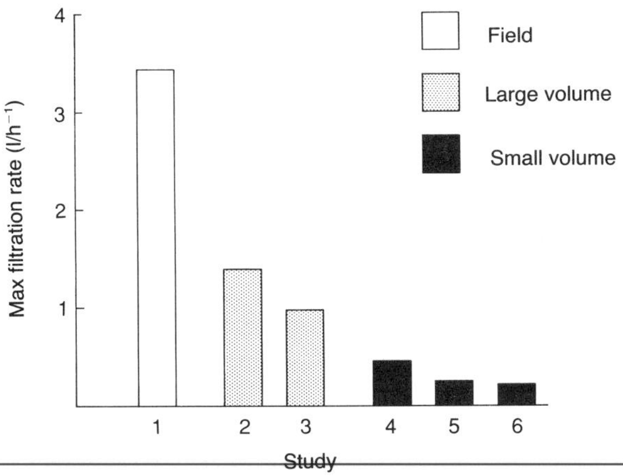

Fig. 1. Comparison of filtration rates measured for *Euphausia superba* with an estimate calculated from field ingestion rate. Studies are: (1) field ingestion rate (Quetin *et al.*, 1988), (2) large chamber, through-flow (Morris, 1984), (3) large chamber (Price *et al.*, 1988), (4–6) small chambers (4, Kato *et al.*, 1979; 5, Antezana *et al.*, 1982; 6, Boyd *et al.*, 1984).

Reproductive effort

Advances in our understanding of reproductive biology and physiology in *Euphausia superba* during the BIOMASS era have been of particular significance. Recent discussion has centred on the duration of the spawning cycle in krill and the level of individual fecundity. These features of krill biology and energetics are so crucial that they warrant a critical review of current evidence and ideas.

General features of spawning in krill

The age at which Antarctic krill spawn is unknown, but based on the size at first spawning, the first reproductive season is believed to be at two years of age during the third summer of existence (Miller & Hampton, 1989). Spawning first occurs when krill are approximately 33 mm in length (Cuzin-Roudy, 1987*a*). Antarctic krill spawn during the summer or late spring (Bargmann, 1945; Marr, 1962). Latitudinal variations in spawning time (Makarov, 1979) suggest that an increased availability of food is essential to ovarian development, and that much of the energy input for reproduction comes from spring and summer phytoplankton blooms.

Observations of external sexual characteristics and histology of small krill have established that differentiation of the gonad occurs before the appearance of secondary sexual characters, just as in mysids and other peracarid crustaceans (Cuzin-Roudy, 1987*a*). Larger immature krill are likely to be postspawn adults that have regressed through successive moults to a subadult appearance following the reproductive season, as shown for adult krill held in the laboratory (Denys & McWhinnie, 1982; Poleck & Denys, 1982) or that postponed spawning to age 3+. Makarov (1975) first suggested that krill spawn during several austral summers, and the ability to resume sexual development after regression to a subadult appearance has been confirmed in the laboratory (McWhinnie *et al.*, 1979; Thomas & Ikeda, 1987).

Energetic cost of reproduction

Little progress has occurred in our understanding of the costs of reproduction in males. We do know that males produce sperm throughout the spring and summer, since Bargmann (1945) found spermatophores developing in males and attached to females from September through March. Mating continues throughout the season because females lose most of the sperm packet when they moult every three weeks or so (Harrington & Ikeda, 1986). Quantitative estimates of the energetic costs of sperm production and of mating are lacking.

The question of energetic costs of reproduction for females devolves into two parts: energy content of reproductive products (eggs), and the duration and frequency of spawning. Several estimates of the energy content of an embryo, either as lipid or carbon, are available (Clarke & Morris, 1983; Clarke, 1984; Ikeda, 1984*b*; Amsler & George, 1985; Ross & Quetin, 1989) and most discussion surrounds questions of total seasonal fecundity, and the effects of environmental conditions on reproductive effort. The available evidence suggests that in regions where primary production is high, reproduction in krill is characterized by multiple spawnings and high fecundity (Denys & McWhinnie, 1982; Ross & Quetin, 1983, 1986), whereas in other regions, fecundity may be significantly lower or reproduction may not occur at all.

Because this view of krill reproductive energetics is not universally accepted we will review the evidence. Critical to this discussion is recent knowledge of the histology of ovarian development.

Histological evidence

The ovary contains a permanent germinal zone with primary and secondary oogonia, and batches of oocytes. Germ cells in the krill ovary go through the same sequence of physiological steps as other crustaceans (Adiyodi & Subramoniam, 1983). After spawning or a winter hiatus, gametogenesis and oogenesis occur; the secondary oogonia (og2) leave the germinal zone and become young oocytes. At this point meiosis is arrested, and the accumulation of yolk (vitellogenesis) in two phases begins (Cuzin-Roudy, 1987*b*) (Fig. 2). During previtellogenesis only proteic yolk is accumulated in the small type 1 oocytes (oc1), whereas during vitellogenesis, lipid yolk is added and the oocytes increase in size (type 2 and type 3 oocytes, oc2 and oc3). Once vitellogenesis is complete, meiosis is resumed, maturation is completed (type 4 oocytes, oc4) and eggs are released. Young oocytes are found in large female krill just prior to winter, suggesting that only vitellogenesis may be delayed until the following spring and summer when yolk accumulation could be quick and repetitive with abundant food (Cuzin-Roudy, 1987*b*).

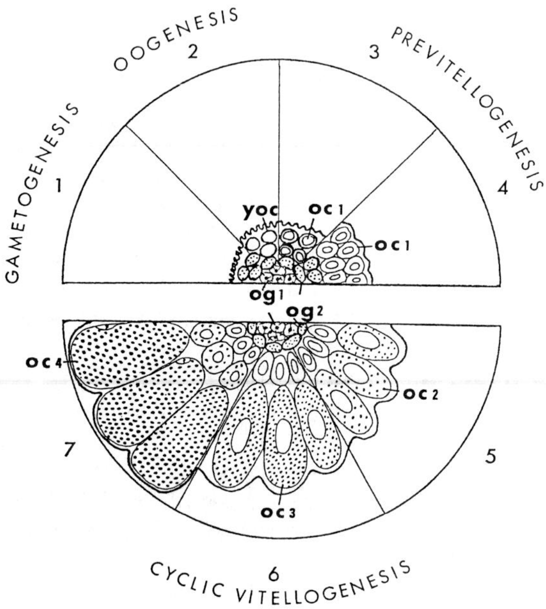

Fig. 2. General representation of the developmental stages of the ovary in adult female *Euphausia superba*. Gametogenesis (1) and oogenesis (2) produce primary oocytes (yoc) which undergo two phases of yolk accumulation, firstly previtellogenesis (3,4) and secondly vitellogenesis (5,6) before reaching maturity (7). Vitellogenesis is cyclic and results in the production of successive spawns. Abbreviations: oc1, primary oocytes accumulating proteic yolk only; oc2, oc3, primary oocytes accumulating lipid yolk; oc4, mature secondary oocytes; og1, og2, primary and secondary oogonia forming the permanent Germinal Zones of the ovary. Redrawn from Cuzin-Roudy & Amsler (1991).

Prior to the past decade, ovarian development was thought to be a simple stepwise process (Bargmann, 1945), and it was assumed that the total seasonal egg production of a female could be assessed by counting either all the oocytes or only the large eggs (oc3 or oc4 stages) in the ovary. Those counting only larger oocytes assumed that the smaller oocytes left in the ovary after spawning were resorbed. The simultaneous presence of oocytes at different stages of development in the ovaries of actively reproducing krill indicates clearly that ovarian development is cyclical rather than stepwise (Kikuno & Kawamura, 1983; Cuzin-Roudy, 1987*b*). Furthermore, a continuous maturation and release of eggs throughout the spawning season is a widespread and common strategy in crustaceans and fish. We can therefore define the total fecundity in a given season (Fec) as

$$\text{Fec} = N_e \times \text{SS/SI},$$

where N_e is the number of eggs per spawning episode, SS is the length of the spawning season and SI is the spawning interval or time between spawning episodes. These variables may change with female age, size or nutritional history, and may be related to factors such as geographical location, interannual variability or time in the season.

Number of eggs per spawning episode

The number of eggs per spawning episode has been estimated from measurements of ovarian volume (Mauchline & Fisher, 1969), counting ovarian eggs (Denys & McWhinnie, 1982; Siegel, 1985), estimating lipid loss (Clarke, 1980), and counting eggs released in a spawning episode. Counting spawned eggs is preferable to counting ovarian eggs because only mature eggs are released and eggs at different physiological stages overlap in size (Cuzin-Roudy & Amsler, 1991).

Estimates of the number of eggs released per spawning episode range from about 100 to 8000 (Denys & McWhinnie, 1982; Kikuno & Kawamura, 1983; Ross & Quetin, 1983; Harrington & Ikeda, 1986). The major cause of variability appears to be female size. We do not yet know whether the number of eggs per spawning episode varies with age of female, or with time during the spawning season. Based on ovarian counts of large eggs Siegel (1985) suggested that there might be differences between years in the number of eggs spawned. Denys & McWhinnie (1982) and Cuzin-Roudy (1987*b*) suggested that the number of ovarian eggs decreased during the season, particularly for the final spawning episode.

Number of spawning episodes

Our view of spawning frequency in *Euphausia superba* has changed significantly during the BIOMASS period. Until recently only one spawning episode per female per year was assumed (Bargmann, 1945), based on the belief of a simple stepwise maturation of eggs in the ovary, and a long period needed for vitellogenesis. Denys & McWhinnie (1982) suggested two spawning episodes per season, based on the observation that spawned eggs were about half the number of ovarian eggs. Most recently the results of shipboard measurements of spawning in freshly sampled krill has led to the concept of multiple spawning episodes, with the number dependent on food availability during the spring and summer (Ross & Quetin, 1983, 1986).

The histological evidence is consistent with the concept of multiple spawning, and suggests that a female krill can successively mature at least three batches of eggs per season (Cuzin-Roudy, 1987*b*). Histological studies showed that oocytes in previtellogenesis were never the only physiological stage in mid-summer, and ovaries either contained 1–3 batches of oocytes in vitellogenesis, or were ready to release the last batch, or were in postspawn condition (that is with only the germinal zone and oogonia) (Cuzin-Roudy, 1987*b*). Similar results were also found in freshly sampled krill by Harrington & Ikeda (1986) and Kikuno & Kawamura (1983). Vitellogenesis is assumed to be rapid, and once oocytes enter the cycle of vitellogenesis they will complete that cycle, mature and be released. Calculations of the energy costs of multiple spawning led Ross & Quetin (1986) to suggest that the number of spawning episodes for an individual female will depend on food availability. The number could range from zero (that is, the oogonia never enter the cycle of vitellogenesis) up to about seven. The higher number implies that the cycle of three batches of eggs is repeated.

Indirect evidence for multiple spawning comes from shipboard experiments with freshly collected krill (Ross & Quetin, 1983, 1986; Harrington & Ikeda, 1986). Slow, short tows ensure that krill are brought aboard in the best possible condition, and shipboard experiments minimize the effects of long-term maintenance in the laboratory. Fifty females are randomly chosen from all those with red thelyca (mature) in a sample of a school of krill, and isolated in 2 l jars. The percentage of the population (of 50) that releases eggs per day is the spawning frequency, and the interval between spawning episodes is the inverse of the spawning frequency. The pattern of spawning frequency in schools with reproductive krill from waters west of the Antarctic Peninsula was similar in the three seasons from 1981–2 through 1983–4, and for krill from Prydz Bay (calculated from data of Harrington and Ikeda (1986)) (Fig. 3). Spawning frequency was zero in early December, increased throughout early December to about 11% (range 4–19%) in January and early February, and

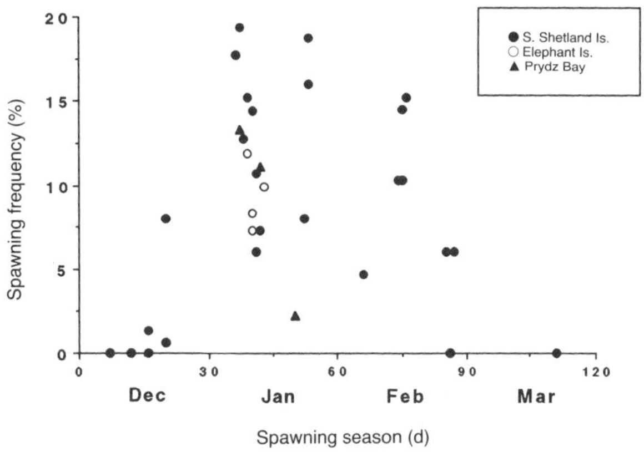

Fig. 3. Spawning frequency in schools of reproductive *Euphausia superba* sampled from waters west of the Antarctic Peninsula from 1981–2 to 1983–4.

decreased in late February. Spawning occurred over a prolonged period of 2.5 months, but was most intense over a 6.5 to 7 week period. In 28 schools sampled that contained mature females, about 11% of the mature females released a batch of eggs daily, or the interval between spawning episodes was about 9 days (Ross & Quetin, 1986), a longer interval than previously calculated from one year of data (Ross & Quetin, 1983). If krill only spawned once per season, then either spawning frequencies of 10–11% per day should be found for only a brief period, about 9 days, or a much lower spawning frequency would be found during a longer spawning season, i.e. 2% for 50 days or 1.3% for 75 days.

The most critical assumption to the interpretation of the results of the spawning frequency experiments is that we are sampling the same population during the entire spawning season. The second assumption is that spawning in any one school is not synchronized, unlike spawning in some decapods (Anderson *et al.*, 1985). If all female krill from a region spawned in synchrony, then very high spawning frequencies would be found only during a brief period, and the variance in estimates of spawning frequency would be much higher than observed.

One argument advanced against multiple spawning is that the second phase of yolk accumulation takes so much energy that it cannot happen on the time scale of days (Siegel, 1985). Evidence for a rapid second phase of yolk accumulation comes both from repeat transects during EPOS (European *Polarstern* Study) during rich spring conditions when females in spawning condition appeared between one transect and the next (Cuzin-Roudy & Schalk, 1989), and from results of analysis of preserved samples with the physiological maturity key of Cuzin-Roudy & Amsler (1991) and spawning frequency experiments on live females from the same schools (Fig. 4). Only 11% of the mature females (with red thelycum, stages 4–9) from a randomly preserved portion of a catch were stage 7, with eggs ready to spawn. Yet in a spawning experiment with a random sample of mature females from the same collection, 7.5%/day spawned. The inference is that the secondary phase of vitellogenesis is very short, less than two days.

Although accumulating evidence supports the concept of multiple spawning by *E. superba* during a spawning season (Denys & McWhinnie, 1982; Ross & Quetin, 1983, 1986; Cuzin-Roudy, 1987*b*), the evidence for more than two or three broods is still indirect.

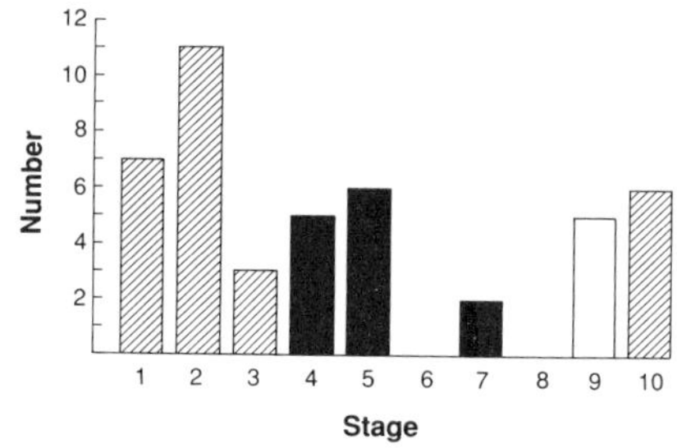

Fig. 4. Reproductive stage of female *Euphausia superba* from a single school sampled on 21 January 1984 from waters west of the Antarctic Peninsula. Females were staged according to Cuzin-Roudy & Amsler (1991). Key: hatched bars, non-reproducing females; solid bars, mature females, clear bar, spent females; stage 10, regressing females.

Length of the spawning season

Very few studies have been able to follow one entire spawning season from beginning to end, but there appears to be considerable interannual variability in both initiation and cessation of spawning even in the same region. For example, spawning did not begin until mid-December in waters west of the Antarctic Peninsula and ended in late February or early March during the three seasons 1981–2, 1982–3, 1983–4 (Fig. 3). Yet in the same region during 1980–1, nearly all females were capable of releasing eggs by the second half of November (Siegel, 1985), and stages capable of spawning dominated from at least mid-November until early January. The 1980–1 spawning season thus both started and ended earlier than usual. Estimation of the total length of the 1980–1 spawning season is difficult because data on the first occurrence of females capable of spawning are lacking. In addition, the assumption that spawning does not begin until spent females begin to appear (Siegel, 1985) is not supported by the histological evidence of Cuzin-Roudy (1987*b*) or the experiments of Harrington & Ikeda (1986).

If food availability is controlling initiation of spawning and timing of the spawning season, then it may also control the length of the spawning season. Low rates of primary production or a delayed spring bloom may affect both timing and length of the spawning season in *Euphausia superba*. The pattern of occurrence of larval stages also supports a prolonged spawning season. The number of larval stages in about equal abundance multiplied by the duration of each stage would yield an estimate of the duration of the period of intense spawning. Brinton & Townsend (1984) found five furcilia stages in about equal abundance in the Scotia Sea. Using a 9–10 day intermoult period for each of the five stages plus one additional intermoult for furcilia 3 to calculate the period of intense spawning leads to an estimate of 54–60 days, very similar to that found with the occurrence of spawning or gravid females. Thus available evidence suggests that a prolonged (2 month) spawning season is both possible and common in the waters west of the Antarctic Peninsula.

Growth

Understanding the growth of *Euphausia superba* has been a problem since the days of the *Discovery* Investigations, and this subject continues to provide difficulties despite intensive work during the BIOMASS period. Much attention continues to be directed at understanding the growth rate and age-at-size of Antarctic krill because of the importance of these parameters in both energetics and population dynamics. The major problems are related to the slow overall growth rate, the marked seasonality of growth, the extended recruitment season, and the likelihood of geographical variations in growth rate. These factors combine to make the traditional approach of analysing length–frequency histograms very difficult. In recent years a number of techniques have been used, including traditional length–frequency analysis of repeated samples, determination of instantaneous growth rates (IGRs) from field samples, and attempts to provide a chemical measure of age. These results have been reviewed extensively by Miller & Hampton (1989). We will limit our comments to salient points.

Length–frequency analysis

The major advance over the early work of the *Discovery* Investigations and Japanese workers (summarized by Mackintosh, 1972; Mauchline, 1980), has been the development of computerized techniques for length–frequency analysis. Both the Macdonald & Pitcher (1979) technique and the ELEFAN (Electronic Length Frequency Analysis) procedure (Pauly & David, 1981) have been used.

Siegel (1986) used the Macdonald & Pitcher (1979) technique to distinguish size classes from complex length–frequency data for samples of krill from the Bransfield Strait, and fitted a von Bertalanffy growth model. His analysis suggests that krill reach a maximum total length of about 60 mm in five years. Although this analysis produces a biologically sensible and internally consistent result, Grant *et al.* (1987) have shown that for size (or age) classes to be assigned with even moderate confidence to length–frequency data exhibiting several peaks, sample sizes considerably greater than normally available are required. Rosenberg *et al.* (1986) analysed krill data collected by the *Discovery* Investigations around South Georgia between 1928 and 1938, and applied a seasonally adjusted von Bertalanffy model. Their analysis suggested a mean daily growth rate for adult krill in summer of 0.148 mm d^{-1}.

McClatchie (1988) used the ELEFAN length–frequency analysis package developed by Pauly & David (1981) to analyse the length–frequency data obtained between December 1978 and March 1980 at Admiralty Bay, King George Island, by Stepnik (1982). The ELEFAN technique optimizes a fit to the complete data set, rather than to the individual samples for each month. This technique was developed for use with organisms where the unequivocal assignment of modal values is difficult, precisely the situation with *Euphausia superba*. The optimized fit for Stepnik's data suggested peak summer growth rates, of 25–55 mm krill, ranging from 0.11 to 0.025 mm d^{-1} and dependent on age.

One difficulty with these results is that the ELEFAN procedure also predicted small positive growth rates in winter, rather than the zero or negative rates predicted by other workers (McClatchie, 1988). However, throughout much of the winter period the predicted mean cohort size from the growth curve suggested by ELEFAN was substantially greater than the sample mean or mode, and in extreme cases (October) fell well outside most of the data. In addition, and also worryingly, the major size-class in the population disappeared from the samples in one month (April). Simple tracing by eye of the dominant modal class suggests that adult krill of about 45 mm in March 1979 shrank during winter, as concluded by Ettershank (1983) in his analysis of the same data set.

Instantaneous growth rates

Another experimental procedure shows promise. The instantaneous growth rate (IGR) of krill can be determined by sampling krill carefully from the wild, holding them in the laboratory for short periods of time, and determining the moulting frequency and size increment per moult. This approach eliminates the assumptions made in length–frequency analysis that the same population is being sampled each time (Quetin & Ross, 1991). In addition, the short experimental protocol that measures only the first intermoult increment after capture yields growth rates representative of those in the wild, as suggested by Buchholz (1985). Krill from Bransfield Strait in late summer and early autumn (1985–6 season) grew 1.75 to 4.40% in length per intermoult period (Quetin & Ross, 1991), equivalent to 0.056 mm d^{-1} for a 40 mm adult.

Growth rates in winter

Although Ikeda & Dixon (1982*b*) showed that adult krill have the ability to withstand winter starvation by shrinkage, there has been considerable debate over the growth of krill in winter. Two sets of data address the question of winter growth rates in postlarval krill. As discussed above, McClatchie (1988) interpreted monthly samples taken by Stepnik (1982) from Admiralty Bay during an unusually ice-free winter (1979) as showing slow positive growth in winter. With the same data set Ettershank (1983) concluded that the adult size class showed clear evidence of negative growth (shrinkage) in winter. Quetin & Ross (1991) used the IGR technique on adult krill collected from the Bransfield Strait in August and September 1985. These krill had negative growth increments (−0.16 to −2.03% length per intermoult period) and long intermoult periods, and length–mass relations showed that they had shrunk in volume as well. Available data thus suggest that, at least in some areas, postlarval krill shrink in winter.

Larval growth rates

Growth rates of larvae maintained in the laboratory with high food availability were 0.06–0.11 mm d^{-1} (Ikeda, 1984*a*, 1985), three to five times those of larvae maintained on winter food concentrations (0.02 mm d^{-1}) (Elias, 1990) at either near-summer or winter temperatures.

Growth rates of larvae in the field have been determined with the same methods as for adults, namely traditional length–frequency analysis and IGRs of freshly collected larvae. Summer growth rates determined either from analysis of preserved samples (Witek *et al.*, 1980) or from a combination of shipboard experiments and sample analysis (Huntley & Brinton, 1987) were of the order of 0.06 mm d^{-1}, similar to those found in the laboratory at 0 °C and with high food concentrations (Ikeda, 1984*a*). In two winter field studies, larvae were grazing on the undersurface of the ice. Daly (1990) measured growth with the length–frequency technique in the Weddell Sea, and Ross & Quetin (1991) worked the waters west of the Antarctic Peninsula with the IGR technique. Larval growth rates in the Weddell Sea study (0.07 mm d^{-1}) were similar to summer growth rates, and four times those in the waters west of the Antarctic Peninsula (0.017 mm d^{-1}).

Significant statistical problems exist for any analysis of growth from repeated samples from the same population (Watkins *et al.*, 1986), and the contrast between these two studies shows the difficulties of measuring growth in the field. Daly (1990) compared the length–frequency distributions of two samples taken 6 weeks and 180 nautical miles apart and attempted to establish that these samples were from the same population by comparing lengths of larvae of a specific stage along a transect, rather than the length–frequency distribution or stage composition of the samples collected along the transect. Unfortunately, stage composition of larval samples often varies significantly along a transect (Brinton & Townsend, 1984). With uncertain verification that these two samples represented the same population, the accuracy of the growth estimate is suspect. As explained previously, the IGR technique circumvents these difficulties.

Growth rate and food concentration

There are no published field data for the effect of food concentration on the growth rate of adult krill, and each of two attempts to model growth in relation to food availability have problems. Holm-

Hansen & Huntley (1984) attempted to calculate lower thresholds for growth, but their analysis was based on a preliminary energy budget that omitted some terms and almost certainly underestimated energy requirements. Their equation related maintenance food concentration (which can be regarded as the threshold for growth) to krill dry mass, and predicted threshold carbon concentrations equivalent to chlorophyll concentrations of 2.64–4.85 μg l^{-1} well above typical oceanic chlorophyll concentrations for the Southern Ocean. Considering that the energy requirements for krill were severely underestimated in this model, the major source of error is undoubtedly the unrealistically low values assumed for the filtration rate of krill (see section on filtration rate). In a second study, McClatchie (1988) compared the growth rates produced from the ELEFAN analysis of krill length–frequency data for Admiralty Bay with chlorophyll concentrations measured at the same time (Lipski, 1987). Chlorophyll concentrations limiting growth were only approached in winter, at 0.06–0.1 μg l^{-1}. These values are more realistic than those of Holm-Hansen & Huntley (1984), but they are based on winter growth rates which may be unrealistically high (see discussion above).

In a laboratory study Ikeda & Thomas (1987) found that when juvenile krill were fed *Phaeodactylum tricornutum* growth was proportional to algal density up to 1000 μg Cl^{-1} (16.7 μg chlorophyll l^{-1} assuming a carbon : chlorophyll ratio of 60). The threshold for growth of juvenile krill was 20–100 μg carbon l^{-1} (0.33–1.67 μg chlorophyll l^{-1}). Such values are unrealistically high for the Southern Ocean, probably because *Phaeodactylum* is not a good food source for krill. Lower threshold values for growth would therefore be expected in the wild.

Moulting

Considerable advances have been made in understanding the process of ecdysis in *Euphausia superba* since the review by Clarke & Morris (1983). Buchholz (1982, 1985, 1991) has established the major features of the moult cycle in euphausiids, provided a definitive terminology and recently reviewed laboratory studies of growth and moulting of *Euphausia superba*. Several laboratory studies have looked at the process of ecdysis in detail and the timing of the various stages (for example Morris & Keck, 1984; Buchholz, 1985, 1991; Maihara & Endo, 1986; Nicol & Stolp, 1990), and detailed field studies have integrated moulting into the general biology of krill (Morris & Priddle, 1984; Morris & Keck, 1984; Morris, 1985; Buchholz *et al.*, 1989). Results from several laboratory studies can be combined to describe the intermoult period (IMP) in relation to temperature, which shows a Q_{10} of 3.5 over the range 0–5 °C, and predicts a mean winter IMP of 93 days for adult krill at −1.8 °C (Fig. 5). Also plotted are the range of field data for krill in the Bransfield Strait in summer and winter (Quetin & Ross, 1991). Both summer and winter moulting rates overlap the laboratory data. The range of field values is large, indicating that the effect of food availability, size and other factors such as synchrony of moulting are superimposed on the effect of temperature.

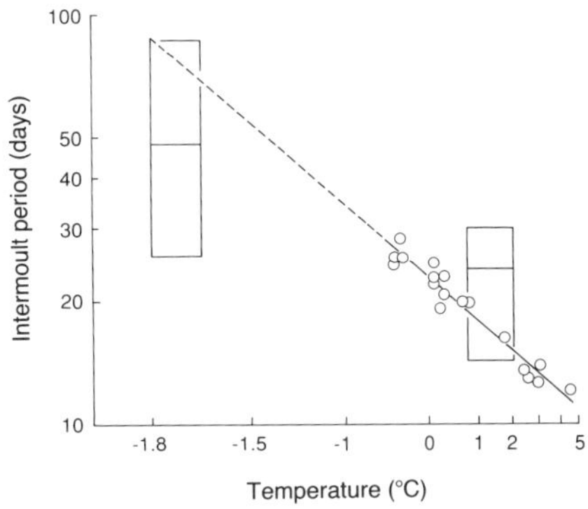

Fig. 5. Relationship between intermoult period (IMP) and temperature for *Euphausia superba*. Open circles show experimental data, with an overall relationship fitted by least-squares regression to logarithmically transformed data and extrapolated (dashed line) to winter temperatures. IMP = 36.47 (T + 2) − 0.587. The addition of 2 is to enable the curve to be fitted by linear regression to logarithmically transformed data. Data plotted from Murano *et al.* (1979), Ikeda & Dixon (1982*a*), Morris & Keck (1984), Buchholz (1985), Ikeda *et al.* (1985) and Nicol & Stolp (1990). The boxes show mean and overall range for field data (LB Quetin & RM Ross, unpublished data) in summer and winter.

Growth rate: summary

Available data on growth rate for *Euphausia superba* show general agreement among the overall growth curve proposed by Ikeda (1985), the summer growth rates calculated by McClatchie (1988), the *in situ* determination of Kanda *et al.* (1982) and the IGR determinations of Quetin & Ross (1991) (Fig. 6). All these exceed the three laboratory determinations (McWhinnie & Denys, 1978; Murano *et al.*, 1979; Poleck & Denys, 1982), emphasizing the difficulties of determining true growth rates from krill held for long periods in aquaria (Buchholz, 1991). The growth rate suggested by Clarke & Morris (1983) for a population of krill near South Georgia in January 1982 is larger than other determinations by a factor of almost three. While growth rates of krill may be higher in the warmer waters around South Georgia (as shown by the growth rate for this area derived by Rosenberg *et al.*, 1986), this difference is so great that the value is suspect. While we are beginning to obtain a working concensus on representative growth rates for postjuvenile krill in the wild, we still have no data about geographical variation of growth rates. These remain major challenges for krill biologists in the post-BIOMASS years.

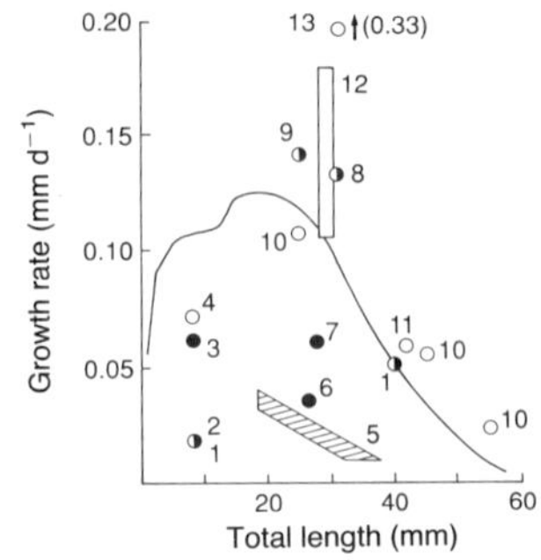

Fig. 6. Growth rate for *Euphausia superba* in relation to Ikeda's (1985) growth model (solid line). Open circles and the open box show field data (usually derived from sequential length–frequency distributions); solid circles and the hatched box show data from krill held in the laboratory. Split symbols show data for krill sampled from the wild and where growth was determined from immediate short-term measurements of growth increment and moulting frequency (IGR) (see text). Key to studies: 1, Quetin & Ross (1991); 2, Elias (1990); 3, Ikeda (1984*a*); 4, Daly (1990); 5 (hatched box), McWhinnie & Denys (1978); 6, Murano *et al.* (1979); 7, Poleck & Denys (1982); 8, Buchholz (1985); 9, Morris & Keck (1984); 10, McClatchie (1988); 11, Kanda *et al.* (1982); 12 (open box), Rosenberg *et al.* (1986); 13, Clarke & Morris (1983).

Respiration

Physiological measurements of free-swimming zooplankton in the open sea are not yet feasible and laboratory studies of confined organisms continue to be necessary. The caveats outlined by Childress (1977) remain important but unresolved.

The swimming costs for euphausiids are believed to be high (Kils, 1981; Torres *et al.*, 1982; Torres & Childress, 1983); Torres (1984) concluded that low swimming efficiencies are responsible for a general underestimation of zooplankton swimming costs. No study of oxygen consumption in *Euphausia superba* has provided for the effect of swimming speed, and only Ikeda & Dixon (1984) have examined the effect of feeding activity on metabolism. All other studies have measured routine oxygen consumption of krill confined to an experimental chamber, and obtained values above basal (as defined by Clarke & Morris, 1983; Clarke, 1991) and below maximum activity. Metabolic costs of activity in the wild remain a major gap in our knowledge of krill energetics.

Ikeda & Dixon (1984) suggested that the oxygen consumption rates on non-feeding krill average only 62.5% of those of krill feeding at realistic food concentrations. With current techniques it is not possible to distinguish the increase in oxygen consumption during feeding due to the mechanical costs of filtration activity and to the costs of growth resulting from the food intake. However, krill probably do not feed adequately during oxygen consumption measurements because of the necessary small size of the experimental chamber (see section on filtration rate).

Most measurements of oxygen consumption in *E. superba* (and all those plotted in Fig. 7) have been made in summer. In krill recently sampled from the wild, growth and/or gamete synthesis will likely continue, at least for a while, during any measurement of oxygen consumption. Summer measurements of metabolism will therefore include a component from the costs of production (R_s and R_g in the energy budget). In theory these costs could be estimated from a knowledge of production rates (Parry, 1983), but we do not yet have sufficient knowledge of growth and reproductive energetics to do this. Nevertheless we can conclude that reduced food availability in winter is likely to reduce the costs

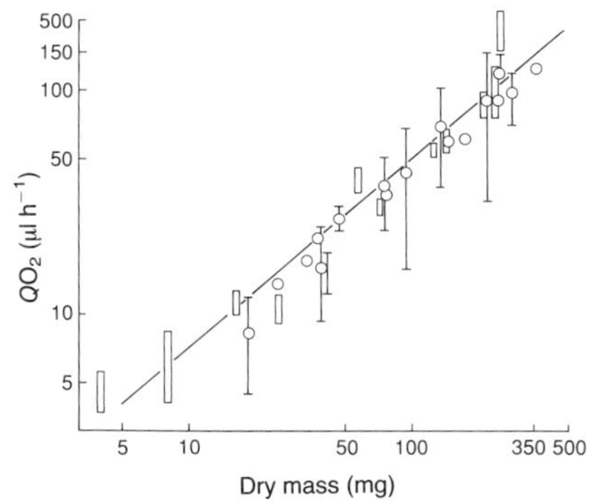

Fig. 7. Laboratory measurements of oxygen consumption (QO_2) as a function of dry body mass (W) in *Euphausia superba*. Published values are plotted as mean (open circle), mean with 95% confidence intervals, or range (open box). The overall relation was fitted by least-squares regression to mean data after logarithmic transformation. $QO_2 = 0.985W^{0.837}$ ($QO_2 = 0.044W^{0.837}$ when expressed in μmol h^{-1}). Data were obtained from Rakusa-Suszczewski & Opalinski (1978), Segawa *et al.* (1979), Kils (1981), Ikeda & Hing Fey (1981), Ikeda & Mitchell (1982), Ikeda (1984*b*), Ikeda & Bruce (1986) and Ishii *et al.* (1987).

of feeding activity as well as those associated with production. We should therefore expect to see a reduced routine metabolism (as would be measured in a respiration chamber) in winter.

The two studies of winter metabolism in *E. superba* have both shown lower rates than in summer. Kawaguchi *et al.* (1986) showed that in krill wintering under fast-ice in Lutzow–Holm Bay the oxygen consumption in August was half that in September, and Quetin & Ross (1991) showed that in postjuvenile krill (total length 30–55 mm) from the Bransfield Strait winter respiration rates based on length (July, −1.5 °C) were only one third of those measured in summer (January, +2 °C). Rakusa-Suszczewski (1990) reported that the oxygen consumption of krill in Admiralty Bay, King George Island, increased by 40% in early spring between early October (−1.8 °C) and late November (−1.0 °C).

In conclusion, although our knowledge of seasonal patterns of krill metabolism has increased in the past decade, incorporation of metabolic data into an energy budget is still hampered severely by our lack of data on the relation between oxygen consumption and swimming speed. Studies of the costs of swimming in *E. superba* are continuing (L B Quetin & R M Ross, unpublished data). At present the overall relation shown in Fig. 7 is the best measure we have for respiratory costs of krill during summer and compares favourably with the relation reported by Clarke & Morris (1983) and that of Ikeda & Mitchell (1982) proposed as a general relation by Miller & Hampton (1989). However, we believe that current values of respiratory costs during summer are underestimates of the real costs in the wild.

Excretion

Excretion rate varies markedly with recent feeding history and so any summary relationship will necessarily be of limited usefulness. Data available on nitrogen excretion in *Euphausia superba* (Fig. 8a) are based on the summary tables in Clarke & Morris (1983) and Miller & Hampton (1989). Some studies

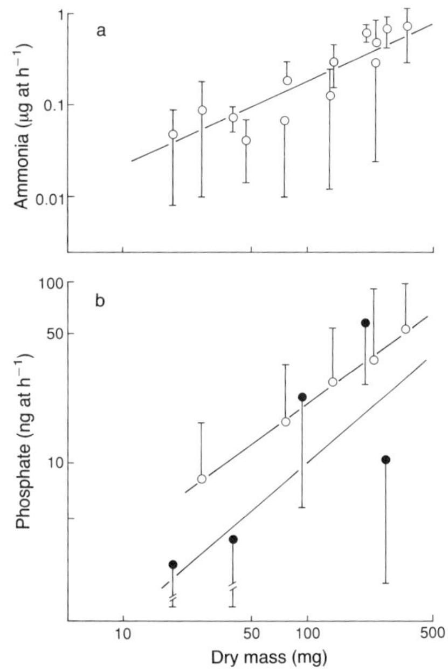

Fig. 8. Excretion in *Euphausia superba*. (a) Ammonia excretion; data from Segawa *et al.* (1982; surface net data only) and Ikeda & Mitchell (1982). (b) Phosphate excretion; data from Ikeda & Mitchell (1982; open circles and upper line) and Ishii *et al.* (1987; closed circles and lower line).

were excluded, however: (1) Johnson *et al.* (1984) for Scotia Sea euphausiids and Biggs (1982) for Ross Sea euphausiids since *E. superba* was a minor component in both studies; (2) Segawa *et al.* (1982) for krill taken by pump sampler, since excretion rates were much higher than krill sampled by surface net; (3) the high data for newly sampled animals reported by George & Fields (1984); and (4) Hirche (1983), whose results were presented in terms of length rather than mass. The data in Fig. 8a suggest that the overall relations determined by Ikeda (1984*b*) or Segawa *et al.* (1982) for krill taken by surface net are representative.

O : N atomic ratios provide information on the relative proportions of lipid, protein and carbohydrate being utilized as metabolic substrate. Representative oxygen consumption (Fig. 7) and ammonia excretion rates for krill (Segawa *et al.*, 1982) suggest an average summer O : N atomic ratio of 20, indicating a mixture of substrates (as would be expected). Experimental studies have determined O : N ratios ranging from 12 to 103 (see recent collation by Miller & Hampton, 1989) emphasizing that O : N ratio will vary with age, season and recent feeding history.

A few investigations have looked at phosphate excretion in *E. superba* (Fig. 8b). Two studies give mean rates that differ by a factor of two, indicating that phosphate excretion probably varies with factors such as recent feeding history. The mean of these two rates suggests an atomic ratio for N : P excretion of 10–20 over the range 50–500 mg dry mass.

Ageing

The difficulties of determining an agreed lifespan for krill by traditional length–frequency analysis have led to an exploration of alternative techniques for ageing krill. Most significant among these has been the attempt to measure accumulated lipofuscin content in krill, a technique first proposed by Ettershank (1984). Lipofuscin is a complex product of peroxidation damage during oxidative metabolism in all cells. Most cells appear unable to excrete lipofuscin, and the granules accumulate with time.

Lipofuscin fluoresces when viewed in histological sections. Laboratory experiments with insects showed that the concentration of extracted fluorescent compounds increased linearly with age (Ettershank *et al.*, 1982). These fluorescent compounds were originally believed to be solubilized lipofuscin but there is now doubt as to their exact identity, and they are better referred to as fluorescent age pigments (FAPs).

Ettershank (1983) extracted FAPs from a sample of formalin-preserved krill and suggested that the frequency distribution of fluorescence values indicated the presence of several age classes in adult female krill. A similar analysis was later extended to immatures and males (Ettershank, 1985). Since the frequency distributions of fluorescence values needed statistical techniques to resolve the putative year classes, this technique represented only a partial advance over traditional analysis of length–frequency distributions. In a further study, Berman *et al.* (1989) measured FAPs from a sample of 252 mature female krill collected from Prydz Bay in January 1985 and concluded from statistical analysis that six classes was the optimum fit for both the FAP and length–frequency distributions.

Nicol (1987) showed that formalin fixation and/or preservation, as used in all the above studies, is likely to have affected the fluorescence values obtained. However, since chemical methods for determining age have such potential for resolving a major problem in krill biology, investigations have continued. These have taken two main approaches. One is to continue looking at krill, and the other is to look at FAP accumulation in organisms that can be aged by independent techniques.

Nicol (1990) collected a sample of juvenile krill from near Vincenness Bay in February 1988, and maintained these in an aquarium until October 1989. The FAP contents of lyophilized samples preserved at the time of collection and killed one year later were compared. The relation between FAP fluorescence and dry mass differed for the two samples of krill, with fluorescence being higher in the older animals. During the course of the experiment the mean dry mass of the krill decreased from 13.3 mg to 9.9 mg and so the concentration of FAP increased with time. There was significant overlap between the two samples in the values of mass-specific FAP fluorescence and so it was not possible to assign all individuals to a specific age class on the basis of FAP fluorescence alone (either as absolute FAP or mass-specific FAP fluorescence). Although krill FAP content increases with age in the laboratory, statistical techniques continue to be necessary to separate age classes. It is not yet clear to what extent the improved analytical technique will enable age classes to be distinguished in field samples of krill where there are no distinct size classes.

An interesting recent development is that Sheehy (1989) has quantified classical lipofuscin granules in the brain of the freshwater crayfish *Cherax cuspidatus* using quantitative image analysis techniques. Although such techniques are unlikely to be of use in the field, they do allow for a definitive investigation of a large, well-controlled sample of krill. If such an approach could be coupled to an aquarium experiment under controlled feeding, photoperiod and temperature conditions that approximate those experienced by krill in the wild, then the problem of krill age-at-size might be resolved.

Clarke *et al.* (1990) looked at the accumulation of FAP with age in the trochid gastropod *Monodonta lineata* whose biology is well understood, and which can be aged unequivocally from checks of shell growth. Total FAP was found to increase monotonically with age in the digestive gland, but FAP concentration showed no relationship, suggesting that total FAP content was simply reflecting growth of the digestive gland. In nervous tissue, however, both total FAP content and FAP concentration varied with age in a complex manner. These data indicate that in *M. lineata* over the age range 1.5 to 11.5 years, FAP does accumulate with time in some tissues but the pattern of accumulation means that a measure of FAP is of little help in ascribing an unequivocal age to any individual.

Recent work thus suggests that FAP may yet provide information on the age structure of krill sampled from the wild, as long as due care is taken with both chemical and statistical analysis. However, work with other organisms indicates that presently available chemical techniques are not likely to provide an unequivocal age for each individual examined (for a recent review see Clarke *et al.*, 1990).

The role of sea-ice in krill energetics

Krill are widely distributed within the area covered by the annual advance and retreat of sea-ice (Marr, 1962) and many recent observations on the distribution and abundance patterns of *Euphausia superba* in relation to pack-ice, and of krill of varying sizes feeding on the under surface of ice, have suggested that the sea-ice community might play an important role in the life history of Antarctic krill. These qualitative observations have led to the uncritical assumption that sea-ice biota must necessarily be of major importance to the energetics of adult and larval krill. More cautious investigators (Stretch *et al.*, 1988) have recognized that these observations do not quantify the relative importance of the sea-ice biota as a food source for krill. In only a few instances have physiological parameters in krill living under the ice been quantified (Daly, 1990; Ross & Quetin, 1991), and the quantitative energetic significance of sea-ice biota to macrozooplankton grazers in winter has yet to be resolved (Clarke & Peck, 1991).

Winter and early spring chlorophyll *a* concentrations in ice floes are 10–100 times those in the water column below (Garrison *et al.*, 1986; Garrison & Buck, 1989; Quetin *et al.*, 1988) and krill are frequently reported as associated with the ice edge and with visible sea-ice microbial communities under the ice from March to December (Fig. 9) (Mackintosh, 1973; Daly & Macaulay, 1988, 1989; Guzman, 1983; O'Brien, 1987; Marschall, 1988; Ainley *et al.*, 1986, 1988; Kottmeier & Sullivan, 1987; Quetin & Ross, 1988; Spiridanov *et al.*, 1985). The pattern is not invariant, and changes with life stage, season, type of ice, and possibly region. In addition adult, juvenile and larval krill have been observed in both field and laboratory effectively grazing on the underside

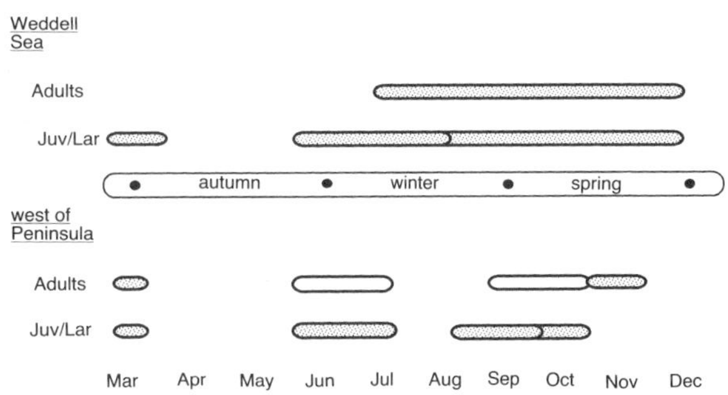

Fig. 9. Observations of *Euphausia superba* associated with sea-ice. The boxes show periods when krill have been observed beneath sea ice: clear boxes indicate non-feeding krill; shaded boxes indicate krill with a green hepatopancreas. Data for adults and larvae are shown separately.

of sea-ice (Stretch *et al.*, 1988; Quetin & Ross, 1988; Kottmeier & Sullivan, 1987; Marschall, 1988; O'Brien, 1987). However, not all life stages have been observed grazing under the ice in all seasons.

The frequency and manner with which the under-ice habitat is utilized by the various life stages of krill may be different (Fig. 9). Larval and juvenile krill are often concentrated at the ice edge or under the pack-ice as compared with open water in both the Weddell Sea and west of the Antarctic Peninsula (Mackintosh, 1973; Daly & Macaulay, 1988; Guzman, 1983). Adult krill may not be as closely associated with sea-ice as are larvae. Adults have been detected at all times of year in open water west of the Antarctic Peninsula (Guzman, 1983; L. B. Quetin & R. M. Ross, personal observation), and in spring in the Weddell Sea krill biomass in open water was comparable to that in the ice zone.

Marschall (1988) found that adult krill in the Weddell Sea were primarily under ice with brine channels or overrafting, rather than under smooth ice; larval and juvenile krill have been reported as primarily associated with first-year ice (Daly, 1990). West of the Antarctic Peninsula, only larval and juvenile krill have been observed grazing the underside of the ice in winter. Adults in winter have clear or opaque digestive glands and are in the water column, not under the ice (Guzman, 1983; L. B. Quetin & R. M. Ross, 1988, personal observation). In both the Weddell Sea and waters west of the Antarctic Peninsula adults are seen grazing on the underside of the ice or close to the ice edge where ice-edge blooms occur in the spring. To put these qualitative observations into perspective, quantitative surveys of the density or distribution of krill in relation to ice type are needed.

Larvae and juveniles may be more dependent on the winter sea-ice biota than adults. Older larvae can starve for only two months (Elias, 1990), whereas adults can tolerate starvation for over 200 days (Ikeda & Dixon, 1982*b*). Thus adults could survive a winter with no food, whereas larvae could not. Of equal importance is the fact that the predators closely associated with the sea-ice probably select the larger, adult krill over the much smaller larvae. Thus the larvae have a refuge in their small size and are subject to lower predation pressure than adults when feeding on the underside of the sea-ice in close proximity to predators. Quetin & Ross (1991) suggest that observed spatial segregation of adults and larvae in relation to winter sea-ice is a result of the differing balance between predation pressure and the need for food in adults and larvae. For adults predation pressure is greater than the need for food in the winter, so unless the underside of the ice is a complex surface that offers some protection from predation, adults remain in the water column away from the ice. For larvae the need for food outweighs predation pressure, and larvae are found grazing on first-year ice. In the spring, however, energetic calculations suggest that a source of food is necessary for krill undergoing gonadal maturation for reproduction that season (Ross & Quetin, 1986), and the observations of adult krill grazing under the ice in the spring support the idea that the balance for adults has shifted to a need for food. Whether predation pressure changes in spring is not known.

It follows from the above discussion that the physiological condition of larvae during a winter with heavy ice and thus good ice habitat should be better than during a winter with no ice. Data on the physiological condition (condition factor, lipid content and growth rate) of larvae from the waters west of the Antarctic Peninsula in two winters, one with light ice throughout the winter (1989) and one with heavy ice (1987) support this prediction (Ross & Quetin, 1991). In the winter with heavy ice, larvae had higher lipid contents, higher condition factors, and were growing. In the winter with little or no ice, larvae were shrinking and had lower lipid contents and condition factors (Ross & Quetin, 1991). These data support the hypothesis that the under-ice habitat plays an important role as a food source in the energetics of larval krill during winter. In our view the quantitative significance of sea-ice as a source of winter or spring food to adult krill remains to be established.

Temperature

In the Southern Ocean the temperature range is from −1.8 °C under ice and in the winter to +3.5 °C in the summer at the northern limits of the distribution of krill. Within this temperature range it has become increasingly apparent that some aspects of the physiology and biology of krill are strongly temperature-dependent. Temperature affects both developmental times and respiration rates of early larvae, but not in a linear fashion (Ross *et al.*, 1988; Quetin & Ross, 1989). Temperature also affects the intermoult period in adults, again not in a linear fashion (see Fig. 5). In these instances, where we have good quantitative data, the effect of temperature on physiological rates is more marked than would be predicted from a typical Q_{10} of 2–3.

Recently a time- and temperature-dependent model was developed that simulates the effect of ambient water temperature on developmental and physiological processes during the descent and ascent of the embryos and early larval stages of *Euphausia superba* (Hofmann *et al.*, 1992). With the observed vertical temperature structures north of the South Shetland Islands, hatching depths ranged from 680 to 790 m, whereas in the Gerlache Strait and eastern Bransfield Strait embryos hit the bottom before hatching. The shallower hatching depths are a result of the interaction between the changing sinking rate with embryological development (Quetin & Ross, 1984), the effect of temperature on embryo and larval developmental times, and the presence of warm (1–2 °C) Circumpolar Deep Water (CDW) between 200 and 700 m. As might be expected from the deeper hatching depth and longer developmental times, carbon reserves were depleted more when embryos and larvae developed in colder waters. These results illustrate the complex interaction between environmental temperature, physiology and biology and demonstrate clearly the need for precise laboratory physiological data for a complete understanding of ecology.

Seasonality

The environment inhabited by krill is characterized by extreme seasonal changes in parameters of importance to krill: food availability, temperature, and ice cover. Four major hypotheses have been advanced to explain how krill survive the long winter period of low food availability in the water column: lipid utilization, shrinkage, switching from primarily herbivory to either carnivory or detritus ingestion, and 'hibernation'.

Initial data (Clarke, 1980) suggested that krill did not synthesize an overwintering lipid store, and the absence of wax esters also implied that lipids were not important as a source of energy over winter. However, more recent data indicate that juvenile and adult krill store significant amounts of lipid for use over winter, primarily in the form of triacylglycerols (Fig. 10). This pattern is also shown clearly in a recent compilation of data by Hagen (1988). The loss of half the total lipid during the low food periods of autumn and winter is evidence that lipid reserves are used as an energy source during the winter. Several analyses of the length–mass relations of krill also suggest shrinkage in volume. Krill of a given length are thinner in winter (Morris *et al.*, 1988; Quetin & Ross, 1991). Shrinkage (loss of mass) over winter has also been clearly demonstrated in several arctic euphausiids (Bamstedt, 1976; Falk-Petersen, 1981).

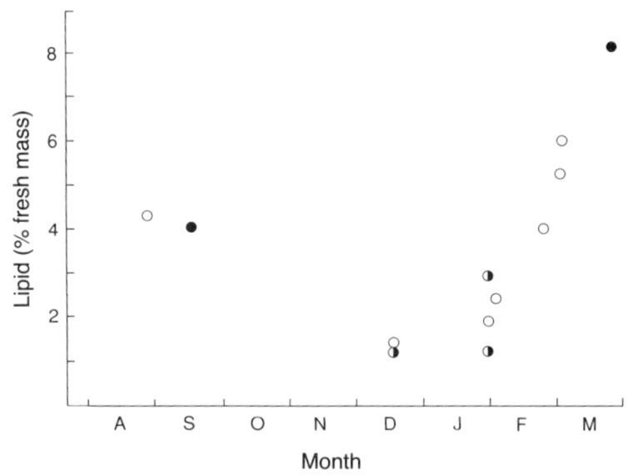

Fig. 10. Lipid content of immature (closed circles) and mature male (open circles) *Euphausia superba*. Data are from Clarke (1980, 1984) and also unpublished data of A. Clarke, R. M. Ross & L. B. Quetin.

Switching of food sources to copepods (Boyd *et al.*, 1984) or to detritus (Kawaguchi *et al.*, 1986), or using the sea-ice microbial community as a food source (Smetacek *et al.*, 1990) during the winter are often-mentioned strategies. The difficulty with many of the observations to date is the lack of quantitative data on the importance of the alternative food resource. In the waters west of the Antarctic Peninsula, both ingestion of plant material and fecal pellet production of adult krill in winter was less than 2–3% of summer rates. Thus, although ingestion is not zero in the winter, for adults from this region energy intake is very low during winter, and switching to other food sources is clearly not the major strategy for winter survival.

There is clear evidence for a lowered oxygen consumption rate in winter from krill west of the

Antarctic Peninsula (Quetin & Ross, 1991), under the fast-ice (Kawaguchi *et al.*, 1986) and from the Weddell Sea (Torres *et al.*, 1988). A lowered oxygen consumption rate is to be expected since the metabolic costs associated with reproduction and growth (R_s, R_g, part of R_a) are greatly reduced. This metabolic reduction has been termed 'hibernation' but this is not advisable because, although hormonal regulation may be involved, the analogy with mammalian hibernation is not valid. A lowered metabolic rate means less requirement for either energy intake or utilization of reserves. Another way to look at this is probably to say that a reduced food availability limits growth and prevents vitellogenesis, and hence metabolic rate is low. To date, whether food availability or another environmental cue is responsible for the reduction in metabolic rate is unknown.

Concluding remarks

The BIOMASS years have resulted in significant advances in our understanding of krill energetics. We now view *Euphausia superba* as an active organism, perhaps more like a small schooling fish than a scaled-up copepod. *E. superba* has a high energy throughput, perhaps 20% of body carbon per day or higher, sustained by a high and effective rate of filtration. Krill thus has the capacity for a high and sustained reproductive output given suitable feeding conditions. Overall growth is limited by a greatly reduced energy throughput in winter; in adults, winter survival is dependent on degrowth and lipid utilization. Larval krill depend critically on sea-ice as a habitat and food source for overwinter survival and condition, whereas in adult krill the quantitative importance of feeding under winter ice has yet to be established but is probably small. Studies of krill during the BIOMASS period have demonstrated the importance of physiological work in gaining a full understanding of the role of Antarctic krill in the food web of the Southern Ocean.

To move forward from demonstrating the importance of physiological studies of krill during the BIOMASS era we must better integrate physiological measurements on freshly caught krill into protocols of quantitative sampling or survey programmes. These integrated physiological–sampling studies will provide data to understand the controls and partitioning of energy flow in krill populations in relation to patterns of physical and biological variables where krill are found. These studies must occur over broad geographical regions and consider seasonal and interannual seasonal differences in patterns of physical and biological variables. New techniques need to be developed to deepen our understanding of the partitioning and controls of energy metabolism, and to simplify existing methodologies so that they can be more easily and thus broadly applied. We need to better estimate the metabolic cost of swimming for krill, the single largest metabolic cost, and relate the estimates to the behaviour of krill in the field. Can we develop behavioural time budgets for krill to which we can apply metabolic estimates? We need to develop techniques and programmes to follow schools of krill that enable us to better understand their behaviour in the field and allow us to make sequential physiological measurements on known discrete populations. We are now at the point where we need to proceed on many fronts and begin to model the flow of energy through krill populations in relation to time and space patterns of the physical and biological variables of their habitat. This process will enable us to understand and eventually predict patterns in variability that we find in krill populations.

Acknowledgements

We thank all of our colleagues who have helped with our studies of krill physiology and biology over the past years. Although some may not agree with all that we have said, the discussions have been stimulating and keep our horizons broadened.

References

Adiyodi R G, Subramoniam T (1983) Arthropoda – Crustacea. In: Adiyodi K G, Adiyodi R G (eds) *Reproductive Biology of Invertebrates*, vol 1 *Oogenesis, Oviposition, and Oosorption*. John Wiley, New York, pp. 443–95.

Ainley D G, Fraser W R, Daly K L (1988) Effects of pack ice on the composition of micronektonic communities in the Weddell Sea. In: Sahrhage D (ed) *Antarctic Ocean and Resources Variability*. Springer-Verlag, Berlin, pp. 140–6.

Ainley D G, Fraser W R, Sullivan C W, Torres J J, Hopkins T L, Smith W O (1986) Antarctic mesopelagic micronekton: evidence from seabirds that pack ice affects community structure. *Science* **232**: 847–9.

Amsler M O, George R Y (1985) Changes in the biochemical composition of *Euphausia superba* Dana embryos during early development. *Polar Biol* **4**: 61–3.

Anderson S L, Clark W H, Chang E S (1985) Multiple spawning and moult synchrony in a free-swimming shrimp (*Sicyonia ingentis*: Penaeoidea). *Biol Bull* **168**: 377–94.

Antezana T, Ray K, Melo C (1982) Trophic behaviour of *Euphausia superba* Dana in laboratory conditions. *Polar Biol* **1**: 77–85.

Bamstedt U (1976) Studies on the deep-water pelagic community of Kongsfjorden, western Norway. Changes in the size and biochemical composition of *Meganyctiphanes norvegica* (Euphausiacea) in relation to its life cycle. *Sarsia* **61**: 15–30.

Bargmann H E (1945) The development and life-history of adolescent and adult krill *Euphausia superba*. *Discovery Rep* **23**: 103–76.

Berman M S, McVey A L, Ettershank G (1989) Age determination of Antarctic krill using fluorescence and image analysis of size. *Polar Biol* **9**: 267–71.

Biggs D C (1982) Zooplankton excretion and NH_4^+ cycling in near-surface waters of the Southern Ocean. 1. Ross Sea, austral summer 1977–1978. *Polar Biol* **1**: 55–67.

Boyd C M, Heyraud M, Boyd C N (1984) Feeding of the Antarctic krill *Euphausia superba*. *J Crust Biol* **4**(Spec No 1): 123–41.

Brinton E, Townsend A W (1984) Regional relationships between development and growth in larvae of Antarctic krill *Euphausia superba*, from field samples. *J Crust Biol* **4**(Spec No 1): 224–46.

Buchholz F (1982) Drach's moult staging system adapted for euphausiids. *Mar Biol* **66**: 301–5.

Buchholz F (1985) Moult and growth in euphausiids. In: Siegfried W R, Condy P R, Laws R M (eds) *Antarctic Nutrient Cycles and Food Webs*. Springer-Verlag, Berlin, pp. 339–45.

Buchholz F (1991) Moult cycle and growth of Antarctic krill *Euphausia superba* in the laboratory. *Mar Ecol Prog Ser* **69**: 217–29.

Buchholz F, Morris D J, Watkins J L (1989) Analyses of field moult data: Prediction of intermoult period and assessment of seasonal growth in Antarctic krill, *Euphausia superba* Dana. *Antarct Sci* **1**: 301–6.

Childress J J (1977) Physiological approaches to the biology of midwater organisms. In: Andersen N R, Zahuranecf B J (eds) *Oceanic Sound Scattering Prediction*. Plenum, New York, pp. 301–24.

Clarke A (1980) The biochemical composition of Antarctic krill, *Euphausia superba* Dana, from South Georgia. *J Exp Mar Biol Ecol* **43**: 221–36.

Clarke A (1984) Lipid content and composition of Antarctic krill, *Euphausia superba* Dana. *J Crust Biol* **4**(Spec No 1): 285–94.

Clarke A (1987) Temperature, latitude and reproductive effort. *Mar Ecol Prog Ser* **38**: 89–99.

Clarke A (1991) What is cold adaptation and how should we measure it? Am Zool **31**: 81–92.

Clarke A, Kendall M A, Gore D J (1990) The accumulation of fluorescent age pigments in the trochid gastropod *Monodonta lineata*. *J Exp Mar Biol Ecol* **144**: 185–204.

Clarke A, Morris D J (1983) Towards an energy budget for krill: the physiology and biochemistry of *Euphausia superba* Dana. *Polar Biol* **2**: 69–86.

Clarke A, Peck L S (1991) The physiology of polar marine zooplankton. In: Sakshaug E, Hopkins C C E, Øritsland N A (eds) *Proc Pro Mare Symposium on Polar Marine Ecology*, Trondheim, 12–16 May, 1990. *Polar Res* **10**(2): 355–69.

Clarke A, Quetin L B, Ross R M (1988) Laboratory and field estimates of the rate of faecal pellet production by Antarctic krill, *Euphausia superba*. *Mar Biol* **98**: 557–63.

Cuzin-Roudy J (1987*a*) Sexual differentiation in the Antarctic krill *Euphausia superba* Dana (Crustacea: Euphausiacea). *J Crust Biol* **7**: 518–24.

Cuzin-Roudy J (1987*b*) Gonad history of the Antarctic krill *Euphausia superba* Dana during its breeding season. *Polar Biol* **7**: 237–44.

Cuzin-Roudy J, Amsler M O (1991) Ovarian development and sexual maturity staging in Antarctic krill, *Euphausia superba* Dana (Euphausiacea). *J Crust Biol* **11**: 236–49.

Cuzin-Roudy J, Schalk P H (1989) Macrozooplankton – biomass, development and activity. In: Hempel I, Schalk P H, Smetacek V (eds) The expedition Antarktis VII/3 (EPOS LEG 2) of RV 'Polarstern' in 1988/89. *Ber Polarforsch* **65**: 146–59.

Daly K L (1990) Overwintering development, growth and feeding of larval *Euphausia superba* in the Antarctic marginal ice zone. *Limnol Oceanogr* **35**: 1564–76.

Daly K L, Macaulay M C (1988) Abundance and distribution of krill in the ice edge zone of the Weddell Sea, austral spring 1983. *Deep-Sea Res* **35**: 21–41.

Daly K L, Macaulay M C (1989) Abundance, distribution, and overwintering strategies of krill in the ice-edge zone. *Ant J US* **24**: 165.

Denys C J, McWhinnie M A (1982) Fecundity and ovarian cycles of the Antarctic krill *Euphausia superba* (Crustacea Euphausiacea). *Can J Zool* **60**: 414–23.

Elias M C (1990) Effects of photoperiod, phytoplankton level and temperature on the growth, development and survival of larval *Euphausia superba* (Dana). MA Thesis, University of California, Santa Barbara, 112 pp.

Ettershank G (1983) Age structure and cyclical annual size change in the Antarctic krill *Euphausia superba* Dana. *Polar Biol* **2**: 189–93.

Ettershank G (1984) A new approach to the assessment of longevity in Antarctic krill *Euphausia superba*. *J Crust Biol* **4**(Spec No 1): 295–305.

Ettershank G (1985) Population age structure in males and juveniles of the Antarctic krill, *Euphausia superba* Dana. *Polar Biol* **4**: 199–201.

Ettershank G G, McDonnell I, Croft R (1982) The estimation of age in the fleshfly *Sarcophaga bullata* using the age pigment lipofuscin. *Aust J Zool* **31**: 131–8.

Falk-Petersen S (1981) Ecological investigations on the zooplankton community of Balsfjorden, northern Norway: seasonal changes in body weight and the main biochemical composition of *Thysanoessa inermis* (Kroyer), *T. raschii* (M. Sars) and *Meganyctiphanes norvegica* (M. Sars) in relation to environmental factors. *J Exp Mar Biol Ecol* **49**: 103–20.

Garrison D L, Buck K R (1989) The biota of the Antarctic pack ice in the Weddell Sea and Antarctic Peninsula regions. *Polar Biol* **10**: 211–19.

Garrison D L, Sullivan C W, Ackley S F (1986) Sea-ice microbial communities in Antarctica. *BioScience* **36**: 243–50.

George R Y, Fields J R (1984) Ammonia excretion in the Antarctic krill *Euphausia superba* in relation to starvation and ontogenetic stages. *J Crust Biol* **4**(Spec No 1): 263–72.

Grant A, Morgan P J, Olive P J W (1987) Use made in marine ecology of methods for estimating demographic parameters from size-frequency data. *Mar Biol* **95**: 1–9.

Guzman O (1983) Distribution and abundance of Antarctic krill (*Euphausia superba*) in the Bransfield Strait. In: Schnack S B (ed) On the biology of krill, *Euphausia superba*. *Ber Polarforsch Sonderheft* **4**: 169–90.

Hagen W (1988) Zur bedeutung der Lipide im antarktischen Zooplankton. *Ber Polarforsch* **49**: 1–129.

Harrington S A, Ikeda T (1986) Laboratory observations on spawning, brood size and egg hatchability of the Antarctic krill *Euphausia superba* from Prydz Bay, Antarctica. *Mar Biol* **92**: 231–5.

Hirche H-J (1983) Excretion and respiration of the Antarctic krill *Euphausia superba*. *Polar Biol* **1**: 205–9.

Hofmann E E, Capella J E, Ross R M, Quetin L B (1992) Models of the early life history of *Euphausia superba*. I. Time and temperature dependence during the descent–ascent cycle. *Deep-Sea Res* **39**: 1177–200.

Holm-Hansen O, Huntley M (1984) Feeding requirements of krill in relation to food sources. *J Crust Biol* **4**(Spec No 1): 156–73.

Huntley M E, Brinton E (1987) RACER: Mesoscale variation in the growth and early development of *Euphausia superba* Dana. *Ant J US* **22**(5): 160–2.

Ikeda T (1984*a*) Development of the larvae of the Antarctic krill (*Euphausia superba* Dana) observed in the laboratory. *J Exp Mar Biol Ecol* **75**: 107–17.

Ikeda T (1984*b*) Sequences in metabolic rates and elemental composition (C, N, P) during the development of *Euphausia superba* Dana and estimated food requirements during its life-span. *J Crust Biol* **4**(Spec No 1): 273–84.

Ikeda T (1985) Metabolic rate and elemental composition (C and N) of embryos and non-feeding early larval stages of Antarctic krill (*Euphausia superba* Dana). *J Exp Mar Biol Ecol* **90**: 119–27.

Ikeda T, Bruce B (1986) Metabolic activity and elemental composition of krill and other zooplankton from Prydz Bay, Antarctica, during early summer (November–December). *Mar Biol* **92**: 545–55.

Ikeda T, Dixon P (1982*a*) Observations on moulting in Antarctic krill (*Euphausia superba* Dana). *Aust J Mar Freshw Res* **33**: 71–6.

Ikeda T, Dixon P (1982*b*) Body shrinkage as a possible over-wintering mechanism of the Antarctic krill, *Euphausia superba* Dana. *J Exp Mar Biol Ecol* **62**: 143–51.

Ikeda T, Dixon P (1984) The influence of feeding on the metabolic activity of Antarctic krill (*Euphausia superba* Dana). *Polar Biol* **3**: 1–9.

Ikeda T, Dixon P, Kirkwood J (1985) Laboratory observations on moulting, growth and maturation in Antarctic krill (*Euphausia superba* Dana). *Polar Biol* **4**: 1–18.

Ikeda T, Hing Fey E (1981) Metabolic activity of zooplankton from the Antarctic Ocean. *Aust J Mar Freshw Res* **32**: 921–30.

Ikeda T, Mitchell A W (1982) Oxygen uptake, ammonia excretion and phosphate excretion by krill and other zooplankton in relation to their body size and chemical composition. *Mar Biol* **71**: 283–98.

Ikeda T, Thomas P G (1987) Moulting interval and growth of juvenile Antarctic krill (*Euphausia superba*) fed different concentrations of the diatom *Phaeodactylum tricornutum* in the laboratory. *Polar Biol* **7**: 339–43.

Ishii H, Omori M, Maeda M, Watanabe Y (1987) Metabolic rates and elemental composition of Antarctic krill *Euphausia superba* Dana. *Polar Biol* **7**: 379–82.

Jobling M (1983) Towards an explanation of specific dynamic action (SDA). *J Fish Biol* **23**: 549–55.

Johnson M A, Macaulay M C, Biggs D C (1984) Respiration and excretion within a mass aggregation of *Euphausia superba*: implications for krill distribution. *J Crust Biol* **4**(Spec No 1): 174–84.

Kanda K, Takagi K, Seki Y (1982) Movement of the larger swarms of Antarctic krill *Euphausia superba* off Enderby Land during 1976–1977 season. *J Tokyo Univ Fish* **68**: 24–42.

Kato M, Murano M, Segawa S (1979) Estimation of the filtering rate of Antarctic krill under laboratory conditions. *Trans Tokyo Univ Fish* **3**: 107–12.

Kawaguchi K, Ishikawa S, Matsuda O (1986) The overwintering of Antarctic krill (*Euphausia superba* Dana) under the coastal fast ice off the Ongul Islands in Lutzow–Holm Bay, Antarctica. *Mem Natl Inst Polar Res Spec Issue* **44**: 67–85.

Kikuno T, Kawamura A (1983) Observations of the ovarian eggs and spawning habits in *Euphausia superba* Dana. *Mem Natl Inst Polar Res Spec Issue* **27**: 104–19.

Kils U (1981) Swimming behaviour, swimming performance and energy balance of Antarctic krill *Euphausia superba*. *BIOMASS Sci Ser* **3**: 1–233.

Kottmeier S T, Sullivan C W (1987) Late winter primary production and bacterial production in sea ice and seawater west of the Antarctic peninsula. *Mar Ecol Prog Ser* **36**: 287–98.

Lipski M (1987) Variations of physical conditions, nutrients and chlorophyll *a* contents in Admiralty Bay (King George Island, South Shetlands, 1979). *Pol Polar Res* **8**: 307–32.

Macdonald P D M, Pitcher T J (1979) Age-groups from size frequency data: a versatile and efficient method of analysing distribution mixtures. *J Fish Res Bd Can* **36**: 987–1001.

Mackintosh N A (1972) Life cycle of Antarctic krill in relation to ice and water conditions. *Discovery Rep* **36**: 1–94.

Mackintosh N A (1973) Distribution of post-larval krill in the Antarctic. *Discovery Rep* **36**: 95–156.

Maihara Y, Endo Y (1986) Laboratory observations on moulting and growth of Antarctic krill *Euphausia superba* Dana. *Mem Natl Inst Polar Res Spec Issue* **1**: 125–7.

Makarov R R (1975) A study of repeated maturation of female euphausiids (Eucaridea, Euphausiacea). *Zool Zh* **54**: 670–81.

Makarov R R (1979) Size composition and conditions of existence of *Euphausia superba* Dana (Crustacea: Euphausiacea) in the eastern part of the Pacific Sector of the Southern Ocean. *Oceanology* **19**: 582–5.

Marr J W S (1962) The natural history and geography of the Antarctic krill (*Euphausia superba* Dana). *Discovery Rep* **32**: 33–464.

Marschall H-P (1988) The overwintering strategy of Antarctic krill under the pack-ice of the Weddell Sea. *Polar Biol* **9**: 245–50.

Mauchline J (1980) The biology of mysids and euphausiids. *Adv Mar Biol* **18**: 681 pp.

Mauchline J, Fisher L R (1969) The biology of euphausiids. *Adv Mar Biol* **7**: 454 pp.

McClatchie S (1988) Food-limited growth of *Euphausia superba* in Admiralty Bay, South Shetland Islands, Antarctica. *Cont Shelf Res* **8**: 329–45.

McWhinnie M A, Denys C J (1978) Antarctic marine living resources with special reference to krill, *Euphausia superba*: Assessment of adequacy of present knowledge. Nat Sci Foundn Wash Rept No NSF/RA-780489, 209 pp.

McWhinnie M A, Denys C J, Parkin R, Parkin K (1979) Biological investigations of *Euphausia superba* Dana. *Antarct J US* **14**: 163–4.

Miller D G M, Hampton I (1989) Biology and ecology of the Antarctic Krill (*Euphausia superba* Dana) a review. *BIOMASS Sci Ser* **9**: 1–166.

Morris D J (1984) Filtration of *Euphausia superba* Dana: under- or over-estimates? *J Crust Biol* **4**(Spec No 1): 185–97.

Morris D J (1985) Integrated model of moulting and feeding of Antarctic krill *Euphausia superba* off South Georgia. *Mar Ecol Prog Ser* **22**: 207–17.

Morris D J, Keck A (1984) The time course of the moult cycle and growth of *Euphausia superba* in the laboratory: a preliminary study. *Meeresforschung* **30**: 94–100.

Morris D J, Priddle J (1984) Observations on the feeding and moulting of the Antarctic krill, *Euphausia superba* Dana, in winter. *Bull Br Antarct Surv* **65**: 57–63.

Morris D J, Watkins J L, Ricketts C, Buchholz F, Priddle J (1988) An assessment of the merits of length and weight measurements of Antarctic krill *Euphausia superba*. *Bull Br Antarct Surv* **79**: 27–50.

Murano M, Segawa S, Kato M (1979) Moult and growth of the Antarctic krill in the laboratory. *Trans Tokyo Univ Fish* **3**: 99–106.

Nicol S (1987) Some limitations on the use of the lipofuscin ageing technique. *Mar Biol* **93**: 609–14.

Nicol S (1990) The age-old problem of krill longevity. *BioScience* **40**: 833–6.

Nicol S, Stolp M (1990) A refinement of the moult staging technique for Antarctic krill (*Euphausia superba*). *Mar Biol* **104**: 169–73.

O'Brien D P (1987) Direct observations of the behavior of *Euphausia superba* and *Euphausia crystallorophias* (Crustacea: Euphausiacea) under pack ice during the Antarctic winter of 1985. *J Crust Biol* **7**: 437–48.

Parry G D (1983) The influence of the cost of growth on ectotherm metabolism. *J Theor Biol* **101**: 453–77.

Pauly D, David N (1981) ELEFAN, a BASIC program for the objective assessment of growth parameters from length frequency data. *Meeresforschung* **28**: 205–11.

Poleck T P, Denys C J (1982) Effect of temperature on the moulting, growth and maturation of the Antarctic krill *Euphausia superba* (Crustacea: Euphausiacea) under laboratory conditions. *Mar Biol* **70**: 255–65.

Price J H, Boyd K R, Boyd C M (1988) Omnivorous feeding behaviour of the Antarctic krill *Euphausia superba*. *Mar Biol* **97**: 67–77.

Quetin L B, Ross R M (1984) Depth distribution of developing *Euphausia superba* embryos predicted from sinking rates. *Mar Biol* **79**: 47–53.

Quetin L B, Ross R M (1988) Summary of WINCRUISE II to the Antarctic Peninsula during June and July 1987. *Ant J US* **23**: 149–51.

Quetin L B, Ross R M (1989) Effects of oxygen, temperature and age on the metabolic rate of the embryos and early larval stages of the Antarctic krill *Euphausia superba* Dana. *J Exp Mar Biol Ecol* **125**: 43–62.

Quetin L B, Ross R M (1991) Behavioral and physiological characteristics of the Antarctic krill, *Euphausia superba*. *Am Zool* **31**: 49–63.

Quetin L B, Ross R M, Amsler M O (1988) Field ingestion rates of *Euphausia superba*. *EOS* **68**: 1785.

Rakusa-Suszczewski S (1990) Seasonal changes in respiration and biomass of *Euphausia superba* from Admiralty Bay (South Shetland Islands, Antarctica). *Polsk Arch Hydrobiol* **37**: 305–11.

Rakusa-Suszczewski S, Opalinksi K W (1978) Oxygen consumption in *Euphausia superba*. *Polsk Arch Hydrobiol* **25**: 633–41.

Rosenberg A A, Beddington J R, Basson M (1986)

Growth and longevity of krill during the first decade of pelagic whaling. *Nature* **324**: 152–4.
Ross R M, Quetin L B (1983) Spawning frequency and fecundity of the Antarctic krill, *Euphausia superba*. *Mar Biol* **77**: 201–5.
Ross R M, Quetin L B (1986) How productive are Antarctic krill? *BioScience* **36**: 264–9.
Ross R M, Quetin L B (1989) Energetic cost to develop to the first feeding stage of *Euphausia superba* and the effects of delays in food availability. *J Exp Mar Biol Ecol* **133**: 103–27.
Ross R M, Quetin L B (1991) Ecological physiology of larval euphausiids, *Euphausia superba* (Euphausiacea). *Mem Queensl Mus* **31**: 321–33.
Ross R M, Quetin L B, Kirsch E (1988) Effect of temperature on developmental times and survival of early larval stages of *Euphausia superba* Dana. *J Exp Mar Biol Ecol* **121**: 55–71.
Segawa S, Kato M, Murano M (1979) Oxygen consumption of the Antarctic krill *Euphausia superba*. *Trans Tokyo Univ Fish* **3**: 113–20.
Segawa S, Kato M, Murano M (1982) Respiration and ammonia excretion rates of the Antarctic krill, *Euphausia superba* Dana. *Trans Tokyo Univ Fish* **5**: 177–87.
Sheehy M R J (1989) Crustacean brain lipofuscin: an examination of the morphological pigment in the fresh-water crayfish *Cherax cuspidatus* (Parastacidae). *J Crust Biol* **9**: 387–91.
Siegel V (1985) On the fecundity of Antarctic krill, *Euphausia superba* (Euphausiacea). *Arch FischWiss* **36**: 185–93.
Siegel V (1986) Untersuchungen zur Biologie des antarktischen Krill, *Euphausia superba*, im Bereich der Bransfield Strasse und angrenzender Gebiete. *Mitt Inst Seefisch Hamburg* **38**: 1–244.
Smetacek V, Scharek R, Nothig E M (1990) Seasonal and regional variation in the pelagial and its relationship to the life history cycle of krill. In: Kerry K R, Hempel G (eds) *Antarctic Ecosystems. Ecological Change and Conservation*. Springer-Verlag, Berlin, pp. 103–14.
Smith W O Jr, Sakshaug E (1990) Polar phytoplankton. In: Smith W O Jr (ed) *Polar Oceanography*, Part B *Chemistry, Biology, and Geology*. Academic Press, New York, pp. 477–525.
Spiridanov V A, Gruzov E N, Pushkin A F (1985) (Investigations of schools of Antarctic *Euphausia superba* under the ice.) *Zool Zh* **64**: 1655–60. (In Russian.)
Stepnik R (1982) All year populational studies of Euphausiacea (Crustacea) in the Admiralty Bay (King George Island, South Shetland Islands, Antarctica). *Pol Polar Res* **3**: 49–68.
Stretch J J, Hamner P P, Hamner W M, Michel W C, Cook J, Sullivan C W (1988) Foraging behaviour of Antarctic krill *Euphausia superba* on sea ice microalgae. *Mar Ecol Prog Ser* **44**: 131–9.
Thomas P G, Ikeda T (1987) Sexual regression, shrinkage, re-maturation and growth in spent female *Euphausia superba* in the laboratory. *Mar Biol* **95**: 357–63.
Torres J J (1984) Relationship of oxygen consumption to swimming speed in *Euphausia pacifica*. II. Drag, efficiency and a comparison with other swimming organisms. *Mar Biol* **78**: 231–8.
Torres J J, Childress J J (1983) Relationship of oxygen consumption to swimming speed in *Euphausia pacifica*. I. Effects of temperature and pressure. *Mar Biol* **74**: 79–86.
Torres J J, Childress J J, Quetin L B (1982) A pressure vessel for the simultaneous determination of oxygen consumption and swimming speed in zooplankton. *Deep-Sea Res* **29**: 631–9.
Torres J J, Hopkins T L, Donnelly J, Lancraft T M (1988) Metabolism and proximate composition of antarctic pelagic crustacea. *EOS* **68**: 1785.
Vahl O (1983) The relationship between specific dynamic action (SDA) and growth in the common starfish *Asterias rubens* L. *Oecologia* **61**: 122–5.
Watkins J L, Morris D J, Ricketts C, Priddle J (1986) Differences between swarms of Antarctic krill and some implications for sampling krill populations. *Mar Biol* **93**: 137–46.
Witek Z, Koronkiewicz A, Soszka G J (1980) Certain aspects of the early life history of krill, *Euphausia superba* Dana (Crustacea). *Pol Polar Res* **1**: 97–115.

Discussant's report: Evaluation of BIOMASS contribution to krill research

D. G. M. MILLER

Introduction

MINDFUL of the injudicious exploitation of whale stocks, and recognizing the key position of krill (*Euphausia superba*) in many important Southern Ocean food chains, the Scientific Committee on Antarctic Research (SCAR) in the early 1970s foresaw a need for a substantial expansion of scientific research on Antarctic marine ecosystem dynamics (Anon, 1977). This initiative culminated in the international BIOMASS (Biological Investigations of Marine Antarctic Systems and Stocks) Programme, primarily cosponsored by SCAR and the Scientific Committee on Oceanic Research (SCOR).

Launched in 1977, BIOMASS was aimed at intensifying marine biological research in the Antarctic, primarily with a view to providing a scientific foundation on which to base future management and rational exploitation of Antarctic marine living resources in general, and krill in particular. The Programme was thus largely geared towards improving knowledge of krill biology, ecology and population dynamics with, among others, the following specific objectives:

1. To develop systems to monitor changes in the krill stock abundance and composition;
2. To make an intensive study of the life history and population dynamics of krill.

Manifestly implicit in these two objectives is that they are applicable to a variety of spatial and temporal scales, a factor greatly complicating their practical realization. Furthermore, it must be stressed that the successful pursuance of studies addressing these objectives was perceived as central to many of the Programme's broader objectives, particularly those dealing with interactions between krill and other components of the Antarctic marine ecosystem (Anon, 1977).

There is little doubt that the results of BIOMASS as a whole were broadly significant (El-Sayed, 1988; this volume), especially insofar as they fostered the scientific expertise necessary for, and utilized in, the drafting of the Convention for the Conservation of Antarctic Marine Living Resources (CCAMLR) (Croxall, this volume). The central importance of the above two objectives, however, requires that their efficacy be adequately and specifically evaluated with respect to the perceived success of BIOMASS in improving the state of knowledge on krill.

I have, therefore, attempted to provide a broad overview of information contained in the three preceding chapters, originally presented at the krill session of the BIOMASS Colloquium. My comments, however, are confined to krill alone and the reader's indulgence is sought in linking these to other research findings reported elsewhere in this volume. Finally, I have structured my review around three spatial scales: the meso- or 'regional' scale, the micro- or 'aggregation' scale and the individual or 'physiological' scale.

Quantifying krill distribution and abundance

On the regional scale, there is little doubt that one of the major achievements of BIOMASS was the implementation of a systematic approach to the problem of quantifying krill distribution and abundance. This was epitomized by the multiship First International BIOMASS Experiment (FIBEX) in 1981.

The definition of 'unit stocks' and the quantification of abundance are unitary principles in the classical fishery paradigm (cf. Dickie, 1979). Hence the identification and quantification of krill stock abundances formed an integral and strategic component of the work of BIOMASS (Miller & Hampton, 1989).

In their chapter, Everson & Miller (this volume) draw attention to the many achievements of BIOMASS in the study of the regional distribution and abundance of krill. They emphasize the key role played by high-frequency echosounders in the direct estimation of krill abundance and indicate that many of the current ideas on factors controlling krill distribution and movement between areas (the so-called krill 'flux') originated during, or as a result of, BIOMASS.

Everson & Miller also highlight the major impact that data accrued during FIBEX have had on work being carried out on management of the krill resource by the CAMLR Commission. In particular, FIBEX results have been used as a basis for establishing a precautionary catch limit for the krill fishery in the west Atlantic sector (SC-CAMLR, 1991). BIOMASS has, therefore, clearly affected issues pertaining to krill resource management, a major goal of the Programme as a whole.

Krill demography and aggregation

As much as FIBEX affected the course of krill stock abundance estimation, so SIBEX (Second International BIOMASS Experiment) contributed to study of the species' small-scale distribution. Siegel & Kalinowski (this volume) provide a thorough review of the impact of BIOMASS on such research, particularly in the elucidation of the proximal causes and ultimate effects of the aggregating behaviour of krill. They highlight the importance of understanding krill aggregation in the context of improving definition of the scales of variance associated with the distribution of krill abundance. It is at such scales that many of the processes associated with krill demography, as well as the cycling of matter and energy through the overall population, are operative. The role of environmental interactions, in particular, still remains unclear.

Krill energetics

The estimation of krill production is as fundamental to assessing the species' potential yield as is knowledge on stock abundance or biomass. Some of the most significant advances in estimating krill production arose during the latter half of BIOMASS and concentrated on refining the parameters necessary to formulate individual energy budgets.

Quetin *et al.* (this volume) have comprehensively reviewed the many significant developments in the study of krill energetics which occurred during the BIOMASS period. Not only do they provide an alternative to more traditional formulations of the energy budget, they also highlight the substantial and significant geographical and interannual variability in krill physiology, biology and energetics. These are complicated by aggregating behaviour and have a direct effect on the extrapolation of energy budgets from one place or season to another.

Conclusions

The results of the three reviews briefly outlined here clearly illustrate the successes as well as certain shortcomings of many BIOMASS-related activities. There is, however, little doubt that many of the results of such activities are directly relevant to the two major objectives of the Programme's krill-directed research. To a large extent this may be attributed to the integrated approach of BIOMASS, and particularly the way it tackled the problems inherent in the scale-dependence of many of the key krill parameters being considered.

It thus remains to be shown whether the influence of BIOMASS was significant based on criteria other

Table 1. *Criteria to evaluate the perceived benefits of research arising from BIOMASS. Based on the criteria used by the South African National Antarctic Research Programme (after Anon, 1987)*

A. Perceived value of BIOMASS-initiated programmes within national and international research priorities, especially the promotion of excellence in self-initiated research;
B. Contribution of such programmes to the creation of new expertise, especially (i) the provision/maintenance of specialized facilities in national laboratories, and (ii) the development of trained manpower;
C. Fostering of both national and international liaison, through (i) the transfer of expertise/technology via national/international collaboration, (ii) ensuring relevance (both national and international) in research, and (iii) fostering national/international collaboration;
D. Facilitating the launching of special research programmes in neglected, emergent or strategically important fields, and
E. Serving decision-makers with the best possible scientific advice available.

than those associated with the scientific relevance of the work reviewed in the three papers highlighted above. I have, therefore, attempted to evaluate BIOMASS-associated krill research according to five criteria used in the assessment of the perceived benefits of research by the South African Antarctic Research Programme (Table 1).

On the basis of the results reported by Everson & Miller (this volume), Siegel & Kalinowski (this volume) and Quetin *et al.* (this volume), as well as in other publications (see, for example, the review by Miller & Hampton, 1989), there is little doubt that BIOMASS has promoted a sustained level of scientific productivity worldwide (i.e. the provisions of Criterion A in Table 1 is met). Furthermore, the standard of a significant number of the publications produced is strong evidence for the 'promotion of excellence in self-initiated research'.

Secondly, BIOMASS-related initiatives have certainly been instrumental in the creation of new expertise internationally (Criterion B). A major component of such expertise is exemplified by the development of acoustic methodology to assess krill abundance, by attempts to define the boundaries and causes of krill aggregation and by the refinement of techniques to measure energetic interchanges.

The fundamental contribution of BIOMASS to the setting of CCAMLR's principal objectives (cf. Knox, 1984; Croxall, this volume) has meant that many scientists have been well placed, by virtue of their involvement in the Programme, to make a significant contribution to the development of a scientifically based management plan for the Antarctic krill fishery (cf. Miller, 1991) (criterion C). In these terms, BIOMASS has also undoubtedly facilitated research in both emergent and strategically important fields (Criterion D).

Finally, the collaborative nature of BIOMASS has undoubtedly facilitated the integration of international research on krill in the Antarctic. The combination of both fundamental scientific dialogue and enhanced organizational skills resulting from various international efforts (especially FIBEX and SIBEX) has certainly improved the standard of scientific knowledge of, and advice on, krill in many of the nations involved in BIOMASS. Not only does this fulfil the final criterion (Criterion E), but it has also provided a useful foundation on which strategic planning of future scientific endeavours in the Antarctic has come to be based (cf. SCAR, 1992; Croxall, this volume).

References

Anon (1977) Biological Investigations of Marine Antarctic Systems and Stocks. *BIOMASS Sci Ser* **1**: 79 pp.

Anon (1987) South African Southern Ocean Research Programme. *S Afr Natnl Sci Prog Rep No* **134**: 58 pp.

Dickie L M (1979) Perspectives on fisheries biology and implications for management. *J Fish Res Bd Can* **36**: 838–44.

El-Sayed S Z (1988) The BIOMASS Programme. *Oceanus* **31**(2): 75–9.

Knox G A (1984) The key role of krill in the ecosystem of the Southern Ocean with special reference to the Convention on the Conservation of Antarctic Marine Living Resources. *Ocean Mgmt* **9**: 113–56.

Miller D G M, Hampton I (1989) Biology and ecology of the Antarctic krill (*Euphausia superba* Dana): A review. *BIOMASS Sci Ser* **9**: 166 pp.

Miller D G M (1991) Exploitation of Antarctic marine living resources: A brief history and a possible

approach to managing the krill fishery. *S Afr J Mar Sci* **10**: 321–39.
SCAR (1992) *SCAR Report*. No. 8: 40 pp. SCAR, Cambridge.
SC-CAMLR (1991) *Report of the Tenth Meeting of the Scientific Committee*. CCAMLR, Hobart, Australia, 427 pp.

Antarctic marine stocks: fish and birds

The contribution of the BIOMASS Programme to Antarctic fish biology

A. KELLERMANN & A. W. NORTH

Introduction: BIOMASS

THE INTERNATIONAL programme Biological Investigations of Marine Antarctic Systems and Stocks (BIOMASS) has been well described (Anon, 1977; El-Sayed, 1991, this volume; Frank, 1983; Heywood, 1985; Hempel, 1987; Hureau & Ślósarczyk, 1990; Tranter & Smith, 1979). The programme originated in the early 1970s when national members of SCAR (Scientific Committee on Antarctic Research) began to develop the idea of an international programme for the study of the living resources of the Southern Ocean (Frank, 1983). In 1972 a subcommittee on the Marine Living Resources of the Southern Ocean was established within the Biology Working Group of SCAR, and after further meetings in the following four years it was elevated to become the Group of Specialists on the Living Resources of the Southern Ocean, co-sponsored by the Scientific Committee on Oceanographic Research (SCOR) as its Working Group 54. The new Group first met at Cambridge in 1975 to begin plans for long-term biological studies of the Antarctic seas. In August 1976, the Group held a major conference at Woods Hole, Massachusetts, where the development of an international programme for the biological investigations of marine Antarctic systems and stocks, which was to run for ten years, was discussed and planned. The principal objective of the programme was:

> to gain a deeper understanding of the structure and dynamic functioning of the Antarctic marine ecosystem for the future management of potential living resources. *(Anon, 1977)*

The fish research programme concentrated attention on 12 species which were believed to be of major importance in the Southern Ocean (Anon, 1977, p 31, part 2.6)

From 1975, the Group of Specialists summarized information on the exploited fish species and identified priority areas for research. Then in 1979, a specialized subgroup, the Working Party on Fish Biology (later the BIOMASS Working Party on Fish Ecology, BWPFE) was created which carried on the work on fish stocks. It voiced concern over the increasing fish catches and made recommendations to improve catch reporting (Anon, 1979, 1980*a*) and produced a list of specific research objectives.

The meeting of the BWPFE in May 1980 (Anon, 1980*a*) was primarily concerned with fishery statistics and population parameters relevant to the status of the fish stocks. It recognized the need for more information on the early life history stages, particularly their identification and early growth rates, to facilitate the interpretation of age from growth patterns in otoliths, bones and scales. Plans for international bio-oceanographic surveys during the First and Second International BIOMASS Experiment (FIBEX and SIBEX) were also discussed. It was decided to improve estimates of krill (*Euphausia*

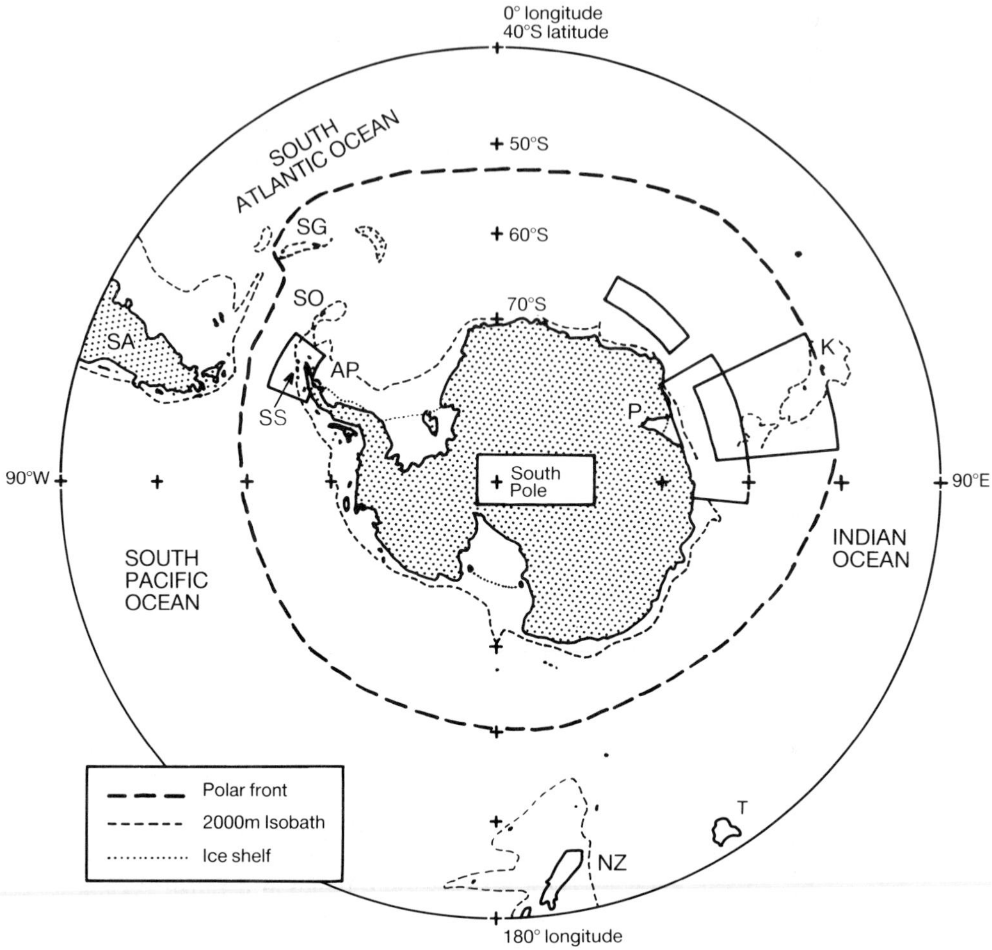

Fig. 1. The Southern Ocean with the areas where fish were sampled during FIBEX and/or SIBEX marked by boxes. Place names are abbreviated thus: South America (SA), South Georgia (SG), South Orkney Islands (SO), South Shetland Islands (SS) (arrowed), Antarctic Peninsula (AP), Prydz Bay (P), Kerguelen Island (K), New Zealand (NZ) and Tasmania (T).

superba) biomass by studies in large areas of the south-west Atlantic, south-east Indian, and Pacific Oceans (Fig. 1), because in these regions some commercial krill fishing had taken place and the marine ecosystem was being studied (Anon, 1980*b*, p 5). The resulting FIBEX and SIBEX surveys were two of the largest internationally coordinated shipborne scientific field investigations in history.

The FIBEX fieldwork during the 1980–1 Antarctic summer season involved 13 ships from 10 countries (Hureau & Ślósarczek, 1990). Krill biomass was estimated from acoustic transects undertaken by up to five ships operating simultaneously to provide a synoptic overview for each region. Krill acoustic target strength was calculated from the size of specimens caught using nets at stations along each transect. The samples also contained the early stages of fish and thereby provided new information about their distribution, particularly in the northern Antarctic Peninsula region, including the South Shetland Islands, Palmer Archipelago, Bransfield Strait and Elephant Island (Asencio & Moreno, 1984; Kellermann, 1986*a*; Kellermann & Kock, 1984; Kellermann & Ślósarczyk, 1984; Kock, 1982; Marin & Moreno, 1982; Mujica & Torres, 1982; Ślósarczk, 1986; Ślósarczyk & Cielniaszek, 1985; Ślósarczyk & Rembiszewski, 1982). Valuable information was also obtained about the early stages of fish from the Prydz Bay region (70–75°E) (Tabeta & Komaki, 1986; Williams *et al.*, 1983; Williams & Duhamel, this volume) and 30–50°E in the southern Indian Ocean (Rasoanarivo & Aboussouan, 1983). The results from FIBEX showed that the pelagic early stages of more than 20 species of notothenioids, and some myctophids and paralepidids, occurred in the Antarctic Peninsula region. The early stages of some species were abundant; those of several species were frequently associated with krill swarms.

Samples of fish otoliths derived from the stomach contents or faeces of predators can be used for prey identification and, if in good condition, for an estimate of prey size. In 1981 the BIOMASS Working Party on Bird Ecology recognized the potential of using otoliths to describe the diet of higher predators and recommended that the BWPFE should produce a guide to otoliths. Scientists within the BIOMASS network contributed material that was utilized by Hecht (1987) to provide a guide to the otoliths of Southern Ocean fishes. Recently, Williams & McEldowney (1990) have produced a guide to fish otoliths from waters off the Australian Antarctic Territory, Heard and Macquarie Islands.

In September 1982 the BWPFE held a meeting in Hamburg (Anon, 1982*a*) where identification guides to Antarctic fish and to fish remains in predators were planned. The Working Party finalized plans for SIBEX which aimed to investigate seasonal variation in two areas where krill were abundant. These regions were around the South Shetland Islands, north-western Antarctic Peninsula, and off Prydz Bay in the southern Indian Ocean (Fig. 1). The SIBEX survey was planned to take place during two austral summer seasons, with a wider seasonal coverage and a greater standardization of methods and sampling locations than FIBEX. The BWPFE also recommended further international collaboration for the planning of bottom trawl surveys to estimate adult fish biomass.

The extensive SIBEX surveys were implemented in 1983–4 (SIBEX-I) and 1984–5 (SIBEX-II). These surveys involved 15 ships from 12 countries. The results from these studies have provided a comprehensive overview of the distribution of the early stages of fish in the north-western Antarctic Peninsula region (Anon, 1987, 1988; Kellermann, 1989*a*; Kellermann & Kock, 1988) and the Prdyz Bay region (Williams & Duhamel, this volume).

Following the signing of the Convention on the Conservation of Antarctic Marine Living Resources by the Antarctic treaty nations in 1981, work on Antarctic fish stock assessment was undertaken by the Commission for the Conservation of Antarctic Marine Living Resources (CCAMLR) and only a small amount of work on fish stocks was performed by the BIOMASS programme. Kock (1986) described the state of fish stocks in the Scotia Arc region during SIBEX, and Skora (1988*a*) reported on the fish catches at Elephant Island during the 1986–7 season. In 1983 the SCAR Group of Specialists on Southern Ocean Ecosystems and their Living Resources formed a committee which worked closely with the BWPFE to compile a 'review of the biology and status of Antarctic fish stocks' (Kock *et al.*, 1985). At the third meeting of Scientific Committee of the CCAMLR (SC-CAMLR), in September 1984 at Hobart (Anon, 1985*a*), the *ad hoc* Working Group on Fish Stock Assessment considered a draft version of the Kock *et al.* (1985) review, and other information on the Polish and Japanese fisheries. The Group identified four fish stocks that had been heavily fished and were in need of conservation measures at South Georgia, and two species that required protection in the Kerguelen Island area. The stocks at Kerguelen Island were already being managed by the French authorities (Duhamel & Hureau, 1981), with annually revised conservation measures that were first implemented in 1980. Proposals that included a 12 mile exclusion zone and restrictions on mesh sizes were submitted by the SC-CAMLR to the Commission to protect the overexploited fish stocks at South Georgia. The proposals were adopted and implemented by the Commission and thereby the CCAMLR began to influence the harvesting of resources in the Southern Ocean.

Advances were made in several areas during the BIOMASS Programme through international workshops. A workshop in 1979 on the age determination of Antarctic fish resulted in several publications (Everson, 1980, 1981) and a further similar workshop (Anon, 1982*b*). The workshop on the identification of the early life history stages in 1981 (Hureau, 1982), produced the first attempt at an identification guide (North & White, 1982); this was followed by a Soviet atlas of Antarctic fish larvae (Efremenko, 1983*a*).

After the completion of the BIOMASS field programme in 1985, the SCAR Executive agreed to propose to the other cosponsors to disband the Group of Specialists on Southern Ocean Ecosystems and their Living Resources; accordingly, the group was effectively disbanded along with its subgroups and Working Parties, including the BWPFE. An executive was formed to manage the BIOMASS Data Centre and plan data analysis workshops (see Thorley & Trathan, this volume).

Several studies were instigated during two workshops in 1986 and 1987 to check and evaluate the SIBEX fish data at the BIOMASS Data Centre in Cambridge (Anon, 1987, 1988). These included a review of the reproduction of notothenioid fishes (Kock & Kellermann, 1991) and several studies on the early stages of Antarctic fish: a catalogue of studies by W. Ślósarczyk, a bibliography by A. Kellermann (Anon, 1987), a field identification key (North & Kellermann, 1989), and a catalogue of the early stages (Kellermann, 1989*b*).

The BIOMASS Programme has provided a framework through which Antarctic scientists gained essential experience of international cooperation; these scientists, through meetings and workshops, have formed a collaborative network. This experience may enhance future international bio-oceanographic studies (e.g. Global Ocean Ecosystems Dynamics, Ecology of the Antarctic Sea Ice Zone) in the Southern Ocean.

Progress in publications on Antarctic fish

The BIOMASS Programme has influenced the management of scientific resources toward the study of Antarctic fish. How studies have been directed may be broadly represented by the number of papers published in each research field. Accordingly, the papers listed in Kock's (1989) bibliography have been subjectively classified according to the field to which they appear to have made the greatest contribution, and cumulatively counted (Fig. 2). The bibliography is largely complete from 1833 to 1987; the trends in publication rate for each subject are illustrated for the decade from 1977, the beginning of BIOMASS, to 1987 (Fig. 2). However, the bibliography also lists papers published in 1988 and 1989, and these are included in the total number of papers given in the text below. The bibliography lists 1430 publications directly on fish and 125 on associated subjects such as fish predators. Of these, 50% were published during 1980–9, in the main period influenced by the BIOMASS Programme. More than 6% (n=46) of these more recent papers were a contribu-

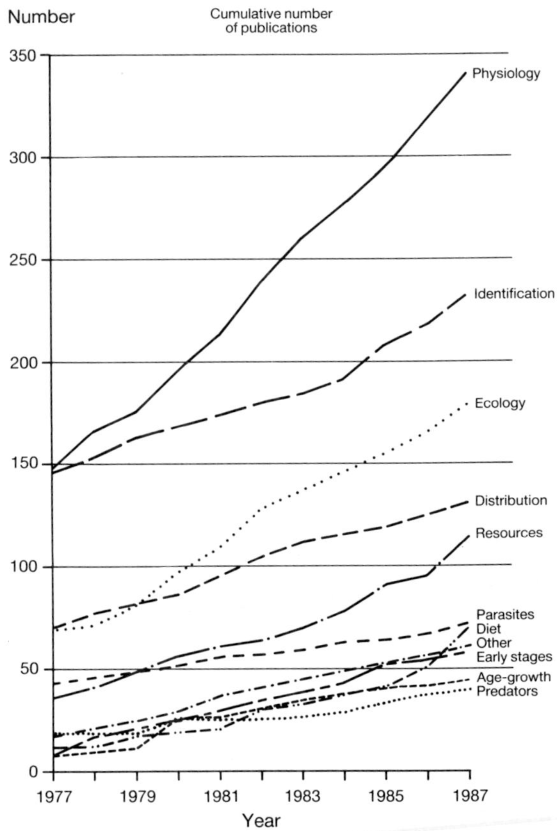

Fig. 2. Cumulative number of publications since 1833 for major Antarctic research fields in the period from 1977 to 1987 (derived from Kock, 1989).

tion to the programme and a further 2.4% (n=18) were linked to it. Thus the BIOMASS Programme made a notable contribution to research on Antarctic fish.

In each of the three decades up to 1987 there was a doubling in the number of publications about Antarctic fish; this increase has occurred in most research fields. Most papers (360) were published in physiology and biochemistry (Fig. 2), an area which was not directly addressed by the BIOMASS Programme. General ecology, with 204 papers, was also a dominant subject (Fig. 2). In both these subject areas there was an increased annual publication rate between 1967 and 1977, and further increases following the FIBEX and SIBEX fieldwork. Another large field was identification (including reports of new species, taxonomic works, and identification guides) with 237 publications, many of which were descriptions by Soviet scientists of new species obtained during their wide-ranging resource surveys. The BIOMASS Programme has influenced reviews concerned with identification, such as the FAO species identification sheets (Fischer & Hureau, 1985, 1987), systematic species accounts (Gon & Heemstra, 1990) and larval fish guides (Efremenko, 1983*a*; Kellermann, 1989*b*; North & Kellermann, 1989; North & White, 1982). Accounts of fish from expeditions published before Norman's (1938) monograph were generally included with papers on identification, whereas later ones were typically counted with those on distribution, a category containing 135 papers. Papers mainly about the early stages (except identification) gave a modest total of 64. Following the BIOMASS workshop on age determination in 1977 (Everson, 1980, 1981) there was a short rise in the publication rate for papers on age and growth, which was not sustained. Between 1984 and 1987 there was an increased publication rate of papers on resources (total 117) which was not matched by those on age and growth (total 64) (Fig. 2). A few papers were published about fish in the diet of predators in 1980, and between 1984 and 1987 there was an increased number of publications on this subject. Since the early reports of an association between the early stages of fish and krill (Rembiszewski *et al.*, 1978; Ślósarczyk & Rembiszewski, 1982) there has been a small number of similar reports, to a total of 14. More detailed trophodynamic studies of larval euphausiids in the diet of larval fish (e.g. Kellermann, 1986*b*) were counted with the early life history papers. There was a consistent output of papers on the parasites of Antarctic fish.

This analysis of papers published cannot accurately reflect the scientific progress made in the various research fields, but it does show that there has been a sustained increase in work on Antarctic fish with much recent interest in their physiology and biochemistry. The analysis also identifies the recent reduction in the ratio of papers on age and growth to those on resources. This may be cause for concern for those relying on accurate demographic data to predict the potential yield of the exploited fish stocks.

Selective summary of present knowledge

At the end of the BIOMASS Programme it is fitting to summarize the present state of knowledge of Antarctic fish. Below is a limited review of the main fields to which the programme has made a substantial contribution. This review is primarily on the members of the dominant notothenioid suborder. There is also a note about the mesopelagic fish species whose importance in the marine ecosystem is becoming increasingly recognized. Because of the limitations of space and other reasons outlined below, some major subject areas are not reviewed here. Information on the distribution of the fish species is not included; this has been reviewed by various authors in Gon & Heemstra (1990). There have been a great many studies of the physiology and biochemistry of Antarctic fish because of the interest in their adaptations to low sea temperatures. Consequently a balanced synthesis of these would be large and is not included here; for reviews of various emphases see: Eastman & DeVries (1986), di Prisco *et al.* (1988, 1991), Macdonald *et al.* (1987), Johnston (1989) and Eastman (1990). The fishery and the status of the fish stocks have been recently reviewed (Kock, 1991) and are not included here.

Antarctic notothenioid fish

Identification

Everson (1977) reviewed the living resources of the Southern Ocean and thereby provided a baseline for what was known before the BIOMASS era. Knowledge of the Antarctic fish fauna was somewhat limited at the beginning of the BIOMASS period. There were only about 120 fish species known from the Antarctic (Andriashev, 1965, 1987; DeWitt, 1971) and the standard guide to their taxonomy and identification was that of Norman (1938), which described only 67 Antarctic species.

During the early meetings of the BWPFE it was soon recognized that there was a need for accurate reporting of species in the fish catches; to facilitate this, a new guide to the species was required. BIOMASS collaborated with FAO and encouraged experts on the taxonomy of the various fish families to produce an identification guide to Antarctic living resources, including fish (Fischer & Hureau, 1985, 1987). The recent book by Gon & Heemstra (1990), endorsed by BIOMASS, was a similar collaborative venture by specialists in each field and has provided a comprehensive guide to the fishes of the Southern Ocean.

The taxonomy of the notothenioids has received much attention in the past decade (Andersen, 1984; Andersen & Hureau, 1979; Balushkin, 1976, 1982, 1984; DeWitt *et al.*, 1990; Eakin, 1981; Hureau, 1985; Iwami, 1985; Voskoboinikova, 1986). In the early 1970s the Antarctic members of the family Nototheniidae were arranged in 6 genera (*Aethotaxis, Dissostichus, Notothenia, Pagothenia, Pleuragramma* and *Trematomus*) (Andriashev & Jakubowski, 1971; DeWitt, 1962, 1964). Systematic revisions have fragmented the genus *Notothenia* so that a further four genera are also recognized (*Gobionotothen, Lepidonotothen, Paranotothenia* and *Patagonotothen*). The status of a further nototheniid genus or sub-genus *Indonotothenia* has not yet been resolved (DeWitt *et al.*, 1990). These revisions have resulted in changes in the Latin names of several species (indeed, some have altered several times in the past decade) and this is a potential hazard to researchers. A list of synonyms is given under each species in DeWitt *et al.* (1990). Since the 1970s two new types of Nototheniidae have been discovered and assigned to new genera *Gvozdarus* and *Cryothenia*.

The Harpagiferidae have been split into two families; members of the four genera of Artedidraconidae all possess a chin barbel, whereas those of Harpagiferidae (containing the single genus *Harpagifer*) have opercular spines (Eakin, 1981, 1990; Hureau, 1990). The nomenclature of genera within the families Channichthydae and Bathydraconidae has remained virtually unchanged.

Reproductive biology (general ecology)

The reproductive strategies of Antarctic notothenioid fish were outlined by Marshall (1953) and compared with those from other realms. From observations on 17 species (largely from Regan, 1916), Marshall noted that they tended to produce few, large yolky eggs, which resulted in large larvae with well-developed jaws. He discussed the advantages of larger larvae, including a greater feeding range during the short austral summer season of elevated secondary production. At the beginning of the BIOMASS Programme the basic reproductive biology of about 11 notothenioid fish was known (Everson, 1977). By the mid-1980s the spawning time, egg diameter and fecundity of about 20 species was known (Everson, 1984; Kock, 1985). In the recent review by Kock & Kellermann (1991) the reproductive parameters are presented for 59 species, including 31 species from the high-Antarctic zone. Notothenioid fish become sexually mature at between 33 and 75% of their maximum length. Generally most species have large eggs (2–5 mm in diameter) and low relative fecundity at around 4–200 eggs per gramme total body mass. Each species has an annual spawning season of 1–2 months' duration. However, sexually mature individuals of some species (e.g. *Champsocephalus gunnari*) may not attain breeding condition every year. The timing of the spawning season varies with locality or stock. Many species spawn during late autumn and/or early winter but spawning periods of different species occur throughout most of the year. Several nearshore demersal species produce a nest of eggs on stones, or in sponges. One or more of the parents may guard the nest. The eggs of other species (e.g. *Notothenia coriiceps* and *C. gunnari*) may be pelagic. The duration of egg development varies from about 2 to 6 months. For example, egg development is 60–75 days for *Pleuragramma antarcticum* and 150–180 days in *Harpagifer antarcticus* (with eggs 2 mm and 2.5 mm in diameter, respectively).

Larval and early juvenile stages

The worldwide interest generated by the BIOMASS Programme provided scientists from several nations with an opportunity to participate in the FIBEX and

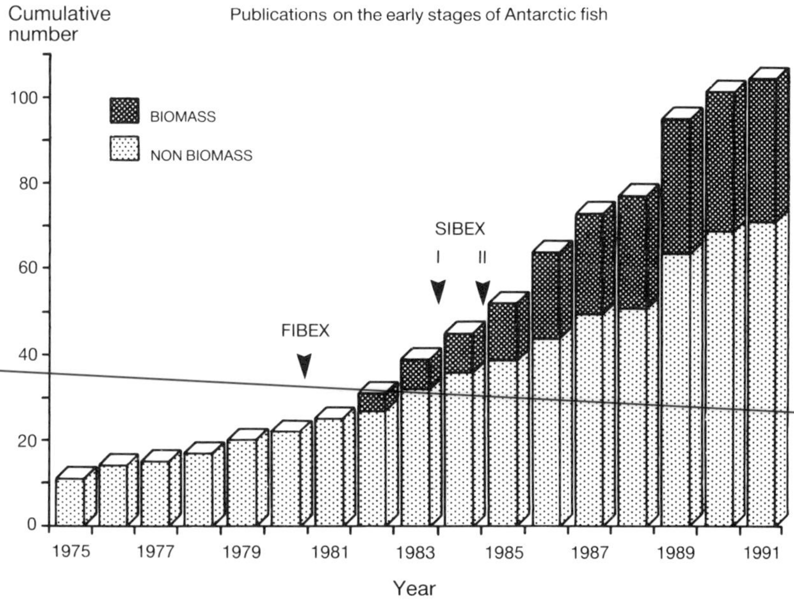

Fig. 3. Cumulative number of publications since 1883 mentioning the early stages of Antarctic fish in the period from 1975 to 1991. Those linked with the BIOMASS Programme and others, and the major BIOMASS phases, are indicated.

SIBEX surveys. The net samples from the surveys contained a by-catch of the pelagic larvae and early juveniles of demersal fish, and these provided much new information on the early life history stages of fish in the northern Antarctic Peninsula and southern Indian Ocean regions of the Antarctic. Since the first descriptions of Antarctic fish larvae early this century (Pappenheim, 1912; Regan, 1916) there were only a few published studies about them until the BIOMASS field programme. From around 20 accounts by the early 1980s, the number of publications with information on the early stages of fish increased rapidly after FIBEX and this increased rate of publication was maintained following SIBEX (Fig. 3). In total, BIOMASS can be directly linked with nearly one third of the studies published by early 1991.

Substantial advances have been made in the identification of larvae. Extensive collections by British and Soviet expeditions were the basis of two publications encouraged by BIOMASS: the key to fish larvae (North & White, 1982) and the *Atlas of Fish Larvae of the Southern Ocean* (Efremenko, 1983*a*). The large amount of material gathered during FIBEX and SIBEX led to a new identification key (North & Kellermann, 1989) and a catalogue of early life stages of notothenioid fishes (Kellermann, 1989*b*) (these were also published by BIOMASS). Similar efforts were undertaken in the Indian Ocean Sector (Gon, 1988; Koubbi *et al.*, 1990, 1991; Rasoanarivo & Aboussouan, 1983). Today, from the total of about 200 demersal and 75 mesopelagic fish species (Andriashev, 1987), the larval stages of 68 species, mostly of the abundant notothenioids, have been described.

Several studies have used the by-catch of larval and juvenile fish from the krill-directed surveys to describe the abundance of larvae and their distribution in space and time. This has led to the description of ichthyoplankton associations and their distribution in relation to bathymetry and to water masses, particularly in the Antarctic Peninsula region (e.g. Asencio & Mujica, 1986; Kellermann & Kock, 1984; Loeb, 1991; Mujica & Asencio, 1985; Sinque *et al.*, 1986*a*, *b*; Skora, 1988*b*; Ślósarczyk, 1986, 1987; Ślósarczyk & Cielniaszek, 1985; Ślósarczyk & Rembiszewski, 1982; White & North, 1987). Larval abundance in some species seems to be greatest in shelf and nearshore waters (North, 1988; Ślósarczyk, 1986). This has to be considered when planning surveys to study them. The temporal distribution pattern that emerged during the BIOMASS field studies was that of a succession of notothenioid hatching periods from late winter through the austral summer to early winter. In some species pelagic development is short, spanning one summer season, but in many others the larvae may remain pelagic during winter and later transform to the demersal stage in their second summer (Kellermann, 1989*a*, *b*).

The FIBEX and SIBEX sampling strategy, based on widely spaced stations covering a large area, was designed to survey the biomass of krill, rather than that of larval fish. The scales of temporal, environmental and biological processes affecting krill distribution and aggregation may be different from those affecting the fish larvae (see Murphy *et al.*, 1988; Ślósarczyk, 1986). A synthesis of larval notothenioid (and mesopelagic) fish distributions recorded during SIBEX and an attempt to relate these to the physical and biological environment was undertaken at two workshops (Anon, 1987, 1988). A quantitative comparison of the data sets provided by the various countries was difficult because of the different sampling methods used. In spite of these shortcomings, owing to the seasonal coverage and the amount of concomitant studies and data, the FIBEX and SIBEX surveys have provided extensive baseline information on fish larvae in the northern Antarctic Peninsula and Prydz Bay regions.

At the onset of exploratory krill fishing, it became apparent that a considerable number of fish species were associated with krill swarms. These included not only adults feeding on krill, but also larval and juvenile stages, especially juvenile channichthyids. This association was interpreted as a fish–krill assemblage that changes during the year with the sequential co-occurrence of the pelagic early stages of different fish species, feeding on krill, until the fish develop to the demersal juvenile–adult phase (Rembiszewski *et al.*, 1978). Perhaps the most significant knowledge emerging from FIBEX and SIBEX was confirmation that large numbers of larval and juvenile fish regularly occur as a by-catch of krill and, therefore, that young fish must be taken by the krill fishery (Kellermann & Kock, 1984; Kellermann & Ślósarczyk, 1984; Kock, 1982; Ślósarczyk, 1983, 1986; Ślósarczyk & Rembiszewski, 1982). The notothenioids typically have a low fecundity; therefore, the stocks of notothenioid fish are particularly vulnerable to additional mortality. The krill fishery has become the largest crustacean fishery in the world (Nicol, 1991). Pre-recruit fishing of Antarctic fish by industrial krill harvesting at South Georgia may have an impact on some fish stocks (Everson *et al.*, 1992). This is especially important for species such as *Champsocephalus gunnari*, which are also heavily exploited as juveniles and adults by the finfish fishery.

Little is known about the predators of Antarctic fish larvae, although many icefish larvae are at least partly piscivorous (Kellermann, 1986*a*, *b*, 1989*b*; North & Ward, 1989, 1990; North & White, 1987; North, 1991; Regan, 1916). Notothenioid larvae feed on copepods and the early stages of krill and other euphausiids (Balbontin *et al.*, 1986; Kellermann, 1986*a*, 1989*b*, 1990; Williams, 1985*a*). The diet of fish larvae in the Antarctic Peninsula region, at first feeding, is usually the eggs, nauplii and small postlarval stages of copepods. In the krill spawning season, the diet of larval fish may include a large proportion of krill eggs. Icefish larvae of several species (e.g. *Chionodraco rastrospinosus* and *Cryodraco antarcticus*) commence feeding on the early juvenile stages of *Euphausia superba* and may also take fish larvae. In years of low krill abundance a smaller euphausiid species, *Thysanoessa macrura*, seems to be an alternative food resource. The functional relationship between krill swarms and the early stages of fish requires further study to quantify the interactions. Of particular relevance are the distribution, trophodynamics, growth rates, and recruitment of the fish larvae and their prey.

There have been few investigations on Antarctic larval fish growth and development. Most estimates of daily growth rates have been interpolated from length–frequency analysis. Estimating larval growth from daily increments in otoliths have been introduced only recently (Radtke *et al.*, 1989). Ślósarczyk (1987) showed that absolute growth in length of larval and early juvenile icefish was typically curvilinear, with slow initial rates and increased rates with development. The maximum summer growth rates of Antarctic fish larvae approach those of temperate marine fish larvae (Clarke & North, 1991; North, 1990, 1991). However, when species of similar initial size were compared, the average daily growth rates of polar larvae were lowest. It was therefore suggested that annual growth was limited by the seasonality of feeding conditions and temperature (Clarke & North, 1991; North, 1991).

The annual temperature range in the surface waters of the Southern Ocean is narrow; however, the development rate of fish may be sensitive to small temperature differences. Eggs of the nototheniid *Notothenia coriiceps* reared at the sub-Antarctic island of South Georgia at water temperatures of 2–3 °C began to hatch after 103 days, while at the more southerly South Orkney Islands, with temperatures of −1.8 to −1.0 °C, they took at least 150 days to develop (White *et al.*, 1982). Thus, intraspecific developmental and growth rates may differ substantially with latitude. A complete explanation of the prolonged incubation periods awaits elucidation. Are the rates of growth and development limited by the adequacy of adaptation to low temperature? Does the timing of spawning reflect seasonal ecological constraints on the energy balance of the adults and/or the larvae? Hatching of the early larvae may be linked with the period when the production of their pelagic prey species is particularly high (as implicated in some invertebrates) (Clarke, 1988).

The paucity of studies on the ecophysiology of larval Antarctic fish is striking: only a few accounts dealing with muscle physiology and swimming abilities of the larvae have been published (Dunn *et al.*, 1989; Johnston, 1989). This field offers considerable opportunities for research.

Age and growth

Olsen (1954, 1955) determined the age of Antarctic fish based on annual rings in otoliths and scales. Similar studies were conducted during the next 20 years (see Everson, 1977). At the beginning of the BIOMASS Programme there was some information on the age and growth rates of about 12 species from work by Olsen (1954, 1955), Wohlschlag (1961), Hureau (1970), Everson (1970), Crisp & Carrick (1975) and Shcherbich (1975). However, it was noted that there were problems associated with reading the age of Antarctic fish from scales and otoliths. For example, Wohlschlag (1961) added two to the number of annual rings he counted in otoliths from *Trematomus bernacchii* to get their age in years. Recognizing the importance of correct age information for estimates of production by the fish stocks, BIOMASS organized a workshop on the age determination of Antarctic fish in 1979. This resulted in a guide to methodology, with an improved and standardized terminology for describing age determination methods, and new information on the age of several species (Everson, 1980, 1981). The guide permitted a more objective appraisal of the published studies on fish age and made it easier to compare between the results obtained by different workers. A similar workshop was held in 1982 (Anon, 1982*b*). A further workshop was recommended by BIOMASS; this was held in 1986 in Moscow under the auspices of CCAMLR. The latter resulted in a CCAMLR otolith and scale exchange study whereby the age of the same material was determined by different scientists. The study on four species found that all readers gave the same age for 0–37% of the specimens depending on species (Kock, 1990). It was concluded that further validation studies were necessary, particularly on exploited notothenioids, to identify the first annuli more precisely. Following the report of daily rings in the otoliths of various Antarctic fish species at the first BIOMASS age workshop (Townsend, 1980), there have been several studies which have used daily rings (Radtke, 1990; Radtke & Hourigan, 1990; Radtke & Kellermann, 1991; Radtke & Targett, 1984; Radtke *et al.*, 1989). Results from daily ring studies largely corroborate those from annual rings (Radtke, 1990).

Annual growth rates of Antarctic notothenioid fish are moderately slow to very slow. The von Bertalanffy growth function *K* (derived from fish length in centimetres and time in years) ranges from 0.32 for the small (maximum length 10 cm) *Harpagifer antarcticus* (Daniels, 1983), to only 0.056 in the large (up to 2 m long) *Dissostichus mawsoni* (Burchett *et al.*, 1984). The former species may reach 10 years old and the latter may live in excess of 30 years. The growth rates of many of the larger sub-Antarctic species caught by the fishery have been studied. In these species, with maximum lengths of about 40–100 cm, the growth rate *K* is around 0.1–0.3 (see North & White, 1987). For example, *Gobionotothen* (formerly *Notothenia*) *gibberifrons*, a moderately large species (*ca.* 50 cm maximum length), has a *K* of 0.095 and may reach 18 years of age (Coggan *et al.*, 1990). There are a few recent growth studies on species that are not presently exploited. Ekau (1988) reported that the growth rates of 10 species, between 12 and 41 cm long from the Weddell Sea were slow ($K = 0.16$ to 0.04). With a maximum size of around 25 cm and longevity of 14+ years, the growth rate of the pelagic *Pleuragramma antarcticum* is slow, with a K of 0.0715 (Hubold & Tomo, 1989). The ice-associated (cryopelagic) *Pagothenia borchgrevinki* (maximum size *ca*. 23 cm and age 7+ years) also grows slowly ($K = 0.070$ to 0.076)(Pankhurst, 1990).

Prey

Information on the diet of juvenile and adult stages of Antarctic fish was very limited when BIOMASS began (e.g. DeWitt & Hopkins 1977; Hoshai, 1979; Hureau, 1970; Linkowski & Rembiszewski, 1978; Permitin, 1970; Permitin & Tarverdiyeva, 1972, 1978; Richardson, 1975; Tarverdiyeva, 1972). Despite several contributions based on work during FIBEX and SIBEX (Balbontin *et al.*, 1986; Barrera Oro & Tomo, 1987; Kawamura, 1981; Kellermann, 1986*a*; Moreno *et al.*, 1986; Williams, 1985*a*, *b*) and the notable work by Targett (1981), there is still only a little known about the geographic, ontogenetic and seasonal variation in the diet of many species. However, it is clear that many notothenioids are opportunistic predators, feeding on a wider range of prey types including gastropods, polychaetes, mysids, tanaids, isopods, decapods, tunicates and fish. Most notothenioids and myctophids feed on copepods, amphipods and euphausiids, especially *E. superba*, at some stage in their life cycle (e.g. Hubold & Ekau, 1990; Kock, 1981; Kozlov *et al.*, 1988; Kozlov & Naumov, 1988; Linkowski & Rembiszewski, 1978; Permitin, 1970; Permitin & Tarverdiyeva, 1972, 1978; Richardson, 1975; Rowedder, 1979*a*; Targett, 1981).

Predators

Many Antarctic penguins and other birds, whales and seals (and probably squid and octopus) feed to some extent on fish (Bonner, 1989; Croxall & Lishman, 1987; Laws, 1985; Prince & Morgan, 1987; Rodhouse *et al.*, 1992). The guides to otoliths (Hecht, 1987; Williams & McEldowney, 1990), other publications (e.g. Hecht & Cooper, 1986; North *et al.*, 1984) and unpublished material have facilitated some studies on the diet of higher predators (e.g. Adams & Klages, 1987; Ainley *et al.*, 1986; Brown & Klages, 1987; Brown *et al.*, 1990; Croxall & Lishman, 1987; Croxall & North, 1988; Croxall *et al.*, 1988; Gales *et al.*, 1990; Green, 1986; Green & Williams, 1986; Green *et al.*, 1989, 1990; Hecht & Cooper, 1986; Klages *et al.*, 1990; North *et al.*, 1983; Plötz, 1986; Ridoux & Offredo, 1989; Testa *et al.*, 1985; Williams, 1989). In general, the common fish prey of animals which forage over the Antarctic continental shelves are notothenioids and epipelagic myctophids; the latter become especially prominent in the open ocean and at the pack-ice edge. Near the coast, in the high Antarctic and in the permanent pack-ice zone, the notothenioid *Pleuragramma antarcticum* is an especially important prey. Fish constitute a large proportion of the diet of some predators (e.g. blue-eyed cormorant, Antarctic petrel, snow petrel, white-chinned petrel, wandering albatross, king penguin, emperor penguin, gentoo penguin and Weddell seal). Thus the status of the fish populations would be expected to be important for such predators. These as well as other predator species with large populations (e.g. macaroni, Adélie and chinstrap penguins; elephant, fur and crabeater seals) which have a lower proportion of fish in their diets may have a considerable impact on the fish populations. This fish predation may be especially pronounced in local areas adjacent to large penguin or fur seal breeding sites. Of particular concern is how little is known about the diet of crabeater seals (Bengtson, 1982; Green & Williams, 1986; Øritsland, 1977), because with a population of about 10^7 individuals they constitute such a large biomass (Erickson & Hanson, 1990).

Mesopelagic fish

Occurrence and biomass

Myctophids (lanternfish), bathylagids (smelts) and representatives of a few other mesopelagic fish families are an important component of the biomass in oceanic waters of the Southern Ocean. Hulley (1981) has produced a monograph on the identification, distribution and biology of Myctophidae (including the genera *Krefftichthys, Electrona, Gymnoscopelus* and *Protomyctophum*) in the Southern Ocean and has recently (Hulley, 1990) provided further details about these fish. The taxonomy of the Bathylagidae has been reviewed (Gon, 1987).

The biomass of mesopelagic fish in the Southern Ocean exceeds that of seals; they are the most prevalent krill predators in the Antarctic oceanic system (Lancraft *et al.*, 1989). The most common species is *Electrona antarctica*, which feeds largely on krill (Rowedder, 1979*a*, *b*). It is common at water depths of over 1000 m and is the most abundant fish in the upper 200 m of the water column, whereas below 200 m depth *Bathylagus antarcticus* (Bathylagidae) contributes a greater biomass (Lancraft *et al.*, 1989, 1991). The myctophid *Electrona carlsbergi* can be the dominant component of the deep (>50 m to 200 m) acoustic scattering layer south of the Polar Frontal Zone in the Pacific Ocean region (Linkowski, 1983), and is one of the most abundant myctophid species in the Atlantic zone (Zasel'sliy *et al.*, 1985).

Many of the mesopelagic fish undertake diel vertical migrations towards the surface at night (Lancraft *et al.*, 1989, 1991). They also have a diurnal feeding cycle; for example, *E. antarctica* feeds more during the day, when it is deeper, than at night (Rowedder, 1979*b*). This may have implications for the rate of vertical carbon flux. It seems likely that, on average, the fish defecate higher in the water column than where they feed and may thus act as a net upward transporter of carbon and other nutrients.

Midwater fish are an important component of the oceanic marginal ice-zone community. The species *E. antarctica, B. antarcticus, Gymnoscopelus braueri* and *G. opisthpterus* were abundant in the marginal ice zone in summer and autumn, when total fish biomass in the upper 100 m was 3.3–4.4 g fresh mass m^2 (Lancraft *et al.*, 1989). At the marginal ice zone in winter, the total fish biomass in the upper 1000 m was 5.4 g fresh mass m^2 and the dominant fish species were *B. antarcticus, E. antarctica, G. braueri* and *Protomyctophum bolini* (Lancraft *et al.*, 1991). In spring, mean fish biomass in the upper 1000 m was 3.3 g fresh mass m^2 and the dominant species were *E. antarctica, G. braueri* and *Notolepis coatsi* (Paralepididae) in the upper 500 m, and *B. antarcticus* and *Cyclothone microdon* (Gonostomatidae) in the 500–1000 m depth layer (Ainley *et al.*, 1986; Torres *et al.*, 1984).

In the Antarctic Peninsula region during FIBEX and SIBEX the most abundant midwater juvenile and adult stages of fish were those of *Krefftichthys anderssoni, E. antarctica, E. carlsbergi, Gymnoscopelus nicholsi, G. braueri, Protomyctophum bolini* and *Notolepis* spp. They represented most of the biomass of pelagic fish in the upper 500 m over oceanic waters. Larval mesopelagic fish also dominated the ichthyoplankton of oceanic waters, especially the larvae of *Electrona* spp., *Gymnoscopelus* spp. and *Notolepis* spp. (Acuna, 1986; Anon, 1987, 1988; Asencio & Moreno, 1984; Efremenko, 1986; Kellermann, 1989*a*; Kellermann & Kock, 1984, 1988; Kellermann & Ślósarczyk, 1984; White & North, 1987). The larvae of these groups also dominate the oceanic ichthyoplankton in the upper 500 m at South Georgia (Efremenko, 1983*a*, 1986; North, 1988, 1990), in the Weddell Sea (Hubold, 1990) and southern Indian Ocean regions (Gon, 1988; Hulley *et al.*, 1989; Rasoanarivo & Aboussouan, 1983; Williams *et al.*, 1983; Williams, 1985*a*, *b*). *B. antarcticus* eggs were found widely distributed at 200–1000 m depth in the warm deep water layer in the Scotia Sea (Efremenko, 1983*b*).

Diet

Antarctic and sub-Antarctic myctophids generally feed on many species of zooplankton, including copepods, euphausiids (especially *Euphausia superba*) and amphipods (Asencio & Moreno, 1984; Gorelova & Efremenko, 1989; Kozlov & Tarverdiyeva, 1989; Rembiszewski *et al.*, 1978; Rowedder, 1979*a*; Oven *et al.*, 1990; Williams, 1985*a*, *b*). Larval *K. anderssoni* and *Electrona antarctica* from the Scotia Sea, sampled during periods representing most of the year, had fed on invertebrate eggs and the early stages of euphausiids, copepods and other zooplankton (Gorelova & Efremenko, 1989). Kozlov & Tarverdiyeva (1989) studied juvenile and adult stages of seven species of myctophids (*E. antarctica, E. carlsbergi, K. anderssoni, G. nicholsi, G. braueri, G. opisthopterus* and *G. (Nasolynchus)* sp.) collected from wide-ranging studies in different parts of the Antarctic seas. They found that copepods, euphausiids and hyperiid amphipods were common in their diets, which included a wide range of zooplankton prey. The paralepidid *N. coatsi* feeds on krill (Post, 1990; Solyanik, 1966).

Predators

The myctophids are consumed by many higher predators in the Southern Ocean, notably king and gentoo penguins, Antarctic fur seals, icefishes and squid (Adams & Klages, 1987; Green *et al.*, 1989; Kozlov *et al.*, 1988; North *et al.*, 1983; Rodhouse *et al.*, 1992; Williams, 1989). A fishery for myctophids has developed in the Southern Ocean. Catches of *E. carlsbergi* from the South Georgia subarea have risen from around 23 000–30 000 tonnes in the 1988–9 and 1989–90 seasons to about 78 500 tonnes in the 1990–1 season (Anon, 1990, pp 164, 185–187; Anon, 1991*a*, p 14).

Growth

Linkowski (1987) determined the von Bertalanffy growth function (K) of four species of myctophids. The growth functions of two common Southern Ocean species were quite different. For *E. antarctica*, K was 0.25 in males and 0.17 in females. K was 0.55 for both sexes of *E. carlsbergi* combined. This value is the same order as that of another common southern myctophid, *G. nicholsi*, with K parameters of 0.48–0.41 depending on the stock (Linkowski, 1985, 1987).

Physiology

The oxygen consumption rates and metabolic enzyme levels of several mesopelagic fishes have been reported by Torres & Somero (1988). Oxygen consumption rates were in the low to intermediate range for Antarctic fish and were generally lower in fish from deeper waters. Mass-specific metabolic enzyme (lactate dehydrogenase and citrate synthase) activity also decreased with increased depth of occurrence. These decreases in metabolic rate and enzyme activity with depth were attributed to the evolutionary adaptations of the fish to the decreasing food supply with depth.

Conclusions

The BIOMASS Programme has made an important contribution to the knowledge of Antarctic fish. The influence of the Programme has already been indicated:

> The main success of the BIOMASS Program has been in fostering and developing interest in research into the Southern Ocean ecosystem; all nations engaged in marine Antarctic research have developed major programmes directly or indirectly due to BIOMASS initiatives. BIOMASS has provided an important framework for international collaboration, instituting multi-ship surveys and multi-national efforts in data handling; it has also contributed greatly to the development, standardization, and intercalibration of methods and instrumentation. Communication developed within the Program, through a wide range of meetings, workshops, and various types of publication, has helped establish an international Antarctic marine science community which has since become the backbone for the work of the Scientific Committee of CCAMLR.
> (Anon, 1985*b*, para. 5.1)

> It was recalled that the BIOMASS Program had been a most successful example of international cooperation in science that had yielded a valuable set of primary data on the Antarctic marine ecosystem.
> (Anon, 1991*b*, para. 12.8)

The BIOMASS Programme has had both successes and difficulties, from which those planning future international studies could learn. Through the annual meetings of the various BIOMASS groups, subgroups and working parties a great amount of new information was communicated, and much cooperation, goodwill and scientific progress was generated. A most useful approach for progress was the identification of problem areas that were blocking further studies (e.g. species identification) or making past studies difficult to utilize (e.g. doubt about age interpretations), and then encouraging a team of experts to tackle these difficulties. Another strategy was to plan scientific sampling and observations which used the resources of different countries to contribute to common objectives. From this latter approach it was learned that standardization of methods and data handling are essential. The data must be integrated within a planned computer database utilizing well supported (business) software (see Thorley & Trathan, this volume). The challenge for the future is to apply the scientific and managerial knowledge gained through the BIOMASS Programme to enhance future international scientific work towards the understanding and conservation of the Antarctic ecosystem as an integral portion of the Earth's biogeosphere.

Acknowledgements

The SCAR/SCOR Group of Specialists on the Living Resources of the Southern Ocean provided the ideas and the management which enabled the BIOMASS Programme to work. A spirit of goodwill and cooperation pervaded the BIOMASS workshops on larval fish identification, age determination and data evaluation, and we acknowledge those who contributed to this, including B Bedford, GR Boltz, J Ciechomski, C Ozouf-Costaz, J Dann, J Dearborn, A DeVries, HH

DeWitt, G Duhamel, P Eberhard, E Everson, G Freytag, O Gon, G Hubold, J-C Hureau, S Koblitz, K-H Kock, C König, B Lamden, CA Moreno, JH Prime, R Radtke, M Rembiszewski, R Rivoharinala, J Ross, C Sinque, KE Skora, W Ślósarczyk, T Targett, AP Tomo, DW Townsend, D Vaughan, MG White, R Williams. We appreciate the important contributions by the late VN Efremenko, (B. H. Ефременко) on the early life history stages of fish. We particularly recognize the diligent contribution by the late Wieslaw Ślósarczyk to the BIOMASS Programme and his significant research on Antarctic fish biology. For constructive criticism of the manuscript we are most grateful to MG White, RA Coggan, A Clarke, RB Heywood and two anonymous reviewers.

References

Acuna E S (1986) Lanternfishes (Family Myctophidae) collected during SIBEX-Phase II cruise in the Bransfield Strait. *Ser Cient INACH* **35**: 87–99.

Adams N J, Klages N T (1987) Seasonal variation in the diet of the king penguin (*Aptenodytes patagonicus*) at sub-Antarctic Marion Island. *J Zool* **212**: 303–24.

Ainley D G, Fraser W R, Sullivan C W, Torres J J, Hopkins T L, Smith W O (1986) Antarctic mesopelagic micronekton: evidence from seabirds that pack ice affects community structure. *Science* **232**: 847–9.

Andersen N C (1984) Genera and subfamilies of the family Nototheniidae (Pisces, Perciformes) from the Antarctic and Subantarctic. *Steenstrupia* **10**: 1–34.

Andersen N C, Hureau J-C (1979) Proposition pour une nouvelle classification des Nototheniinae (Pisces, Perciformes, Nototheniidae). *Cybium* **1979**(6): 47–53.

Andriashev A P (1965) A general review of the Antarctic fish fauna. In: Mieghen I van, Oye P van (eds) *Biogeography and Ecology in Antarctica*. Junk, The Hague, pp 491–550.

Andriashev A P (1987) A general review of the Antarctic bottom fish fauna. In: Kullander S O, Fernholm B (eds) *Proc V Congr europ Ichthyol, Stockholm* 1985. Swedish Museum of Natural History, Stockholm, pp 357–72.

Andriashev A P, Jakubowski M (1971) Morphological grounds for generic separation of the Antarctic broadhead fishes (*Trematomus borchgrevinki* Boulenger and *T. brachysoma* Pappenheim) and a new status of the genus *Pagothenia* Nicols et LaMonte (Nototheniidae). *Zool Zh* **50**: 1041–55. (In Russian with English summary.)

Anon (1977) Biological Investigations of Marine Antarctic Systems and Stocks. *BIOMASS Sci Ser* **1**: 1–79.

Anon (1979) Working Party on Antarctic Fish Biology. Report of the First Meeting, Hamburg, January 1979. *BIOMASS Rep Ser* **3**: 1–36.

Anon (1980*a*) Working Party on Antarctic Fish Biology. Report of the 2nd Meeting, Dammarie-les-Lys, France, 27–31 May 1980. *BIOMASS Rep Ser* **12**: 1–47.

Anon (1980*b*) Working Party on Acoustic Krill Estimation. Report of the 2nd Meeting, Horten, Norway, September 1979. *BIOMASS Rep Ser* **9**: 1–12.

Anon (1982*a*) Meeting of the BIOMASS Working Party on Fish Ecology, Hamburg, F.R.G., 20–25 September 1982. *BIOMASS Rep Ser* **28**: 1–24.

Anon (1982*b*) Working Party on Antarctic Fish Biology. Report of the 2nd Workshop on the Ageing of Antarctic Fishes, Orono, Maine USA, 7–10 June 1982. *BIOMASS Rep Ser* **26**: 1–7.

Anon (1985*a*) Commission for the Conservation of Antarctic Marine Living Resources. Report of the third meeting of the scientific committee (Hobart, Australia, 3–13 September 1984). SC-CAMLR-III, Hobart, 238 pp.

Anon (1985*b*) Meeting of the Group of Specialists on Southern Ocean Ecosystems and their Living Resources (Dammarie-les-Lys, France 26–28 June 1985). *BIOMASS Rep Ser* **43**: 1–32.

Anon (1987) Report of the Post-SIBEX Fish Data Evaluation Workshop: Phase 1, Validation, Cambridge UK, 6–17 October 1986. *BIOMASS Rep Ser* **52**: 1–46.

Anon (1988) Report of the Post-SIBEX Fish Data Evaluation Workshop: Phase 2, Evaluation, Cambridge UK, 17–28 August 1987. *BIOMASS Rep Ser* **54**: 1–38.

Anon (1990) Commission for the Conservation of Antarctic Marine Living Resources. Report of the Ninth Meeting of the Scientific Committee. CCAMLR, Hobart, 345 pp.

Anon (1991*a*) Commission for the Conservation of Antarctic Marine Living Resources. Report of the Working Group on Fish Stock Assessment (Hobart, Australia, 8–17 October 1991). SC-CAMLR-X/5, Hobart, 142 pp.

Anon (1991*b*) Commission for the Conservation of Antarctic Marine Living Resources. Report of the Tenth Meeting of the Commission (Hobart, Australia, 21 October–1 November 1991). CCAMLR-X, Hobart, 101 pp.

Asencio G, Moreno C A (1984) Dieta y selectividad alimentaria de *Protomyctophum bolini* Fraser-Brunner (Pisces; Myctophidae) en el Paso Drake (Antartica). *Ser Cient INACH* **31**: 85–96.

Asencio G, Mujica A (1986) Ictioplancton del estrecho Bransfield (SIBEX Fase II-Chile). *Ser Cient INACH* **35**: 105–10.

Balbontin F, Garreton M, Neuling J (1986) Composicion de alimento y tamano de las presas en larvas de peces del estrecho Bransfield (SIBEX-Fase II). *Ser Cient INACH* **35**: 125–44.

Balushkin A V (1976) Kratykaya reviziya nototenii (*Notothenia* Richardson i blizkie rody) iz semeistva Nototheniidae. In: Korovina V M (ed) *Zoogeografiya i Sistematika Ryb.* Izd ZIN AN SSSR, Leningrad, pp 118–34. (In Russian.)

Balushkin A V (1982) Klassifikatsiya trematomovykh ryb. *Biol Morya* **5**: 70–7. (In Russian.)
Balushkin A V (1984) *Morfologicheskie Osnovy Sistematiki i Filogenii Nototenievykh Ryb.* Akademiya Nauk SSR Zoologicheskii Institut, Leningrad, 141 pp. (*Translated from Russian*: Balushkin A V (1989) *Morphological Bases of the Systematics and Phylogeny of the Nototheniid Fishes.* National Science Foundation, Amerind, New Delhi, 153 pp.)
Barrera Oro E R, Tomo A P (1987) Feeding and Ecology of *Notothenia larsenii* Lönnberg. In: El-Sayed S Z (ed) *Antarctic Aquatic Biology. BIOMASS Sci Ser* **7**: 99–106.
Bengtson J L (1982) Reproductive ecology of crabeater and leopard seals along the Antarctic Peninsula. *Ant J US* **17**: 185.
Bonner W N (1989) *The Natural History of Seals.* Christopher Helm, London, 196 pp.
Brown C R, Klages N T (1987) Seasonal and annual variation in the diets of macaroni (*Eudyptes chrysolophus chrysolophus*) and southern rockhopper (*E. chrysocome chrysocome*) penguins at sub-Antarctic Marion Island. *J Zool* **212**: 7–28.
Brown C R, Klages N T, Adams N J (1990) Short and medium term variations in the diet of penguins at Marion Island. *S Afr J Antarct Res* **20**: 13–20.
Burchett M S, DeVries A, Briggs A J (1984) Age determination and growth of *Dissostichus mawsoni* (Norman, 1937) (Pisces, Nototheniidae) from McMurdo Sound (Antarctica). *Cybium* **8**(1): 27–31.
Clarke A (1988) Seasonality in the Antarctic marine environment. *Comp Biochem Physiol* **90B**(3): 461–73.
Clarke A, North A W (1991) Is the growth of polar fish limited by temperature? In: di Prisco G, Maresca B, Tota B (eds) *Biology of Antarctic Fish.* Springer-Verlag, Berlin, pp 54–69.
Coggan R, Skora K, Murray A, White M (1990) A comparison between age determinations of the Antarctic fish *Notothenia gibberifrons* Lönnberg using scales and otoliths. *Cybium* **14**(1): 43–55.
Crisp D T, Carrick S M (1975) Some observations on the growth and length: weight relationship of the South Georgia cod *Notothenia rossii marmorata* Fischer during the first four years of life. *J Fish Biol* **7**: 407–9.
Croxall J P, Lishman G S (1987) The food and feeding ecology of penguins. In: Croxall J P (ed) *Seabirds: Feeding Ecology and Role in Marine Ecosystems.* Cambridge University Press, pp 101–33.
Croxall JP, North AW (1988) Fish prey of Wilson's storm petrel *Oceanites oceanicus* at South Georgia. *Br Antarct Surv Bull* **78**: 37–42.
Croxall J P, North A W, Prince P A (1988) Fish prey of the Wandering Albatross *Diomedea exulans* at South Georgia. *Polar Biol* **9**: 9–16.
Daniels R A (1983) Demographic characteristics of the Antarctic plunderfish, *Harpagifer bispinis antarcticus. Mar Ecol Prog Ser* **13**: 181–7.
DeWitt H H (1962) A new Antarctic nototheniid fish with notes on two recently described nototheniiformes. *Copeia* **1962**: 826–33.
DeWitt H H (1964) A redescription of *Pagothenia antarctica*, with remarks on the genus *Trematomus* (Pisces, Nototheniidae). *Copeia* **1964**: 683–6.
DeWitt H H (1971) *Coastal and Deepwater Benthic fishes of the Antarctic.* American Geophysical Union, Antarctic Map Folio Series, Folio 15, 10 pp.
DeWitt H H, Heemstra P C, Gon O (1990) Nototheniidae: notothens. In: Gon O, Heemstra P C (eds) Fishes of the Southern Ocean. J L B Smith Institute of Ichthyology, Grahamstown, pp 279–331.
DeWitt H H, Hopkins T L (1977) Aspects of the diet of the Antarctic silverfish, *Pleuragramma antarcticum.* In: Llano G A (ed) *Adaptations within Antarctic Ecosystems.* Smithsonian Institution, Washington DC, pp 557–67.
di Prisco G, Maresca B, Tota B (1988) Marine biology of Antarctic organisms: proceedings of an international conference on the marine biology of the Antarctic. *Comp Biochem Physiol* **90B**: 459–60.
di Prisco G, Maresca B, Tota B (eds) (1991) *Biology of Antarctic Fish.* Springer-Verlag, Berlin, 292 pp.
Duhamel G, Hureau J C (1981) La situation de la pêche aux îles Kerguelen en 1981. *La Pêche Maritime* **1238**: 272–9.
Dunn J F, Archer S D, Johnston I A (1989) Muscle fibre types and metabolism in post-larval and adult stages of Notothenioid fish. *Polar Biol* **9**: 213–23.
Eakin R R (1981) Osteology and relationships of the fishes of the Antarctic family Harpagiferidae (Pisces, Notothenioidei). *Antarct Res Ser* **1**: 81–147.
Eakin R R (1990) Artedidraconidae: plunderfishes. In: Gon O, Heemstra P C (eds) *Fishes of the Southern Ocean.* J L B Smith Institute of Ichthyology, Grahamstown, pp 332–56.
Eastman J T (1990) The biology and physiological ecology of notothenioid fishes. In: Gon O, Heemstra PC (eds) *Fishes of the Southern Ocean.* J L B Smith Institute of Ichthyology, Grahamstown, pp 34–51.
Eastman J T, DeVries AL (1986) Antarctic fishes. *Sci Amer* **255**(5): 96–103.
Efremenko V N (1983*a*) *Atlas of Fish Larvae of the Southern Ocean. Cybium* **7**(2): 1–74; and BIOMASS Handbook No 22: 74 pp.
Efremenko V N (1983*b*) A finding of *Bathylagus antarcticus* (Bathylagidae) eggs, their characteristics and distribution in the Scotia Sea. *J. Ichthyol* **22**(2): 135–7.
Efremenko V N (1986) Distribution of eggs and larvae of Myctophidae in the Southern Atlantic. *J Ichthyol* **26**(6): 141–7.
Ekau W (1988) Ecomorphology of nototheniid fish from the Weddell Sea, Antarctica. *Ber Polarforsch* **51**: 1–140.
El-Sayed S Z (1991) The contribution of the BIOMASS Program to Antarctic marine research. In: di Prisco G, Maresca B, Tota B (eds) *Biology of Antarctic Fish.* Springer-Verlag, Berlin, pp 131–42.
Erickson A W, Hanson MB (1990) Continental estimates

and population trends of Antarctic ice seals. In: Kerry K R, Hempel G (eds) *Antarctic Ecosystems: Ecological Change and Conservation*. Springer-Verlag, Berlin, pp 253–64.

Everson I (1970) The ecology of inshore fish at Signy Island (South Orkneys). PhD thesis, Univ London, 231 pp.

Everson I (1977) *The Living Resources of the Southern Ocean*. FAO, Rome, GLO/SO/77/1, 176 pp.

Everson I (1980) Age determination methods for Antarctic fish. *Cybium* (1980) **8**: 5–51.

Everson I (ed) 1981 *Antarctic Fish Age Determination Methods*. BIOMASS Handbook No 8, 24 pp.

Everson I (1984) Fish biology. In: Laws RM (ed) *Antarctic Ecology*, vol II. Academic Press, London, pp 491–532.

Everson E, Néyelov A, Permitin Yu E (1992) Bycatch of fish in the South Atlantic krill fishery. *Antarct Sci* **4**: 389–92.

Fischer W, Hureau J-C (eds) (1985) FAO species identification sheets for fishery purposes. Southern Ocean (Fishing areas 48, 58 and 88) (CCAMLR Convention Area). Rome, FAO, Vol 2: 233–470.

Fischer W, Hureau J-C (eds) (1987) Fiches FAO d'identification des espèces pour les besoins de la pêche. Océan Austral (Zones de la pêche 48, 58 et 88) (Zone de la Convention CCAMLR). Rome, FAO 2: 235–479.

Frank R F (1983) The Convention on the Conservation of Antarctic Marine Living Resources. *Ocean Development and International Law Journal* **13**(3): 291–330.

Gales N J, Klages N T W, Williams R, Woehler E J (1990) The diet of the emperor penguin, *Aptenodytes forsteri*, in Amanda Bay, Princess Elizabeth Land, Antarctica. *Antarct Sci* **2**(1): 23–8.

Gon O (1987) *The Fishes of the Genus* Bathylagus *of the Southern Ocean*. JLB Smith Institute of Ichthyology, Special Publication No 43, 22 pp.

Gon O (1988) The fishes collected during the South African SIBEX I and II expeditions to the Indian Ocean sector of the Southern Ocean (60–66° S, 48–64° E). *S Afr J Ant Res* **18**: 55–70.

Gon O, Heemstra P C (eds) (1990) *Fishes of the Southern Ocean*. JLB Smith Institute of Ichthyology, Grahamstown, 462 pp.

Gorelova T A, Efremenko VN (1989) On the food composition of the larvae of two species of lantern anchovies (Myctophidae) from the Scotia Sea. *J Ichthyol* **29**(5): 106–9.

Green K (1986) Food of the emperor penguin *Aptenodytes forsteri* on the Antarctic fast ice edge in late winter and early spring. *Polar Biol* **6**: 187–8.

Green K, Burton H R, Williams R (1989) The diet of Antarctic fur seals *Arctocephalus gazella* (Peters) during the breeding season at Heard Island. *Antarct Sci* **1**(4): 317–24.

Green K, Williams R (1986) Observations on food remains in faeces of elephant, leopard and crabeater seals. *Polar Biol* **6**: 43–5.

Green K, Williams R, Woehler E J, Burton H R, Gales N J, Jones R T (1990) Diet of the Heard Island cormorant *Phalacrocorax atriceps nivalis*. *Antarct Sci* **2**(2): 139–41.

Hecht T (1987) A guide to the otoliths of Southern Ocean fishes. *S Afr J Antarct Res* **17**(1): 1–87.

Hecht T, Cooper J (1986) Length/mass relationships, energetic content and the otoliths of the Antarctic cod *Paranotothenia magellanica* (Nototheniidae: Pisces) at sub-Antarctic Marion Island. *S Afr J Zool* **21**: 294–6.

Hempel G (1987) BIOMASS. *Environmental International* **13**: 27–31.

Heywood R B (1985) Antarctica: Offshore marine research. *Biologist* **32**(3): 139–44.

Hoshai T (1979) Feeding behavior of juvenile *Notothenia rossii marmorata* Fischer at South Georgia station. *Antarct Rec* **66**: 25–36.

Hubold G (1990) Seasonal patterns of ichthyoplankton distribution and abundance in the southern Weddell Sea. In: Kerry K R, Hempel G (eds) *Antarctic Ecosystems: Ecological Change and Conservation*. Springer-Verlag, Berlin, pp 149–58.

Hubold G, Ekau W (1990) Feeding patterns of post-larval and juvenile notothenioids in the southern Weddell Sea (Antarctica). *Polar Biol* **10**: 255–60.

Hubold G, Tomo A P (1989) Age and growth of the Antarctic silverfish *Pleuragramma antarcticum* Boulenger, 1902, from the southern Weddell Sea and Antarctic Peninsula. *Polar Biol* **9**: 205–12.

Hulley P A (1981) Results of the research cruises of FRV 'Walter Herwig' to South America. LVIII. Family Myctophidae (Osteichthys, Myctophiformes). *Arch FischWiss* **3**(1): 1–300.

Hulley P A (1990) Family Myctophidae. In: Gon O, Heemstra P C (eds) *Fishes of the Southern Ocean*. JLB Smith Institute of Ichthyology, Grahamstown, pp 146–78.

Hulley P A, Camus P, Duhamel G (1989) Ichthyological results of cruise MD-42/SIBEX II. Part I. Fishes from RMT8 stations, with additional records of lanternfishes (Myctophidae: Osteichthyes) from the Indian and Atlantic sectors of the Southern Ocean. *Cybium* **13**(1): 83–99.

Hureau J C (1970) Biologie comparée de quelques poissons antarctiques (Nototheniidae). *Bull Inst Océanogr Monaco* **68**(1391): 244 pp.

Hureau J C (1982) Methods for studying early life history stages of Antarctic fishes. *Cybium* **6**(1): 43–51.

Hureau J C (1985) Nototheniidae. In: Fischer W, Hureau J-C (eds) FAO species identification sheets for fishery purposes. Southern Ocean (Fishing areas 48, 58 and 88) (CCAMLR Convention Area) Vol 2. Rome, FAO, pp 323–85.

Hureau J C (1990) Harpagiferidae: spiny plunderfishes. In: Gon O, Heemstra P C (eds) *Fishes of the Southern Ocean*. J L B Smith Institute of Ichthyology, Grahamstown, pp 357–63.

Hureau J C, Ślósarczyk W (1990) Exploitation and conservation of Antarctic fishes and recent ichthyological research in the Southern Ocean. In:

Gon O, Heemstra P C (eds) *Fishes of the Southern Ocean.* JLB Institute of Ichthyology, Grahamstown, pp 52–63.

Iwami T (1985) Osteology and relationships of the family Channichthyidae. *Mem Natl Inst Pol Res* (Ser E) **36**: 1–69.

Johnston I A (1989) Antarctic fish muscles – structure, function and physiology. *Antarct Sci* **1**(2): 97–108.

Kawamura A (1981) Stomach contents of fish caught during exploratory trawl fishing for krill. In: El-Sayed S Z (ed) *BIOMASS 2.* SCAR, Cambridge, pp 69–71.

Kellermann A (1986*a*) On the biology of the early life stages of notothenioid fishes (Pisces) off the Antarctic Peninsula. *Ber Polarforsch* **31**: 1–149.

Kellermann A (1986*b*) Trophic relationships between early stages of icefish (Notothenioidei; Channichthyidae) and euphausiids and ichthyoplankton during FIBEX 1981. *ICES CM* 1986 **L6**: 1–7.

Kellermann A (1989*a*) The larval fish community in the zone of the seasonal pack-ice cover and its seasonal and interannual variability. *Arch FischWiss* **39**(1): 81–109.

Kellermann A (1989*b*) Catalogue of early life stages of Antarctic notothenioid fishes. *BIOMASS Sci Ser* **10**: 45–136. ((1990) *Ber Polarforsch* **67**: 45–136.)

Kellermann A (1990) Food and feeding dynamics of the larval Antarctic fish *Nototheniops larseni. Mar Biol* **106**: 159–67.

Kellermann A, Kock K-H (1984) Postlarval and juvenile notothenioids (Pisces, Perciformes) in the Southern Scotia Sea and Northern Weddell Sea during FIBEX 1981. *Meeresforschung* **30**(2): 82–93.

Kellermann A, Kock K-H (1988) Patterns of spatial and temporal distribution and their variation in early life stages of Antarctic fish in the Antarctic Peninsula region. In: Sahrhage D (ed) *Antarctic Ocean and Resources Variability.* Springer-Verlag, Berlin, pp 147–59.

Kellermann A, Ślósarczyk W (1984) Distribution of postlarval and juvenile Notothenioidei in the Atlantic sector of the Southern Ocean during FIBEX 1981. *BIOMASS Rep Ser* **36**: 1–27.

Klages N, Gales R, Pemberton D (1990) The stomach contents of Antarctic petrels *Thalassoica antarctica* feeding young chicks at Scullin Monolith, Mawson Coast, Antarctica. *Polar Biol* **10**: 545–7.

Kock K-H (1981) Fischereibiologische Untersuchungen an drei antarktischen Fischarten: *Champsocephalus gunnari* Lönnberg, 1905, *Chaenocephalus aceratus* (Lönnberg, 1906) und *Pseudochaenichthys georgianus* Norman, 1937 (Notothenioidei, Channichthyidae). *Mitt Inst Seefisch Hamburg* **32**: 1–226.

Kock K-H (1982) Fische aus RMT8- und Krillschwimmschleppnetzfängen während FIBEX 1981. *Arch FischWiss* **33**(1): 97–112.

Kock K-H (1985) Marine habitats – Antarctic fish. In: Bonner W N, Walton D W H (eds) *Key Environments Antarctica.* Pergamon Press, Oxford, pp 173–92.

Kock K-H (1986) The state of exploited Antarctic fish stocks in the Scotia Arc region during SIBEX (1983–1985). *Arch FischWiss* **37**(1): 129–86.

Kock K-H (1989) *Bibliography of Antarctic Fish.* Bundesforschungsanstalt für Fischerei. Information/Documentation, Hamburg. Literaturliste No 9, 136 pp.

Kock K-H (1990) Results of the CCAMLR Antarctic fish otoliths/scales/bones exchange scheme. *Scientific Committee of the Convention for the Conservation of Antarctic Marine Living Resources Selected Scientific Papers 1989* (SC-CAMLR-VII/BG/46), pp 197–226.

Kock K-H (1991) The state of the exploited fish stocks in the Southern Ocean – A review. *Arch FischWiss* **41**(1): 1–66.

Kock K-H, Duhamel G, Hureau J C (1985) Biology and status of exploited Antarctic fish stocks: a review. *BIOMASS Sci Ser* **6**: 1–143.

Kock K-H, Kellermann A (1991) Reproduction in Antarctic Notothenioid fish. *Antarct Sci* **3**(2): 125–50.

Koubbi P, Duhamel G, Camus P (1990) Early life stages of Notothenioidei from the Kerguelen Islands. *Cybium* **14**(3): 225–50.

Koubbi P, Ibanez F, Duhamel G (1991) Environmental influences on spatio-temporal distribution of ichthyoplankton around the Kerguelen Islands (Southern Ocean). *Mar Ecol Prog Ser* **72**: 225–38.

Kozlov A N, Naumov A G (1988) Feeding of young of three notothenoid (Notothenioidei) species in the Indian Ocean sector of the Antarctic. *J Ichthyol* **28**(1): 70–3.

Kozlov A N, Pinskaya I A, Podrazhanskaya S G, Tarverdieva M I (1988) Feeding of glassfishes in different regions of the Atlantic sector of Antarctica. *J Ichthyol* **28**(6): 137–45.

Kozlov A N, Tarverdiyeva M I (1989) Feeding of different species of myctophidae in different parts of the Southern Ocean. *J Ichthyol* **29**(3): 160–7.

Lancraft T M, Hopkins T L, Torres J J, Donelly J (1991) Oceanic micronektonic/macrozooplanktonic community structure and feeding in ice covered Antarctic waters during the winter (AMERIEZ 1988). *Polar Biol* **11**: 157–67.

Lancraft T M, Torres J J, Hopkins T L (1989) Micronekton and macrozooplankton in the open waters near antarctic ice edge zones (AMERIEZ 1983 and 1986). *Polar Biol* **9**: 225–33.

Laws R M (1985) The ecology of the Southern Ocean. *Am Sci* **73**: 26–40.

Linkowski T B (1983) *Electrona carlsbergi* (Taning, 1932), the principal component of a deep scattering layer in the Pacific sector of the Antarctic Ocean. *Pol Polar Res* **4**(1–4): 71–8.

Linkowski T B (1985) Population biology of the myctophid fish *Gymnoscopelus nicholsi* (Gilbert, 1911) from the western South Atlantic. *J Fish Biol* **27**: 683–98.

Linkowski T B (1987) Age and growth of four species of *Electrona* (Teleostei, Myctophidae). In: Kullander S O, Fernholm B (eds) *Proc V Congr europ Ichthyol,*

Stockholm 1985. Swedish Museum of Natural History, Stockholm, pp 435–42.

Linkowski T B, Rembiszewski J M (1978) Ichthyological observations off the South Georgia coasts. *Polsk Arch Hydrobiol* **25**(3): 697–704.

Loeb V J (1991) Distribution and abundance of larval fishes collected in the western Bransfield Strait region, 1986–87. *Deep-Sea Res* **38**(8/9): 1251–60.

Macdonald J A, Montgomery J C, Wells R M G (1987) Comparative physiology of Antarctic fishes. In: Blaxter J H S, Southward A J (eds) *Advances in Marine Biology* 24. Academic Press, London, pp 321–88.

Marin EE, Moreno CA (1982) Peces recolectados durante la expedicion FIBEX (Chile) en la Antartica. *Ser Cient INACH* **38**: 213–16.

Marshall N B (1953) Egg size in Arctic, Antarctic and deep-sea fish. *Evolution* **7**: 328–41.

Moreno C A, Rueda T, Asencio G (1986) Nicho trofico de *Pleuragramma antarcticum* en la region del estrecho Bransfield con una comparacion cuantitativa con otras areas del Océano Antartico. *Ser Cient INACH* **35**: 145–69.

Mujica A R, Asencio V (1985) Larvas de peces, eufausidos y estructura comunitaria del zooplancton del Estrecho Bransfield (Crucero SIBEX-INACH 1984). *Ser Cient INACH* **33**: 159–86.

Mujica A R, Torres A G (1982) Analisis cualitativo y cuantitativo del zooplancton Antartico. *Ser Cient INACH* **28**: 181–91.

Murphy E J, Morris D J, Watkins J L, Priddle J (1988) Scales of interaction between Antarctic krill and the environment. In: Sahrhage D (ed) *Antarctic Ocean and Resources Variability.* Springer-Verlag, Berlin, pp 120–30.

Nicol S (1991) CCAMLR and its approaches to management of the krill fishery. *Polar Rec* **27**: 229–36.

Norman J R (1938) Coast fishes part III. The Antarctic zone. *Discovery Rep* **18**: 1–104.

North A W (1988) Distribution of fish larvae at South Georgia: horizontal, vertical and temporal distribution and early life history relevant to monitoring year-class strength and recruitment. *Scientific Committee of the Convention for the Conservation of Antarctic Marine Living Resources Selected Scientific Papers 1987* (WG-FSA-87/16). pp 105–42.

North A W (1990) Ecological studies on Antarctic fish with emphasis on early development of inshore stages at South Georgia. PhD thesis, CNAA. 319 pp.

North A W (1991) Review of the early life history of Antarctic Notothenioid fish. In: di Prisco G, Maresca B, Tota B (eds) *Biology of Antarctic Fish.* Springer-Verlag, Berlin, pp 70–86.

North A W, Burchett M S, Gilbert C J, White M G (1984) Identification of fish from the Southern Ocean by means of otoliths. *Br Antarct Surv Bull* **62**: 83–94.

North A W, Croxall J P, Doidge D W (1983) Fish prey of the Antarctic fur seal *Arctocephalus gazella* at South Georgia. *Br Antarct Surv Bull* **61**: 27–37.

North A W, Kellermann A (1989) Key to the early stages of Antarctic fish. *BIOMASS Sci Ser* **10**: 1–44. ((1990) *Ber Polarforsch* **67**: 1–44.)

North A W, Ward P (1989) Initial feeding by Antarctic fish larvae during winter at South Georgia. *Cybium* **13**: 357–64.

North A W, Ward P (1990) The feeding ecology of larval fish in an Antarctic fjord, with emphasis on *Champsocephalus gunnari.* In: Kerry K R, Hempel G (eds) *Antarctic Ecosystems. Ecological Change and Conservation.* Springer-Verlag, Berlin, pp 299–307.

North A W, White M G (1982) Key to fish postlarvae from the Scotia Sea, Antarctica. *Cybium* **6**(1): 13–32.

North A W, White M G (1987) Reproductive strategies of Antarctic fish. In: Kullander S O, Fernholm B (eds) *Proc V Congr Europ Ichthyol, Stockholm 1985.* Swedish Museum of Natural History, Stockholm, pp 381–90.

Olsen S (1954) South Georgia cod (*Notothenia rossii marmorata* Fischer). *Norsk Hvalffangsttid* **7**(43): 373–82.

Olsen S (1955) A contribution to the systematics and biology of chaenichthyid fishes from South Georgia. *Nytt Mag Zool* **3**: 79–93.

Øritsland T (1977) Food consumption of seals in the Antarctic pack ice. In: Llano GA (ed) *Adaptations within Antarctic Ecosystems* (Proceedings of the third SCAR symposium on Antarctic biology). Smithsonian Institution, Washington, pp 749–68.

Oven L S, Konstantinova M P, Shevchenko N F (1990) Aspects of reproduction and feeding of myctophids (Myctophidae) in the southwest Atlantic. *J Ichthyol* **30**(2): 229–37.

Pankhurst N W (1990) Growth and reprduction of the Antarctic nototheniid fish *Pagothenia borchgrevinki. Polar Biol* **10**: 387–91.

Pappenheim P (1912) Die Fische der Deutschen Südpolar-Expedition 1901–1903. I. Die Fische der Antarktis und Subantarktis. *Deutsche Südpol.-Exped. 1901–1903*, XIII, *Zool.* **5**: 161–82.

Permitin Yu E (1970) The consumption of krill by Antarctic fishes. In: Holdgate MW (ed) *Antarctic Ecology*, vol 1. Academic Press, London, pp. 177–82.

Permitin Yu E, Tarverdieva M I (1972) Feeding habits of some species of Antarctic fishes near South Georgia island. *Vopr Ikhtiol* **12, 1, 72**: 120–32. (In Russian.)

Permitin Yu E, Tarverdieva M I (1978) Feeding of fishes of the families Nototheniidae and Chaenichthyidae in the South Orkney Islands. *Sov J Mar Biol* **4**(2): 619–22.

Plötz J (1986) Summer diet of Weddell seals (*Leptonychotes weddelli*) in the eastern and southern Weddell Sea, Antarctica. *Polar Biol* **6**: 97–102.

Post A (1990) Paralepididae. In: Gon O, Heemstra P C (eds) *Fishes of the Southern Ocean.* J L B Smith Institute of Ichthyology, Grahamstown, pp 138–41.

Prince P A, Morgan (1987) Diet and feeding ecology of Procellariiformes. In: Croxall J P (ed) *Seabirds: Feeding Ecology and Role in Marine Ecosystems.* Cambridge University Press, pp 135–71.

Radtke R L (1990) Age determination of the Antarctic

fishes *Champsocephalus gunnari* and *Notothenia rossii marmorata* from South Georgia. *Polar Biol* **10**: 321–7.

Radtke R L, Hourigan T F (1990) Age and growth of the Antarctic fish *Nototheniops nudifrons*. *Fish Bull US* **88**: 557–71.

Radtke R L, Kellermann A (1991) Microstructural analysis of growth patterns in the early life history of Antarctic fishes. In: di Prisco G, Maresca B, Tota B (eds) *Biology of Antarctic Fish*. Springer-Verlag, Berlin, pp 101–15.

Radtke R L, Targett T E (1984) Rhythmic structural and chemical patterns in otoliths of the Antarctic fish *Notothenia larseni*: their application to age determination. *Polar Biol* **3**: 203–10.

Radtke R L, Targett T E, Kellermann A, Bell J L, Hill K T (1989) Antarctic fish growth: profile of *Trematomus newnesi*. *Mar Ecol Prog Ser* **57**: 103–17.

Rasoanarivo R, Aboussouan A (1983) Larves de *Electrona antarctica* (Günther, 1878) (Teleostei, Myctophidae) récoltées durant la campagne FIBEX-MD/25 dans le Sud-Ouest de l'ocean Indien. *Cybium* **7**(2): 75–86.

Regan C T (1916) British Antarctic ('Terra Nova') Expedition, 1910. Larval and post-larval fishes. *Terra Nova Rep* (Zool) **1**(4): 125–56.

Rembiszewski J M, Krzeptowski M, Linkowski T B (1978) Fishes (Pisces) as by-catch in fisheries of krill *Euphausia superba* Dana (Euphausiacea, Crustacea) *Polsk Arch Hydrobiol* **25**(3): 677–95.

Richardson M G (1975) The dietary composition of some Antarctic fish. *Bull Br Antarct Surv* **41/42**: 113–20.

Ridoux V, Offredo C (1989) The diets of five summer breeding seabirds in Adélie Land, Antarctica. *Polar Biol* **9**: 137–45.

Rodhouse P G, White M G, Jones M R R (1992) Trophic relations of the cephalopod *Martialia hyadesi* (Teuthoidea: Ommastrephidae) at the Antarctic Polar Front, Scotia Sea. *Mar Biol* **114**: 415–22.

Rowedder U (1979*a*) Feeding ecology of the myctophid *Electrona antarctica* (Gunther, 1878) (Teleostei). *Meeresforschung* **27**: 252–63.

Rowedder U (1979*b*) Some aspects of the biology of *Electrona antarctica* (Gunther, 1878) (Family Myctophidae). *Meeresforschung* **27**: 244–51.

Shcherbich L V (1975) Method of determining age and onset of sexual maturity in the marbled Antarctic cod *Notothenia rossii marmorata*. *J Ichthyol* **15**(1): 82–8.

Sinque C, Koblitz S, Costa L M (1986*a*) Ichthyoplankton of Bransfield Strait – Antarctica. *Neritica* **1**(3): 91–102.

Sinque C, Koblitz S, Costa L M (1986*b*) Distribution of larval and post-larval Antarctic fishes around Elephant Island and Bransfield Strait – Antarctica. *Neritica* **1**(3): 103–10.

Skora K E (1988*a*) Benthic fishes of the Elephant Island shelf (BIOMASS III, October–November 1986 and January 1987). *Pol Polar Res* **9**(2–3): 385–98.

Skora K E (1988*b*) Fishes in pelagic catches in the South Shetlands area (BIOMASS III, October–November 1986 and January 1987). *Pol Polar Res* **9**(2–3): 367–83.

Ślósarczyk W (1983) Preliminary estimation of abundance of juvenile *Nototheniidae* and *Channichthidae* within krill swarms east of South Georgia. *Acta Ichthyol Pisc* **13**(1): 3–11.

Ślósarczyk W (1986) Attempts at a quantitative estimate by trawl sampling of distribution of postlarval and juvenile notothenioids (Pisces, Perciformes) in relation to environmental conditions in the Antarctic Peninsula region during SIBEX 1983–84. *Mem Natl Inst Polar Res Spec Issue* **40**: 299–315.

Ślósarczyk W (1987) Contribution to the early life history of Channichthyidae from the Bransfield Strait and South Georgia (Antarctica). In: Kullander S O, Fernholm B (eds) *Proc V Congr Europ Ichthyol Stockholm 1985*. Swedish Museum of Natural History, Stockholm, pp 427–33.

Ślósarczyk W, Cielniaszek Z (1985) Postlarval and juvenile fish (Pisces, Perciformes and Myctophiformes) in Antarctic Peninsula region during BIOMASS-SIBEX 1983/84. *Pol Polar Res* **6**(1–2): 159–65.

Ślósarczyk W, Rembiszewski J M (1982) The occurrence of juvenile Notothenioidei (Pisces) within krill concentrations in the region of the Bransfield Strait and the southern Drake Passage. *Pol Polar Res* **3**(3–4): 299–312.

Solyanik G A (1966) Distribution and nutrition of bathypelagic fish *Notolepis coatsi* (Paralepididae family). *Sov Antarct Exped Inform Bull No 5* **5**: 365–6.

Tabeta O, Komaki Y (1986) Distribution and abundance of pelagic fishes in the epipelagic layers of the Indian Sector of the Southern Ocean in summer, 1983–84. *Mem Natl Inst Polar Res Spec Issue* **40**: 316–22.

Targett T E (1981) Trophic ecology and structure of coastal Antarctic fish communities. *Mar Ecol Prog Ser* **4**: 243–63.

Tarverdiyeva MI (1972) Daily food consumption and feeding pattern of the Georgian cod (*Notothenia rossi marmorata* Fischer) and the Patagonian toothfish (*Dissostichus eleginoides* Smitt) (Fam. Notothenidiida) in the South Georgia area. *J Ichthyol* **12**(4): 684–92.

Tresta J W, Siniff D B, Ross M J, Winter J D (1985) Weddell seal – Antarctic cod interactions in McMurdo Sound, Antarctica. In: Siegfried W R, Condy P R, Laws R M (eds) *Antarctic Nutrient Cycles and Food Webs*. Springer-Verlag, Berlin, pp 561–5.

Torres J J, Lancraft T M, Weigle B L, Hopkins T L (1984) Distribution and abundance of fishes and salps in relation to the marginal ice zone of the Scotia Sea, November and December 1983. *Ant J US* **19**(5): 117–19.

Torres J J, Somero G N (1988) Metabolism, enzymic activities and cold adaptation in Antarctic mesopelagic fishes. *Mar Biol* **98**: 169–80.

Townsend D W (1980) Microstructural growth

increments in some Antarctic fish otoliths. *Cybium* **8**: 17–22.

Tranter D J, Smith J E (1979) The Antarctic 'BIOMASS' program. *Search* **10**(9): 304–7.

Voskoboinikova O S (1986) Evolution of the visceral skeleton and problems of phylogeny of the nototheniid fishes. *Proc Zool Inst Leningrad* **153**: 46–66. (In Russian.)

White M G, North A W (1987) Postlarval notothenioidei and midwater fish collected during the SIBEX cruise by British Antarctic Survey, 1985. In: Kullander S O, Fernholm B (eds) *Proc V Congr Europ Ichthyol Stockholm 1985.* Swedish Museum of Natural History, Stockholm, pp 405–11.

White M G, North A W, Twelves E L, Jones S (1982) Early development of *Notothenia neglecta* from the Scotia Sea, Antarctica. *Cybium* **6**(1): 43–51.

Williams R (1985*a*) Trophic relationships between pelagic fish and euphausiids in Antarctic waters. In: Siegfried W R, Condy P R, Laws R M (eds) *Antarctic Nutrient Cycles and Food Webs.* Springer-Verlag, Berlin, pp 452–9.

Williams R (1985*b*) The potential impact of a krill fishery upon pelagic fish in the Prydz Bay area of Antarctica. *Polar Biol* **5**: 1–4.

Williams R (1989) Australian research on Antarctic bird and seal diets. *Scientific Committee of the Convention for the Conservation of Antarctic Marine Living Resources Selected Scientific Papers 1988* Part II (SC-CAMLR-VII/BG/32), pp 231–49.

Williams R, Kirkwood J M, O'Sullivan D (1983) FIBEX cruise zooplankton data. *ANARE Res Notes* **7**: 1–163.

Williams R, McEldowney A (1990) A guide to the fish otoliths from waters off the Australian Antarctic Territory, Heard and Macquarie Islands. *ANARE Res Notes* **75**: 1–173.

Wohlschlag D E (1961) Growth of an Antarctic fish at freezing temperatures. *Copeia* **1961**(1): 11–18.

Zasel'sliy V S, Kudrin B D, Poletayev V A, Chechenin S Ch (1985) Some features of the biology of *Electrona carlsbergi* (Taning) (Myctophidae) in the Atlantic sector of the Antarctic. *J Ichthyol* **25**(2): 163–6.

Studies on fish of the Indian Ocean sector of the Southern Ocean during the BIOMASS Programme

R. WILLIAMS & G. DUHAMEL

Introduction

THE PRINCIPAL objective of BIOMASS was to 'gain a deeper understanding of the structure and dynamic functioning of the Antarctic marine ecosystem as a basis for the management of potential living resources' (Anon, 1977). The specific objectives emphasized krill (*Euphausia superba*) as the major potential food resource of Antarctic waters and as a key element of the marine food web. However, other Antarctic marine studies were recognized as important, including autecological studies of fishes, assessment of their biomass and production in selected areas of the Southern Ocean, and description of the major food chains.

This type of approach was particularly important in the context of this chapter because, before BIOMASS, virtually nothing was known about fish in the Indian Ocean sector, except for some early taxonomy and distribution data and information from the early phases of the commercial fishery around Iles Kerguelen. As few data have been published since BIOMASS fieldwork started, this paper deals firstly with the basic information such as distribution and abundance before proceeding to more general topics.

In line with the BIOMASS objectives, nearly all the fieldwork in the Indian Ocean sector was primarily for assessment of krill, and sampling was confined to pelagic zones, generally with plankton nets. Exceptions were some of the South African SIBEX data, which were obtained with a commercial-sized midwater trawl, and the French bottom trawls in Prydz Bay on SIBEX II. Data on fish were thus acquired as a by-product in most cases from open ocean epipelagic and mesopelagic zones. This resulted in the vast majority of data being on ichthyoplankton.

BIOMASS activities in the Indian Ocean sector were conducted by Australia, France, Japan and South Africa during FIBEX and SIBEX (Table 1). Australia also conducted some surveys between the major experiments.

Sources and treatment of data

Data for analysis of species composition, distribution and abundance have been obtained from the results of the Post-SIBEX Fish Data Evaluation Workshop (Anon, 1988), unpublished Australian records, and accounts of the South African (Gon, 1988), Japanese (Tabeta & Komaki, 1986; Iwami & Kubodera, 1990) and French (Camus, 1990; Duhamel, 1987; Duhamel and Williams, 1990; Hulley *et al.*, 1989; Hureau, 1986; Rasoanarivo & Aboussouan, 1983) BIOMASS fieldwork.

SIBEX ichthyoplankton data have been analysed by the Post-SIBEX Fish Data Evaluation Workshop (Anon, 1988). Very few data on fish were available

Table 1. *Summary of BIOMASS work in the Indian Ocean sector*

Expt.	Country	Ship	Dates	Positions	Gear	Haul types	Depth range	No. hauls	Ref.
FIBEX (1981)	France	*Marion Dufresne*	25 Jan	61–64°S	RMT	oblique	0–235 m	23	Hureau, 1986
			5 Mar	30–50°E	ORI	oblique	0–225 m	28	
	Australia	*Nella Dan*	17 Jan	62–69°S	RMT	oblique/horizontal	0–800 m	61	Williams *et al.*, 1983
			15 Mar	62–90°E	bongo			41	
ADBEX I (1982)	Australia	*Nella Dan*	19 Nov	59–69°S	RMT	oblique/horizontal	0–500 m	64	Williams *et al.*, 1986
			27 Dec	63–90°E	bongo			16	
SIBEX I (1984)	S. Africa	*Agulhas*	15 Mar	62–66°S	bongo	double oblique	0–250 m	48	Gon, 1988
			3 May	52–64°E					
	Japan	*Kaiyo Maru*	Dec 83	46–61°S	KYMT	oblique	0–135 m	84	Tabeta & Komaki, 1986
			Feb 84	65–75°E	ORI	oblique	0–110 m	66	
	Japan	*Hakuho Maru*		Aust–Ant, 110–115°E	10′ IKMT	oblique	0–1050 m	9	Miya *et al.*, 1986
SIBEX II (1985)	S. Africa	*Africana*	20 Feb	60–67°S	bongo	double oblique	0–300 m	13	Gon, 1988
			23 Mar	48–63°E	krill trawl		0–500 m	6	
	Australia	*Nella Dan*	4 Jan	60–69°S	RMT1+8	Oblique	0–1029 m	75	Ikeda *et al.*, 1986
			1 Feb	58–93°E					
	France	*Marion Dufresne*	12 Jan	Heard Is	RMT8	Oblique (fish)	0–1002 m	31	Hulley *et al.*, 1989
			6 Feb	Prydz Bay	RMT8	Oblique/horiz (krill)	0–281 m	30	
					beam trawl	benthic	150–2000 m	22	Duhamel & Williams, 1990
					bongo	oblique	0–200 m	45	Camus, 1990

from FIBEX, except from Australia. Many kinds of nets were used by the various nations in different fieldwork phases. These included ring nets, bongos, RMT1 and 8, and krill trawls. The Workshop studied the selectivity between these different nets in catching ichthyoplankton, and concluded that for the Prydz Bay area, the RMT8 nets did not catch the smallest individuals of *Electrona antarctica* and *Notolepis coatsi* that appeared in the bongo nets. For *Pleuragramma antarcticum* the difference between the two net types was much smaller. Abundance data are therefore treated for each net type separately, while distribution and similar data are not separated because, in most cases, bongo or RMT1 nets were used at the same locations as RMT8 nets, and no bias in these results is anticipated.

Results and discussion

Ichthyoplankton

Distribution

Two obvious spatial distribution patterns exist.

(1) Most species or family groupings are confined to the oceanic waters, with a clearly defined southern boundary at the Antarctic continental shelf edge. Fish with this type of distribution include *Electrona antarctica* and all other myctophid species (Fig. 1), *Notolepis coatsi, Bathylagus antarcticus*, and *Cyclothone* sp. (Fig. 2).

(2) Other species groups are mostly confined to the shelf zones. In the case of the Antarctic continental shelf, most groups also occur well to seaward of the shelf edge. These include *Pleuragramma antarcticum*, and pelagic juveniles of species from the families Artedidraconidae, Bathydraconidae, Nototheniidae and Channichthyidae, which are usually benthic when adult (Figs 3 and 4). This group is rather under-represented in the data because of the low number of hauls on the shelf, and are in general abundant within their range.

The bathydraconid and nototheniid juveniles are also associated with the Kerguelen Plateau, although in general represented by different species than in coastal Antarctica. Camus (1990) showed that species such as *Gobionotothen acuta, Notothenia rossii* and *Lepidonotothen mizops* are confined to the northern part of the Kerguelen Plateau, whereas *Lepidonotothen kempi, Pleuragramma antarcticum* and *Trematomus* spp. are found in the Antarctic continental shelf area.

Size frequency and distribution by size

Sufficient data exist for the length frequency of four species to be analysed (Fig. 5).

> *Electrona antarctica.* The size range from all nets combined covers the entire known size range (Hulley, 1981). Although they probably each represent more than one age class, there are three distinct size groups.
> *Notolepis coatsi.* Only juvenile sizes were caught. Within this size range, probably two size classes exist.
> *Bathylagus* sp. The specimens captured range from premetamorphosed larvae to adults. At least two preadult size groups exist, as well as a wide polymodal adult group.
> *Pleuragramma antarcticum.* Most specimens of this species fall into two small size groups, which correspond with ages 0+ and 1+ and possibly some 2+ of Hubold & Ekau (1987).

The oceanic species *Electrona antarctica, Notolepis coatsi* and *Bathylagus antarcticus* appear to have no horizontal distribution patterns relative to size (Figs 1 and 2). Of the shelf species, only the smallest two groups of *Pleuragramma* were caught off the shelf (Fig. 3). The larger specimens (2+ age group and older) were caught only on the shelf, but even there they were rare. This may be a result of the larger fish avoiding the nets or being below the depth range of sampling, as Hubold & Ekau (1987) found that older fish only occurred at depths greater than 400 m in the Weddell Sea. However, as far more samples were taken off the shelf, one would expect some of the larger group to be caught if they were there, just as on the shelf. This therefore indicates that *P. antarcticum* juveniles are carried off the shelf but either do not survive or return to the shelf waters after their second year.

Very few specimens of Channichthyidae longer than 50 mm standard length (SL) were caught. With such a small size range, it would be surprising to see any distribution patterns. Channichthyidae are not dispersed off the shelf as widely as is *P. antarcticum* (Fig. 4). Hubold (1990) described similar off shelf extensions to the range of *P. antarcticum* and Channichthyidae juveniles in the Weddell Sea, which he suggested was an adaptation to extend the geo-

A

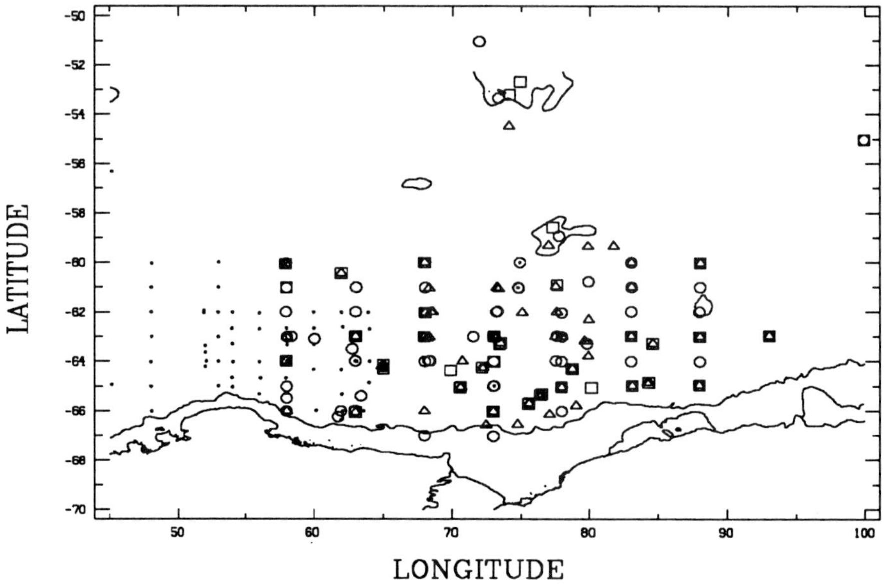

Fig. 1. Distribution of species of Myctophidae based on BIOMASS data. (A) *Electrona antarctica*, not measured (·), <25 mm SL (○), 25–45 mm SL (□), >45 mm SL (△). (B) *Krefftichthys anderssoni* (○), *Protomyctophum bolini* (□), *Electrona carlsbergi* (X). (C) *Gymnoscopelus nicholsi* (○), *G. braueri* (□), *G. opisthopterus* (X).

B

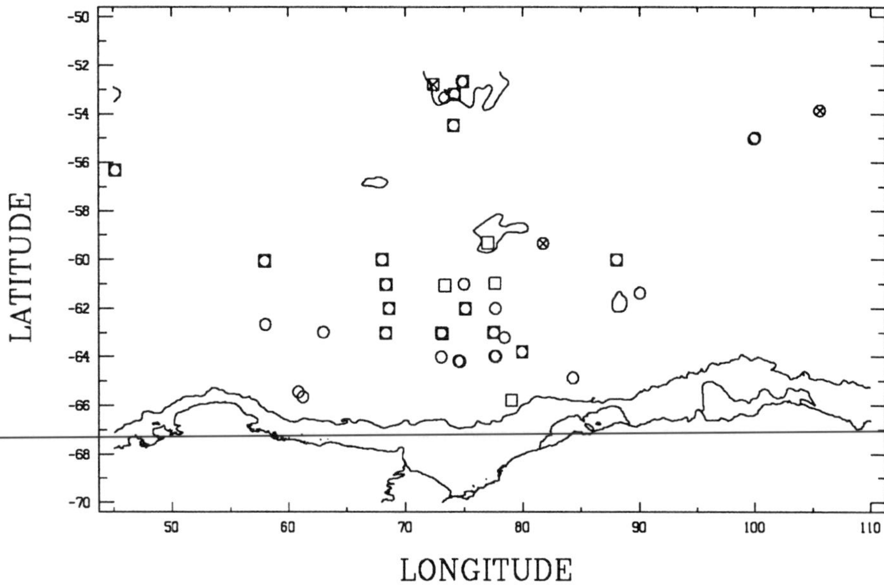

C

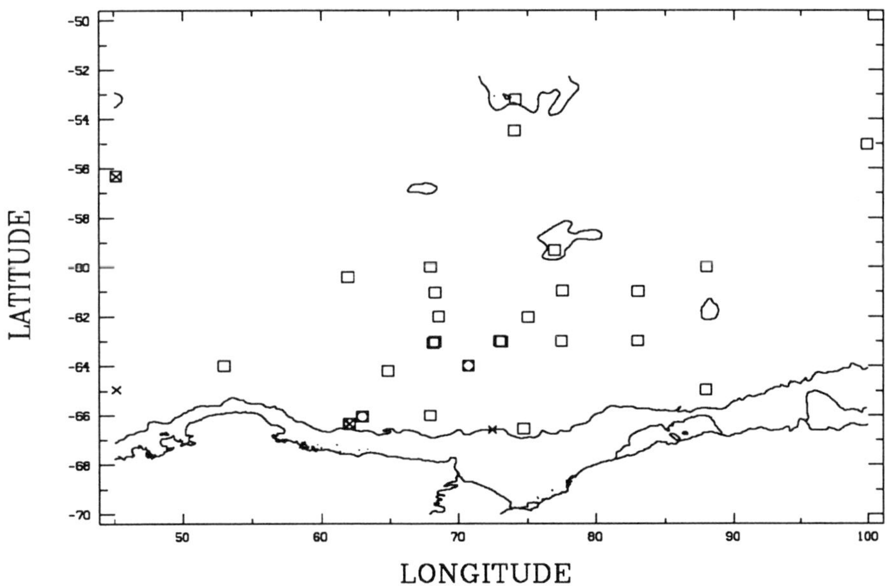

A

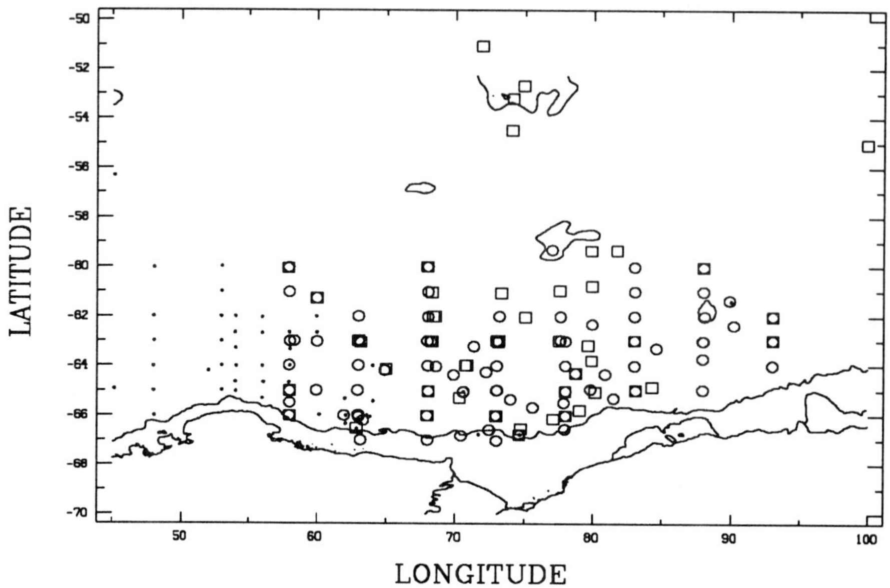

Fig. 2. Distribution of oceanic species based on BIOMASS data. (A) *Notolepis coatsi*, not measured (·), <50 mm SL (○), >50 mm SL (□). (B) *Bathylagus antarcticus* <25 mm SL (○), 25–54 mm SL (□), >55 mm SL (△). (C) *Cyclothone* sp.

B

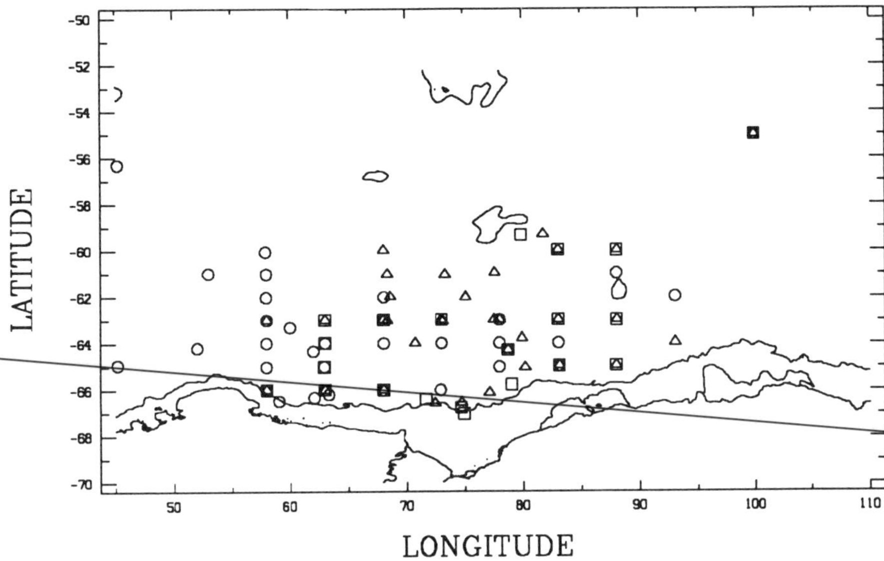

C

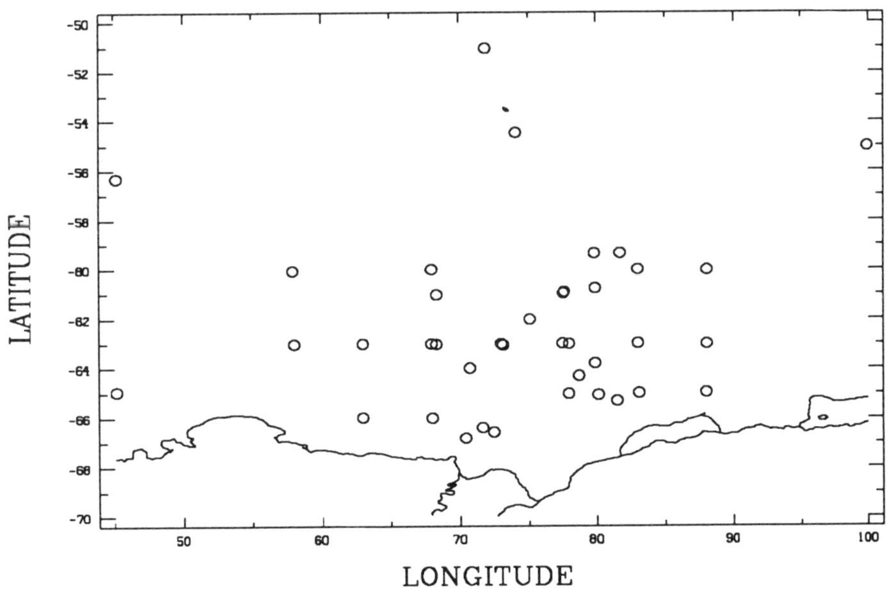

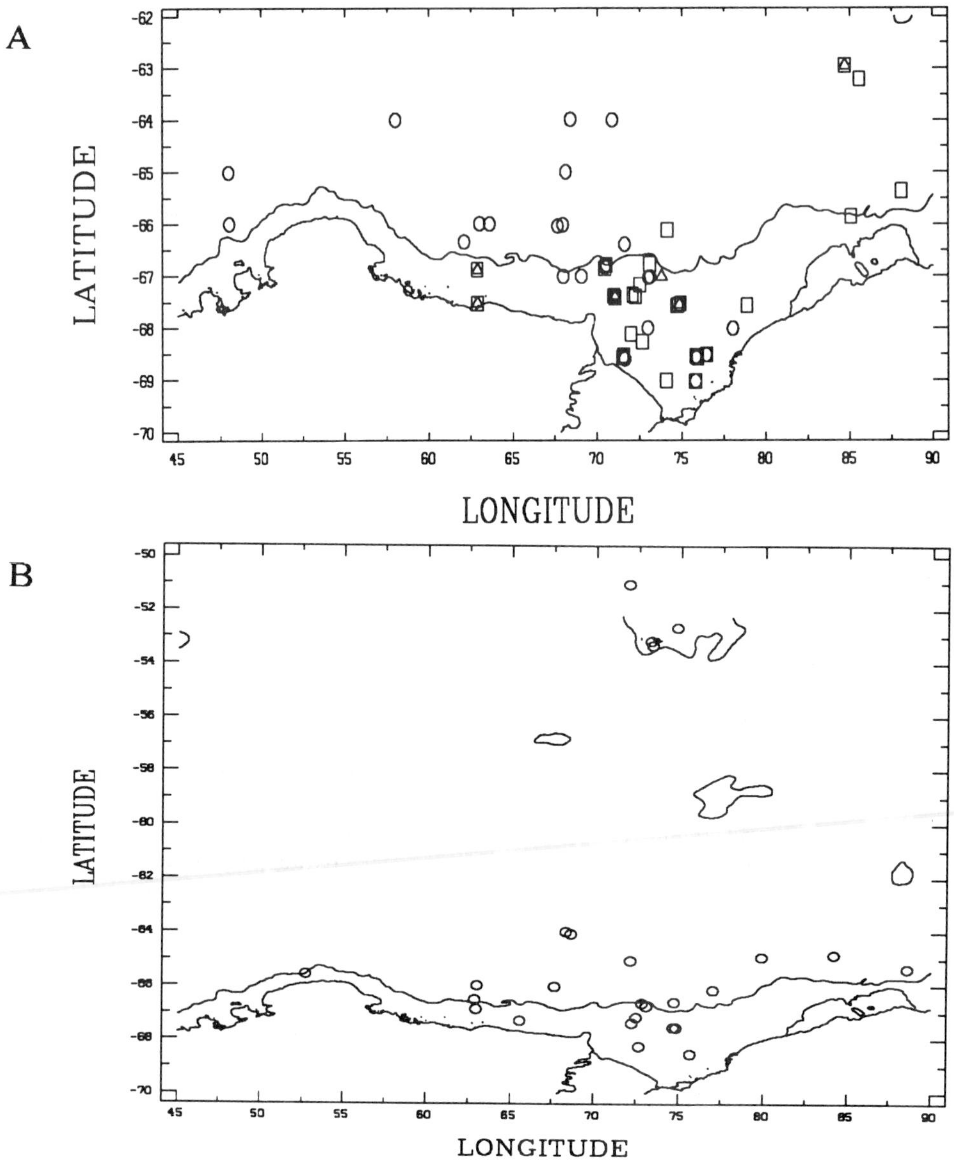

Fig. 3. Distribution of species of Nototheniidae based on BIOMASS data. (A) *Pleuragramma antarcticum*, <25 mm SL (○), 25–65 mm SL (□), >65 mm SL (△). (B) Notothenid juveniles (except *P. antarcticum*).

graphical basis of the larval drift beyond the limited continental shelf area. It is interesting that this also happens off the much wider shelf of Prydz Bay, where such habitat extension should be less important in the species' life history.

As nearly all hauls during the BIOMASS Programme were oblique hauls with non-closing nets, it is difficult to obtain meaningful data on the vertical distribution of species. However, plots of fish length versus maximum depth of hauls do give some information (Fig. 6).

For *Notolepis coatsi*, the small sizes which constitute the bulk of the specimens are found in both deep and shallow hauls, but the larger sizes (>80 mm SL) tend to be found in the shallower hauls. Because the larger sizes are quite rare, this could merely mean that all sizes are usually found near the surface, but very few larger specimens occur in the hauls with large depth maxima.

The depth distribution of *Electrona antarctica* is interesting in that no specimen comes from a haul with a maximum depth of less than 180 m, implying that this species does not occur in depths shallower than this. Hauls of less than 600 m maximum depth had a very high proportion of fish less than 20 mm SL, whereas deeper hauls had a very even range of size, from 5 to greater than 90 mm SL. Myctophid larvae show no real pattern.

Bathylagus antarcticus shows a similar pattern to that of *E. antarctica*, with only small individuals in the shallow trawls, but the full range of sizes in the hauls with deeper maximum depths.

Pleuragramma antarcticum occurs virtually only in hauls with a maximum depth less than 400 m, which reflects its distribution over the continental shelf. Most of the larger size range occurs in the hauls with maximum depth less than 100 m.

The Artedidraconidae, Nototheniidae and Bathydraconidae all have a similar pattern, being mostly in shallower hauls. This reflects their shelf distribution, with the few specimens in deeper hauls indicating the occurrence of off-shelf strays, which probably were still mostly present in the upper waters.

The Channichthyidae are slightly different in being more confined to shallower trawls, indicating their less extensive dispersal off the shelf.

Influence of water circulation on fish distribution

Oceanographic data taken on SIBEX II indicate a possible mechanism to explain the distribution of some groups of ichthyoplankton. A plot of geostrophic flow (Fig. 7) indicates the presence of a gyre in Prydz Bay, and shows a north-westerly flow off the shelf in the western part of the bay, which could explain the presence of shelf-based species, e.g. *Pleuragramma antarcticum* and channichthyid species, in adjacent oceanic waters. These individuals could be lost to the system, or return to the shelf in the central and eastern part of the bay if they stay in the gyre system.

North of 63°S, the strong easterly flow of the West Wind Drift encounters the Kerguelen Plateau. This may induce some upwelling, consequently increasing productivity in the area, which could explain the relatively high abundances of *E. antarctica* there.

Abundance

Few data on absolute abundance of any species are available. The BIOMASS Post-SIBEX Fish Data Evaluation Workshop (Anon, 1988) could find no overall pattern of abundance of the commonest species, and no difference between months. However, Camus (1990) has shown that *Electrona antarctica* larvae are most abundant in the southern part of their range, particularly over the southern end of the Kerguelen Plateau (Fig. 8).

Benthic fish

Duhamel *et al.* (1983) confirmed that the major zoogeographic groupings of fish in the Indian Ocean sector, were:

> the continental fauna (including Bouvet Island);
> the Kerguelen–Heard fauna, dominated by the sub-Antarctic nototheniids (*Notothenia rossii, Lepidonotothen squamifrons*) and the presence of channichthyids (*Champsocephalus gunnari, Channichthys rhinoceratus*); and
> the islands north of the Antarctic Convergence (Marion, Prince Edward, Crozet, etc.), which lack channichthyids.

Williams (1988) surveyed the inshore fish fauna at Davis Station in Prydz Bay, and established that it was similar to those at other coastal sites in East Antarctica such as Dumont d'Urville and Syowa.

A

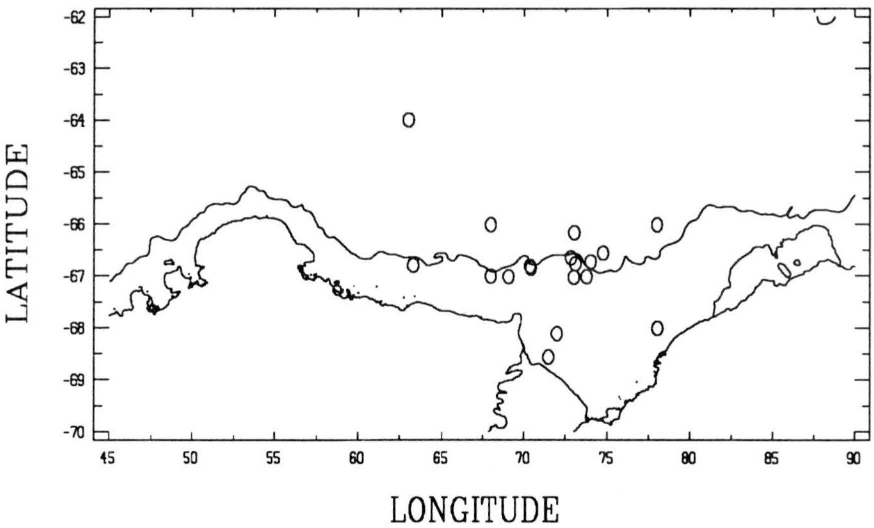

Fig. 4. Distribution of species of (A) Artedidraconidae, (B) Bathydraconidae and (C) Channichthyidae <55 mm SL (○), >55 mm SL (□) based on BIOMASS data.

B

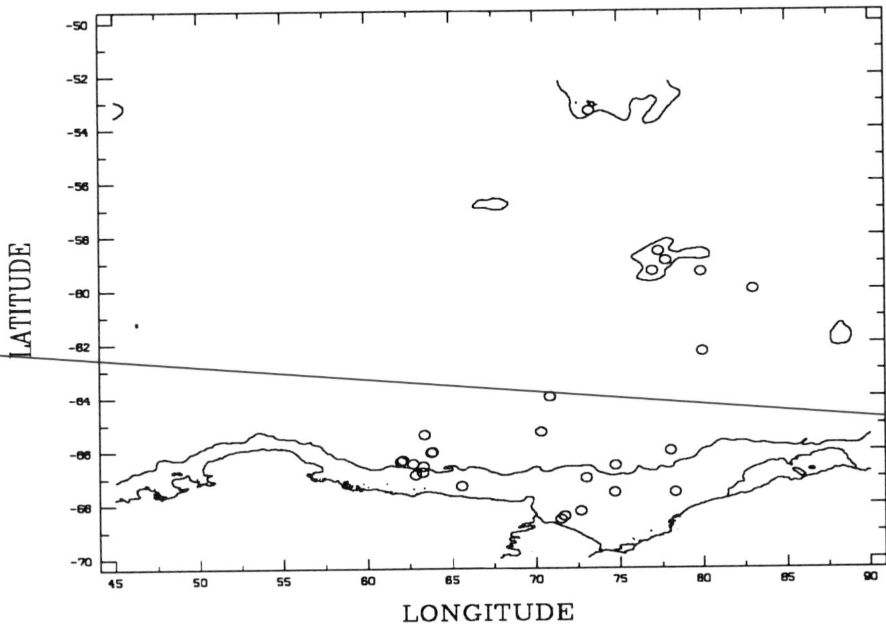
LATITUDE
-50
-52
-54
-56
-58
-60
-62
-64
-66
-68
-70
45
50
55
60
65
70
75
80
85
90
LONGITUDE

C

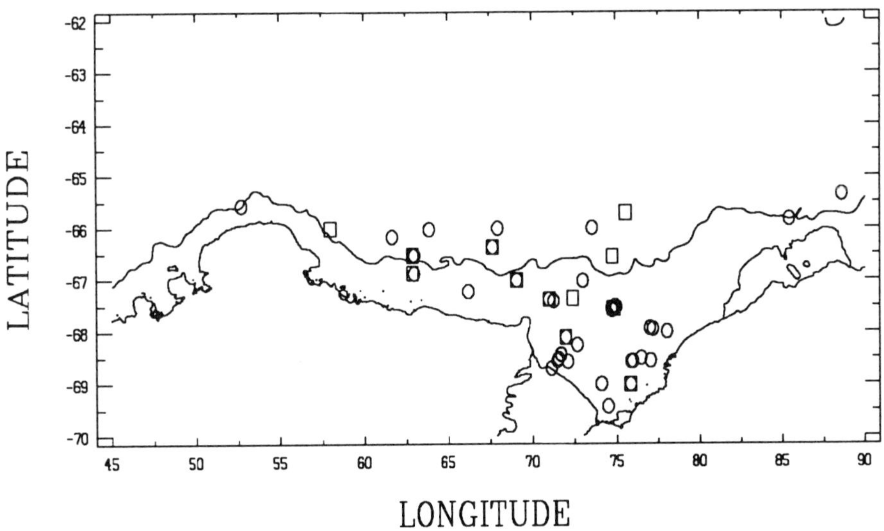
LATITUDE
-62
-63
-64
-65
-66
-67
-68
-69
-70
45
50
55
60
65
70
75
80
85
90
LONGITUDE

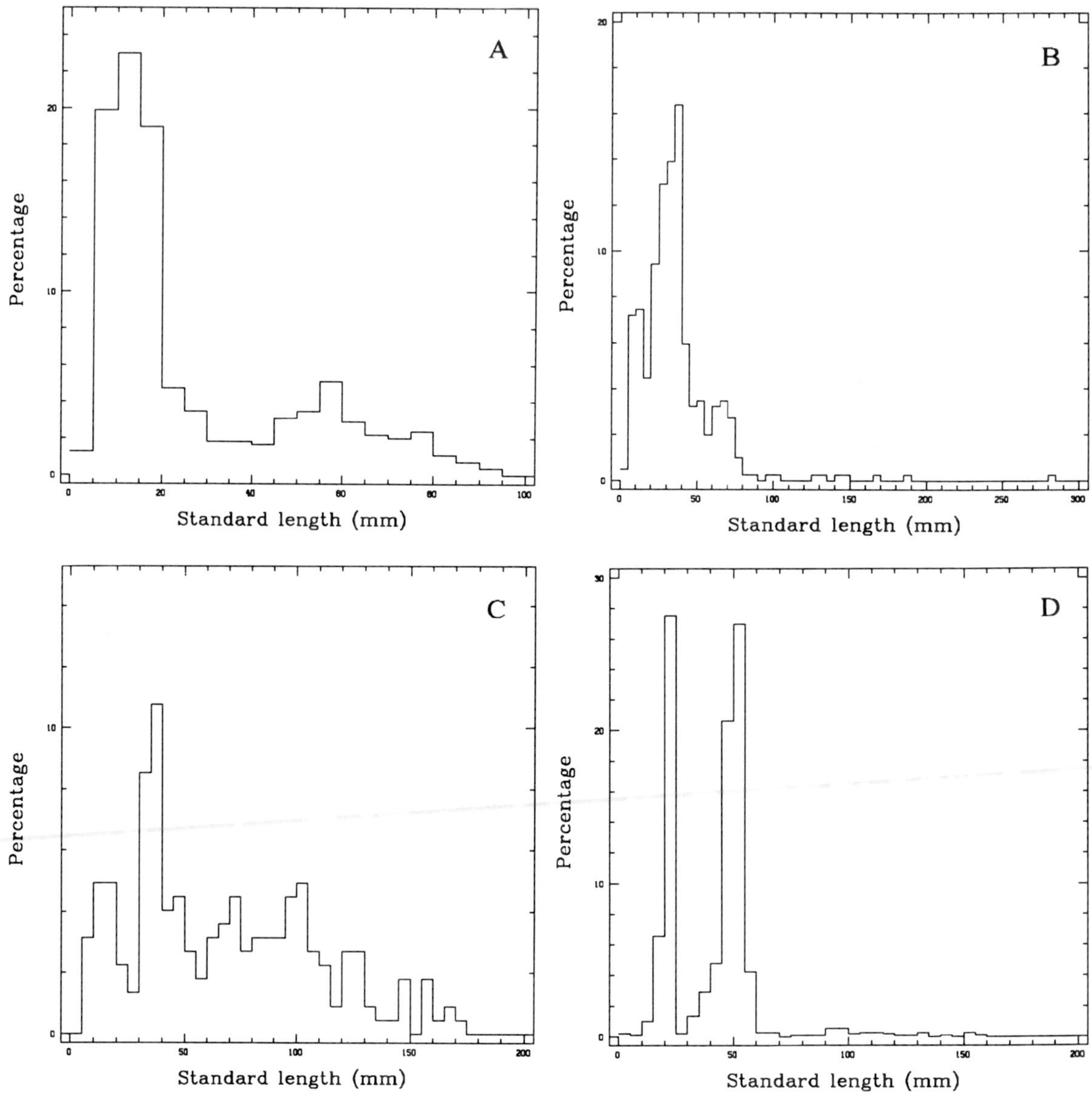

Fig. 5. Standard length frequencies of (A) *Electrona antarctica*, (B) *Notolepis coatsi*, (C) *Bathylagus antarcticus* and (D) *Pleuragramma antarcticum* based on BIOMASS data.

Caption for Fig. 6 (*opposite*). Standard length versus maximum depth of haul of (A) *Notolepis coatsi*, (B) *Electrona antarctica*, (C) *Bathylagus antarcticus*, (D) *Pleuragramma antarcticum*, (E) Artedidraconidae juveniles and (F) Channichthyidae juveniles based on BIOMASS data.

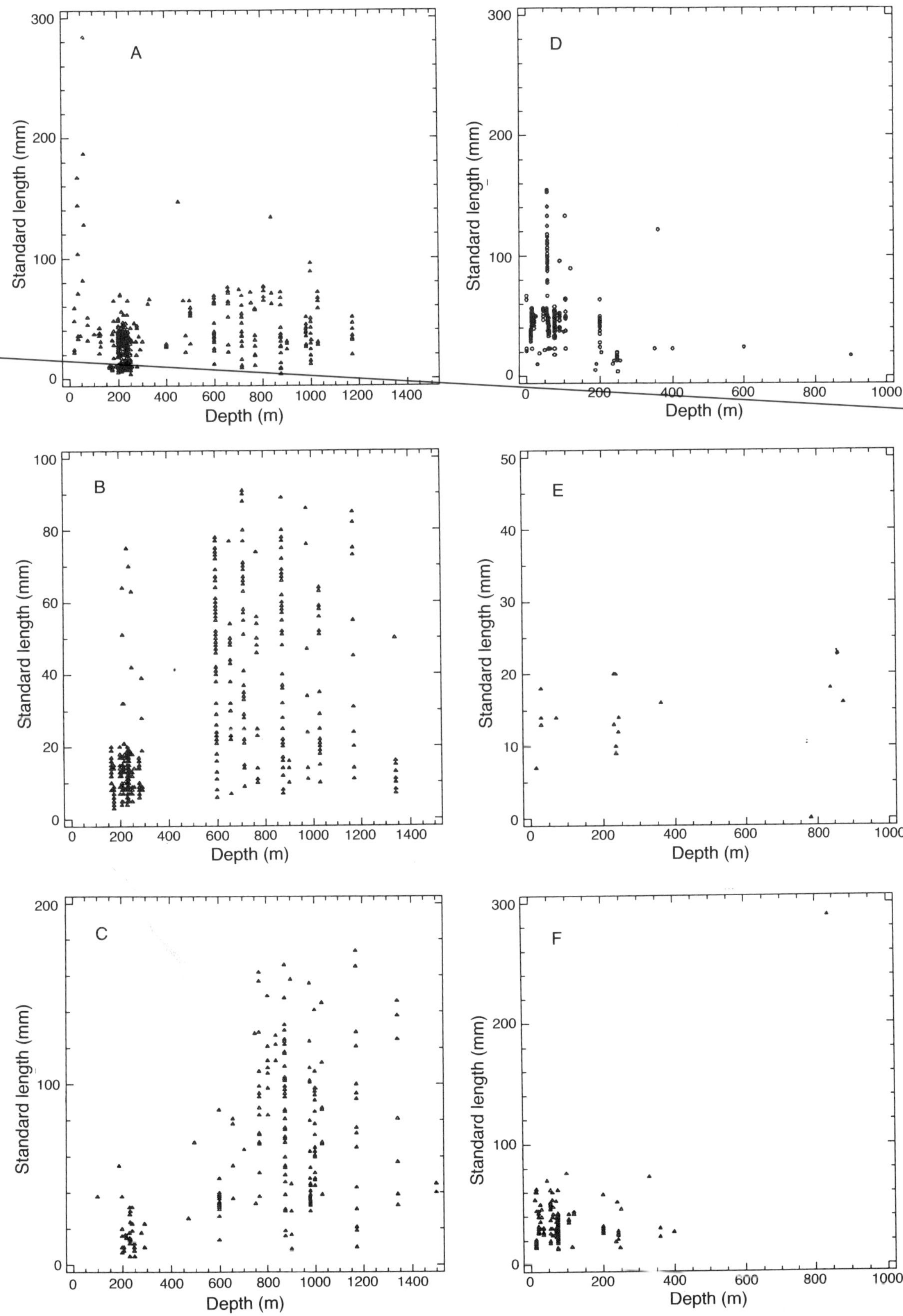
A
B
C
D
E
F
Standard length (mm)
Depth (m)
0
100
200
300
0
20
40
60
80
100
0
10
20
30
40
50
0
200
400
600
800
1000
1200
1400

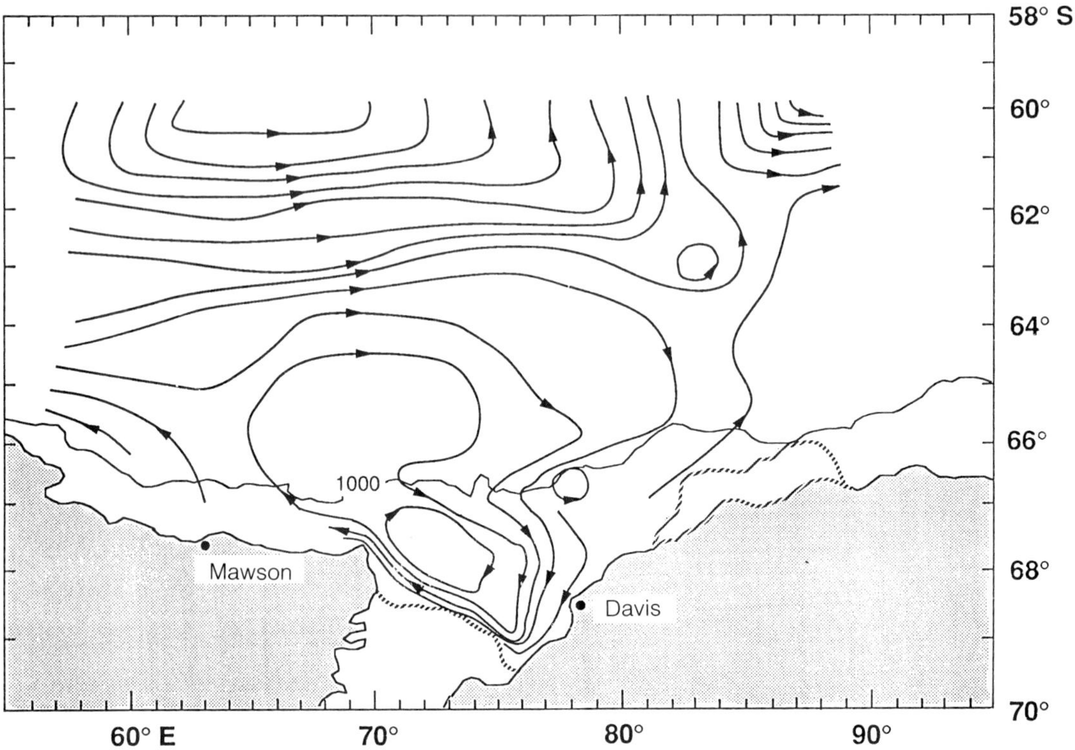

Fig. 7. Geostrophic flow redrawn from the geopotential anomaly contours of RA Nunes-Vaz (personal communication, 1990).

Thus *Trematomus bernacchii* and *Pagothenia borchgrevinki* dominated the shallow waters less than 20 m deep, and *Trematomus hansoni* and *Chionodraco hamatus* were dominant in waters greater than 20 m deep. This demonstrated the important point that species widespread over the adjacent continental shelf (e.g. *C. hamatus*) can also be important members of the coastal fauna, implying that factors affecting the coastal fish fauna may also affect the community in the wider shelf area.

During the French SIBEX II cruise, a series of benthic beam trawls were made on the Kerguelen–Heard Banks, around Heard Island, on the southern part of the Kerguelen Plateau (BANZARE Bank), and on the continental shelf of Prydz Bay (Duhamel & Williams, 1990). The main conclusions were:

(1) The ichthyofauna of Heard Island is similar to that from Kerguelen, and exchanges via the intermediate shallow banks probably facilitate this.

(2) The ichthyofauna of Prydz Bay is typically Antarctic. The circum-Antarctic nature of several species (e.g. *Lepidonotothen kempi, Chionodraco myersi*) was confirmed, along with the first record of a ray (family Rajidae) in East Antarctica. The species diversity is high compared with the Kerguelen–Heard region.

(3) The great depth of the BANZARE Bank (minimum depth greater than 600 m) compared with other sea mounts in this sector precluded the discovery of a fauna intermediate between the sub-Antarctic Kerguelen–Heard fauna and the Antarctic shelf fauna. Most species belonged to the typically deep water groups Macrouridae, Liparididae and Zoarcidae.

(4) Abundances of all species within the respect-

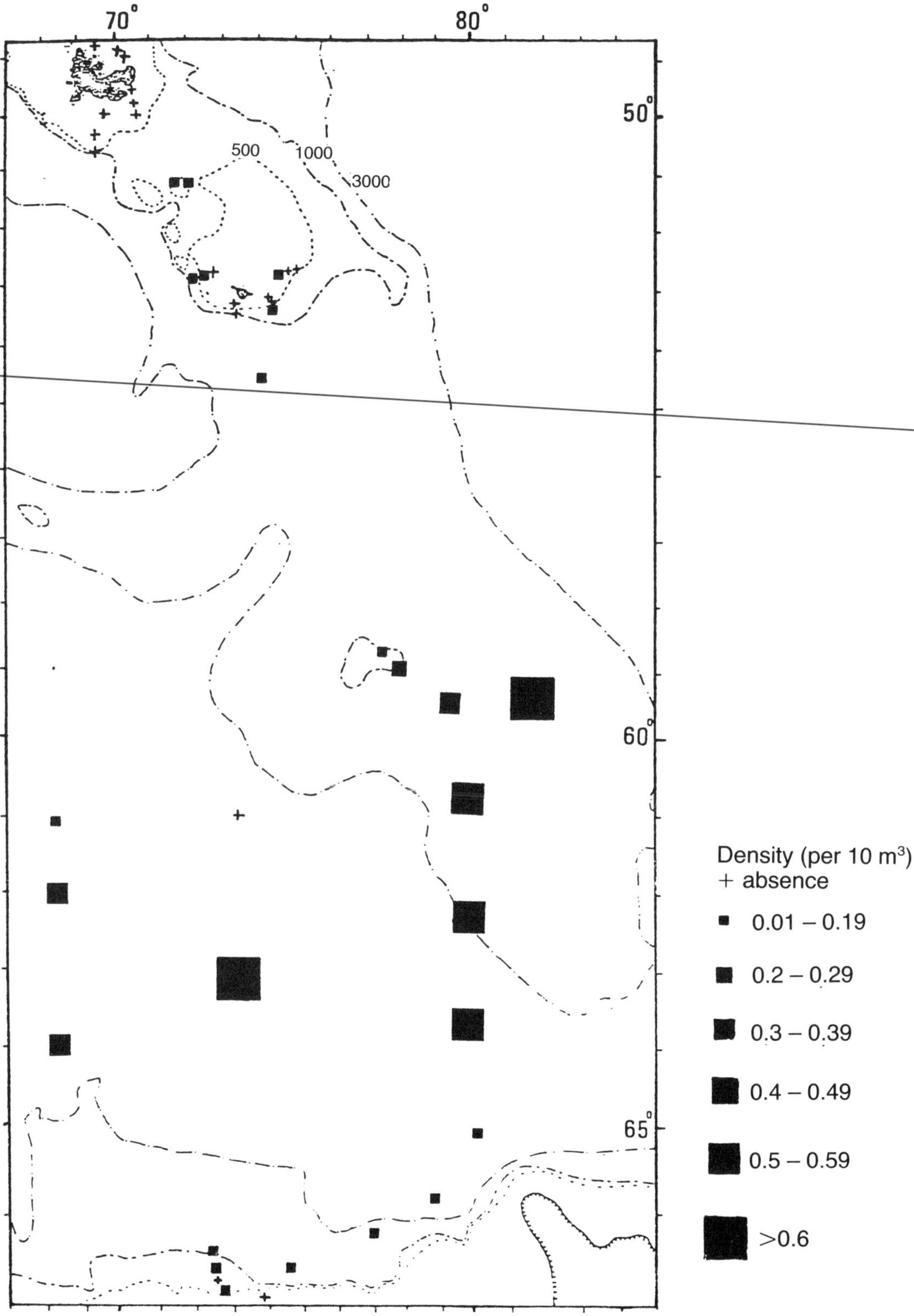

Fig. 8. Abundance of *Electrona antarctica* larvae during the French SIBEX II (from Camus, 1990).

Table 2. *Summary of mean percentage composition (by mass) of euphausiids and fish in the diets of vertebrate predators in Prydz Bay*

Species	*E. superba*	*E. crystallorophias*	Fish
Adélie penguin[a]	35.1	33.8	23.7
Emperor penguin	<1	—	96.7
Cape petrel	80.7 (nearly all *E. Superba*)		18.6
Antarctic petrel	22.3	—	77.5
Antarctic fulmar	36.2	—	63.2
Weddell seal	—	—	77.1

[a] Posthatch period only.
Source: From Williams (1989).

ive zones were: Kerguelen–Heard Banks, 44–455 kg km^{-2}; Heard Island, 114–1129 kg km^{-2}; banks and slopes >600 m deep, 377–1889 kg km^{-2}; Prydz Bay, 349–717 kg km^{-2}. Some of these figures are from a very few trawls and should be treated with caution. Points of interest include the relatively high biomass in the deep waters.

Trophic relations

Many species of ichthyoplankton caught on the Prydz Bay shelf during FIBEX had euphausiids as an important part of their diet (Williams, 1985*a*). These included holopelagic species such as *Notolepis coatsi* (diet mostly adult krill, 96% by mass) and *Pleuragramma antarcticum* (diet mostly *Euphausia superba* and *E. crystallorophias* larvae, 25% by mass) and pelagic juvenile stages of fish which are benthic when adult (such as the channichthyids *Cryodraco antarcticus, Pagetopsis macropterus* and *Chaenodraco wilsoni*) in which more than 90% of the diet was euphausiids.

In addition, many of the adult bottom-living fish caught in the French beam trawls in Prydz Bay at 450–550 m depth had a high proportion of *E. superba* in their diet. These species included *Trematomus eulepidotus* (60% by mass), *T. scotti* (70%), *T. lepidorhinus* (60%), *Racovitzia glacialis* (59% and *Artedidraco orianae* (82%) (Duhamel & Williams, 1990). The observed vertical distribution of *E. superba* in Prydz Bay, as elsewhere, is within the top 200 m, and mostly within the top 100 m. Some of these fish, notably *T. eulepidotus*, are known to be benthopelagic (Ekau, 1990) and could migrate to the upper waters to feed, but many of the others are probably strictly benthic, and the presence of krill in their diet may indicate a wider vertical distribution of krill than hitherto thought. The poster presented by Gutt in this Colloquium (Gutt, 1992), demonstrating the presence of benthic krill in the Weddell Sea in a habitat almost identical to that where the Prydz Bay fish were taken, supports this.

Long-term work by Australian scientists at Davis Station and elsewhere in Prydz Bay has demonstrated the importance of fish, in particular *P. antarcticum*, in the diets of many birds and seals (Table 2). Given the large number of predators in the region (Fig. 9), a significant proportion of the *P. antarcticum* biomass is taken each year (R Williams, in preparation). This situation appears to be consistent in all areas studied in East Antarctica (for example, see Franeker, 1992), and contrasts sharply with the diets of predators in the Atlantic Ocean sector, where *Euphausia superba* is the major prey item for nearly all top predators.

Allied with the extensive feeding on krill by juvenile and adult fish is the presence in krill catches of juvenile fish. Williams (1985*b*), on the basis of RMT hauls, established that up to 18% by mass of krill hauls on the shelf is composed of fish (mainly juvenile *P. antarcticum*). The proportion of fish in krill catches beyond the shelf is 1.2%, mostly myctophids. Ślósarczyk (1983) found similar results off the Balleny Islands using a commercial-sized krill trawl.

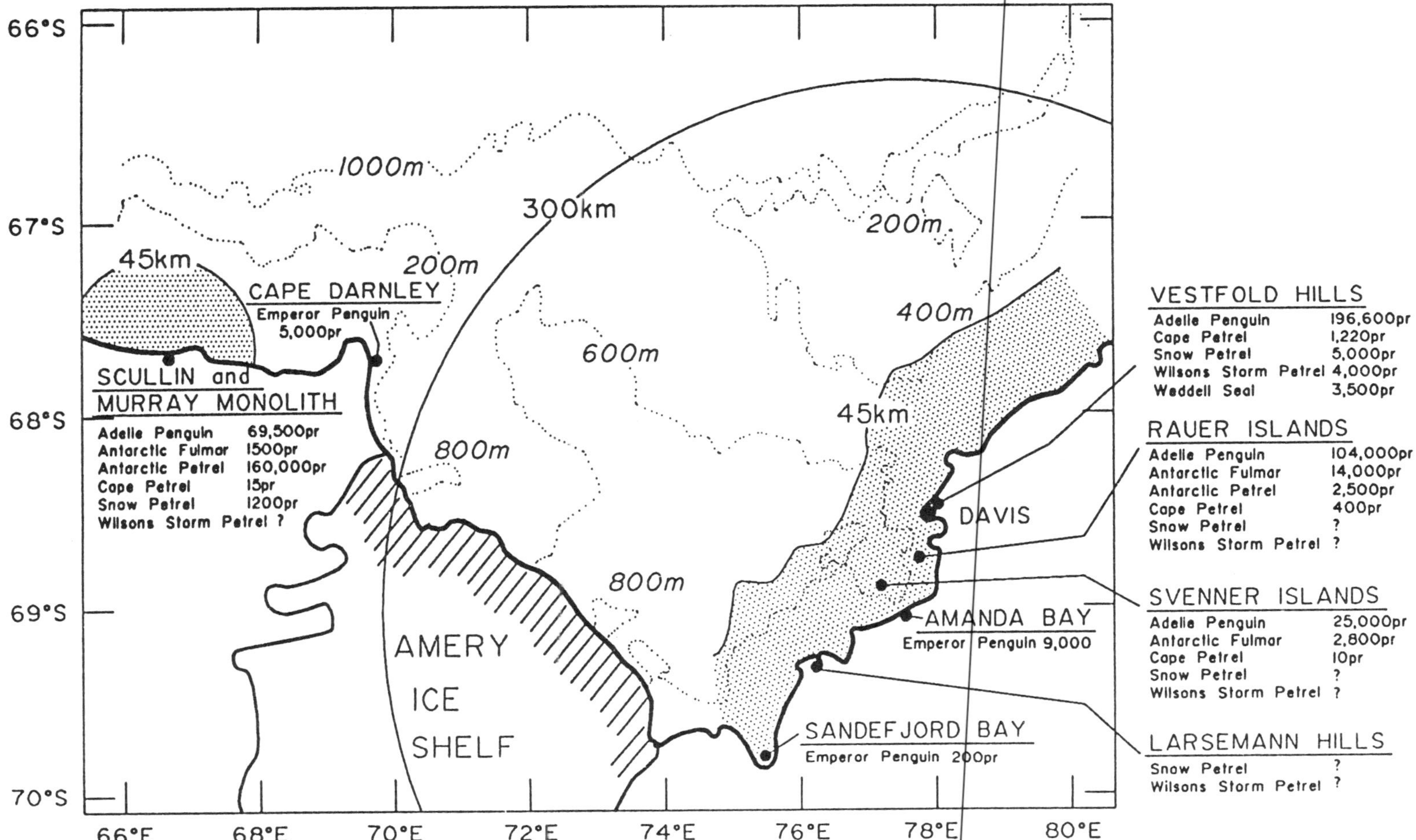

Fig. 9. Summary of the populations in summer of predators known to have fish as a significant part of their diet. The shaded area with 45 km from the coast represents the foraging range of Adélie penguins when feeding chicks. The 300 km radius represents the approximate foraging range of emperor penguins and petrel species. Bird data from Woehler & Johnstone (1991); seal data from K Green (personal communication, 1991).

He found a mean number of 1818 *T. bernacchii* and 44 *Pagothenia brachysoma* juveniles per hour of fishing.

The extensive feeding on, and association with, krill by fish have important consequences for CCAMLR in its requirement to manage any krill fishery in the area on an ecosystem basis.

Fish in the Indian Ocean sector therefore appear to occupy an important intermediate role in the food chain. This is in contrast to the Atlantic Ocean sector, where krill is the clearly dominant food source for top predators; one of the outcomes of BIOMASS is that a number of widespread sites need to be studied to gain a representative picture of the Antarctic marine ecosystem. The important legacy of BIOMASS in relation to fish, especially to CCAMLR, is that the trophic relations of fish, and fish incidence in krill trawls, are fundamental to the ecosystem approach to resource management.

References

Anon (1977) *Biological Investigations of Marine Antarctic Systems and Stocks (BIOMASS).* Volume 1: *Research Proposals*, pp 1–79.

Anon (1988) Post-SIBEX Fish Data Evaluation Workshop: Phase 2, Evaluation. *BIOMASS Rep Ser* **58**: 1–41.

Camus P (1990) Ichtyoplancton (oeufs et larves de poissons mesopélagiques et benthiques). In: Mayzaud P, Gautier S (eds) MD 42/SIBEX à bord du 'Marion Dufresne' 3 janvier – 18 fèvrier 1985. *Rapports Campagnes Mer* **85–01**: 133–48.

Duhamel G (1987) Ichtyofaune des secteurs indien occidental et atlantique oriental de l'océan austral: biogeographie, cycles biologiques et dynamique des populations. Thèse de Doctorat d'Etat, Université Paris 6, 687 pp. Microfiche SN87 200 447, Museum National d'Histoire naturelle, 512 pp.

Duhamel G, Hureau JC, Ozouf-Costaz C (1983) Ecological survey of the notothenioid fishes in the Southern Ocean from Bouvet to Kerguelen Islands. In: Nemoto T, Matsuda T (eds) Proceedings of the BIOMASS Colloquium 1982. *Proc Nat Inst Polar Res Special Issue* **27**: 176–82.

Duhamel G, Williams R (1990) Campagne MD42/SIBEX ichtyofaune demersale: analyse des captures de chalut à perche. In: Mayzaud P, Gautier S (eds) MD 42/SIBEX à bord du 'Marion Dufresne' 3 janvier–18 fèvrier 1985. *Rapports Campagnes Mer* **85–1**: 149–72.

Ekau W (1990) Demersal fish fauna of the Weddell Sea, Antarctica. *Antarct Sci* **2**(2): 129–37.

Franeker J A van (1992) Diet of fulmarine petrels in the Windmill Islands, Wilkes Land, Antarctica. In: BIOMASS Colloquium Poster Abstracts. Alfred-Wegener-Institut für Polar- und Meeresforschung, Bremerhaven, p 13.

Gon O (1988) The fishes collected during the South African SIBEX I+II expeditions to the Indian Ocean sector of the Southern Ocean (60–60°S, 48–64°E). *S Afr J Antarct Res* **18**: 55–70.

Gutt J (1992) Bentho-pelagic krill. In: *BIOMASS Colloquium Poster Abstracts*. Alfred-Wegener-Institut für Polar- und Meeresforschung, Bremerhaven, p 16.

Hubold G (1990) Seasonal patterns of ichthyoplankton distribution and abundance in the southern Weddell Sea. In: Kerry K R, Hempel G (eds) *Antarctic Ecosystems: Ecological Change and Conservation*. Springer-Verlag, Berlin, pp 149–58.

Hubold G, Ekau W (1987) Midwater fish fauna of the Weddell Sea, Antarctica. *Proc V Congr Europ Ichthyol, Stockholm 1985*, pp 391–6.

Hulley P A (1981) Results of the research cruises of FRV 'Walther Herwig' to South America. LVIII. Family Myctophidae (Osteichthyes, Myctophiformes). *Arch FischWiss* **31**: 1–300.

Hulley P A, Camus P, Duhamel G (1989) Ichthyological results of cruise MD42/SIBEX-II. Part 1. Fishes from RMT-8 stations, with additional records of lanternfishes (Myctophidae:Osteichthyes) from the Indian and Atlantic sectors of the Southern Ocean. *Cybium* **13**: 83–99.

Hureau J C (1986) MD 24/BIOMASS (9 août–19 septembre 1980); MD 25/FIBEX (25 janvier–5 mars 1981) à bord du 'Marion Dufresne'. *Rapports Campagnes Mer* (Mission de Recherche des Terres Australes et Antarctiques Francaises) **81**: 1–101.

Ikeda T, Hosie G, Stolp M (1986) SIBEX II cruise krill/zooplankton data. *ANARE Res Notes* **32**: 1–70.

Iwami T, Kubodera T (1990) Mesopelagic fishes collected with 10-foot IKPT in the Indian Sector of the Antarctic Ocean and its neighbouring waters during the JARE-28 cruise, 1987. *Proc NIPR Symp Polar Biol* **3**: 64–70.

Miya M, Okiyama M, Nemoto T (1986) Midwater fishes of the Southern Ocean south of Australia (extended abstract). *Mem Natl Inst Polar Res Spec Issue* **40**: 323–4.

Rasoanarivo R, Aboussouan A (1983) Larves de *Electrona antarctica* (Gunther, 1878) (Teleostei, Myctophidae) recoltées durant la campagne FIBEX-MD/25 dans le Sud-Ouest de l'Océan Indien. *Cybium* **7**: 75–86.

Ślósarczyk W (1983) Juvenile *Trematomus bernacchii* and *Pagothenia brachysoma* (Pisces, Nototheniidae) within krill concentrations off Balleny Islands (Antarctic). *Polsk Polar Res* **4**: 57–69.

Tabeta O, Komaki Y (1986) Distribution and abundance of pelagic fishes in the epipelagic layers of the Indian Sector of the Southern Ocean in summer 1983–84. *Mem Natl Inst Polar Res Spec Issue* **40**: 316–22.

Williams R (1985*a*) Trophic relationships between pelagic fish and euphausiids in Antarctic waters. In: Siegfried W R, Condy P R, Laws RM (eds) *Antarctic*

Nutrient Cycles and Food Webs. Springer-Verlag, Berlin, pp 452–9.

Williams R (1985*b*) The potential impact of a krill fishery upon pelagic fish in the Prydz Bay area of Antarctica. *Polar Biol* **5**: 1–4.

Williams R (1988) The inshore marine fishes of the Vestfold Hills region, Antarctica. *Hydrobiologia* **165**: 161–7.

Williams R (1989) Australian research on Antarctic bird and seal diets. *Scientific Committee for the Conservation of Antarctic Marine Living Resources: Selected Scientific Papers 1988.* SC-CCAMLR-SSP/**5**: 321–49.

Williams R, Kirkwood J M, O'Sullivan D B (1983) FIBEX cruise zooplankton data. *ANARE Res Notes* **7**: 1–163.

Williams R, Kirkwood J M, O'Sullivan D B (1986) ADBEX I cruise zooplankton data. *ANARE Res Notes* **31**: 1–107.

Woehler E J, Johnstone G W (1991) Status and conservation of seabirds of the Australian Antarctic Territory. In Croxall J P (ed) *Seabird Status and Conservation: a Supplement.* ICBP Technical Publication **11**: 279–308.

Marine ornithology in the southern Drake Passage and Bransfield Strait during the BIOMASS Programme

G. L. HUNT, Jr., J. P. CROXALL & P. N. TRATHAN

Introduction

ONE GOAL of the international Biological Investigations of Marine Antarctic Systems and Stocks (BIOMASS) Programme was to study the pelagic distributions and abundances of marine birds to understand better their role as consumers in the Antarctic marine ecosystem (BIOMASS, 1977). Subsequently, more specific aims were defined (BIOMASS, 1985) as to:

1. obtain detailed information on the spatial and temporal distribution of avian species, their species diversity, biomass and density;
2. investigate correlations between the distribution of birds at sea and features of the physical and biological environment;
3. determine whether seabirds can be used as indicators of the distribution and abundance of selected prey stocks, especially krill and cephalopods.

Here we review the accomplishments of BIOMASS-related marine ornithological research in the Atlantic sector of the Southern Ocean. The ornithological accomplishments of the BIOMASS Programme include not only the marine bird observations taken during FIBEX (1980–1) and SIBEX (1983–4; 1984–5), but also the development of a set of methods for conducting marine bird observations (BIOMASS, 1982, 1985; BIOMASS Working Party on Bird Ecology, 1992) that facilitate comparison of results, regardless of where or by whom data were gathered. Such standardization of methodology is taken for granted in many mature fields of science; in the relatively new area of marine ornithology, this attempt at standardization meant that many cooperating nations had to change their methods of data collection. In our review of BIOMASS we discuss the ecological and management implications of ornithological observations in southern Drake Passage and Bransfield Strait. We also review what we have learned about the influence of methods on the quality and interpretation of the available data, as well as limitations in how the data may be used.

Pelagic studies of marine birds in southern Drake Passage and Bransfield Strait, Antarctica, prior to the BIOMASS Programme were few. Among the early studies were those of Tickell & Woods (1972) and Kock & Reinsch (1978) in the southern Drake Passage and Bransfield Strait, Brown *et al.* (1975) and Linkowski & Rembiszewski (1978) in the Drake Passage, and Cline *et al.* (1969) in the Weddell Sea. These studies are primarily useful as sources of information on the distributions of species, and to a lesser extent on their relative abundances. Data on the location and abundance of breeding seabirds in the area (Croxall & Kirkwood, 1979; Jablonski, 1984; Croxall *et al.*, 1984*a*; Myrcha *et al.*, 1987; Poncet & Poncet, 1987; Peter *et al.*, 1988; Shuford & Spear, 1988) provide important background

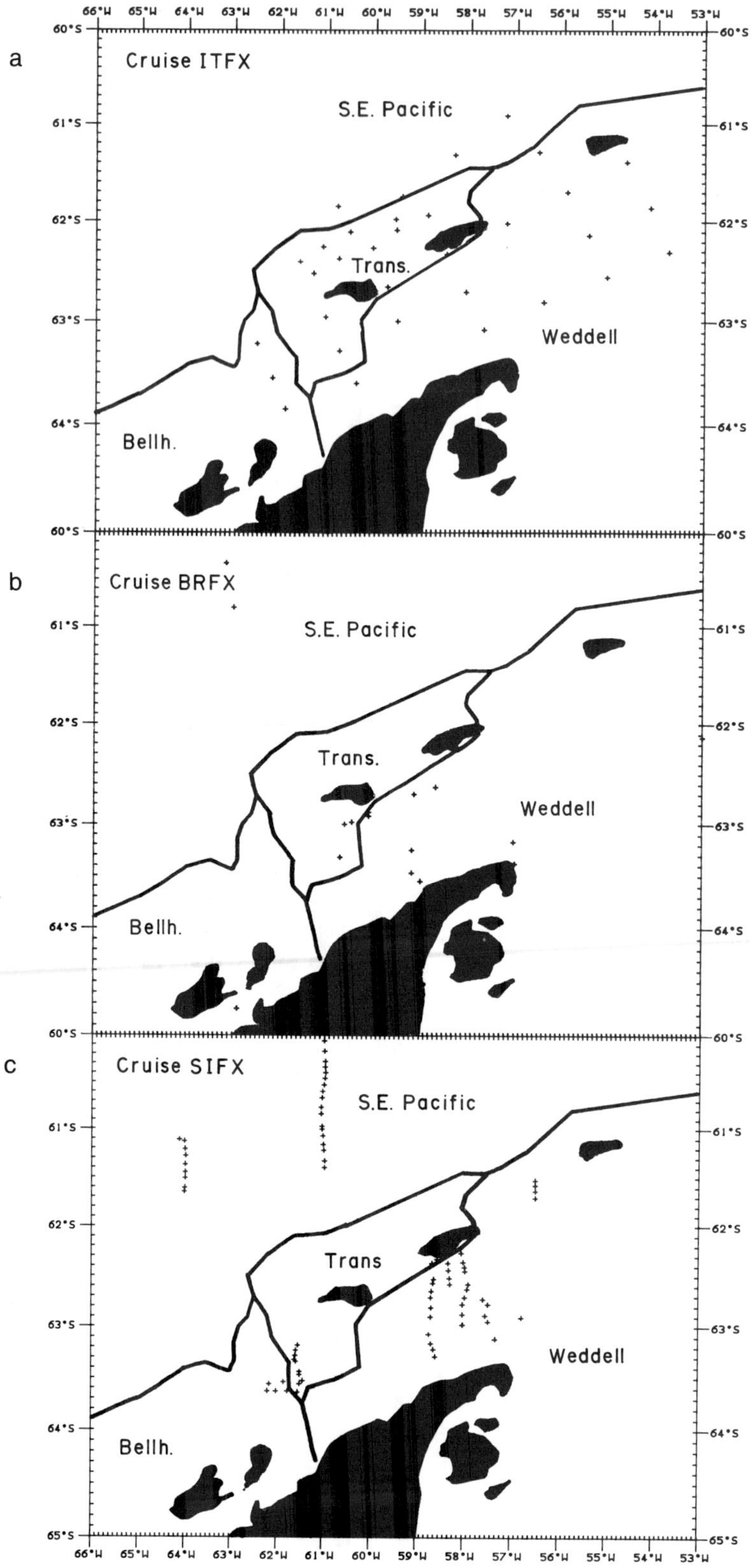

a
Cruise ITFX
S.E. Pacific
Trans.
Weddell
Bellh.
b
Cruise BRFX
S.E. Pacific
Trans.
Weddell
Bellh.
c
Cruise SIFX
S.E. Pacific
Trans
Weddell
Bellh.
66°W 65°W 64°W 63°W 62°W 61°W 60°W 59°W 58°W 57°W 56°W 55°W 54°W 53°W
60°S 61°S 62°S 63°S 64°S 65°S

Table 1. *Seabird data cards submitted to the BIOMASS Data Centre and used in our analysis.*

Cards were judged unsuitable for analysis if the vessel was not underway, if the observations were not made of birds in the foreward quarter, and if the transect was not limited to 300 m width.

Country	Cruise	Duration	Total cards	Cards suitable for analysis
FIBEX				
Chile	ITFF	3–21 Feb 1981	33	29
Poland	SIFX	19 Feb–12 Mar 1981	338	65
U.K.	BRFX	3–29 Mar 1981	296	21
Total FIBEX			667	115
SIBEX				
Poland	SIS1	21 Dec 1983–8 Jan 1984	220	220
Japan	KMS2	10–14 Dec 1984	233	20
U.K.	JBS2	16 Jan–6 Feb 1985	618	583
Total SIBEX			1071	823
Total SIBEX and FIBEX			1783	938

information for the interpretation of the pelagic records. Likewise, colony-based studies of penguin foraging ecology at Admiralty Bay (Trivelpiece *et al.*, 1987, 1990; Volkman *et al.*, 1980) are useful for assessing the impact of these birds on prey populations.

Methods and data

Standardized instructions for recording seabirds at sea were developed for FIBEX (BIOMASS, 1982). However, the participants of a workshop to analyze the FIBEX data concluded that considerably improved and more detailed instructions were needed to provide a uniform methodology for recording quantitatively data on seabirds at sea in the Antarctic (BIOMASS, 1985). These methods (BIOMASS Working Party on Bird Ecology, 1992) are based on recording birds in 10 min time periods (cards). During SIBEX, Hunt *et al.* (1990*b*) and Heinemann *et al.* (1989) recorded birds to the nearest 0.1 min by entering observations as they occurred directly into a hand-held computer. These records were subsequently partitioned into 10 min segments for analysis. Tasker *et al.* (1984) recommended a 'snapshot' technique. JA van Francker (unpublished manuscript) compared the 'snapshot' technique with the continuous count technique employed by Hunt *et al.* (1990*b*) and Heinemann *et al.* (1989) and concluded that the snapshot technique avoided the systematic over-counts of flying birds that result from the continous count technique. These overcounts are on average 1.8 times greater than counts obtained by the 'snapshot' technique. The snapshot method is thus particularly useful if the objective of the research is to obtain estimates of avian biomass which include that of flying birds.

During FIBEX and SIBEX, ornithological observations were made on seven cruises that visited the study area (60–65°S; 53–66°W) (Table 1). Ornithological effort varied considerably on these cruises (Figs 1 and 2), but overall the study area received considerable coverage (Fig. 3). During FIBEX some observers failed to record zero records or transect width, thereby preventing calculation of avian densities or biomass at sea (BIOMASS, 1985). BIOMASS (1985) also documented the substantial

Fig. 1. Location of bird sighting efforts during FIBEX by Chile (a), U.K. (b) and Poland (c), in southern Drake Passage and Bransfield Strait. Names refer to water masses: Trans, Transition Water; Bellh., Bellingshausen Water.

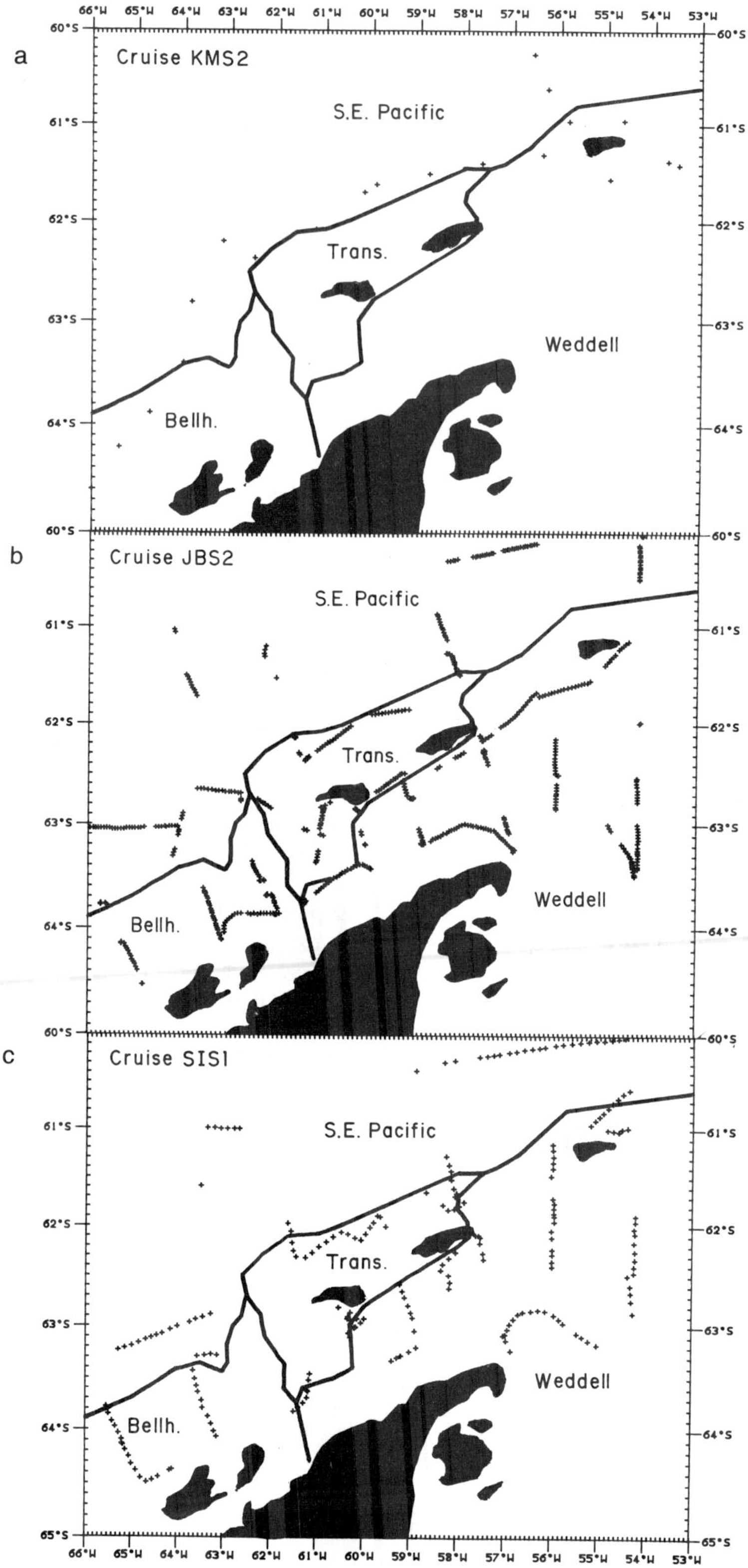

a
Cruise KMS2
S.E. Pacific
Trans.
Weddell
Bellh.
b
Cruise JBS2
S.E. Pacific
Trans.
Weddell
Bellh.
c
Cruise SIS1
S.E. Pacific
Trans.
Weddell
Bellh.
66°W 65°W 64°W 63°W 62°W 61°W 60°W 59°W 58°W 57°W 56°W 55°W 54°W 53°W
60°S 61°S 62°S 63°S 64°S 65°S

Table 2. *Geographical distribution of seabird cards selected for analysis within Bransfield Strait and southern Drake Passage*

Cruise	Total no. of cards	SE Pacific	Weddell Sea	Bellingshausen Sea	Transition Zone
SIFX	63	21	26	4	12
SIS1	220	58	98	28	36
JBS2	583	156	236	92	99
Total	866	235	360	124	147

number of birds associated with, or following, an observer's ship and indicated that, if ship-following birds were not excluded from counts, it would be impossible to provide realistic estimates of seabird density and biomass. The impact of ship-following birds on reported counts remains a difficult feature of the interpretation of multi-investigator data sets when documentation for assessing how ship-following birds were identified and recorded is not available.

In our evaluations of the multiple data sets available in the BIOMASS Data Centre, we have only used records for which observations were made within a transect width of 300 m and a viewing arc of 90° off the bow (three cruises) (Table 2). These restrictions severely limit the number of records for our analyses. Additionally, for the comparison of species abundances within water mass categories, only data from the two SIBEX cruises (SISI, JBS2) were used; data for SIFX were too sparse to permit analysis.

The extent of ornithological effort varies greatly between research cruises, and it is therefore important to understand the influence of sampling effort in obtaining observations of 'uncommon' species, which may be indicative of seasonal or water-mass changes (e.g. Hunt *et al.* 1992*b*), or of rare large flocks. We examined the effect of sample size on the likelihood of observing all bird species actually present, and of observing rare large flocks by using a Monte Carlo analysis. For estimates of species richness, each species was recorded in terms of presence or absence on each of 618 cards generated from 10 min transect segments on the JBS2 cruise. For each species, these presence–absence scores were then randomized independently and assigned back to the cards. A second level of randomization was carried out to reorder the cards. The cumulative number of species was determined by taking the cards in the new order. This process was repeated independently three times. A similar process was used to generate curves depicting the maximum 'flock' size 'seen' as a function of the number of cards. Flock size was defined as the number of birds of a given species recorded on a card. The maximum 'flock' size was determined by taking the cards in the new order and determining the largest 'flock' size, irrespective of species. The process was repeated independently ten times.

Results and discussion

Because the two cruises with the largest data sets occurred at different seasons and in different years, and the majority of the data came from only three cruises (SIFX, SISI and JBS2), there were insufficient data to separate the effects of year and season. Additionally, inspection of the data suggests that ship-following birds may have been treated differently on each cruise.

Large samples are required if the full suite of species or rare large flocks are to be encountered with certainty by a survey (Fig. 4). Although fewer than 100 of 618 ten-minute counts were required to record 50% of the bird species present in the survey

Fig. 2. Location of bird sighting efforts during SIBEX by Japan (a), U.K. (b) and Poland (c), in southern Drake Passage and Bransfield Strait. (Abbreviations: as Fig. 1.)

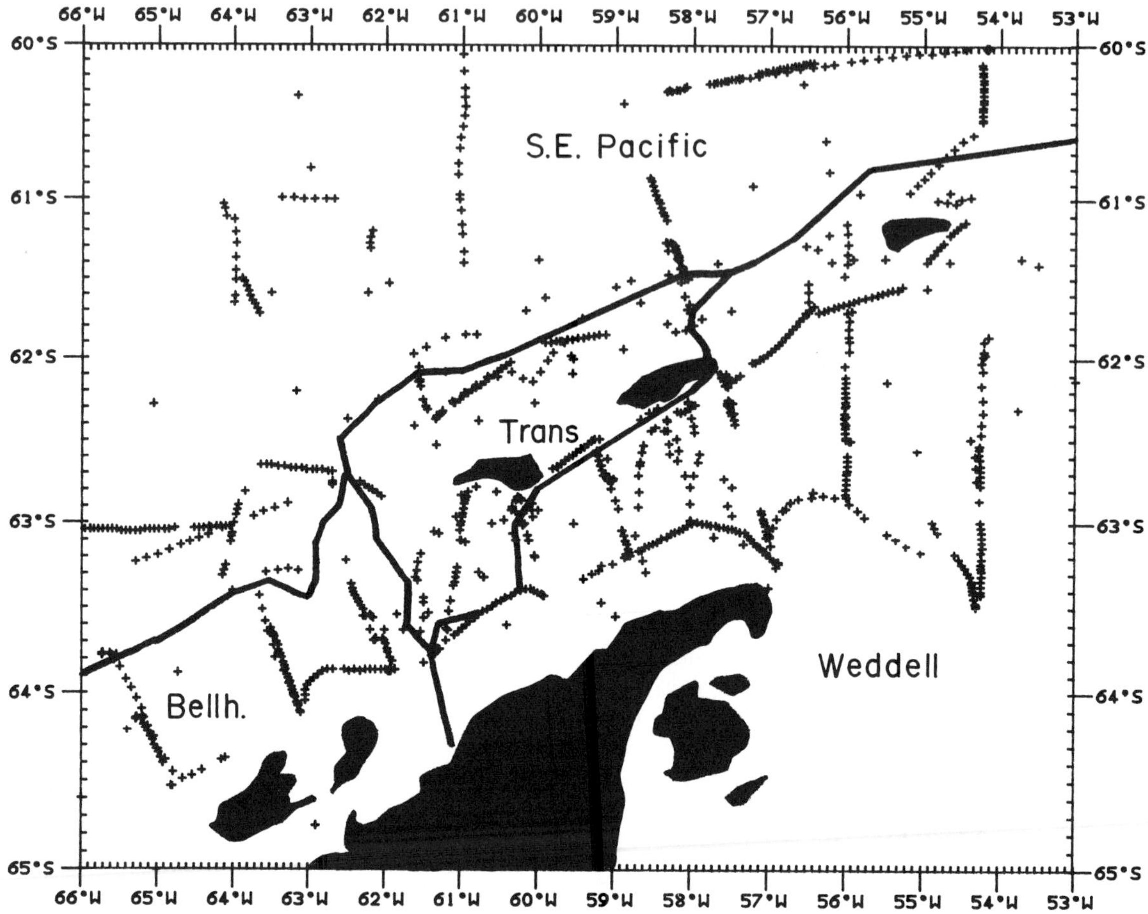

Fig. 3. Bird sighting efforts for FIBEX and SIBEX in southern Drake Passage and Bransfield Strait.

area during JBS2, between 400 and 500 counts were required to find up to 90% of the species present. Seven bird species were recorded on 100 or more cards, but another twelve were seen on five or fewer cards each. Similarly, on average, 300 ten-minute observation periods would be needed to ensure encountering all but the two largest flocks, if flocks were distributed randomly in space. In fact, the largest flock was observed after just over 100 observations. If large foraging flocks assemble owing to prey interacting with geographically fixed physical processes (see Hunt, 1991, for a review), then the inclusion of these flocks on a survey will be extremely sensitive to survey design. In regions where flocks are randomly distributed, extensive surveys are necessary if the largest flocks are to be found.

Distribution patterns

Large-scale maps of seabird distributions based on the FIBEX cruises (Starck & Wyrzykowski, 1982; BIOMASS, 1985) and more detailed mapping of distribution and relative abundance of birds in the study area during SIBEX (Hunt *et al.*, 1990*b*) provide useful descriptions of where species are likely to occur and variation in relative densities of the species present. From these maps it is evident that, even within the confines of the Bransfield Strait,

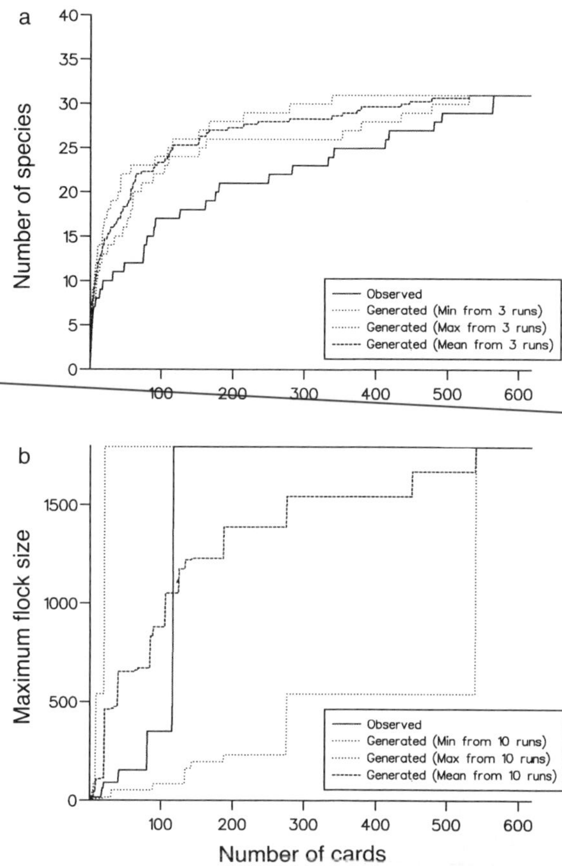

Fig. 4. (a) Number of bird species appearing on cards as a function of observation effort, which provides a comparison of observed values with those generated by randomization of cards and species (see methods). These results suggest that after the most common species are observed, sightings of additional species are roughly proportional to effort, with new species being added even after 500 ten-minute survey periods were completed. (b) Maximum flock size appearing on cards as a function of the observation effort, which provides a comparison of observed values with those generated by a randomization of cards and flock size (see methods). The great variation in the shape of these curves suggests that it will be very difficult to predict the effort needed to ensure adequate sampling of large aggregations. Since the aggregations are almost certainly not randomly distributed, the likelihood of their occurrences in a sample will also be sensitive to the design of the cruise track.

Table 3. *Frequency of occurrence (% cards on which recorded) and abundance when present (mean abundance per card on which recorded) of seabirds on cruises SIFX, SIS1 and JBS2*

The + indicates that penguins were seen in small numbers, but not identified to species.

	Frequency of occurrence			Abundance when present			Rank sum		
Species	SIFX	SIS1	JBS2	SIFX	SIS1	JBS2	SIFX	SIS1	JBS2
Pygoscelis papua	—	—	0.5	—	—	3.0	—	—	15
P. adeliae	+	7.2	6.3	+	21.8	42.9	+	4	1
P. antarctica	+	23.1	27.4	+	13.6	11.3	+	2	2
Diomedea exulans	28.5	1.3	0.3	1.9	1.0	1.0	4	19	21.5
D. melanophris	6.3	59.5	29.1	1.0	2.1	2.0	11.5	5	7
D. chrysostoma	6.3	1.8	6.8	1.0	1.0	3.6	11.5	17.5	9
Phoebetria palpebrata	—	3.1	1.3	—	1.0	1.0	—	16	16.5
Macronectes giganteus	38.0	43.6	17.6	2.0	1.5	1.2	3	7	11
M. halli	26.9	5.0	0.3	1.5	1.0	1.0	6	14	21.5
Fulmarus glacialoides	17.4	59.5	36.7	2.5	3.0	8.4	5	3	3
Thalassoica antarctica	1.5	5.0	1.0	1.0	1.3	1.0	16.5	13	18
Daption capense	65.0	89.5	40.8	9.9	13.6	7.1	1	1	4
Pagodroma nivea	4.7	0.4	0.3	1.0	1.0	1.0	13	20.5	21.5
Pachyptila spp.	4.7	19.5	16.4	1.0	1.4	2.7	14.5	9	8
Halobaena caerulea	—	—	7.0	—	—	1.2	—	—	12
Procellaria aequinoctialis	3	8.1	0.3	—	1.3	1.0	—	11	21.5
Oceanites oceanicus	66.6	53.1	66.3	7.5	1.8	2.9	2	6	5
Fregetta tropica	6.3	19.5	35.1	2.0	1.1	1.9	8	10	6
Pelecanoides spp.	—	—	1.0	—	—	1.3	—	—	16.5
Phalacrocorax atriceps	9.7	0.4	—	5.25	1.0	—	7	20.5	—
Catharacta spp.	3.1	3.1	1.8	1.0	1.1	1.0	14.5	15	14
Sterna spp.	3.1	5.4	2.5	3.0	1.3	21.9	10	12	13
Larus dominicanus	7.9	12.2	0.8	1.4	2.9	1.0	9	8	19
Chionis alba (at sea)	1.5	1.8	—	1.1	1.0	—	16.5	17.5	—

there was considerable spatial pattern both in where species occurred and in their abundances. Temporal shifts in distribution and abundance have not been analyzed statistically (see above), but some temporal shifts are evident based on the analyses of individual cruises. For instance, Hunt *et al.* (1990*b*) recorded blue petrels (*Halobaena caerulea*) on 41 of 583 cards during late January 1985, whereas Starck (1985) working in the same area recorded them on only 6 of 396 cards in late December 1983 and early January 1984. Blue petrels were recorded on 0 of 491 cards obtained from mid-February to mid-March 1981 (Starck & Wyrzykowski, 1982). Starck (1985) reported a greater than two-fold increase between abundances recorded in the autumn (February–March) 1981 cruise and in the early summer (December–January) cruise in SIBEX 1. These latter observations of differences are particularly useful in describing the magnitude of short-term changes because methodology and observers were the same on both cruises.

Bird species varied greatly in the patchiness of their distributions (Table 3). Some species (e.g. black-browed albatross (*Diomedea melanophris*), southern giant petrel (*Macronectes giganteus*), Antarctic fulmar (*Fulmarus glaciloides*), Cape petrel (*Daption capense*) and Wilson's storm-petrel (*Oceanites oceanicus*)) were relatively widespread throughout the study area, as judged by the relatively large percentage of the data cards on each cruise that contained records of their presence. However, one must be cautious in the interpretation of these records because most of these widespread species are also known to follow ships (Griffiths, 1982). In the case of penguins, which do not follow ships, Adélie penguins (*Pygoscelis adeliae*) were

Table 4. *Grouping of seabird species.*

(See Hunt *et al.* (1990*b*), Figs 9 and 10). Data from the two very large foraging flocks were excluded from the cluster analyses because they were statistical outliers.

Dominant bird species in clusters of oceanographic regions based on bird species abundance		Groups with 70% similarity based on a cluster analysis using bird species abundances at the scale of single oceanographic regions	
Water mass clusters	Dominant bird species	Bird species cluster	Dominant bird species
Bellingshausen Water	*D. melanophris* *D. chrysostoma* *F. glacialoides*	Group 1	*D. melanophris* *D. chrysostoma* *F. glacialoides*
South-east Pacific Water	*H. caerulea* *F. tropica* *Pachyptila* spp.	Group 2	*H. caerulea* *F. tropica* *Pachyptila* spp.
Transition Water (group 1)	*P. antarctica* *D. capense* *F. tropica*	Group 3	*P. antarctica* *D. capense* *F. tropica*
Transition Water (group 2)	*O. oceanicus*	Group 4	*O. oceanicus*
Weddell Water	*P. adeliae*	Group 5	*P. adeliae*

abundant, but recorded on a small number of cards compared with chinstrap penguins (*P. antarctica*) which were both abundant and widespread.

Many of the penguin records came from the inshore portions of survey lines where surveys were within the foraging ranges of nearby colonies. Had cruise tracks approached land in areas lacking penguin colonies, or passed closer to larger numbers of colonies, the number of penguins recorded would have varied significantly. Thus, when surveys are conducted near land where and when penguins are breeding, slight variations in cruise tracks can greatly affect the number of penguins seen.

Seabird distribution in relation to physical environment

For the study area, BIOMASS (1985) characterized the physical environment in which each of 24 seabird species or species groups had been recorded. Differences in preferred physical environments primarily reflected the latitudinal distribution of bird species. For instance, species commonly found at low latitudes were typically in areas where water temperature, salinity, and wind speed were relatively high, and ice cover was absent (BIOMASS 1985) (Tables 3–5, Fig. 4). Fraser & Ainley (1986) and DG Ainley *et al.* (unpublished data) working in the Weddell Sea have documented that there are strong differences between seabird species in terms of their affinity for pack ice. These different bird species assemblages shift north and south over a period of days to weeks as the ice edge moves in response to weather. Adélie penguins were usually associated with ice, chinstrap penguins preferred open water to the north of the pack-ice zone. Elsewhere, Ryan & Cooper (1989) showed that the distribution of bird species in the Prydz Bay region of Antarctica reflected the distribution of sea-surface temperatures.

Cluster analyses of bird species groupings within the Atlantic Sector of FIBEX revealed no clusters with similarities greater than 50%, but at the 30% similarity level, 11 of the 24 taxa examined fell into three groupings (BIOMASS, 1985). These three groupings were representative of the birds occupying different latitudinal ranges, with Antarctic petrel (*Thalassoica antarctica*) and snow petrel (*Pagodroma nivea*) forming a high-latitude group; soft-plumaged petrels (*Pterodroma mollis*) and grey-headed albatrosses (*D. chrysostoma*) formed the lowest-latitude (northernmost) group. A third group included species occupying intermediate latitudes.

Table 5. *Abundance ranks of the most abundant seabirds in each of four oceanographic regions*
Giant petrels (*Macronectes* spp.) are excluded because of the frequency of their attraction to ships.

Bellingshausen Water

Species	Rank SIS1	Rank JBS2	Combined rank
D. melanophris	2	2	1.5
F. glacialoides	3	1	1.5
D. capense	1	6	3.5
O. oceanicus	4	3	3.5
P. antarctica	5	4	5
T. antarctica	6	8.5	6
D. chrysostoma	—	5	—

Southern Pacific Water

Species	Rank SIS1	Rank JBS2	Combined rank
D. capense	1	4	1.5
O. oceanicus	4	1	1.5
P. antarctica	3	5	3.5
Pachyptila spp.	6	2	3.5
D. melanophris	2	7	5
F. tropica	7	3	6
F. glacialoides	5	6	7
H. caerulea	—	8	—

Transition Water

Species	Rank SIS1	Rank JBS2	Combined rank
P. antarctica	2	1	1
D. capense	1	3	2
O. oceanicus	5	2	3
D. melanophris	3	5	4
F. glacialoides	4	6	5
F. tropica	8	4	6
D. chrysostoma	—	7	—
H. caerulea	—	8	—

Weddell Water

Species	Rank SIS1	Rank JBS2	Combined rank
P. adeliae	2	1	1.5
D. capense	1	2	1.5
F. glacialoides	3	3	3
P. antarctica	4	4	4
O. oceanicus	5	5	5
D. melanophris	6	7	6
F. tropica	8	6	7

Within the southern Drake Passage and Bransfield Strait, Hunt *et al.* (1990*b*) described 11 water masses based on temperature, salinity, and silicate characteristics. They identified five clusters (regions) of these water masses, based on bird species assemblages, using the ten most abundant species. In each of these five regions, different bird species, or groups of species, predominated (Table 4). When bird species were clustered using a correlation matrix, a grouping of species similar to that based on the oceanographic regions was achieved.

Using the SIBEX data for the SISI and JBS2 cruises, we ranked the most abundant bird species recorded in each of the four major oceanographic regions (Table 5). Despite there being a difference in season and observers, the results of the two cruises were remarkably similar. Species abundance rankings between the two cruises were almost identical in Weddell Sea Water, where Adélie penguins were a predominant species. In Transition Water, prions (*Pachyptila* spp.) and storm petrels (*O. oceanicus* and *Fregetta tropica*) had higher abundances than elsewhere. Blue petrels were more abundant in the Transition Water and Southeast Pacific Water during the JBS2 cruise than in other regions. Bellingshausen Water was distinguished by the presence of Antarctic petrels and large numbers of albatrosses, particularly black-browed albatrosses.

Estimates of seabird biomass at sea

Estimates of seabird biomass at sea in southern Drake Passage and Bransfield Strait were calculated by Hunt *et al.* (1990*b*), and were averaged for the study area as a whole (Table 6). Birds present in two large aggregations (see below) made up 32% of the mean density and 44% of the total seabird biomass recorded. Penguins accounted for 84% of the total seabird biomass, and 75% of the biomass when

Table 6. *Estimated mean density and biomass for some seabirds in the southern Drake Passage and Bransfield Strait during JBS2.*

Values in parentheses are means calculated with data from two exceptionally large aggregations excluded.

Species	Mean abundance per nautical mile ± s.d.	Mean density of birds per km^2	Mean biomass (kg per km^2)
Pygoscelis adeliae	3.10±47.37	1.72	7.40
	(0.83±6.30)	(0.46)	(1.98)
P. antarctica	1.72±6.33	0.95	3.61
Diomedea melanophris	0.34±1.04	0.19	0.65
D. chrysostoma	0.16±1.70	0.09	0.33
Fulmarus glacialoides	1.55±9.37	0.86	0.67
	(1.03±4.23)	(0.57)	(0.44)
Daption capense	1.56±8.71	0.87	0.39
	(1.06±3.63)	(0.59)	(0.27)
Pachyptila spp.	0.30±0.94	0.17	0.03
Halobaena caerulea	0.07±0.37	0.04	0.01
Oceanites oceanicus	1.00±1.71	0.55	0.02
Fregetta tropica	0.41±0.84	0.23	0.01
Total	10.21±50.60	5.67	13.21
	(6.95±11.42)	(3.86)	(7.36)

Source: From Hunt *et al.* (1990*b*); values of seabird mass taken from Jouventin & Mougin (1981).

birds in the two large aggregations were excluded. Because penguins are much heavier than flying seabirds, they have a disproportionately large impact on the estimation of avian biomass at sea. Other species such as Antarctic fulmars, Cape petrels and Wilson's storm-petrels together accounted for 40% of the numbers of birds seen, but only 8.2% of the total seabird biomass observed in the study area.

Biomass estimates from at-sea observations have sometimes been used to assess energy or carbon fluxes to seabirds (e.g. Schneider *et al.*, 1986; van Franeker, 1993). In the present case, the spatial heterogeneity in bird distributions caused by water-mass differences and proximity to colonies, and the lack of stratified surveys that would have adequately sampled the study area, precluded a similar analysis. Assessments of prey consumption by Antarctic seabirds have also been derived from estimates of breeding populations (e.g. Croxall *et al.*, 1984*b*, 1985; Croxall & Prince, 1987). However, at the small spatial scale being considered here, knowledge of breeding populations of seabirds other than penguins is grossly inadequate to do this. Critical comparisons of colony- and ship-based estimates of seabird abundance and the resulting estimates of prey consumption are badly needed, but must be based on appropriately designed studies.

Seabird distribution in relation to prey distribution

Heinemann *et al.* (1989) used the results of simultaneous surveys of marine birds and hydroacoustic estimates of krill during SIBEX to examine two statistically independent components of correlation: spatial concordance or association; and numerical concordance, the rank correlation of seabird and krill numbers when both were present. They found a positive spatial association between each of two species of birds (Antarctic fulmar and Cape petrel) and krill. Only two bird species, Adélie penguin (r= 0.60) and Cape petrel (r=22), had statistically significant ($p \leq 0.05$) positive numerical correlations with prey abundance, in nautical miles where both predator and prey were present. At spatial scales larger than a nautical mile, positive correlations between predators and their prey appeared stronger for some species, but for others, correlations became negative, possibly because of the inclusion of oceanographic regions with high krill densities in which the birds were not abundant.

Other studies of the spatial and numerical concordance of seabirds and planktonic prey have also had mixed success in demonstrating strong positive correlations, particularly at small spatial scales, when scale dependence has been investigated (e.g. Woodby, 1984; Obst, 1985; Ryan & Cooper, 1989; Hunt *et al.*, 1990*a*; Hunt *et al.*, 1992*a*). There are methodological reasons why the explained variance at small spatial scales is often weak (0.21 to 0.59%) (Hunt *et al.*, 1990*a*), such as the lack of night-time sampling and a lack of acoustical data for the top 10 m of the water column. Additionally, it may be that background levels of prey are sufficiently high that it is not of advantage for birds to seek the densest concentrations of plankton (Woodby, 1984; Hunt *et al.*, 1990*a*). Birds may have considerable difficulty in locating the areas of highest prey abundance, but are able to locate areas where prey is generally at sufficient densities to meet energetic needs. Within these larger areas, the choice of a foraging locale may be a chance event or may be directed by social attraction to previously and/or currently feeding individuals.

The importance of the presence of other species of birds (and marine mammals) is illustrated by the work of Harrison *et al.* (1991) who found black-browed and grey-headed albatrosses foraging on krill in small mixed-species flocks in the daytime near South Georgia. Nineteen species of birds were recorded associated with these flocks; the flocks were usually accompanied by penguins, or Antarctic fur seals (*Arctocephalus gazella*), or both. The pursuit-diving species apparently drove krill to the surface where the non-pursuit-diving species were able to surface seize them or plunge-dive for them. When birds were actively feeding at the surface, other birds quickly joined the flock. Similar observations have been recorded in the Gulf of Alaska (Hoffman *et al.*, 1981) and the northern Bering Sea (Obst & Hunt, 1990).

During the cruises discussed here, a significant proportion of seabird feeding activity involved large aggregations of birds associated with the large patches of krill, which may have important implications for management of krill resources. Hunt *et al.* (1985) described two foraging flocks encountered in the Bransfield Strait during SIBEX. One flock contained 73% of all Adélie penguins recorded during the survey; the other flock contained 32% (not 62% as reported by Hunt *et al.*, 1990*b*) of all Cape petrels and Antarctic fulmars recorded. These flocks were associated, respectively, with the largest and third largest concentrations of krill encountered during the survey (Heinemann *et al.*, 1989). Similarly, on a circumnavigation of Antarctica in 1983, Veit & Hunt (1991) found nearly 45% of all birds recorded during their cruise in two foraging flocks. One flock off Enderby Land (65°41′S, 25°51′E) was estimated to contain as many as one million Antarctic petrels and the second, off Wilkes Land (66°6′S, 110°50′E), was dominated by 100 000 short-tailed shearwaters (*Puffinus tenuirostris*). Although no samples of the prey taken were obtained, there was evidence that one and possibly both flocks were associated with very large aggregations of krill. Others have also commented that the largest aggregations of birds encountered were flocks associated with unusually large aggregations of krill. Thus, Starck (1985) reported that flocks of Cape petrels and Antarctic fulmars were associated with the only large, dense krill aggregations found in the region. Similarly, Starck & Wyrzykowski (1982) reported flocks of Cape petrels and Antarctic fulmars associated with large, dense aggregations of krill. In the Bering Sea, short-tailed shearwaters show similar behaviours in that they form flocks of the tens of thousands of birds when foraging on aggregations of the euphausiid *Thysanoessa raschii* (Guzman & Myers, 1986; G L Hunt *et al.*, unpublished data).

The finescale spatial and temporal distributions of krill aggregations are probably unpredictable because there are so many potential influences operating at a wide variety of temporal and spatial scales. If very large aggregations of krill persist for several days or longer, they may serve as a predictable resource over a short period of time. These large aggregations therefore may be disproportionately important for the foraging success of birds because the aggregation, once located, can be followed and foraged upon repeatedly. These very large aggregations are likely to be the focus for commercial harvesting for the same reason. Thus, the potential for harm to seabird populations would exist if the required large krill aggregations were not available within foraging range of seabird colonies. In penguins, this range is quite restricted. In future studies, it would be profitable to learn more about the importance of these large krill aggregations to foraging birds. In addition, virtually all of our observations of foraging seabirds have occurred in daytime (but see Fraser *et al.*, 1989). Thus, there is also a need to develop means of comparing daytime foraging activities of birds with those that occur at night when krill distributions near the surface are likely to differ greatly from those found during the day (Kalinowski & Witek, 1980; Everson, 1982; Miller & Hampton, 1989).

Regardless of the reasons for the weak correlations between avian predators and planktonic prey, the available data suggest that the pelagic surveys of birds in these BIOMASS studies did not provide a consistent, accurate index of the relative, let alone the absolute, size of krill populations. More generally, although the presence of large flocks of foraging birds is likely to indicate the presence of krill, not all krill swarms are so accompanied, and the number of birds does not correlate well with the amount of krill present (Woodby, 1984; Obst, 1985; Heinemann *et al.*, 1989; but see Ryan & Cooper (1989) for a positive result). For these reasons, even the large flocks are unlikely to be useful indicators of the status of krill stocks, although they are obviously good indicators of the presence of local concentrations of krill. An alternative approach, monitoring of marine birds on colonies, provides data which may more directly reflect variations in their ability to locate prey near the colony (Croxall *et al.*, 1988). However, to interpret these results we need to learn more about the areas over which birds from monitored colonies forage, and how changes in parameters measured at colonies relate to changes in local and regional krill stocks (Hunt *et al.*, 1991).

Acknowledgements

We thank the observers who made the field observations during FIBEX and SIBEX, the scientists who processed and validated these data and submitted them to the BIOMASS Data Centre, and Mark Thorley, the Data Centre Manager, for help and guidance in analysis. John Cooper and Eric Woehler and two reviewers made helpful comments on a draft manuscript. The research of GLH has been supported by the Division of Polar Programs, National Science Foundation.

References

BIOMASS (1977) *Biological Investigations of Marine Antarctic Systems and Stocks*, vol 1: *Research Proposals*. SCAR/SCOR, Cambridge. 79 pp.

BIOMASS (1982) *Recording Observations of Birds at Sea*. BIOMASS Handbook no. 18: 14 pp.

BIOMASS (1985) FIBEX Seabird Data Interpretation Workshop, Cape Town, South Africa. *BIOMASS Rep Ser* **44**: 1–40.

BIOMASS Working Party on Bird Ecology (1992) Recording distribution and abundance of seabirds at sea in the Southern Ocean: methods used in the BIOMASS programmes. *Mar Ornithol* (in press).

Brown R G B, Cook F, Kinnear P K, Mills E L (1975) Summer seabird distributions in Drake Passage, the Chilean fjords and off southern South America. *Ibis* **117**: 339–56.

Cline D R, Siniff D B, Erickson A W (1969) Summer birds of the pack ice in the Weddell Sea, Antarctica. *Auk* **86**: 701–16.

Croxall J P, Kirkwood E D (1979) *The Distribution of Penguins on the Antarctic Peninsula and Islands of the Scotia Sea*. British Antarctic Survey, Cambridge, 188 pp.

Croxall J P, Prince P A (1987) Seabirds as predators on marine resources, especially krill, at South Georgia. In: Croxall J P (ed) *Seabirds: Feeding Ecology and Role in Marine Ecosystems*. Cambridge University Press, pp 347–68.

Croxall J P, Prince P A, Hunter I, McInnes S J, Copestake P G (1984*a*) The seabirds of the Antarctic Peninsula, islands of the Scotia Sea and Antarctic Continent between 80°W and 20°W; their status and conservation. In: Croxall J P, Evans P G H, Schreiber RW (eds) *Status and Conservation of the World's Seabirds*, International Council for Bird Preservation, Cambridge, England.

Croxall J P, Ricketts C, Prince P A (1984*b*) Impact of seabirds on marine resources, especially krill, of South Georgia waters. In: Whittow G C, Rahn H (eds) *Seabird Energetics*. Plenum Publishing Corporation, New York, pp 285–318.

Croxall J P, Prince P A, Ricketts C (1985) Relationships between prey life-cycles and the extent, nature and timing of seal and seabird predation in the Scotia Sea. In: Siegfried W R, Condy P T, Laws R M (eds) *Antarctic Nutrient Cycles and Food Webs*. Springer-Verlag, Berlin, pp 516–33.

Croxall J P, McCann T S, Prince P A, Rothery P (1988) Reproductive performance of seabirds and seals at South Georgia and Signy Island, South Orkney Islands, 1967–1987. Implications for Southern Ocean monitoring studies. In: Sahrhage D (ed) *Antarctic Ocean and Resources Variability*. Springer-Verlag, Berlin, pp 261–85.

Everson I (1982) Diurnal variations in mean volume backscattering strength of an Antarctic krill *Euphausia superba* patch. *J Plankt Res* **4**: 155–62.

Fraser W R, Ainley D A (1986) Ice edges and seabird occurrence in Antarctica. *BioScience* **36**: 258–63.

Fraser W R, Pitman R L, Ainley D G (1989) Seabird and fur seal responses to vertically migrating winter krill swarms in Antarctica. *Polar Biol* **10**: 37–41.

Fraser W R, Trivelpiece W L, Ainley D G, Trivelpiece S G (1992) Increases in Antarctic penguin populations:

reduced competition with whales or loss of sea ice due to environmental warming? *Polar Biol* **11**: 525–31.

Griffiths A M (1982) Reaction of some seabirds to a ship in the Southern Ocean. *Ostrich* **53**: 228–35.

Guzman J R, Myers M T (1987) Ecology and behaviour of southern hemisphere shearwaters (Genus *Puffinus*) when over the outer continental shelf of the Gulf of Alaska and Bering Sea during the northern summer (1975–1976). *Outer Continental Shelf Environmental Assessment Program, Final Reports of Principal Investigators* **54**: 571–682. NOAA, Washington DC.

Harrison N M, Whitehouse M J, Prince P A, Heinemann D, Hunt G L Jr, Veit R R (1991) Observations of multispecies seabird flocks around South Georgia. *Auk* **108**: 801–10.

Heinemann D, Hunt G L Jr, Everson I (1989) Relationships between the distributions of marine avian predators and their prey, *Euphausia superba*, in Bransfield Strait and southern Drake Passage, Antarctica. *Mar Ecol Prog Ser* **58**: 3–16.

Hoffmann W, Heinemann D, Wiens J A (1981) The ecology of seabird feeding flocks in Alaska. *Auk* **98**: 437–56.

Hunt G L Jr (1991) Marine ecology of seabirds in polar oceans. *Amer Zool* **31**: 131–42.

Hunt G L Jr, Harrison N M (1990) Foraging habitat and prey taken by least auklets at King Island, Alaska. Mar Ecol Prog Ser **65**: 141–50.

Hunt G L Jr, Everson I, Veit R R, Heinemann D (1985) Distribution of marine birds and their prey in Bransfield Strait and southern Drake Passage. *Antarct J US* **19**: 166–8.

Hunt G L Jr, Harrison N M, Cooney T (1990*a*) Foraging of least auklets: the influence of hydrographic structure and prey abundance. *Studies Avian Biol* **14**: 7–22.

Hunt G L Jr, Heinemann D, Veit R R, Heywood R B, Everson I (1990*b*) The distribution, abundance and community structure of marine birds in southern Drake Passage and Bransfield Strait, Antarctica. *Continental Shelf Res* **10**: 237–57.

Hunt G L Jr, Heinemann D, Everson I (1992*a*) The distributions and predator-prey interactions of macaroni penguins, Antarctic fur seals, and Antarctic krill near Bird Island, South Georgia. *Mar Ecol Prog Ser* **86**: 15–30.

Hunt G L Jr, Piatt J F, Erikstad K E (1991) How do foraging seabirds sample their environment? *Int Ornithol Congr Acta* **20**: 2272–9.

Hunt G L, Priddle J, Whitehouse M J, Veit R R, Heywood RB (1992*b*) Observations of changes in seabird species abundance near South Georgia during a period of rapid change in sea surface temperature. *Antarctic Sci* **4**: 15–22.

Jablonski B (1984) Distribution and numbers of penguins in the region of King George Island (South Shetland Islands) in the breeding season 1980/1981. *Pol Polar Res* **5**: 17–30.

Jouventin P, Mougin J-L (1981) Les strategies adaptives des oiseaux de mer. *Rev Ecol* **35**: 217–72.

Kalinowski J, Witek Z (1980) Diurnal vertical distribution of krill aggregations in the western Antarctic. *Pol Polar Res* **1**: 127–46.

Kock K-H, Reinsch HH (1978) Ornithological observations during the 'German Antarctic Expedition 1975/76'. *Beitr Vogelkd* **24**: 305–28.

Linkowski T B, Rembiszewski J M (1978) Distribution of seabirds of Argentina coast and the feeding habits of the bird fauna in the Drake Passage and Scotia Sea. *Polsk Arch Hydrobiol* **25**: 717–27.

Miller D G M, Hampton I (1989) Biology and ecology of the Antarctic krill (*Euphausia superba* Dana): a review. *BIOMASS Sci Ser* **9**: 1–166.

Myrcha A, Tatur A, Del Valle R (1987) Numbers of Adélie penguins breeding at Hope Island and Seymour Island rookeries (West Antarctica) in 1985. *Pol Polar Res* **8**: 411–22.

Obst B S (1985) Densities of Antarctic seabirds at sea and the presence of the krill *Euphausia superba*. Auk **102**: 540–9.

Obst B S, Hunt G L Jr (1990) Foraging of marine birds at Gray Whale mud-plumes in the Bering Sea. *Auk* **107**: 678–88.

Peter H-U, Kaiser M, Gebauer A (1988) Untersuchungen an Vogeln und Robben auf King George Island (South Shetland Islands, Antarktis). *Geodat Geophys Veroffentlich* **1**(14): 1–128.

Poncet S, Poncet J (1987) Censuses of penguin populations of the Antarctic Peninsula, 1983–87. *Br Antarct Surv Bull* **77**: 109–29.

Ryan P G, Cooper J (1989) The distribution of aerial seabirds in relation to Antarctic krill in the Prydz Bay region, Antarctica, during late summer. *Polar Biol* **10**: 199–209.

Schneider D C, Hunt G L Jr, Harrison N M (1986) Mass and energy transfer to seabirds in the southeastern Bering Sea. *Continental Shelf Res* **5**: 241–57.

Shuford W D, Spear L B (1988) Surveys of breeding Chinstrap penguins in the South Shetland Islands, Antarctica. *Br Antarct Surv Bul* **1**: 18–30.

Starck W (1985) Seabird observations in the region of the South Shetland Islands and South Orkney Islands during BIOMASS-SIBEX (December 1983–January 1984). *Pol Polar Res* **6**: 167–74.

Starck W, Wyrzykowski R (1982) Seabird observations in southern Drake Passage and the Bransfield Strait (BIOMASS-FIBEX programme) in February–March 1981. *Pol Polar Res* **3**: 313–32.

Tasker M L, Hope Jones P, Dixon T, Blake B F (1984) Counting seabirds at sea from ships: a review of methods employed and a suggestion for a standardized approach. *Auk* **101**: 567–77.

Tickell W L N, Woods R W (1972) Ornithological observations at sea in the South Atlantic Ocean, 1954–64. *Br Antarct Surv Bull* **31**: 63–84.

Trivelpiece W Z, Trivelpiece S G, Geupel G R, Kjelmyr J, Volkman N J (1990) Adélie and Chinstrap penguins: their potential as monitors of the Southern Ocean marine ecosystem. In: Kerry K R, Hempel G (eds) *Ecological Change and the Conservation of Antarctic Ecosystems (Proc 5th SCAR Symp Antarct*

Biol, Hobart, August–September 1988). Springer-Verlag, Berlin, pp 191–202.

Trivelpiece W Z, Trivelpiece S G, Volkman N J (1987) Ecological separation of Adélie, gentoo and chinstrap penguins at King George Island, Antarctica. *Ecology* **68**: 351–61.

Van Franeker J A (1993) Top predators as indicators for ecosystem events in the confluence zone and marginal ice zone of the Weddell and Scotia Seas, Antarctica, November 1988 to January 1989 (EPOS leg 2). *Polar Biol* (in press).

Veit R R, Hunt G L Jr (1991) Broadscale density and aggregation of pelagic birds from a circumnavigational survey of the Antarctic Ocean. *Auk* **108**: 790–800.

Volkman N J, Presler P, Trivelpiece W Z (1980) Diets of pygoscelid penguins at King George Island, Antarctica. *Condor* **82**: 373–8.

Woodby D (1984) The April distribution of murres and prey patches in the southeastern Bering Sea. *Limnol Oceanogr* **29**: 181–8.

Consumption of Antarctic krill (*Euphausia superba*) by seabirds during summer in the Prydz Bay region, Antarctica

J. COOPER & E. J. WOEHLER

Introduction

THE PRYDZ BAY region of Antarctica has been identified as an area of high abundance of Antarctic krill, *Euphausia superba* (e.g. Marr, 1962; Everson, 1977). This elevated abundance may be due to the presence of an anticyclonic gyre (Smith *et al.*, 1984), although a gyre is not always present and krill may not always be abundant (Miller, 1985, 1986*a*; Hosie *et al.*, 1988). Higginbottom *et al.* (1988) estimated the standing stock of Antarctic krill in the Prydz Bay region to be approximately 3.7 × 10^6 tonnes in 1985, a resource which has been fished commercially (Williams, 1985).

Biomass of breeding seabirds (Woehler, 1990), at-sea densities in the region of foraging seabirds (Montague, 1988) and whale populations (Best & Butterworth, 1980) are thought to be relatively high, owing to a high krill biomass. The Prydz Bay region contains numerous, often large, colonies of nine species of seabirds (Woehler & Johnstone, 1991), whose diets include Antarctic krill (Williams, 1989). Until recently, little quantitative information has been available on the distribution and abundance of seabirds at sea in the region (e.g. Johnstone & Kerry, 1976; Ohyama & Naito, 1982; Woehler *et al.*, 1990).

The Biological Investigations of Marine Antarctic Systems and Stocks (BIOMASS) Programme selected the Prydz Bay region as one of the recommended study areas for the Second BIOMASS Experiment (SIBEX) (Anon, 1982), after the First BIOMASS Experiment (FIBEX) had found high krill abundances in the region (Anon, 1981; Hampton, 1983). Part of the BIOMASS Programme was an attempt to determine whether observations of seabirds at sea could be used as indicators of the distribution and abundance of selected prey stocks, especially krill and cephalopods (BIOMASS Working Party on Bird Ecology, 1985). Ryan & Cooper (1989*a*) found strong correlations between Antarctic krill and seabird abundances in the Prydz Bay region, lending credence to the usefulness of at-sea observations of seabirds as krill indicators (see also Obst, 1985; Heinemann *et al.*, 1989; Plötz *et al.*, 1991).

Several estimates of prey consumption by seabirds in various regions of the Southern Ocean have been based on censuses of breeding birds at colonies. Consumption estimates exist for the waters around South Georgia (Croxall & Prince, 1982, 1987; Croxall *et al.*, 1984); the Scotia and Weddell Seas (Croxall *et al.*, 1985; Joiris, 1991; van Franeker, 1992), near the South Orkney Islands (Hunt, 1985), in the African Sector of the Southern Ocean (Abrams, 1985; Miller *et al.*, 1985), and for four species of penguins at the sub-Antarctic Prince Edward Islands (Adams, 1990). No similar estimate has as yet been made for the Prydz Bay region (Woehler, 1990).

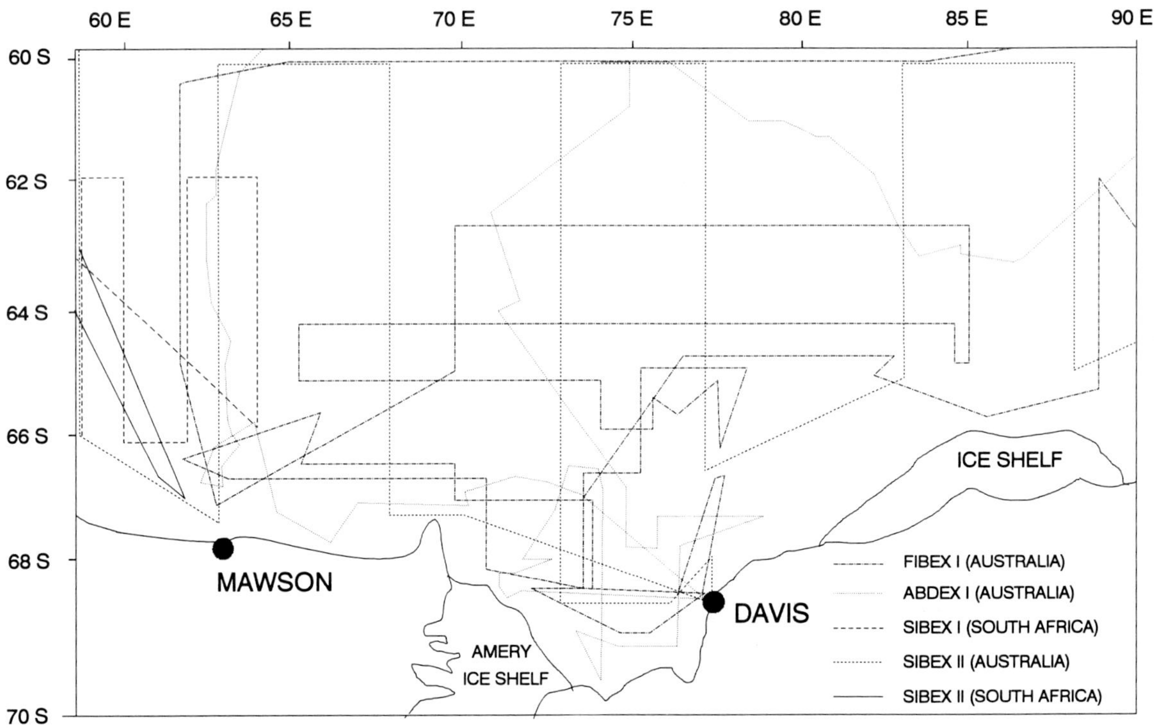

Fig. 1. Cruise tracks of Australian and South African BIOMASS cruises to the Prydz Bay Region. (See Table 1 for sources.)

In this chapter we estimate the consumption of Antarctic krill by seabirds (including penguins) in the Prydz Bay region during summer, based on data from at-sea observations, supported by dietary studies and censuses conducted at breeding colonies, and discuss problems and biases in such an approach.

Study area and data set

The Prydz Bay region is situated off MacRobertson and Princess Elizabeth Lands, Antarctica, and is defined for the purposes of this paper as that part of the Southern Ocean from 58 to 90°E and south of 60°S, an area larger than the confines of Prydz Bay itself and totalling 1.294×10^6 km^2 (Fig. 1). Observations of seabirds were made during three Australian and two South African BIOMASS cruises to the Prydz Bay region during the period from 1981 to 1985 (Williams *et al.*, 1983; Allanson & Boden, 1985; Ikeda *et al.*, 1986; Miller, 1986*b*; Montague, 1988; Ryan & Cooper, 1989*a*; Woehler *et al.*, 1990) (Fig. 1; Table 1). Data coverage was unequal, owing to differences in cruise lengths and ship speeds and the fact that sections of the two South African BIOMASS cruises took place west of 58°E. Most data were collected during the Australian SIBEX II cruise (Table 1).

Methods

Colony census data were taken from Woehler & Johnstone (1991) and Woehler (1993). Summer includes the period 1 October to 30 April, when nearly all species of Antarctic seabirds breed (Watson, 1975). All BIOMASS cruises to the Prydz Bay region took place during this period (Table 1).

Observations of seabirds at sea were made following procedures laid down for the BIOMASS Pro-

Table 1. *Summary of ornithological observations on five Australian (Aus) and South African (RSA) BIOMASS cruises to the Prydz Bay region, Antarctica, 1981–5*

Cruise name (and vessel)	Duration	No. 10 min cards	%	No. birds	%	Transect width (m)	Mean speed (kmh^{-1})	Area surveyed (km^2)	Sources
FIBEX (Aus) (MV *Nella Dan*)	20 Jan–13 Mar 1981	468	27.4	25 233	49.9	300	15.5	362.5	Williams *et al.*, 1983; BIOMASS Working Party on Bird Ecology, 1985
ADBEX I (Aus) (MV *Nella Dan*)[a]	19 Nov–19 Dec 1982	358	20.9	15 543	30.7	1000*a*	11.1	?	Montague, 1988
SIBEX I (RSA) (MV *S.A. Agulhas*)	28 Mar–11 Apr 1984	250	14.6	3 560	7.0	300	22.2	276.9	Allanson & Boden, 1985; Ryan & Cooper, 1989*a*
SIBEX II (Aus) (MV *Nella Dan*)	4–26 Jan 1985	599	35.0	5 993	11.9	300	21.6	658.4	Ikeda *et al.*, 1986
SIBEX II (RSA) (RV *Africana*)	7–8 Mar 1985	36	2.1	220	0.4	300	20.6	36.9	Miller, 1986*b*; Ryan & Cooper, 1989*a*
Totals		1 711		50 549				1334.7[b]	

[a] Antarctic Division BIOMASS Experiment, Phase I; transect width is a 1 km radius from ship.
[b] Represents total for four cruises only.

gramme (BIOMASS Working Party on Bird Ecology, 1982, 1984), except for data collected by Montague (1988). Counts were recorded on standard 10 min cards, normally in one-hour blocks, while cruising at between 2 and 24 km h^{-1}. Ship-followers and birds recorded on 'stern counts' were not included. Observations were made from the bridge, scanning forward in a 90° quadrant with dimensions of either 300 m to one side of the bow or in a one-kilometre radius (Table 1). Seabird density is expressed as a function of the number of birds seen and the area surveyed (transect length × width). Because of differences in the area surveyed, Montague's (1988) data were excluded from density and subsequent calculations.

We used an assimilation efficiency of 75% (Jackson, 1986, 1990) combined with estimates of Daily Energy Expenditure (DEE) for each species from an allometric equation for field metabolic rate. This was calculated from 12 seabird species (Koteja, 1991). We used mass data in Cramp & Simmons (1982) and Croxall (1984). Consumption of Antarctic krill was estimated by multiplying gross consumption by the proportion of krill in each species' diet. This proportion for all nine locally breeding species was obtained from dietary studies conducted during summer within the Prydz Bay region (Montague, 1984; Green, 1986*a–d*; Green & Johnstone, 1988; Puddicombe & Johnstone, 1988; Thomas & Green, 1988; Williams, 1989; Gales *et al.*, 1990; Klages *et al.*, 1990; Norman & Ward, 1990; Arnould & Whitehead, 1991; Whitehead, 1991; P Hodum & MD Whitehead, in litt.; G Robertson, in litt.) The proportions of krill in the diets of non-breeding species have been less accurately estimated from data given by Kerry *et al.* (1983), Schramm (1986), Croxall & Prince (1987) and Zotier (1990), assuming in some cases that the crustacean portion of the diet equates to consumption of krill. Antarctic krill was taken to have an energetic content of 4.35 kJ g^{-1} fresh mass (Croxall *et al.*, 1984).

Results

Breeding populations and consumption estimates

Nearly one and a quarter million seabirds of nine species are estimated to breed annually in the Prydz Bay region. Of this number, 86% are emperor (*Aptenodytes forsteri*) and Adélie (*Pygoscelis adeliae*) penguins (Table 2). In terms of biomass, penguins represent 99% of a total of 7621 t (Table 2). Of the seven non-penguin species, Woehler & Johnstone (1991) consider census data to be underestimates or even 'gross underestimates' for five. However, because of their relatively small body mass (Table 2), more accurate censuses are unlikely

Table 2. *Minimum population sizes and biomass of breeding seabirds in the Prydz Bay region, Antarctica*

Data sources given in text; see Table 5 for scientific names.

Species	No. individuals	%	Individual mass (g)	Biomass (kg)	% Total biomass
emperor penguin	103 196	8.32	32 550	3 359 030	43.70
Adélie penguin	968 736	78.08	4 400	4 262 438	55.46
southern giant petrel	46	<0.01	4 500	207	<0.01
Antarctic fulmar	36 940	2.98	775	28 629	0.37
Antarctic petrel	43 256	3.49	680	29 414	0.38
pintado petrel	3 474	0.28	425	1 476	0.02
snow petrel	4 606	0.37	270	1 244	0.02
Wilson's storm petrel	80 000	6.45	36	2 880	0.04
South Polar skua	436	0.04	1 000	436	0.01
Total	1 240 690			7 685 754	

Table 3. *Antarctic krill (*Euphausia superba*) in the diets of breeding seabirds in the Prydz Bay region, Antarctica*

Species	Proportion of Antarctic krill by mass (%)	Sampling period	No. samples	Source
emperor penguin	0.1	Aug–Oct 1986	45	Gales *et al.*, 1990
Adélie penguin	32.1	Dec 1982–Feb 1983; Oct 1983–Feb 1984; Oct 1984–Jan 1985	706	Green & Johnstone, 1988; Puddicombe & Johnstone, 1988
southern giant petrel	0.0	Feb 1984; Feb 1985	60*	Green, 1986*a*
Antarctic fulmar	36.0	Jan–Feb 1988	35	Arnould & Whitehead, 1991
Antarctic petrel	17.9	Feb 1987; Jan–Feb 1988	79	Klages *et al.*, 1990; Arnould & Whitehead, 1991
pintado petrel	79.1	Jan–Feb 1988	113	Green, 1986*b*; Arnould & Whitehead, 1991
snow petrel	6.0	Feb 1990	35	Whitehead, 1991; P Hodum & MD Whitehead, in litt.
Wilson's storm petrel	25.0	Feb 1990	41	Whitehead, 1991; P Hodum & MD Whitehead, in litt.
South Polar skua	0.0	Summer 1983–84; Dec 1988–Feb 1989	1037*	Green, 1986*c*; Norman & Ward, 1990

[a] Regurgitated casts; marine prey thought under-represented by authors.

Table 4. *Estimated summer consumption of Antarctic krill (*Euphausia superba*) by breeding seabirds in the Prydz Bay region, Antarctica*

Species	Daily Energy Demand (DEE × 1.33) (kJd^{-1})	Estimated individual consumption (gd^{-1})	Population daily consumption ($kg\ d^{-1}$)	Population summer consumption (t)	%
emperor penguin	14 898	3.4	350	74	0.11
Adélie penguin	4 386	323.7	313 580	66 479	97.13
southern giant petrel	4 447	0.0	0	0	0.00
Antarctic fulmar	1 518	125.6	4 640	984	1.44
Antarctic petrel	1 402	57.7	2 496	529	0.77
pintado petrel	1 052	191.3	665	141	0.21
snow petrel	797	11.0	51	12	0.02
Wilson's storm petrel	233	13.4	1 070	227	0.33
South Polar skua	1 774	0.0	0	0	0.00
Totals	30 507	726.1	322 852	68 446	

to alter significantly the overwhelming predominance of penguins in terms of biomass.

The percentages by mass of Antarctic krill in the diets of seabirds breeding in the Prydz Bay region are given in Table 3, showing that in only the pintado or Cape petrel *Daption capense* does krill predominate. The diet of the Adélie penguin has been studied in most detail, including seasonal, geographical and year-to-year variations (briefly summarized by Williams, 1989) The mean percentage by mass of krill in this species' diet in the Prydz Bay region varied from 24.2% to 72.8% in two different studies (Green & Johnstone, 1988; Puddicombe & Johnstone, 1988). There is a tendency for the proportion of Antarctic krill in the diet to increase seasonally, mainly by replacing *Euphausia crystallorophias* or amphipods (Green & Johnstone, 1988; Thomas & Green, 1988; Whitehead *et al.*, 1990).

Estimates of the quantity of Antarctic krill consumed by the breeding seabird populations of the Prydz Bay region during summer are given in Table 4. These figures do not include the elevated demands of moulting and breeding. Antarctic krill forms a very small part of the diet of emperor penguins in the Prydz Bay region (Green, 1986d; Gales *et al.*, 1990; G Robertson in litt.) (Table 3), and therefore nearly all krill (97.1%) is consumed by Adélie penguins (Table 4). The estimated total consumption of Antarctic krill by breeding Adélie penguins is sensitive to the proportion of krill in the species' diet. At the higher figure (72.8% by mass) the Adélie penguin population consumes 686 t daily or 145 400 t during the seven-month summer season, 2.3 times the mean figure (Table 4) and representing 98.6% of krill consumption by all breeding species.

At-sea observations and consumption estimates

The two South African cruises occurred late in the season and did not travel south of 66°S (Ryan & Cooper, 1989*a*) (Fig. 1; Tables 1 and 5). This is in contrast to the three Australian cruises, which took place earlier in the season and reached south of 66°S.

Twenty-three seabird taxa were identified on five cruises in the Prydz Bay region. Of these, nine breed within the region, the rest being primarily non-breeding visitors from sub-Antarctic and southern cool-temperate islands (Watson, 1975; Woehler & Johnstone, 1991) (Table 5). A total of 50 549 birds was counted when all cruises were combined, of which approximately one third were Antarctic petrels, *Thalassoica antarctica*. Species breeding within the Prydz Bay region made up 55.5% (Table 5). The total number of birds seen per cruise varied greatly, from 220 to 25 233, as did the number of birds seen per 10 min card (Table 5). The total avian biomass recorded at sea was 67 t. Emperor and Adélie penguins formed 49% of the total biomass (Table 5).

Details of the diet of non-breeding seabirds occurring in the Prydz Bay region are given in Table 6.

Table 5. *Numbers and density (birds km^{-2}) of seabirds observed during Australian and South African BIOMASS cruises to the Prydz Bay region, Antarctica, 1981–5*

Taxon	FIBEX		ADBEX I	SIBEX I		SIBEX II		SIBEX II		All cruises combined		Overall relative abundance (%)	Overall biomass (kg)	Overall relative biomass (%)
	Nos	Density	Nos	Nos	Density	Nos	Density	Nos	Density	Nos	Density[a]			
emperor penguin *Aptenodytes forsteri*	47	0.13	595	0	–	1	<0.01	1	0.03	644	0.04	1.27	20 962	30.90
Adélie penguin *Pygoscelis adeliae*	480	1.32	2267	0	–	60	0.09	9	0.24	2 816	0.41	5.57	12 390	18.50
wandering albatross *Diomedea exulans*	78	0.22	1	1	<0.01	5	0.01	0	–	85	0.06	0.17	742	1.11
black-browed albatross *D. melanophris*	6	0.02	0	0	–	6	0.01	0	–	12	0.01	0.02	45	0.07
grey-headed albatross *D. chrysostoma*	0	–	0	1	<0.01	0	–	0	–	1	<0.01	<0.01	4	0.01
light-mantled sooty albatross *Phoebetria palpebrata*	368	1.01	13	113	0.41	112	0.17	6	0.16	612	0.45	1.21	1 738	2.60
northern giant petrel *Macronectes halli*	45	0.12	0	0	–	9	0.01	0	–	54	0.04	0.11	243	0.36
southern giant petrel *M. giganteus*	0	–	58	6	0.02	15	0.02	2	0.05	81	0.02	0.16	365	0.55
Antarctic fulmar *Fulmarus glacialoides*	1339	3.69	1964	284	1.03	104	0.16	34	0.92	3 725	1.32	7.37	2 887	4.31
Antarctic petrel *Thalassoica antarctica*	6671	18.40	7264	216	0.78	3101	4.71	32	0.87	17 284	7.51	34.19	11 753	17.55
pintado petrel *Daption capense*	1585	4.37	702	101	0.36	126	0.19	6	0.16	2 520	1.36	4.99	1 071	1.60
snow petrel *Pagodroma nivea*	1506	4.15	2019	92	0.33	71	0.11	57	1.54	3 745	1.29	7.41	1 011	1.51

blue petrel *Halobaena caerulea*	3	0.01	0	199	0.72	159	0.24	0	–	361	0.27	0.71	472	0.70
prions *Pachyptila* spp.	1336	3.69	103	303	1.09	1268	1.93	11	0.30	3 021	2.19	5.98	489	0.73
Kerguelen petrel *Pterodroma brevirostris*	0	–	0	22	0.08	10	0.02	0	–	32	0.02	0.06	10	0.01
white-headed petrel *P. lessonii*	14	0.04	0	10	0.04	0	–	0	–	24	0.02	0.05	18	0.03
mottled petrel *P. inexpectata*	0	–	0	0	–	3	0.01	0	–	3	<0.01	0.01	1	<0.01
white-chinned petrel *Procellaria aequinoctialis*	3072	8.47	187	77	0.28	326	0.50	1	0.03	3 663	2.60	7.25	4 652	6.95
sooty/short-tailed shearwaters *Puffinus griseus/tenuirostris*	7865	21.70	2	2098	7.58	67	0.10	0	–	10 032	7.51	19.85	7 895	11.79
Wilson's storm petrel *Oceanites oceanicus*	347	0.96	324	26	0.09	126	0.19	48	1.30	871	0.41	1.72	31	0.05
black-bellied storm petrel *Fregetta tropica*	18	0.05	0	0	–	0	–	0	–	18	0.01	0.04	1	
South Polar skua *Catharacta Maccormicki*	57	0.16	38	1	<0.01	6	0.01	2	0.05	104	0.05	0.21	104	0.16
Antarctic/Arctic terns *Sterna vittata/paradisaea*	396	1.09	6	10	0.04	418	0.63	11	0.30	841	0.63	1.66	88	0.13
No. of species	19		15	17		20		13		22				
Totals	25 233		15 543	3560	12.86	5993		220	5.96	50 549	26.23		66 972	
Total 10 minute cards	468		358	250	–	–		36	–	1 711	–		–	
Mean no. birds/card	48.4		–	13.2	–	–		6.1	–	–	–		–	

[a] Not including ADBEX I cruise.

Table 6. *Antarctic krill (*Euphausia superba*) in the diets of non-breeding seabirds in the Prydz Bay region, Antarctica*

Taxon	Individual mass (g)[a]	Estimated proportion of Antarctic krill by mass (%)[b]	Daily Energy Demand (DEE × 1.33) (kJd^{-1})	Estimated individual consumption (gd^{-1})	Source of diet data
wandering albatross	8730	0	6662	0.0	Croxall & Prince, 1987
black-browed albatross	3790	38	4004	349.7	Croxall & Prince, 1987
grey-headed albatross	3790	15	4004	138.1	Croxall & Prince, 1987
light-mantled sooty albatross	2840	37	3357	285.5	Croxall & Prince, 1987
northern giant petrel	4500	15	4447	153.3	Croxall & Prince, 1987
blue petrel	200	82	664	125.2	Croxall & Prince, 1987
prions	162	58	583	77.7	Croxall & Prince, 1987
Kerguelen petrel	300	(24)	850	46.9	Schramm, 1986
white-headed petrel	750	(30)	1488	102.6	Zotier, 1990
mottled petrel	323	(1)	889	2.0	Ainley *et al.*, 1984
white-chinned petrel	1270	27	2063	128.1	Croxall & Prince, 1987
sooty/short-tailed shearwaters	787	(93)	1532	327.5	Kerry *et al.*, 1983
black-bellied storm petrel	56	(40)	305	28.1	Croxall & Prince, 1987
Antarctic/Arctic terns	105	(15)	447	15.4	Croxall & Prince, 1987

[a] See text for sources of masses.
[b] Figures in parentheses are estimated values from qualitative studies or from localities to the north of the Antarctic Polar Front.

Shearwaters (*Puffinus* sp.) are estimated to consume relatively large amounts of krill. Based on mean at-sea densities from four BIOMASS cruises (Table 5), species' population estimates for the whole Prydz Bay region are given in Table 7. Antarctic petrels formed 28.6% of the estimated total of 33.9×10^6 birds; penguins formed only 1.7%. Total Daily Energy Demand was estimated at 38 423 kJ km^{-2} d^{-1}. Estimated daily krill consumption was 4.3 kg km^{-2} for all birds combined, with Antarctic petrels and shearwaters together consuming 66.7%. Based on extrapolations of at-sea observations, the population of seabirds of the Prydz Bay region consumed 5.4 kt of krill per day, equivalent to 1.141 Mt in a seven-month summer season (Table 7). Of this total, Antarctic petrels and shearwaters consumed an estimated 794 kt. Penguins consumed only an estimated 37 kt of Antarctic krill during summer.

In a comparison of data from the four FIBEX and SIBEX cruises (Table 8), the estimated seabird population of the Prydz Bay region, and therefore its consumption of Antarctic krill, varies widely from 7.7×10^6 to 90.3×10^6 birds and from 0.106 to 3.199 Mt of krill in a summer, respectively. The highest estimate comes from the Australian FIBEX cruise.

Discussion

Seabird consumption of Antarctic krill in relation to standing stock in the Prydz Bay region

Published estimates of the standing stock of Antarctic krill for the Prydz Bay region range from 1.6 Mt during the Australian FIBEX cruise to 3.7 Mt during the Australian SIBEX II cruise (Higginbottom *et al.*, 1988). Estimates of krill consumed in a seven-month summer by seabirds, based on counts made during these two cruises (Table 8), represent 199.9% and 5.2% of these standing stock figures, respectively. Higginbottom *et al.* (1988) consider the SIBEX II figure to be the most reliable of the three given (an estimate for the Australian ADBEX II cruise of 3.5 Mt is the third, similar to that for SIBEX II). Using the figure of 3.7 Mt for the

Table 7. *Estimated at-sea populations of seabirds and their consumption of Antarctic krill in the Prydz Bay region*
Data based on Australian FIBEX and SIBEX II and South African SIBEX I and II Cruises.

Taxon	Population (thousands)	%	Daily Energy Demand (DEE × 1.33) (kJ km^{-2} d^{-1})	Consumption of krill (g km^{-2} d^{-1})	Population daily consumption (t d^{-1})	Population summer consumption (kt per 212 d)	%
emperor penguin	47	0.14	547	0.1	0.16	0.03	<0.01
Adélie penguin	532	1.57	1804	133.1	172.19	36.50	3.20
wandering albatross	81	0.24	419	0.0	0.00	0.00	0.00
grey-headed albatross	1	<0.01	3	0.1	0.13	0.03	<0.01
black-browed albatross	12	0.04	36	3.1	4.07	0.86	0.08
light-mantled sooty albatross	580	1.71	1 507	128.1	165.75	35.14	3.08
northern giant petrel	52	0.15	180	6.2	8.02	1.70	0.15
southern giant petrel	22	0.06	77	0.0	0.00	0.00	0.00
Antarctic fulmar	1706	5.03	2 003	165.8	214.39	45.45	3.98
Antarctic petrel	9708	28.62	10 525	433.1	560.18	118.76	10.41
pintado petrel	1762	5.19	1 433	260.6	337.01	71.45	6.26
snow petrel	1673	4.93	1 031	14.2	18.39	3.90	0.34
blue petrel	350	1.03	180	210.4	43.79	9.28	0.81
prions	2828	8.34	1 275	169.9	219.81	46.60	4.08
Kerguelen petrel	31	0.09	20	1.1	1.45	0.31	0.03
white-headed petrel	23	0.07	28	1.8	2.39	0.51	0.04
mottled petrel	3	0.01	2	<0.1	0.01	<0.01	<0.01
white-chinned petrel	3368	9.93	5 373	333.5	431.32	91.44	8.01
sooty/short-tailed shearwaters	9720	28.66	11 513	2461.3	3 183.50	674.90	59.14
Wilson's storm petrel	530	1.56	95	5.5	7.10	1.51	0.13
black-bellied storm petrel	17	0.05	4	0.4	0.49	0.10	0.01
South Polar skua	64	0.19	88	0.0	0.00	0.00	0.00
Antarctic/Arctic terns	809	2.39	280	9.6	12.47	2.64	0.23
Totals	33 919					1 141.12	

Table 8. *Estimates of the Prydz Bay region seabird population and its consumption of Antarctic krill from four BIOMASS cruises*

Cruise	Area surveyed (km^2)	No. birds counted	Mean no. birds km^{-2}	Population (millions)	Consumption of krill (kt)	
					daily	summer
FIBEX (Aus)	362.5	25 233	69.61	90.03	15.09	3199
SIBEX I (RSA)	276.9	3 560	12.86	16.6	3.97	840
SIBEX II (Aus)	658.4	5 993	9.10	11.8	0.91	193
SIBEX II (RSA)	36.9	220	5.96	7.7	0.50	106
Totals/Means	1334.7	35 006	26.23	33.92	5.38	1141

standing stock of krill for all four cruises for which krill consumed by birds have been calculated (mean 1.1 Mt) gives a mean consumption of 29.8% (range 2.9–86.5%). However, the lowest estimate is based on the South African SIBEX II cruise, which counted only 220 birds in an area of 36.9 km^2, so more credence should be placed on figures for the other three cruises.

Comparison with other Southern Ocean at-sea studies

Several studies have estimated densities, biomass or consumption of seabirds from at-sea counts in Antarctic waters south of 60°S, including from BIOMASS cruises, summarized by the BIOMASS Working Party on Bird Ecology (1985).

Abrams (1985) reported energy and food requirements of aerial seabirds (i.e. excluding penguins) in the African Sector of the Southern Ocean. His densities south of 60°S for aerial birds ranged from 3.3 to 4.5 birds km^{-2}, lower than the Prydz Bay region figures reported here. He obtained densities for 'planktivores' south of 60°S of 0.1–2.6 birds km^{-2}. Abrams (1985) assumed that planktivores fed principally on Antarctic krill, but his category excluded a number of species which are known to feed on krill to a greater or lesser extent, so no detailed comparisons with our data from the Prydz Bay region are possible.

Ainley (1985) calculated density and biomass figures for birds in the Ross Sea, obtaining an average density of 16.3 birds km^{-2}. Only nine species were recorded by him, compared with 23 in the Prydz Bay region. Antarctic petrels predominated in terms of numbers in the Ross Sea, but Adélie and emperor penguins predominated in terms of biomass (82.1%) to a greater extent than in the present study (49.4%). This may be ascribed, at least partially, to the fact that Ainley's survey spent more time among sea ice where roosting penguins were far more conspicuous than when swimming at the sea surface.

Hunt (1985) obtained densities of 62.3 birds km^{-2} in the Scotia Sea near the South Orkney Islands in summer, whereas Hunt *et al.* (1990) reported much lower densities of 5.7 birds km^{-2} in the southern Drake Passage and in the Bransfield Strait, off the Antarctic Peninsula. The difference is probably due to the nearby presence of large penguin colonies in the former study. It is apparent that closeness to penguin colonies, and the presence of sea ice (e.g. Ainely & Jacobs, 1981; Joiris, 1991; van Franeker, 1992), can markedly influence at-sea densities of seabirds, and therefore calculations of consumption of Antarctic krill.

Miller *et al.* (1985) estimated the amount of Antarctic krill consumed by seabirds (excluding penguins) in the South African FIBEX region (62–69°S, 15–30°E) to be 45 kt in a 150-d period in an area of 620 000 km^2, which was 49.4% of the estimated standing stock of 91 kt. Converting to a seven-month summer gives a consumption equivalent to 69.9% of standing stock, higher than the mean figure of 29.8% for the Prydz Bay region. Lack of inclusion of penguins and an inadequate knowledge of the proportions of krill in the diets of the species studied during the South African FIBEX cruise reduce the value of this comparison.

In a Weddell Sea study, Joiris (1991) obtained densities of 25.81 birds km^{-2}, with highest densities in areas with ice present. Van Franeker (1992) obtained mean densities of 11.4 birds km^{-2} in the Weddell and Scotia Seas and estimated mean Daily Energy Demand to be 21 900 kJ day^{-1} km^{-2}. He recorded highest densities and energy demands in areas with partial ice cover.

Problems and biases

As previously pointed out (e.g. Everson, 1977; Mougin & Prévost, 1980; Prévost, 1981; Croxall, 1984), penguins form the largest proportion of Antarctic avian biomass and therefore consume the greatest proportion of prey. This is clearly also the case for breeding seabirds in the Prydz Bay region. Thus, although non-penguin censuses of breeding seabirds in the Prydz Bay region are considered to be underestimates (Woehler & Johnstone, 1991), more complete or accurate surveys would probably not alter the situation of Adélie and emperor penguins dominating the biomass, or the Adélie penguin consuming by far the largest proportion of Antarctic krill in the region. It is therefore noticeable that when at-sea records of seabirds in the Prydz Bay region are considered, penguins do not predominate in either biomass or krill consumption. Penguins are difficult to observe at sea (personal observations), so some analyses of at-sea distribution and consumption by seabirds in the Southern Ocean have ignored penguins, restricting themselves to aerial birds only (e.g. Abrams, 1985; Miller *et al.*, 1985). Given that penguins dominate the breeding avian biomass, such restrictions are likely to cause serious errors in estimating avian consumption at sea in the Southern Ocean. However, including penguins, as here, will also lead to biases because of inevitable undercounting. At present this bias cannot easily be resolved.

Other inaccuracies inherent in estimating the consumption of Antarctic krill by seabirds in the Prydz Bay region are related to assumptions made about the proportion of krill in their diets. Very few quantitative studies of the diets of Southern Ocean seabirds while at sea have been made (Griffiths (1983) and Ainley *et al.* (1984) are exceptions). In fact, for only one species of non-breeding seabird, the short-tailed shearwater *Puffinus tenuirostris*, were dietary data available from the Prydz Bay region itself (Kerry *et al.*, 1983). Because it was not possible to distinguish this species from the similar and closely related sooty shearwater *P. griseus* in the data sets, and because the dietary data were not properly quantified and were further based on a small sample, little trust can be placed in the figure of 93% used for Antarctic krill in the taxon's diet. In addition, it has been suggested that sooty shearwaters may be migrating through the region rapidly and thus may not be undertaking much feeding activity in the area (Phillips, 1963; Ryan & Cooper, 1989*a*). Although shearwaters were estimated to represent only 11.8% of the biomass at sea in the Prydz Bay region, they were estimated to consume 59.1% of the total consumption of Antarctic krill by birds. If this taxon is discounted (see above), then the total consumption of krill in the Prydz Bay region based on data from four BIOMASS cruises is more than halved to 47 kt during a seven-month summer.

Large differences in densities, biomass and krill consumption estimates exist among the four FIBEX and SIBEX cruises. At least some of these differences may be ascribed to year-to-year and seasonal variations, and probably to differences in the presence of sea-ice (Ryan & Cooper 1989*a*) and distances from breeding colonies. These differences may also be affected by differences in ship speeds. The MV *Nella Dan* travelled at a mean speed of 15.5 km h^{-1}, and as slowly as 2 km h^{-1} at times during the Australian FIBEX cruise, whereas during the other three cruises for which population estimates have been made vessels travelled at mean speeds of 21–22 km h^{-1}. At slow speeds birds attracted to the ship (Griffiths, 1981, 1982) are better able to 'keep up' and therefore counts may be artificially inflated (Tasker *et al.*, 1984; personal observations). It was noticeable during the South African SIBEX I cruise that on northward legs moving away from out of sea-ice there were more species that are normally associated with sea-ice following the vessel (such as Antarctic and snow petrels) (Ainley & Jacobs, 1981; Griffiths, 1983) than on southward legs (JC, personal observations). The inference that the vessel was attracting birds from distances greater than transect widths is therefore strong with the inevitable tendency being to overestimate the true densities of such species. This bias is likely to be exacerbated if the vessel is moving relatively slowly. In this regard it is noticeable that densities of most fulmarine petrel species and the white-chinned petrel (*Procellaria aequinoctialis*) were highest on the relatively slow-travelling Australian FIBEX cruise. These species are known to be attracted to ships (Griffiths, 1982; Ryan & Cooper, 1989*b*). Antarctic petrels were estimated to consume 10.4% of the total krill consumed by birds in the Prydz Bay region, so an overestimate of their true density would have an effect on estimates of total consumption.

The extrapolated at-sea population estimates for the Prydz Bay region are very high for a number of species, given what is known of their local breeding populations. It is therefore considered that the at-sea estimates (except for penguins) are too large, and that the attraction effect of the vessel could be an important biasing factor for several species.

Lastly, the data collected on the five BIOMASS cruises to the Prydz Bay region were conducted by a number of different persons. Differences between observers can affect numbers of birds counted at sea (Ryan & Cooper, 1989*b*) and therefore any subsequent calculations

Conclusions

We believe that penguins were undercounted during the cruises, whereas some aerial species were probably overestimated, especially those attracted to vessels. The figures for total consumption of Antarctic krill by seabirds during summer in the Prydz Bay region (ranging from 1.1 to 3.2 Mt on four different cruises) must therefore be regarded as preliminary estimates only.

Future research needs to solve the two problems of knowing what the birds are actually eating in the area during the time censuses are undertaken (e.g. Ainley *et al.*, 1984) and obtaining more accurate estimates of at-sea densities of all species of birds, including penguins and species attracted to ships. The latter will prove particularly hard to accomplish.

Acknowledgements

We acknowledge financial and logistic support received from the Australian Antarctic Division, the BIOMASS Executive and the Scientific Committee on Antarctic Research through SZ El-Sayed, and the South African Department of Environment Affairs through the auspices of the South African Committee for Antarctic Research. We thank our colleagues for help with at-sea counts and the captains and crews of the RV *Africana*, MV *Nella Dan* and MV *S.A, Agulhas*. P Hodum, GL Hunt Jr, G Robertson, PG Ryan, JA van Franeker, DJ Watts, MD Whitehead and anonymous referees helped variously with unpublished data, in-press manuscripts, data analyses and comments on drafts. We thank them all.

References

Abrams R W (1985) Energy and food requirements of pelagic aerial seabirds in different regions of the African Sector of the Southern Ocean. In: Siegfried W R, Condy P R, Laws R M (eds) *Antarctic Nutrient Cycles and Food Webs*. Springer-Verlag, Berlin, pp 466–72.

Adams N J (1990) Feeding biology and energetics of king *Aptenodytes patagonicus* and gentoo *Pygoscelis papua* penguins at sub-Antarctic Marion Island. Unpublished PhD Thesis, University of Cape Town.

Ainley D G (1985) Biomass of birds and mammals in the Ross Sea. In: Siefried W R, Condy P R, Laws R M (eds) *Antarctic Nutrient Cycles and Food Webs.* Springer-Verlag, Berlin, pp 498–515.

Ainley D G, Jacobs SS (1981) Sea-bird affinities for ocean and ice boundaries in the Antarctic. *Deep-Sea Res* **28**A: 1173–85.

Ainley D G, O'Connor E F, Boekelheide R J (1984) The marine ecology of birds in the Ross Sea, Antarctica. *Orn Monogr* **32**: 1–97.

Allanson B R, Boden B P (1985) Preliminary results of the South African SIBEX I cruise to the Prydz Bay region, Antarctica, 15 March–3 May 1984: overall resumé. *S Afr J Antarct Res* **15**: 3–5.

Anon (1981) Post-FIBEX Data Interpretation Workshop, Hamburg, F.R.G., 21 September–9 October 1981. *BIOMASS Rep Ser* **20**: 1–33.

Anon (1982) Technical Group on Programme Implementation and Coordination–Third Meeting. *BIOMASS Rep Ser* **29**: 1–28.

Arnould J P Y, Whitehead M D (1991) The diet of Antarctic petrels, cape petrels and southern fulmars rearing chicks in Prydz Bay. *Antarct Sci* **3**: 19–27.

Best P B, Butterworth D S (1980) Report of the Southern Hemisphere minke whale assessment cruise 1978/79. *Rep Int Whal Commn* **30**: 257–83.

BIOMASS Working Party on Bird Ecology (1982) Recording observations of birds at sea. *BIOMASS Handbk* **18**: 1–14.

BIOMASS Working Party on Bird Ecology (1984) Recording observations of birds at sea (rev edn). *BIOMASS Handbk* **18**: 1–20.

BIOMASS Working Party on Bird Ecology (1985) Report on the FIBEX Seabird Data Interpretation Workshop, Cape Town, South Africa, 10–18 April 1985. *BIOMASS Rep Ser* **44**: 1–40.

Cramp S, Simmons K E L (eds) (1982) *Birds of the Western Palearctic*, vol 3. Oxford University Press.

Croxall J P (1984) Seabirds. In: Laws R M (ed) *Antarctic*

Ecology, vol 2. Academic Press, London, pp 533–618.

Croxall J P, Prince P A (1982) A preliminary assessment of the impact of seabirds on marine resources at South Georgia. *Com Nat Franc Recherch Antarct* **51**: 501–9.

Croxall J P, Prince P A (1987) Seabirds as predators on marine resources, especially krill, at South Georgia. In: Croxall J P (ed) *Seabirds. Feeding Ecology and Role in Marine Ecosystems.* Cambridge University Press, pp 347–68.

Croxall J P, Prince P A, Ricketts C (1985) Relationships between prey life-cycles and the extent, nature and timing of seal and seabird predation in the Scotia Sea. In: Siegfried W R, Condy P R, Laws R M (eds) *Antarctic Nutrient Cycles and Food Webs*. Springer-Verlag, Berlin, pp 515–33.

Croxall J P, Ricketts C, Prince P A (1984) Impact of seabirds on marine resources, especially krill, of South Georgia waters. In: Whittow G C, Rahn H (eds) *Seabird Energetics*. Plenum, New York, pp 285–317.

Everson I (1977) *The Living Resources of the Southern Ocean.* Southern Oceans Survey Programme GLO/SO.77.1. FAO, UNDP, Rome.

Franeker J A van (1992) Top predators as indicators for ecosystem events in the confluence zone and marginal ice zone of the Weddell and Scotia seas, Antarctica, November 1988 to January 1989 (EPOS Leg 2). *Polar Biol* **12**: 93–102.

Gales N J, Klages, N T W, Williams R, Woehler E J (1990) The diet of the emperor penguin, *Aptenodytes forsteri*, in Amanda Bay, Princess Elizabeth Land, Antarctica. *Antarct Sci* **2**: 23–8.

Green K (1986*a*) Observations on the food of the southern giant petrel near Davis, Antarctica. *Notornis* **33**: 90–4.

Green K (1986*b*) Food of the cape pigeon (*Daption capense*) from Princess Elizabeth Land, East Antarctica. *Notornis* **33**: 151–4.

Green K (1986*c*) Observations on the food of the South Polar skua, *Catharacta maccormicki* near Davis, Antarctica. *Polar Biol* **6**: 185–6.

Green K (1986*d*) Food of the emperor penguin *Aptenodytes forsteri* on the Antarctic fast ice edge in late winter and early spring. *Polar Biol* **6**: 187–8.

Green K, Johnstone G W (1988) Changes in the diet of Adélie penguins breeding in East Antarctica. *Aust Wildl Res* **15**: 103–10.

Griffiths A M (1981) Biases in censuses of pelagic seabirds at sea in the Southern Ocean. In: Cooper J (ed) *Proceedings of the Symposium on Birds of the Sea and Shore, 1979.* African Seabird Group, Cape Town, pp 189–96.

Griffiths A M (1982) Reactions of some seabirds to a ship in the Southern Ocean. *Ostrich* **53**: 228–35.

Griffiths A M (1983) Factors affecting the distribution of the snow petrel (*Pagodroma nivea*) and the Antarctic petrel (*Thalassoica antarctica*). *Ardea* **71**: 145–50.

Hampton I (1983) Preliminary report on the FIBEX acoustic work to estimate the abundance of *Euphausia superba. Proc Natl Inst Polar Res Spec Issue* **27**: 165–75.

Heinemann D, Hunt G, Everson I (1989) Relationships between the distributions of marine avian predators and their prey, *Euphausia superba*, in Bransfield Strait and southern Drake Passage, Antarctica. *Mar Ecol Prog Ser* **58**: 3–16.

Higginbottom I R, Kerry K R, Wayte S E (1988) Hydroacoustic surveys of the distribution and abundance of krill: Prydz Bay region – FIBEX, ADBEX II and SIBEX II, MV *Nella Dan. ANARE Res Notes* **62**: 1–46.

Hosie G W, Ikeda T, Stolp M (1988) Distribution, abundance and population structure of the Antarctic krill (*Euphausia superba* Dana) in the Prydz Bay region, Antarctica. *Polar Biol* **8**: 213–24.

Hunt G L Jr (1985) A preliminary comparison of marine bird biomass and food consumption between the southeastern Bering Sea and parts of the Southern Ocean. In: Siegfried W R, Condy P R, Laws R M (eds) *Antarctic Nutrient Cycles and Food Webs*. Springer-Verlag, Berlin, pp 487–97.

Hunt G L Jr, Heinemann D, Veit R R, Heywood R N B, Everson I (1990) The distribution, abundance and community structure of marine birds in southern Drake Passage and Bransfield Strait, Antarctica. *Continental Shelf Res* **10**: 243–57.

Ikeda T, Hosie G, Stolp M (1986) SIBEX II cruise krill/zooplankton data. *ANARE Res Notes* **32**: 1–70.

Jackson S (1986) Assimilation efficiencies of whitechinned petrels (*Procellaria aequinoctialis*) fed different prey. *Comp Biochem Physiol* **35**A: 301–3.

Jackson S (1990) Seabird digestive physiology in relation to foraging ecology. PhD thesis, University of Cape Town.

Johnstone G W, Kerry K R (1976) Ornithological observations in the Australian Sector of the Southern Ocean. *Proc 16th Int Orn Congr,* pp 725–38.

Joiris C R (1991) Spring distribution and ecological role of seabirds and marine mammals in the Weddell Sea, Antarctica. *Polar Biol* **11**: 415–24.

Kerry K R, Horne R S C, Dorward D F (1983) Records of the short-tailed shearwater *Puffinus tenuirostris* in Antarctic waters. *Emu* **83**: 35–7.

Klages N, Gales R, Pemberton D (1990) The stomach contents of Antarctic petrels *Thalassoica antarctica* feeding young chicks at Scullin Monolith, Mawson Coast, Antarctica. *Polar Biol* **10**: 545–7.

Koteja P (1991) On the relation between basal and field metabolic rates in birds and mammals. *Funct Ecol* **5**: 56–64.

Marr J W S (1962) The natural history and geography of the Antarctic krill (*Euphausia superba* Dana). *Discovery Rep* **32**: 33–464.

Miller D G M (1985) The South African SIBEX I cruise to the Prydz Bay region, 1984: IX. Krill (*Euphausia superba* Dana). *S Afr J Antarct Res* **15**: 33–41.

Miller D G M (1986*a*) Results from biological investigations of krill (*Euphausia superba*) in the southern Indian Ocean during SIBEX I. *Mem Natl Inst Polar Res Spec Issue* **40**: 117–39.

Miller D G M (1986*b*) SIBEX-II: report of the South African study in the sector (48–64°E) of the Southern Ocean. *S Afr Nat Sci Prog Rep* **132**: 1–47.

Miller D G M, Hampton I (1989) Biology and ecology of the Antarctic krill (*Euphausia superba* Dana): a review. *BIOMASS Sci Ser* **9**: 1–166.

Miller D G M, Hampton I, Henry J, Abrams RW, Cooper J (1985) The relationship between krill food requirements and phytoplankton production in a sector of the southern Indian Ocean. In: Siegfried W R, Condy P R, Laws R M (eds) *Antarctic Nutrient Cycles and Food Webs.* Springer-Verlag, Berlin, pp 362–71.

Montague T L (1984) The food of Antarctic petrels (*Thalassoica antarctica*). *Emu* **84**: 244–5.

Montague T L (1988) Birds of Prydz Bay, Antarctica: distribution and abundance. *Hydrobiologia* **165**: 227–37.

Mougin J-L, Prévost J (1980) Evolution annuelle des effectifs et des biomasses des oiseaux antarctiques. *Rev Ecol (Terre Vie)* **34**: 101–33.

Norman F J, Ward S J (1990) Foods of the South Polar Skua at Hop Island, Rauer Group, East Antarctica. *Polar Biol* **10**: 489–93.

Obst B S (1985) Densities of antarctic seabirds at sea and the presence of krill *Euphausia superba. Auk* **102**: 540–9.

Ohyama Y, Naito Y (1982) Visual observations of the Antarctic sea birds during the FIBEX cruise of the Kaiyo Maru. *Mem Natl Inst Polar Res Spec Issue* **23**: 87–93.

Phillips J H (1963) The pelagic distribution of the sooty shearwater, *Procellaria grisea. Ibis* **105**: 340–53.

Plötz J, Weidel H, Bersch M (1991) Winter aggregations of marine mammals and birds in the north-eastern Weddell Sea pack ice. *Polar Biol* **11**: 305–9.

Prévost J (1981) Population, biomass and energy requirements of Antarctic birds. *BIOMASS Sci Ser* **2**: 125–37.

Puddicombe R A, Johnstone G W (1988) The breeding season diet of Adélie penguins at the Vestfold Hills, East Antarctica. *Hydrobiologia* **165**: 239–53.

Ryan P G, Cooper J (1989*a*) The distribution and abundance of aerial seabirds in relation to Antarctic Krill in the Prydz Bay region, Antarctica, during late summer. *Polar Biol* **10**: 199–209.

Ryan P G, Cooper J (1989*b*) Observer precision and bird conspicuousness during counts of birds at sea. *S Afr J Mar Sci* **8**: 271–6.

Schramm M (1986) The diet of chicks of greatwinged, Kerguelen and softplumaged petrels at the Prince Edward Islands. *Ostrich* **57**: 9–15.

Smith N R, Dong Z, Kerry K R, Wright S (1984) Water masses and circulation in the region of Prydz Bay, Antarctica. *Deep-Sea Res* **31**: 114–16.

Tasker M L, Hope Jones P, Blake B F, Dixon T J (1984) Counting seabirds at sea from ships: a review of methods employed and a suggestion for a standardized approach. *Auk* **101**: 567–77.

Thomas P G, Green K (1988) Distribution of *Euphausia crystallorophias* within Prydz Bay and its importance to the inshore marine ecosystem. *Polar Biol* **8**: 327–31.

Watson G E (1975) *Birds of the Antarctic and sub-Antarctic.* American Geophysical Union, Washington, DC.

Whitehead M D (1991) Food resource utilization by seabirds breeding in Prydz Bay, Antarctica. *Acta XX Int Orn Congr*, pp 1384–92.

Whitehead M D, Johnstone G W, Burton H R (1990) Annual fluctuations in productivity and breeding success of Adélie penguins and fulmarine petrels in Prydz Bay, East Antarctica. In: Kerry K R, Hempel G (eds) *Antarctic Ecosystems, Ecological Change and Conservation.* Springer-Verlag, Berlin, pp 214–23.

Williams R (1985) The potential impact of a krill fishery upon pelagic fish in the Prydz Bay Region. *Polar Biol* **5**: 1–4.

Williams R (1989) Australian research on Antarctic bird and seal diets. *Sci Com Conserv Antarct Mar Living Resources 1988 Select Sci Pap* Part II, VII/BG. **32**: 231–47.

Williams R, Kirkwood J M, O'Sullivan D (1983) FIBEX cruise zooplankton data. *ANARE Res Notes* **7**: 1–163.

Woehler E J (1990) The distribution of seabird biomass in the Australian Antarctic Territory: implications for conservation. *Environ Conserv* **17**: 256–61.

Woehler E J (1993) *The Distribution and Abundance of Antarctic and Subantarctic Penguins.* Scientific Committee on Antarctic Research, Cambridge.

Woehler E J, Hodges C L, Watts D J (1990) An atlas of the pelagic distribution and abundance of seabirds in the southern Indian Ocean, 1981 to 1990. *ANARE Res Notes* **77**: 1–406.

Woehler E J, Johnstone G W (1991) Status and conservation of the seabirds of the Australian Antarctic Terrritory. *Int Council Bird Preserv Tech Publ* **11**: 279–308.

Zotier R (1990) Breeding ecology of the white-headed petrel *Pterodroma lessoni* in the Kerguelen Islands. *Ibis* **132**: 525–34.

Discussant's Report: Antarctic marine stocks – fish and birds

J.-C. HUREAU

FOUR CHAPTERS are presented in this section. Two report on fish (Kellermann & North) and birds (Hunt *et al.*) from the Antarctic Peninsula and Southwest Atlantic Region of the Southern Ocean; the other two report on fish (Williams & Duhamel) and birds (Cooper & Woehler) from the Prydz Bay Region in the Indian Sector of the Southern Ocean. These authors have given comprehensive overviews of the top predators from the two main regions where the majority of the BIOMASS activities took place (during FIBEX, SIBEX I and SIBEX II cruises).

In their chapter, Williams & Duhamel point out that data on fish were acquired in most cases as a by-product of the assessment of krill population, and were confined to the epi- and mesopelagic zones. As a result, most of the data on fish obtained during BIOMASS concerned ichthyoplankton, i.e. eggs and larval stages. Nevertheless, the BIOMASS Programme has made a substantial contribution in coordinating research on the ecology of Antarctic fish in the two areas of the Southern Ocean mentioned above.

The two chapters on fish are somewhat different in their approach. Williams & Duhamel concentrated their effort on describing the results obtained in the Prydz Bay region during FIBEX and SIBEX; they explained that most of the results obtained dealt with pelagic fish, summarizing the numerous new data acquired during BIOMASS in Antarctic ichthyoplanktology and pelagic ichthyology: geographical distribution, length frequency distributions, spatial and depth distribution, and abundance in relation to the water masses' circulation. During SIBEX II, a few studies were made on benthic fish. Most of the investigations on adult benthic fish were carried out as part of CCAMLR activities. The main results of the fish studies in the Prydz Bay region during BIOMASS focused on the trophic relations in the food web and on the importance of the mesopelagic family Myctophidae. It has been shown that most juvenile stages of fish have more than 90% of their diet composed of euphausiids, and that many fish species are particularly important in the diet of many birds and seals. Fish, at least in the Indian Sector of the Southern Ocean, appear to occupy an important intermediate role in the food web. Williams & Duhamel point out that 'This is in contrast to the South Atlantic sector, where krill is clearly the dominant food source for the top predators and one of the outcomes of BIOMASS is that a number of widespread sites need to be studied to gain a representative picture of the Antarctic marine ecosystem'. One has to agree with the authors in stating that 'Perhaps the important legacy of BIOMASS in relation to fish, especially to CCAMLR fish study programmes, is in that the trophic relations of fish, and fish incidence in krill trawls, are fundamental to the ecosystem approach to management'.

Kellermann & North, on the other hand, underscored the role of the BIOMASS Programme in promoting international research and cooperation in the study of Antarctic ichthyology. They emphasized the role of the various meetings and workshops in stimulating and furthering research in the various fields of fish ecology, distribution, reproduction, early life history stages, age and growth determination, identification, diets, physiology and predator–prey relationships. They reported that 51% of the papers published on Antarctic fish (1430) from 1833 to 1989 were published between 1977 and 1989, i.e. during the duration of the BIOMASS Programme. About one tenth of the papers on fish published during that period were directly related to the BIOMASS Programme. These included the publication of key works on fish identification and methodology which, in turn, promoted further studies.

In their chapter, Cooper & Woehler describe the consumption of Antarctic krill by seabirds during summer in the Prydz Bay region. Based on population and dietary data collected during the BIOMASS Programme, the quantity of Antarctic krill consumed by the breeding seabird populations of the Prydz Bay region during summer was estimated to be 68 446 t for nine species, including penguins. For the non-breeding populations occurring in the Prydz Bay region, the consumption was estimated at 2358 Mt during summer for 22 species. The authors contend that some species (penguins) were undercounted during FIBEX and SIBEX, whereas others were probably overestimated (species attracted to vessels) owing to biases inherent in the number and variety of observers involved in this research. It was also pointed out that these estimates should be considered only as preliminary and that more accurate observations need to be carried out in the future. Nevertheless, this was the first time that such estimates had been made in the Prydz Bay region.

The chapter by Hunt *et al.* describes the pelagic marine observations of birds in the southern Drake Passage and Bransfield Strait made during the BIOMASS Programme. The chapter does not include the results of the national programmes which were initiated as a result of the BIOMASS Programme. FIBEX and SIBEX cruises provided new information on species distribution, abundance and association of bird species with both physical and biological components of their environment. Marine bird distributions were correlated with the hydrography of the water masses, and their abundance was a function of Antarctic krill concentration, at least on large spatial scales. Comparisons of the results between cruises were hampered by difference in methodology, and by seasonal and interannual variations in distribution and abundance. Nevertheless, the overall similarity in the distributional patterns of birds suggests that multiyear, multinational studies of the biology of marine birds are useful, provided that seabird data are collected by rigorous methods.

Undoubtedly the BIOMASS Programme has succeeded in establishing strong ties between scientists within their own disciplines. Nevertheless, because of the heavy emphasis of BIOMASS on krill studies and pelagic ecosystems, there are still huge gaps in our knowledge of both the ichthyo- and avifauna. For instance, regarding the fish, except for the myctophids, very little is known about the abundance and ecology of other mesopelagic groups, e.g. the salmoniformes (Bathylagidae, Alepocephalidae, Searsliidae) and the stomiiformes (Gonostomatidae, Stomiidae), which probably play important ecological roles in the midwater environment. Much less is known about the deep-sea fishes. Further, the deep waters of the Antarctic Ocean have been little explored, and our knowledge of their faunal composition is meagre. Yet a vast field of research is there to be explored: for instance, what is the role of the deep-sea fishes within that specific biotope; what is their impact on the turnover of organic matter that originates in the surface layers of the Southern ocean and what is their adaptation strategy to that biotope? Additionally, future research using modern tools and state-of-the-art techniques in molecular biology and cytogenetics might contribute to a better identification and characterization of these organisms and to unravelling their evolutionary relationships.

More work must also be done on the biology of fish and bird fauna within the Antarctic sea-ice zone. The EASIZ programme has identified several key questions where fish and birds are important components of that zone. For instance, what is the composition (species/major taxa/trophic level and role) of the sea-ice assemblages? At what life history stages and times of year do fish and birds inhabit or utilize sea-ice? What proportion, in terms of number, biomass and age distribution, of the populations of Antarctic fish and birds are found in the sea-ice zone?

Despite the fact that intensive fieldwork has been carried out for more than thirty years at some shore-stations (as part of national programmes), more emphasis needs to be placed on the Antarctic coastal and shelf zone systems. This is mainly due to the fact that the Antarctic coastal and near-shore ecosystem is considerably different from that dominated by the oceanic seasonal pack-ice: the presence of anchored ice, the increased amount of frazil ice and the presence of icebergs that modify and/or disturb the surrounding waters. The presence of shore-stations adjacent to the coast provides the ideal locale for experimental research in the Antarctic coastal and shelf zone.

The BIOMASS Programme has certainly helped to enhance our knowledge of fish and bird distribution and ecology, and has contributed to a better understanding of the trophic relations between them and other faunal groups. By fostering and coordinating joint international research programmes, BIOMASS has been the catalyst in the development and exploration of Antarctic marine ecosystems. As a result, many national and international programmes are now being developed, especially in the fields of biochemistry, cytogenetics and behaviour of Antarctic fish and birds. These studies, now more than ever before, are needed, at a time when there is a world-wide concern regarding the maintenance of the Southern Ocean's biodiversity and the protection of its living resources.

Antarctic marine systems

Interactions at the base of the Antarctic food web

H. J. MARCHANT & E. J. MURPHY

Introduction

THE STATED aim of the BIOMASS Programme was to gain a deeper understanding of the structure and dynamic functioning of the Southern Ocean marine ecosystem to provide a basis for any future management of its living resources. It is difficult in retrospect to ascribe the contributions of a particular programme, such as BIOMASS, to the development of a research field. Perhaps the best testimony for a programme comes from the extent to which it was seen to be addressing the key scientific questions of the day. Reviews during the active phase of the BIOMASS Programme show that it was seen to be at the centre of Antarctic marine research.

> BIOMASS (Biological Investigations of Marine Antarctic Systems and Stocks) is now yielding a more detailed and realistic picture of the distribution and abundance of some of the key elements of Southern Ocean food webs. (Hempel, 1985)

At the time of the development of this programme the pivotal role of krill (*Euphausia superba*) in the ecosystem was emphasized. Gaining an understanding of the processes and controls of primary production and the links to the higher trophic levels were a major component of BIOMASS. An aspect of this, and one of the fundamental questions being asked at the time of the inception of the BIOMASS Programme, was 'how productive is the Southern Ocean?'. There was the widely held view at the time that it was among the most productive marine ecosystems on earth. This was in part based on the high harvesting rates of fish, seals and whales and the implied need for high levels of primary productivity to support them. Extensive studies in the 1970s and 1980s demonstrated that the primary productivity of the Southern Ocean is highly variable and comparable with the annual production rates found in oligotrophic tropical waters (Smith & Sakshaug, 1990).

Here we consider the major developments that have taken place in the last decade and give an overview of the status of knowledge of some of the key groups of organisms. We then discuss the spatial and temporal variability and dynamics of the lower levels of the ecosystem, the larger-scale role of the Antarctic marine ecosystem, and some of the planned future directions for research.

Development of an understanding of the lower trophic levels

Since the inception of the BIOMASS Programme in 1976 there have been dramatic developments in the way we view the functioning of the Southern Ocean ecosystem. A simple linear food chain of large diatoms–krill–predators used to be considered as characteristic of the Southern Ocean. However, the complexity of the Antarctic marine food web has

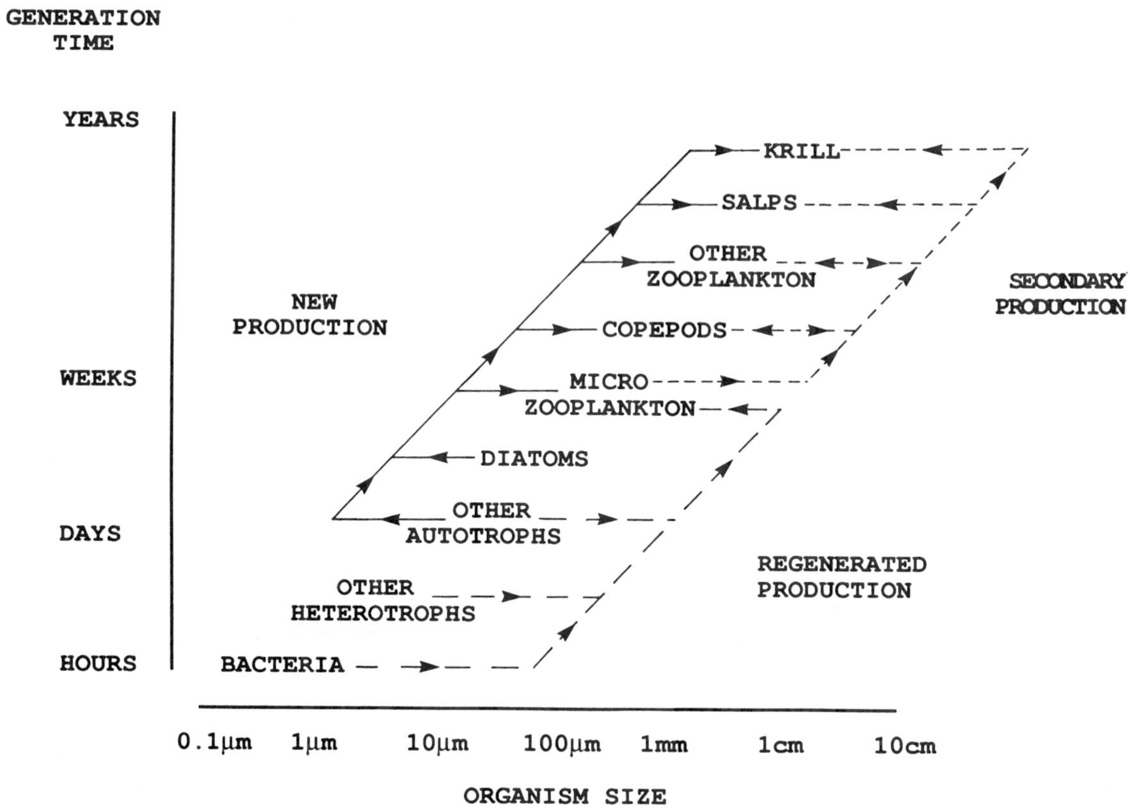

Fig. 1. Size-based view of the lower trophic level interactions of the Southern Ocean ecosystem. New, regenerated and secondary production pathways are illustrated based on a relation between organism size and generation time.

been increasingly emphasized and is now considered to be as high as that of many food webs in lower-latitude ecosystems (Garrison, 1991). Arguably, the most important development in marine science has been the re-evaluation of the microbial components of marine ecosystems. The 'microbial loop' hypothesis was developed (Azam *et al.*, 1983) and shown to be as applicable to the Southern Ocean as to other aquatic ecosystems (Hewes *et al.*, 1985). This has been part of a wider re-evaluation of the pathways of material and energy flux through Southern Ocean food webs (Murphy *et al.*, 1988; Smetacek *et al.*, 1990).

Rather than attempting to list all possible pathways of material and energy flux we show a size-based view of the lower trophic levels of the Southern Ocean ecosystem in Fig. 1. This is based on a generalized relation of organism size to generation time (Peters, 1983) and current thinking on the major pathways of energy flux in microbial communities. The new production system is characterized by the traditional view of the Southern Ocean ecosystem with nitrate-based production dominated by the large diatoms, which are grazed by the large herbivorous zooplankton such as krill. These krill are, in turn, available as prey to the larger predators such as fish, squid, seabirds and marine mammals.

The regeneration systems can be characterized by the microzooplankton such as ciliates and flagellates. These organisms consume the smaller autotrophic and heterotrophic components. Much of the autotrophic production of these smaller categories is based on ammonia produced by other biotic components of the ecosystem. The heterotrophic production is based on the consumption of dissolved or particulate organic matter generated by other biotic components of the ecosystem. The microzooplankton provide a potential food source for the larger zooplankton. Thus these regenerative systems convey energy to higher levels of the food web, providing important maintenance systems in areas where new production is low. These pathways mean that much of the production in some areas is being recycled rather than dissipated.

Within these overview pathways there are many complex interactions taking place, some of which have been carefully elucidated (Hopkins, 1987; Vincent, 1988). In attempting to list all pathways of flux we obviously greatly increase the complexity of our view of the operation of the ecosystem. The fact that pathways occur does not make them important for larger-scale questions of energy flux. However, apparently unimportant pathways in terms of the magnitude of flux can have major effects in complex networks (Ulanowicz, 1991; DeAngelis & Post, 1991). Such pathways can be limiting or may have indirect effects on other pathways. Pathways of flux that are unimportant in one area at a particular time may become the key controlling steps elsewhere. Such problems raise questions of tradeoff between complexity and simplicity in analyses of the Southern Ocean ecosystem. Addressing these questions requires carefully integrated field and modelling programmes to continuously assess data requirements.

We cannot hope to detail all pathways and fluxes so we do not attempt to comprehensively review the groups in the lower food web in this paper. Rather, we highlight those groups which are major components or whose role has been reassessed or recognized during the past decade, and highlight something of their role in the Southern Ocean. There have been a number of key reviews and papers in recent years which have brought together much of the detailed knowledge on the lower trophic levels of the Southern Ocean ecosystem. In particular El-Sayed (1988), Garrison (1991), Smith & Sakshaug (1990) and Vincent (1988) describe the microbial and primary production processes. The zooplanktonic components of the system are described by Smith & Schnack-Schiel (1990). More general reviews of relevance are Hempel (1983, 1985, 1991).

Photoautotrophs

Diatoms have been recognized as the major constituent of the Antarctic marine phytoplankton since the time of the earliest explorations (Hart, 1942; Hasle, 1969; El-Sayed, 1984). Much of the information derived in the past decade has consolidated earlier data and attempted to consider the larger scale role of diatoms in the Southern Ocean ecosystem (Priddle, 1990; Jacques & Panouse, 1991). The early studies, when collections were made by net and samples preserved in fixatives for examination months or years later, were selective for the larger and more robust species. As a consequence the idea of the phytoplankton being composed principally of large diatoms emerged. More recent investigations indicate that net-phytoplankton constitute a highly variable proportion of the phytoplanktonic biomass (Wilson *et al.*, 1986; Gieskes & Elbrächter, 1986; Weber & El-Sayed, 1987; Nöthig *et al.*, 1991). Weber & El-Sayed (1987) discuss the contributions of pico-, nano- and net-phytoplankton to the standing crop and primary productivity from a variety of locations and seasons; the indications are that Antarctic marine phytoplankton is, for the most part, dominated by nanoplankton (see also von Bröckel, 1981). These authors also report that picoplankton generally account for a minor proportion of phytoplankton biomass and primary production although some extremely high values (74% and 42% of chlorophyll *a* in the Bransfield Strait and Indian Ocean, respectively) have been recorded (Weber & El-Sayed, 1987). The nanoplankton biomass is more constant both in geographical distribution and throughout the year and as a consequence becomes more important at times of lower total biomass (Perrin *et al.*, 1987; Rivkin *et al.*, 1990). Diatoms, however, usually make the major contribution to overall biomass (Heywood & Priddle, 1987).

The abundance of dinoflagellates in Antarctic waters is highly variable. The diverse array of species include both autotrophic and heterotrophic forms. Although recognized in the early expeditions, they have so far received little attention and their role in the Southern Ocean is not clear. Hada (1970) and Balech (1975) provide a descriptive key to the array of these organisms from Antarctic waters; an ecological investigation of the dinoflagellates near South Georgia was reported by Dodge & Priddle (1987). They found that, whereas diatom community structure had distinct associations of species, this was not the case with the dinoflagellates and that, probably because they contain both autotrophic and heterotrophic taxa, it was difficult to identify environmental factors which would act on the different taxa. During the winter Garrison *et al.* (1991) found that autotrophic dinoflagellates, mostly the athecate *Gymnodinium*, constituted some 38% of the autotrophic biomass in the ice-edge zone of the Weddell Sea. These organisms are also occasionally dominant in summer populations. As many of the athecate autotrophic dinoflagellates are extremely fragile, their abundance is likely to be underestimated.

The prymnesiophyte *Phaeocystis* can occur in very high concentrations (>10^7 cells l^{-1}) and form extensive blooms (El-Sayed, 1984). This alga is one of the principal components of the phytoplankton of the marginal ice-edge zone (Fryxell & Kendrick, 1988). As has been found in other parts of the world, few other algae bloom during the peak of the *Phaeocystis* bloom (Davidson & Marchant, 1992*b*). The mucilage of this alga binds manganese (Davidson & Marchant, 1987; Lubbers *et al.*, 1990) and it appears likely that this alga is important in the cycling and transport of this micronutrient. *Phaeocystis* produces β-dimethylsulphoniopropionate (DMSP), presumably to control osmotic balance. This solute breaks down to form acrylic acid, reported to be a bacteriocide (Sieburth, 1960) and dimethyl sulphide (DMS). The implications of this are considered below.

The widespread abundance in the world's oceans of picoplanktonic cyanobacteria was reported by Johnson & Sieburth (1979) and Waterbury *et al.* (1979). These organisms occur in high concentration and have specific growth rates of about 1.6 doublings per day. They reportedly contribute substantially to the oceanic food web with 30–40% of the standing crop being grazed per day (Iturriaga & Mitchell, 1986). Across the Southern Ocean their abundance decreases by four orders of magnitude from 10^7 cells l^{-1} to 10^3 cells l^{-1} with decreasing water temperature (Marchant *et al.*, 1987; Letelier & Karl, 1989).

Microheterotrophs

At the start of the BIOMASS Programme the understanding of the role of microheterotrophs in aquatic systems was vastly inferior to our present concentration on the importance of the microbial loop. Until about 15 years ago the main flow of energy in the marine environment was considered to be a food chain directly from phytoplankton to herbivores and carnivores. Pomeroy (1974) focused attention on the significance of microorganisms as movers of energy and materials in aquatic environments; this significance is related to their small size and higher metabolic rates per unit mass. Later, Azam *et al.* (1983) elaborated the complexity of the microbial food chain as the 'microbial loop' hypothesis, whereby bacteria exploit dissolved organic matter and are preyed upon by heterotrophic flagellates, which in turn are predated by ciliates. In this way energy is rather inefficiently recycled to Metazoa. The importance of the microbial loop in the Southern Ocean is not clear. Heterotrophic flagellates and ciliates can comprise over 50% of the total nano- and microplankton biomass, implying that they are playing an important role (Garrison & Buck, 1989*a*). However, few studies have been conducted to ascertain the feeding rates of these heterotrophs.

Investigations on microheterotrophs in the Southern Ocean have not been comprehensive in terms of the groups of organisms and the areas and times of the year in which studies have been conducted. Choanoflagellates have been investigated in some detail whereas naked amoebae, for example, have been substantially overlooked (see, however, Hara *et al.*, 1986). In some broad-scale surveys choanoflagellates and naked amoebae have been reported as the dominant heterotrophs. The variety and abundance of Antarctic protozooplankton has been recently comprehensively reviewed by Garrison & Gowing (1992) and the roles of these organisms have been considered (Garrison, 1991).

Considerable attention has been given to the choanoflagellates, especially the loricate forms, of the Southern Ocean (Thomsen *et al.*, 1990). At present some 30 species are recognized and there have been a number of investigations into their ecology. Rather than being an ecologically homogeneous group of bactivores they also ingest other particulate material including cyanobacteria, detritus and high-molecular-mass dissolved organic material (DOM) (Sherr, 1988). The various species exhibit differences in their ability to ingest different sized particles and different molecular masses of DOM (Marchant & Scott, 1993). Their role in the Southern Ocean is still not clear but it is apparent that their numbers and species composition probably reflect differences in the distribution of food (Buck & Garrison, 1988).

In spring and winter heterotrophic dinoflagellates can account for a large proportion of the microprotozoan biomass (Garrison, 1991). At the ice edge of the Weddell Sea in winter, Garrison *et al.* (1991) found that around 73% of the heterotrophic biomass was composed of dinoflagellates. Some of these organisms can feed on prey as large as or even larger than themselves. On the other hand, the clearance rate of *Protoperidinium antarcticum* was found to be about an order of magnitude less for algal cells than for bacteria by Lessard & Rivkin (1986), who concluded that the heterotrophic dinoflagellates were feeding on bacteria and other heterotrophs rather than diatoms. Yet Bjørnsen & Kuparinen (1991), on the basis of microcosm experiments, indicated that heterotrophic dinoflagellates may contribute significantly to the maintenance of low phytoplankton biomass in the Southern Ocean. During the winter heterotrophic protozooplankton may be a sink for winter production (Garrison *et al.*, 1991). The significance of dinoflagellates in the diet of krill and other grazers is at present unknown.

The dominant ciliates in terms of biomass in Antarctic waters are the non-tintinnid oligotrichs. However, they have received less attention than the loricate tintinnids. What is known of the role of these organisms is discussed by Garrison (1991) who indicates that their abundance in Antarctic waters is within the range reported for temperate waters including coastal and upwelling areas. Recent investigations have highlighted the abundance, depth distribution and diet of large Protozoa, including phaeodarian and polycystine radiolarians, Foraminifera, Heliozoa and acantharians (Gowing, 1989; Nöthig & Gowing, 1991). Although their biomass is small compared with flagellates and ciliates, in some situations they constitute a significant link between nano- and microplankton with metazoan zooplankton (Nöthig & Gowing, 1991).

Bacterioplankton

The concentration of bacteria in the Southern Ocean is roughly comparable with that found in warmer marine environments (Fuhrman & Azam, 1980; Hanson *et al.*, 1983; Painting *et al.*, 1985). However, their concentration reportedly differs substantially with location and season. Gibson *et al.* (1990) reported that at a coastal site near the Australian Antarctic station of Davis bacterial concentration remained at around 1×10^5 cells ml^{-1} between May and the end of September. Bacterial numbers increased to around 4×10^5 cells ml^{-1} in October, coinciding with increased detrital material in the water column, then declined to fewer than 5×10^4 cells ml^{-1}, the lowest concentration for the year, in early November before increasing to the seasonal maximum of over 1×10^6 cells ml^{-1} in January. This bimodal abundance of bacteria at this coastal site has been confirmed by Davidson & Marchant (1992*a*). In contrast, in a seasonal study of the bacterial abundance under the sea ice near Syowa station, Satoh *et al.* (1989) reported that the bacterial concentration began to increase in October and reached a single peak in December. They concluded that bacterial growth was linked with the growth of the under-ice algal community. In addition, Rivkin *et al.* (1989) found that the bacterial abundance in McMurdo Sound showed seasonal variability and was around 1–2 orders of magnitude lower than the concentration in temperate waters. Detailed investigations of bacterial biomass and productivity in marginal ice-edge zones in all seasons have been reported by Cota *et al.* (1990) and Kottmeier & Sullivan (1990). These studies found that bacterial biomass and production mirrored the phytoplankton biomass and production and that bacterial productivity is a significant source of production. Estimates of bacterial productivity based on measurements of bacterial biomass and specific growth rate are higher than those estimated from tritiated thymidine assimilation (Karl *et al.*, 1991*a*). Both methods indicate that generation times of Antarctic marine bacteria are longer than for these organisms in temperate waters. In the Weddell–Scotia Sea bacterial production ranged from 11% of primary production in spring (Sullivan *et al.*, 1990) to 76% of primary production in autumn (Cota *et al.*, 1990).

Larger zooplankton and krill

Krill (*Euphausia superba*) are one of the central elements of the Southern Ocean ecosystem, providing the major link to many of the higher predators in the system. The past decade has seen a vast amount of research effort aimed at this species, a major component of which was the BIOMASS Programme. The biology and ecology of this organism has been extensively reviewed (Priddle *et al.*, 1988; Miller & Hampton, 1989) and is further developed in this volume (see Everson & Miller, this volume; Siegel & Kalinowski, this volume; Quetin *et al.*, this volume).

Much of the comprehensive work on other zooplankton species and groups was carried out in the early studies of the Southern Ocean (Hardy & Gunther, 1935). Studies over the past decade have been more specialized and regionally restricted (Atkinson & Peck, 1988; Hopkins & Torres, 1988; Huntley & Escritor, 1991; Brinton & Townsend, 1991). There has been little study of many of the key groups. The importance of spatial (horizontal and vertical) and temporal variation in zooplankton community structure has been emphasized in regional studies (Hopkins & Torres, 1989; Smith & Schnack-Schiel, 1990). Translating some of that distributional information into rate process studies has also begun (Huntley *et al.*, 1991*a*). Studies of patterns of development and ontogenic changes and overwintering strategies have still to be comprehensively tackled.

There are some key species which, for a number of reasons, not least sampling practicalities, have received particularly little attention. Of these perhaps salps, such as *Salpa thompsoni* (Foxton, 1966), are likely to play a major role in microbial biogeochemical cycling. They are a major component of the ice-free zone communities (see below) away from the krill areas. They can occur in extremely large aggregations (Fig. 2). However, they are not organisms that are well studied using traditional techniques of net sampling or use of predator stomach contents. The knowledge of the rate processes of these organisms and their biology does not allow us to consider as yet their importance in biogeochemical cycles or their role in the ecosystem (Duhamel & Hureau, 1985).

With the view of the increased complexity of the Southern Ocean ecosystem, studies are now turning again to investigating the role of many of the less conspicuous zooplankton. The pathways of larval and adult fish grazing are not simply krill-based. Copepods are known to be major components in the diet of many Antarctic fish (Duhamel & Hureau, 1985). Squid and fish interact, probably consuming each other at various ontogenic stages, and both are major components of the diets of marine mammals and land-based predators (Croxall *et al.*, 1985).

Key dynamic links

Factors limiting autotrophic production

In general, macronutrients such as phosphate, nitrate, and for diatoms, silicate apparently do not play the same dominant role in limiting the growth of phytoplankton in Antarctic waters. They can be locally important at particular times (Holm-Hansen & Mitchell, 1991; Owens *et al.*, 1991; Lizotte & Sullivan, 1992) but they do not limit growth to the extent that is observed in the Arctic and in more temperate and tropical regions (Sakshaug & Holm-Hansen, 1984). The limitation of growth of autotrophs is now thought to be dependent on a combination of physical and biological processes (Jacques, 1991). That no one single element or process can be said to be the dominant factor is a reflection of the high spatial and temporal variability of this ecosystem at all scales. The inputs of variability over such a wide range of spatiotemporal scales produces wide variation in the dominance of biological processes and operation of the marine ecosystem.

Tilzer *et al.* (1986) have shown that the phytoplankton light utilization efficiencies in Antarctic regions are reduced at low temperatures and are generally lower than in lower latitudes. The relatively low light climate and short season result in a limited period for phytoplankton blooms to occur. In some areas the growth and accumulation of phytoplankton appear principally determined by the stability of the water column (Smith & Nelson, 1985). The high wind stress across much of the Southern Ocean results in a relatively deep mixed water column even during the summer months. The result is that average light levels are low within the mixed layer and thus can limit the growth of the phytoplankton.

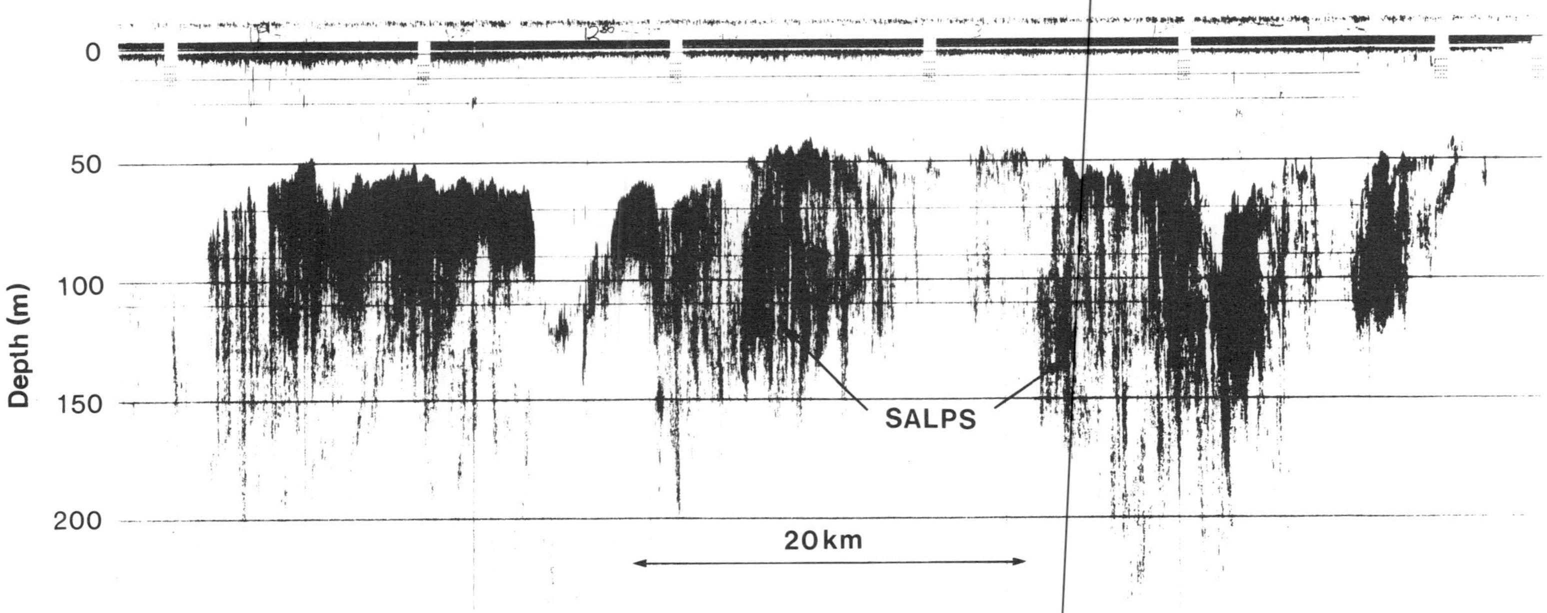

Fig. 2. A planktonic aggregation detected hydroacoustically in the northern Scotia Sea in the region of the Antarctic Convergence. Restricted sampling indicated that the major scatterers were salps. (British Antarctic Survey, unpublished data; L Madureira, personal communication.)

The limitation of phytoplankton growth by trace elements was first discussed in relation to the Southern Ocean by Hart (1934). This aspect of nutrient cycling has received a lot of attention recently; some investigations indicate that iron limits primary productivity (Martin *et al.* 1990). Other investigators (e.g. De Baar *et al.*, 1990), however, indicate that iron is not solely limiting for phytoplankton in the Weddell and Scotia Seas. They find that sediments in the coastal areas supply sufficient iron for phytoplankton growth but that in the more remote areas iron is more likely to occasionally limit phytoplankton productivity.

With the apparent size-dependency of autotrophic production, another aspect of biogeochemical cycling is emphasized, one which has also been considered to be involved in the limitation of production in the Southern Ocean. Ammonia appears to be preferentially taken up by the nano- and picoautotrophic components of the community (Probyn & Painting, 1985). The possibility that ammonia is the major limiting factor for primary production has been suggested (Biggs *et al.*, 1985; Owens *et al.*, 1991) based on the fact that in some areas it is taken up preferentially over nitrate. Biggs *et al.* (1985) caution against considering all ammonia to be routed directly through biotic components along short chains. There is some evidence that in some marine ecosystems high ammonia concentrations may inhibit the growth of larger phytoplankton (Paasche, 1988). The microheterotrophic and microzooplankton appear to be important in ammonia generation (Koike *et al.*, 1986). However, the relative importance of these various aspects of the nitrogen flux pathways are not known. Flux of ammonia across the pycnocline, transport out from under ice where heterotrophic processes may be important or indeed out of the melting ice are all possible mechanisms for ammonia replenishment (Biggs *et al.*, 1985). The possibility of large krill swarms releasing ammonia in areas where the large diatoms have been grazed by the krill probably results in a spatial mosaic of size based interactions taking place.

High levels of the herbivorous zooplankton may limit production through grazing (Smetacek *et al.*, 1990) although this may be spatially and temporally restricted (Lancelot *et al.*, 1992). There are numerous reports of krill and other grazers feeding on the underside of sea-ice (e.g. Daly & Macaulay, 1988; Stretch *et al.*, 1988), and it has been proposed that the sea-ice community plays an important role in the overwintering survival of larval krill (Daly, 1990). However, the extent to which the ice-associated community contributes to the diet of krill and other grazers is not clear. Krill can apparently feed over a wide range of particle size. The smallest particles they can ingest are 2–3 μm in diameter but for optimal feeding efficiency particles of 20–40 μm in diameter are required. Quetin & Ross (1985) concluded that size of prey species rather than chemical composition is likely to influence the clearance rate of krill. Examination of gut contents and faecal material from net-sampled krill by transmission electron microscopy reveals that they are omnivorous. As well as phytoplankton that can be recognized by resistant parts such as diatom frustules, silicoflagellate skeletons, and the periplast scales and spines diagnostic of particular species of phytoflagellates, some protozoa also can be recognized (Marchant & Nash, 1986). Discharged trichocysts of dinoflagellates and the siliceous costal strips that make up the lorica of choanoflagellates resist digestion by krill. Analysis of the fatty acids and amino acids of krill faecal material and electron microscopy have shown that choanoflagellates can constitute a substantial component of their diet (Tanoue & Hara, 1986). These authors propose that this grazing by krill on choanoflagellates provides an avenue for particulate organic carbon to be transferred via faecal pellets to deep-water consumers.

The question of limitation of production at the various trophic levels requires careful analyses, model development and parameterization. It is likely that models will only be appropriate over restricted spatial and temporal scales. In the Southern Ocean many different situations occur, so to determine the relative importance of these different factors a range of models will be required, the output of which will need to be cross-referenced to check for consistency between different model structures.

Other production pathways

The pathways of possible heterotrophic production are the same as those considered for most other areas of the oceans. As well as the classical 'food chain' and the microbial loop, there are a number of variations. For example, choanoflagellates take up bacteria, autotrophs, particulate organic matter

(POM) and DOM. Recently the ingestion of colloidal sized particles by planktonic tunicates was demonstrated by Flood *et al.* (1992). The uptake of DOM has been suggested to supply a significant input to the nutrition of larval and embryonic invertebrates in McMurdo Sound (Manahan *et al.*, 1990).

The extent to which mixotrophy plays a role in the trophic interactions of the organisms of the Southern Ocean has yet to be elucidated. Mixotrophy occurs in two principal ways. Some heterotrophs retain the chloroplasts of ingested autotrophs in such a way as to permit the continuation of photosynthesis, and some autotrophs are capable of feeding by phagocytosis. In temperate and tropical waters a significant contribution to gross primary productivity is made by chloroplasts sequestered by ciliates, especially *Strombidium* sp. (Stoecker *et al.*, 1987). Although *Strombidium* is a component of the ciliate fauna of the Southern Ocean (Hada, 1970), whether these organisms harbour chloroplasts, and, if so, the extent to which they contribute to primary productivity has not been reported (Garrison, 1991). In addition, *Myrionecta (Mesodinium) rubrum*, a ciliate with a cryptomonad endosymbiont, has been reported to occur in high numbers in Antarctic coastal waters and to be photosynthetically active (Satoh & Watanabe, 1991). Although the importance of algal phagotrophy has been clearly demonstrated in freshwater environments (Bird & Kalff, 1986, 1987), few detailed studies on algal phagotrophy have been reported from marine habitats (Estep *et al.*, 1986; Sanders & Porter, 1988; Sanders, 1991). Species of *Chrysochromulina*, a genus of phytoplankton that has a wide distribution in the world's oceans, have been shown to be able to graze on bacteria, other phytoplankton, particulate and dissolved organic material (see Estep & McIntyre (1989) for review). Several species of *Chrysochromulina* have been found in the Southern Ocean and preliminary evidence indicates that they are mixotrophic (HJ Marchant, unpublished data). However, the extent to which their grazing contributes to carbon and energy flux has yet to be ascertained. Potentially they are of considerable importance as scales of *Chrysochromulina* have been found in the gut contents and faeces of krill at different stages of maturity collected at various times of the year (HJ Marchant, unpublished data). If this alga is able to grow during the Antarctic winter by feeding phagotrophically when light is limiting for photosynthesis, it may provide an important source of food for grazers.

The effects of grazing, faecal pellet production and export of the material from surface layers have all been extensively studied during the last decade (see Karl *et al.*, 1991*b*) and Abelmann & Gersonde (1991) for reviews). Recently the importance of faecal material produced by protozoa has been shown (Nöthig & Bodungen, 1989). The membrane-bound faecal pellets produced by the abundant, unusual athecate dinoflagellate reported by Buck *et al.* (1990) from the sea-ice of the Weddell Sea are likely to be contributing substantially to this vertical flux, which would constitute a significant food source for detritivores, such as phaeodarian radiolarians, inhabiting the lower photic zone at a depth of 100–200 m.

Ecosystem dynamics and spatial and temporal variability

The above comments on the current status of knowledge of the biology of the major groups serve to illustrate the importance of those organisms recognized in the past decade and to put in context the earlier work. This applies in particular to the study of the large siliceous diatoms (Priddle, 1990), the major components of the zooplankton community (Smith & Schnack-Schiel, 1990), and krill (Miller & Hampton, 1989). One of the major elements stressed over the past decade has been the spatial and temporal variability of this ecosystem (Murphy *et al.*, 1988). There has been a realization that the system cannot be characterized in a simple homogeneous fashion. There have also been major developments in our understanding of the spatial and temporal interactions of biological and physical processes. Perhaps the most significant of these has been the increased recognition of the role of the marginal ice zone. Interactions within the microbial system show high variability over spatial scales of a few metres and temporal scales of hours–days as a result of the small size of the organisms and their rapid rates of turnover. At spatial scales of tens or hundreds of kilometres and temporal scales of weeks to months, some general microbial–microplankton ecosystem features can be identified. Four ecosystem types can be identified, three of which are based on zones which characterize zooplankton

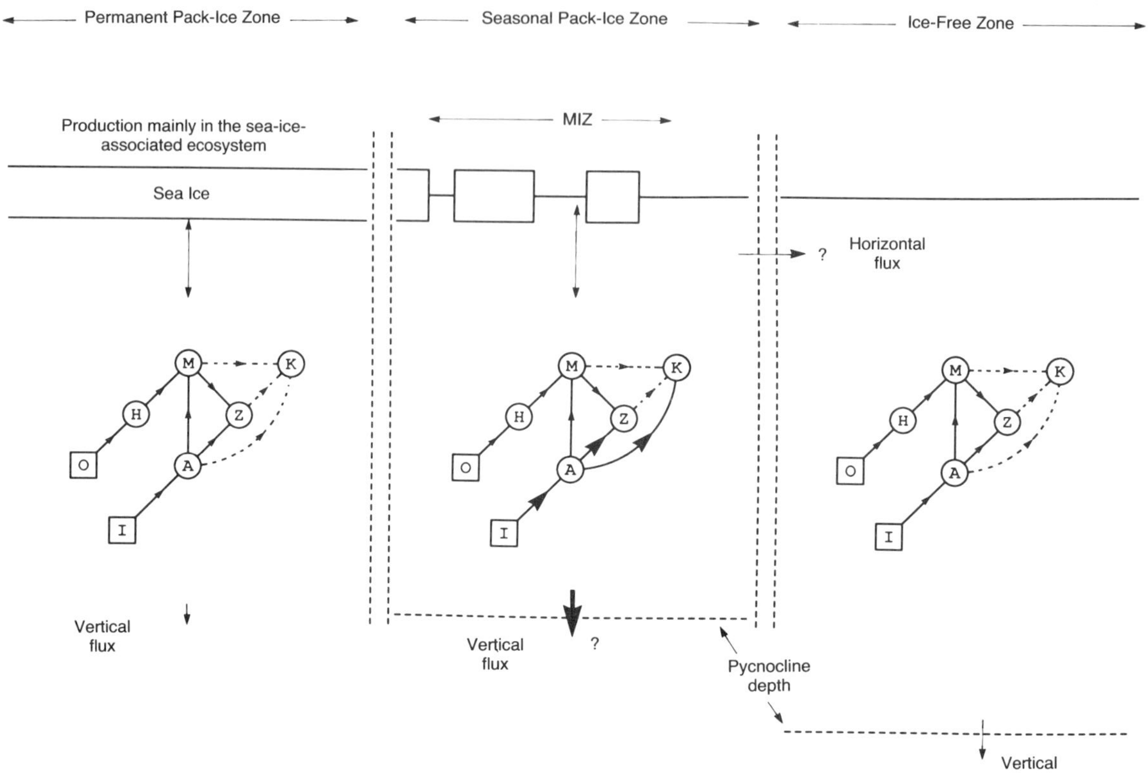

Fig. 3. Schematic representation of the three major zones of the Southern Ocean which characterize much of the large-scale variation in operation of the lower trophic level networks. O, organics; I, inorganics; A, autotrophs; H, heterotrophs; M, microzooplankton; Z, zooplankton other than krill; K, krill; MIZ, Marginal Ice Zone.

communities (Hempel, 1985; Smith & Schnack-Schiel, 1990). These microbial systems are illustrated in Fig. 3. This schematic exercise also serves to illustrate our lack of knowledge of the spatial and temporal variations and connections of the marine ecosystems of the Antarctic.

Northern Zone (Ice-Free Zone, IFZ)

In this region, the production is generally low although macronutrient levels are high. The pycnocline probably remains deep in this area of high wind stress. Frequent storms pass over most of this area, producing deep wind mixing even at the time of the season when thermal energy input is high. This can result in low overall light levels for the phytoplankton present in the surface waters. Of the other factors discussed above which may restrict production in this area, those such as low levels of trace elements, low ammonia levels or restrictive grazing pressure may all occur. Copepods, salps, small euphausiids, chaetognaths and amphipods dominate the zooplankton in these areas with herbivorous and carnivorous species at varying stages of their life cycle producing a complex web of interactions (Smith & Schnack-Schiel, 1990). Much of the production is probably produced and recycled within the smaller components of the community. When krill occur in these areas they are probably operating at maintenance levels owing to the low production (Ross & Quetin, 1986). It is difficult to establish the seasonal progression of the production processes in these regions. In the winter the water column is homogeneous with a deep thermocline. In the summer stratification occurs but the thermocline tends to remain relatively deep (Jacques, 1989). It is not clear from the studies carried out so far whether there are spring blooms of diatoms in these regions. Much of the production throughout the summer is probably dominated by the smaller microbial components of the system. An interesting problem is to establish to what extent this system is dependent on regenerated and advected production throughout the whole season. The influence of the various factors limiting growth in these regions is still the focus of a lot of research.

Intermediate Zone (Seasonal Pack-Ice Zone, SPIZ)

For most of the year the sea around Antarctica is frozen. From the time of the earliest cruises to the Antarctic, the biological richness of the sea-ice has been noted. Much of the harvesting of the great whales was carried out in these regions. At the winter maximum, in September, the area covered by sea-ice can exceed 20×10^6 km^2. During spring and early summer the ice retreats to a summer minimum of less than 4×10^6 km^2 (Jacka, 1983). As a consequence of the low angle of the sun and the attenuation by the sea-ice and its frequent coverings by snow, the amount of light reaching the water column is very low. During this time of year the greatest biomass of the Southern Ocean is concentrated within or directly under the sea-ice. This zone has been the target of extensive research programmes over the past decade emphasizing its importance as an area of enhanced primary productivity which supports a complex web of higher trophic level interactions.

In this region, trailing the receding ice edge in spring, the surface waters are stabilized by a layer of less saline water released from the melting sea-ice (see Sullivan *et al.*, 1988). This nutrient-rich, high-light, shallow mixed layer provides optimal conditions for phytoplankton growth (Smith, 1987). Primary productivity in this region is elevated against the background of low productivity and biomass of the Southern Ocean. There is now strong evidence that the ice-edge zone plays a major role in the enhancement of biological activity at all trophic levels (Ainley *et al.*, 1986). Not surprisingly, this sea-ice associated community has received considerable attention and has been the subject of a number of recent reviews (Smith & Nelson, 1986; Garrison & Buck, 1989*b*; Horner, 1990). The mechanisms of incorporation of organisms into this community, the growth of the various components and the interactions between them have been discussed extensively elsewhere (see Horner (1990) for review).

Melting of the sea-ice during spring and summer generates a pycnocline at about 20 m depth above which phytoplankton bloom in the nutrient-rich, high-light environment. Seeding of this marginal ice-edge zone (MIZ) by the sea-ice community is likely to contribute to the developing planktonic population (Garrison *et al.*, 1987) although the extent to which this occurs has been questioned (Riebesell *et al.*, 1991). This southward-moving region of high phytoplankton productivity (Smith & Nelson, 1986) is utilized by the higher trophic levels (Ainley *et al.*, 1986) providing much of the carbon required to sustain the large populations of zooplankton, birds and mammals for which the Southern Ocean is noted (Sakshaug & Skjoldal, 1989). The massive deposits of diatomaceous ooze in Southern Ocean sediments are thought to be due to reduced coupling of production and consumption in the marginal ice-edge zone, compared with other oceanic areas, and a substantial amount of the biogenic production sinking rapidly from the euphotic zone (Smith & Nelson, 1986; Karl *et al.*, 1991*b*; Riebesell *et al.*, 1991). Wind forcing is thought to be responsible for the breakup of the shallow mixed depth of the marginal ice-edge zone to a mixed depth of 100–200 m in the open ocean.

The seasonal pack-ice zone provides a highly variable environment in time and space. The extreme seasonality of Southern Ocean systems is emphasized in these areas. The ice-edge is a transitory phenomenon; physical conditions suitable for high production occur only during a short period of time following the passage of the ice-edge. The variability of this region is high over all scales from the interannual changes in sea-ice extent (Smith *et al.*, 1988; Eicken, 1992; EJ Murphy *et al.*, in preparation), through seasonal changes in ice-edge position, to the high variability in the MIZ produced by such factors as floes affecting local light climate on a spatial scale of tens of metres and a temporal scale of minutes to hours. This environmental variability is intricately linked to the variability in the ecosystem. Thus, the pathways of energy flux will show high variability in the MIZ. In the region where the ice-edge has retreated, blooms can develop, but this depends on the patchiness generated in the retreat of the MIZ. The meltwater stability of the area provides an environment where large blooms can develop. Other factors may also be important in the development of these blooms, such as local seeding, nutrient releases and the local effects of grazing (Lancelot *et al.*, 1992). Given the very different scales of variability characterizing this environment compared with the ice-free zone, there may be very different spatial distribution of the major components of the plankton than in other regions of the Southern Ocean (Weber *et al.*, 1986; EJ Murphy, unpublished data). We would also expect to see a progressive increase in the importance of recycled production further from the bloom regions.

It is likely that this high spatial and temporal variability results in highly variable vertical flux rates from the MIZ. This probably reflects the patchiness of the production as well as the patchiness in the grazing effects. The faecal pellet production of the herbivorous zooplankton may be rapidly recycled although there is evidence that faecal material sediments rapidly out of the euphotic zone (Bodungen, 1986; Abelmann & Gersonde, 1991).

Southern Zone (Permanent Pack Ice Zone, PPIZ)

There is little production associated with the pelagic environment under the ice in those regions where it persists from year to year. In addition, there is little vertical structure in the water column under the ice except very close to the ice produced by melting or brine expulsion. The high light absorption properties of the ice results in a rapid attenuation of light. The majority of the production is in the sea-ice community. Diatoms and new production appear to be a major element of the productivity in these areas (Vincent, 1988). Productivity can become limited by nutrient availability in these environments, so recycling processes are likely to be important. The sea-ice ecosystem shows a high variability over small scales, resulting in patchiness in the microbial community interactions (Murphy *et al.*, 1988; Vincent, 1988; Eicken, 1992).

Zone-independent regions

Although the major zones described above characterize much of the Southern Ocean production, there are regions which do not fit these categories. These are often areas where enhanced production can occur associated with fronts (Lutjeharms *et al.*, 1985), upwelling (Grindley & David, 1985), islands (Allanson *et al.*, 1985), shelf regions and eddy effects. In these areas the local pathways of material flux can be very different from those of the surrounding water masses. In the Antarctic many of these regions produce large diatom blooms. Some of these blooms appear to be of extended duration, indicating that nutrient replenishment or regeneration may be taking place (Mitchell & Holm-Hansen, 1991). Frequent changes in the depth of the pycnocline may occur, producing a replenishment of nutrients. Grazing effects combined with these changes may also produce conditions favourable for recycling and high turnover.

Zone-independent regions are of major importance in the Southern Ocean. The Bransfield Strait, for example, is thought to be one of the major krill production regions (Siegel, 1988; Brinton, 1991). Many of the predator colonies are associated with these regions, indicating their importance as areas where energy transfer to higher trophic levels can occur.

The larger-scale interactions of the Southern Ocean ecosystem

The change in the position of the ice-edge in summer produces a spatially distributed input of energy into the Southern Ocean ecosystem. There is, however, little information on the spatial transfer of material between regions. Various studies have suggested that spatial and temporal changes occur in the distribution of krill (Marr, 1962; Everson, 1977; Murphy *et al.*, 1988; Siegel, 1988; Miller & Hampton, 1989; Smetacek *et al.*, 1990: Everson & Goss, 1991). Such krill immigration and emigration between regions may be important in the development of management procedures (CCAMLR, 1991). The quantification of horizontal as well as vertical fluxes is likely to be extremely important in gaining an understanding of the operation of this spatially variable large marine ecosystem. Huntley *et al.* (1991*a*) included horizontal fluxes of particulate carbon in their carbon flux model of the Bransfield Strait. Lancelot *et al.* (1992) considered the development and passage of marginal ice-edge blooms. The further development of our knowledge of the spatial and temporal dissipation of energy in the Antarctic marine ecosystems requires further model developments which couple models of the biological and physical processes. A key aspect in the spatial and temporal transfer of ecosystem conditions is the overwintering strategies of the various planktonic components. The information available so far is limited and there is little understanding of how these are coupled. This aspect is of crucial importance because models of seasonal planktonic succession which include herbivore dynamics are generally sensitive to the spring conditions.

Elucidating the role of the Southern Ocean in global climate processes has become a major area of research in the last few years. During the time of the BIOMASS Programme, concern has been widely voiced about the increase in atmospheric carbon dioxide and the global climatic changes that are a likely consequence of this increase. The extent to which the Southern Ocean is a sink for atmospheric CO_2 is equivocal. Recent investigations indicate that the pCO_2 of surface waters south of 60°S is higher than in the air and lower to the north (Inoue & Sugimura, 1986). Substantial interest is now focused on the role of the Southern Ocean in processes that affect global climate. The primary productivity of the Southern Ocean is in the order of 3.5×10^9 t of carbon per year, approximately 15% of the global total (Huntley *et al.*, 1991*b*). Huntley *et al.* (1991*b*) have suggested that the Southern Ocean may be highly inefficient as a sink for CO_2 because of the potentially large flux of carbon respired directly to the atmosphere by birds and mammals. Clarifying the importance of such pathways will require careful analyses of the ecosystem operation over appropriate scales.

Another aspect of the role of the Southern Ocean ecosystem in climatic processes which has been studied in the past decade is the planktonic production of DMS. DMS is a principal source of sulphate aerosol particles, which have been reported to function as cloud condensation nuclei (CCN) over oceanic areas. Increased cloud cover increases albedo; a feedback loop has been proposed so that DMS production can influence global climate (Bates *et al.*, 1987; Charlson *et al.*, 1987). *Phaeocystis* in Antarctic coastal waters reportedly produces substantially more DMS per cell than this alga from other regions and it has been calculated that this alga is responsible for some 10% of the global biogenic production of atmospheric sulphur aerosol particles (Gibson *et al.*, 1989).

Since the discovery of the springtime depletion of stratospheric ozone over Antarctica there has been concern that the resultant increase in UV-B irradiation could have an adverse effect on the phytoplankton of the Southern Ocean. A consequence of UV damage would be to limit the food availability for higher trophic levels and reduce the magnitude of vertical carbon flux. The effect of elevated levels of UV-B irradiation on Antarctic marine phytoplankton remains equivocal. Based on growth studies, El-Sayed *et al.* (1990) concluded that Antarctic marine phytoplankton are presently under UV stress, that any further increase in UV irradiance would be deleterious to phytoplankton populations, and that there would be consequential effects on the trophodynamics of the Southern Ocean. The penetration of UV irradiation in Antarctic waters has been investigated for different irradiances and different wavelengths (see Karentz (1991) for review). Because of the coincidence in the timing of seasonal stratospheric ozone depletion and the development of the marginal ice-edge zone which trails the retreating sea-ice, organisms blooming in this shallow mixed region are likely to be exposed to levels of UV-B as high as those occurring at the summer solstice (Frederick & Snell, 1988). Recent evidence indicates that primary production is diminished by at least 6–12% as consequence of ozone depletion (Smith *et al.*, 1992). Different species are highly variable in their tolerance to UV-B exposure (Calkins & Thordardottir, 1980; Karentz *et al.*, 1991; Marchant *et al.*, 1991, Bramich *et al.*, (1993) and the most likely impact of elevated UV irradiance on the Antarctic phytoplankton assemblage is a shift in species composition (Bidigare, 1989; Karentz, 1991; Marchant, 1993). The consequences of this to trophodynamics and vertical carbon flux are presently being explored.

As well as the potential effects of climate change, another anthropogenic impact on the Southern

Ocean is the disturbance of the ecosystem by harvesting various trophic levels. The living resources are presently being exploited with the fishery for Antarctic krill (*Euphausia superba*) taking some 400 kt per annum. In 1988–9 the Antarctic krill fishery was the largest single-species crustacean fishery in the world; in terms of tonnage landed it ranked as number 30 of all fisheries in the world. Some 100 kt of finfish are also caught annually. The international Convention for the Conservation of Antarctic Marine Living Resources (CCAMLR) came in to force in 1982 to regulate the exploitation of the living resources of the Southern Ocean. Attempting to monitor and manage a large marine ecosystem (Sherman & Gold, 1990; Sherman, this volume) in the way that is being carried out in the Antarctic is pioneering work. Human exploitation of lower trophic levels such as krill, which are of major importance in the foodweb, means that understanding primary production processes remains a fundamental requirement of future research effort.

Concluding remarks

During the time of the BIOMASS Programme our knowledge of the physical and biological structure of the Southern Ocean increased dramatically. This has led to a significant increase in our understanding of the processes occurring in Antarctic waters. There are perhaps two basic concepts developed during the course of the BIOMASS Programme that are central to our present understanding of the dynamics of the Southern Ocean. One is the recognition of the complexity of the Southern Ocean food web; the other is the realization of the importance of the marginal ice-edge zone as the most productive region of the Southern Ocean. In simplified views of the food chain in the Southern Ocean, krill feed on diatoms and are themselves the principal prey of the higher trophic levels. Perhaps one of the most significant outcomes of work during the BIOMASS Programme was the finding that the food web is vastly more complex than originally envisaged, with microheterotrophs constituting a major component of the plankton and being reported as an element in the diet of krill. The past decade has seen an increasing emphasis on the need to measure flux rates of marine ecosystems. This must, however, be balanced by the need to gain an understanding of the variability in ecosystem operation. It will always be possible that further interaction effects will be elucidated in the future. However, the importance of such pathways does not come simply from their existence, but from their influence in the whole ecosystem.

It is likely, with the developing JGOFS Programme, that the next decade will see some major advances in our knowledge of the interactions occurring at the base of the food web in the Antarctic marine ecosystem. The key to many of these interactions are the zooplankton, including all the other groups as well as krill. Of these interactions much more information is required on the importance of grazing against sedimentation and other euphotic zone loss processes. The indirect effects of the contribution of zooplankton to the recycling systems are also likely to be of major importance and require much detailed study. The development of the GLOBEC Programme promises much in terms of elucidating the primary–secondary production links. With the next generation of remote sensing systems becoming available covering the Southern Ocean we are likely to see rapid improvement in our knowledge of the operation of the lower trophic levels of the Southern Ocean marine ecosystem.

Understanding the environmental variation at all scales, understanding the way in which this determines the ecosystem structure, and elucidating the spatial and temporal scales of interaction of biological and physical processes now encompass the goals of much of the research in the Southern Ocean. Investigating the role of the sea-ice variability over all scales is likely to be a key aspect of this work.

So, in the time of the BIOMASS Programme, the Southern Ocean has emerged as a region of major fisheries and is now regarded as one of the most sensitive regions of the world in determining global climate change. The role of krill in the Southern Ocean is still viewed as central, but simple food chain descriptions are no longer useful. The research programmes of the 1980s, of which BIOMASS was crucial, now show that krill are a key component of an intricate network of spatially and temporally varying lower trophic level interactions.

Acknowledgements

The authors thank Sayed Z El-Sayed, Graham Hosie, Steve Nicol, Julian Priddle and Simon Wright for useful discussions and comments on the manuscript.

References

Abelmann A, Gersonde R (1991) Biosiliceous particle flux in the Southern Ocean. *Mar Chem* **35**: 503–36.

Ainley D G, Fraser W R, Sullivan C W, Torres J J, Hopkins T L, Smith W O (1986) Antarctic mesopelagic micronekton: Evidence from seabirds that pack ice affects community structure. *Science* **232**: 847–9.

Allanson B R, Boden B, Parker L, Duncombe Rae C (1985) A contribution to the oceanology of the Prince Edward Islands. In: Siegfried W R, Condy P R, Laws R M (eds) *Antarctic Nutrient Cycles and Food Webs.* Springer-Verlag, Berlin, pp 38–45.

Atkinson A, Peck J M (1988) Summer–winter differences in copepod distributions around South Georgia. *Polar Biol* **8**: 463–73.

Azam F, Fenchel T, Field J G, Gray J S, Meyer-Reil L A, Thingstad F (1983) The ecological role of water column microbes in the sea. *Mar Ecol Prog Ser* **10**: 257–63.

Balech E (1975) *Clave Ilustrada de Dinoflagelados Antarticos.* Instituto Antartico Argentino, Buenos Aires.

Bates T S, Charlson R J, Gammon R H (1987) Evidence for the climatic role of marine biogenic sulphur. *Nature* **329**: 319–21.

Bidigare R R (1989) Potential effects of UV-B radiation on marine organisms of the Southern Ocean: Distributions of phytoplankton and krill during austral spring. *Photochem Photobiol* **50**: 469–77.

Biggs D C, Amos A F, Holm-Hansen O (1985) Oceanographic studies of epi-pelagic ammonium distributions: the Ross Sea NH_4^+ experiment. In: Siegfried WR, Condy PR, Laws RM (eds) *Antarctic Nutrient Cycles and Food Webs.* Springer-Verlag, Berlin, pp 93–103.

Bird D F, Kalff J (1986) Bacterial grazing by planktonic lake algae. *Science* **231**: 493–5.

Bird D F, Kalff J (1987) Algal phagotrophy: Regulating factors and importance relative to photosynthesis in *Dinobryon (Chrysophyceae). Limnol Oceanogr* **32**: 277–84.

Bjørnsen P K, Kuparinen J (1991) Growth and herbivory by heterotrophic dinoflagellates in the Southern Ocean, studied by microcosm experiments. *Mar Biol* **109**: 397–405.

Bodungen B von (1986) Phytoplankton growth and krill grazing during spring in the Bransfield Strait, Antarctica – implications from sediment trap collections. *Polar Biol* **6**: 153–60.

Bramich D, Davidson A T, Marchant H J, McMinn A (1993) Effects of UV-B irradiation on growth and survival of Antarctic marine diatoms. *Mar Biol* (submitted).

Brinton E (1991) Distribution and population structures of immature and adult *Euphausia superba* in the western Bransfield Strait region during the 1986–1987 summer. *Deep-Sea Res* **38**: 1169–93.

Brinton E, Townsend A W (1991) Developmental rates and habitat shifts in the Antarctic neritic euphausiid *Euphausia crystallorophias*, 1986–1987. *Deep-Sea Res* **38**: 1195–211.

Bröckel K von (1981) The importance of nanoplankton within the pelagic Antarctic ecosystem. *Kieler Meeresforsch* **5**: 61–7.

Buck K R, Bolt P A, Garrison D L (1990) Phagotrophy and fecal pellet production by an athecate dinoflagellate in Antarctic sea ice. *Mar Ecol Prog Ser* **60**: 75–84.

Buck K R, Garrison D L (1988) Distribution and abundance of choanoflagellates (Acanthoecidae) across the ice edge zone in the Weddell Sea, Antarctica. *Mar Biol* **98**: 263–9.

Calkins J, Thordardottir T (1980) The ecological significance of solar UV radiation on aquatic organisms. *Nature* **283**: 563–6.

CCAMLR (1991) *Report of the Third Meeting on the Working Group on Krill, USSR, July 1991.* SC-CCAMLR-X/4. 98 pp.

Charlson R L, Lovelock J E, Andreae M O, Warren S G (1987) Oceanic phytoplankton, atmospheric sulfur, cloud albedo and climate. *Nature* **326**: 655–61.

Cota G F, Kottmeier S T, Robinson D H, Smith W O Jr, Sullivan C W (1990) Bacterioplankton in the marginal ice zone of the Weddell Sea: biomass, production and metabolic activities during austral autumn. *Deep-Sea Res* **37**: 1145–67.

Croxall J P, Prince P A, Ricketts C (1985) Relationships between prey life-cycles and the extent, nature and timing of seal and seabird predation in the Scotia Sea. In: Siegfried W R, Condy P R, Laws R M (eds) *Antarctic Nutrient Cycles and Food Webs.* Springer-Verlag, Berlin, pp 516–33.

Daly K L (1990) Overwintering development, growth, and feeding of larval *Euphausia superba* in the Antarctic marginal ice zone. *Limnol Oceanogr* **35**: 1564–76.

Daly K L, Macaulay M C (1988) Abundance and distribution of krill in the ice edge zone of the Weddell Sea, austral spring 1983. *Deep-Sea Res* **35**: 21–41.

Davidson A T, Marchant H J (1987) Binding of manganese by antarctic *Phaeocystis pouchetii* and the role of bacteria in its release. *Mar Biol* **95**: 481–7.

Davidson A T, Marchant H J (1992*a*) Protist interactions and carbon dynamics of a *Phaeocystis*-dominated bloom at an Antarctic coastal site. *Polar Biol* **12**: 387–95.

Davidson A T, Marchant H J (1992*b*) The biology and ecology of *Phaeocystis* (Prymnesiophyceae). In: Round F E, Chapman D J (eds) *Progress in Phycological Research* **8**: 1–45. Biopress, Bristol.

De Baar H J W, Buma A G I, Nolting R F, Cadée G C, Jacques G, Trégeur P J (1990) On iron limitation of the Southern Ocean: experimental observations in the Weddell and Scotia Seas. *Mar Ecol Prog Ser* **65**: 105–22.

DeAngelis D L, Post W M (1991) Positive feedback and ecosystem organization. In: Higashi M, Burns T (eds) *Theoretical Studies of Ecosystems: the Network Perspective.* Cambridge University Press, pp 155–78.

Dodge J D, Priddle J (1987) Species composition and ecology of dinoflagellates from the Southern Ocean near South Georgia. *J Plank Res* **9**: 685–97.

Duhamel G, Hureau J C (1985) The role of zooplankton in the diets of certain sub-antarctic marine fish. In: Siegfried W R, Condy P R, Laws R M (eds) *Antarctic Nutrient Cycles and Food Webs.* Springer-Verlag, Berlin, pp 421–9.

Eicken H (1992) The role of sea ice in structuring Antarctic ecosystems. *Polar Biol* **12**: 3–13.

El-Sayed S Z (1984) Productivity of the Antarctic waters – a reappraisal. In: Holm-Hansen O, Bolis L, Giles R (eds) *Marine Phytoplankton and Productivity. Lecture Notes on Coastal and Estuarine Studies.* Springer-Verlag, New York, pp 19–34.

El-Sayed S Z (1988) Seasonal and interannual variabilities in Antarctic phytoplankton with reference to krill distribution. In: Sahrhage D (ed) *Antarctic Ocean and Resources Variability.* Springer-Verlag, New York, pp 101–19.

El-Sayed S Z, Stephens F C, Bidigare R R, Ondrusek M E (1990) Effect of ultraviolet radiation on Antarctic marine phytoplankton. In: Kerry K R, Hempel G (eds) *Antarctic Ecosystems. Ecological Change and Conservation.* Springer-Verlag, Berlin, pp 379–85.

Estep K W, Davis P G, Hargraves P E, Sieburth J McN (1986) How important are algal nanoflagellates in bactivory? *Limnol Oceanogr* **31**: 646–50.

Estep K W, MacIntyre F (1989) Taxonomy, life cycle, distribution and dasmotrophy of *Chrysochromulina*: a theory accounting for scales, haptonema, muciferous bodies and toxicity. *Mar Ecol Prog Ser* **57**: 11–21.

Everson I (1977) *The Living Resources of the Southern Ocean.* GLO/SO/77/1, FAO, Rome. 156 pp.

Everson I, Goss M (1991) Krill fishing activity in the south west Atlantic. *Antarct Sci* **3**: 351–8.

Flood P R, Deibel D, Morris C C (1992) Filtration of colloidal melanin from sea water by planktonic tunicates. *Nature* **355**: 630–2.

Foxton P (1966) The distribution and life history of *Salpa thompsoni* Foxton with observations on a related species *Salpa gerlachei* Foxton. *Discovery Rep* **34**: 1–116.

Frederick J E, Snell H E (1988) Ultraviolet radiation levels during the Antarctic spring. *Science* **241**: 438–40.

Fryxell G A, Kendrick G A (1988) Austral spring microalgae across the Weddell Sea ice edge: spatial relationships found along a northward transect during AMERIEZ 83. *Deep-Sea Res* **35**: 1–20.

Fuhrman J A, Azam F (1980) Bacterioplankton secondary production estimates for coastal waters of British Columbia, Antarctica, and California. *Appl Environ Microbiol* **39**: 1085–95.

Garrison D L (1991) An overview of the abundance and role of protozooplankton in Antarctic waters. *J Mar Syst* **2**: 317–31.

Garrison D L, Buck K R (1989*a*) Protozooplankton in the Weddell Sea, Antarctica: abundance and distribution in the ice-edge zone. *Polar Biol* **9**: 341–51.

Garrison D L, Buck K R (1989*b*) The biota of Antarctic pack ice in the Weddell Sea and Antarctic Peninsula regions. *Polar Biol* **10**: 211–19.

Garrison D L, Buck K R, Fryxell G A (1987) Algal assemblages in the Antarctic pack ice and in ice-edge plankton. *J Phycol* **23**: 564–72.

Garrison D L, Buck K R, Gowing M M (1991) Plankton assemblages in the ice-edge zone of the Weddell Sea during the austral winter. *J Mar Syst* **2**: 123–30.

Garrison D L, Gowing M M (1992) Protozooplankton. In: Friedman EI (ed) *Antarctic Microbiology* (in press).

Gibson J A E, Garrick R C, Burton H R (1990) Seasonal fluctuation of bacterial numbers near the Antarctic continent. *Proc NIPR Symp Polar Biol* **3**: 16–22.

Gibson J A E, Garrick R C, Burton H R, McTaggart A R (1989) Dimethylsulfide and the alga *Phaeocystis pouchetii* in Antarctic coastal waters. *Mar Biol* **104**: 339–46.

Gieskes W W C, Elbrächter M (1986) Abundance of nanoplankton-size chlorophyll-containing particles caused by diatom disruption in surface waters of the Southern Ocean (Antarctic Peninsula region). *Neth J Sea Res* **20**: 291–303.

Gowing M M (1989) Abundance and feeding ecology of Antarctic phaeodarian radiolarians. *Mar Biol* **103**: 107–18.

Grindley J R, David P (1985) Nutrient upwelling and its effects in the lee of Marion Island. In: Siegfried W R, Condy P R, Laws R M (eds) *Antarctic Nutrient Cycles and Food Webs.* Springer-Verlag, Berlin, pp 46–51.

Hada Y (1970) *The protozoan plankton of the Antarctic and subantarctic seas.* JARE Scientific Reports, Series E, No. 30.

Hanson R B, Lowery H K, Shafer D, Sorocco R, Pope D H (1983) Microbes in Antarctic water of the Drake Passage: Vertical patterns of substrate uptake, productivity and biomass in January 1980. *Polar Biol* **2**: 179–88.

Hara S, Tanoue E, Zenimoto M, Komaki Y, Takahashi E (1986) Morphology and distribution of heterotrophic protists along 75°E in the Southern Ocean. *Mem Natl Inst Polar Res Spec Issue* **40**: 69–80.

Hardy A C, Gunther E R (1935) The plankton of the South Georgia whaling grounds and adjacent waters, 1926–1927. *Discovery Rep* **11**: 1–456.

Hart T J (1934) On the phytoplankton of the South-West Atlantic and the Bellingshausen Sea. *Discovery Rep* **8**: 1–268.

Hart T J (1942) Phytoplankton periodicity in Antarctic surface waters. *Discovery Rep* **21**: 261–356.

Hasle G R (1969) An analysis of the phytoplankton of the Pacific Southern Ocean: abundance, composition and distribution during the Brategg Expedition 1947–48. *Hvalråd Skr* **52**: 1–168.

Hempel G (1983) FIBEX – An international survey in the Southern Ocean: review and outlook. *Mem Natl Inst Polar Res Spec Issue* **27**: 1–15.
Hempel G (1985) Antarctic marine food webs. In: Siegfried W R, Condy P R, Laws R M (eds) *Antarctic Nutrient Cycles and Food Webs.* Springer-Verlag, Berlin, pp 266–70.
Hempel G (1991) Life in the Antarctic sea ice zone. *Polar Rec* **27**: 249–54.
Hewes C D, Holm-Hansen O, Sakshaug E (1985) Alternative carbon pathways at lower trophic levels in the Antarctic food web. In: Siegfried W R, Condy P R, Laws R M (eds) *Antarctic Nutrient Cycles and Food Webs.* Springer-Verlag, Berlin, pp 277–83.
Heywood R B, Priddle J (1987) Retention of phytoplankton by an eddy. *Continental Shelf Res* **7**: 937–55.
Holm-Hansen O, Mitchell B G (1991) Spatial and temporal distribution of phytoplankton and primary production in the western Bransfield Strait region. *Deep-Sea Res* **38**: 961–80.
Hopkins T L (1987) Midwater foodweb in McMurdo Sound, Ross Sea, Antarctica. *Mar Biol* **96**: 93–106.
Hopkins T L, Torres J J (1988) The zooplankton community in the vicinity of the ice edge, western Weddell Sea, March 1986. *Polar Biol* **9**: 79–87.
Hopkins T L, Torres J J (1989) Midwater food web in the vicinity of a marginal ice zone in the western Weddell Sea. *Deep-Sea Res* **36**: 543–60.
Horner R (1990) Ice associated ecosystems. In: Medlin L K, Priddle J (eds) *Polar Marine Diatoms.* British Antarctic Survey, Cambridge, pp 9–14.
Huntley M E, Escritor F (1991) Dynamics of *Calanoides acutus* (Copepoda: Calanoida) in Antarctic coastal waters. *Deep-Sea Res* **38**: 1145–67.
Huntley M E, Karl D M, Niler P, Holm-Hansen O (1991*a*) Research on Antarctic coastal ecosystem rates (RACER): an interdisciplinary field experiment. *Deep-Sea Res* **38**: 911–41.
Huntley M E, Lopez M D G, Karl D M (1991*b*) Top predators in the Southern Ocean: a major leak in the biological carbon pump. *Science* **253**: 64–6.
Inoue H, Sugimura Y (1986) Distribution of pCO_2 and $\delta^{13}C$ in the air and surface seawater in the Southern Ocean, south of Australia. *Mem Natl Inst Polar Res Spec Issue* **40**: 454–61.
Iturriaga R, Mitchell B G (1986) Chroococcoid cyanobacteria: a significant component in the food web dynamics of the open ocean. *Mar Ecol Prog Ser* **28**: 291–7.
Jacka T H (1983) *A computer database for Antarctic sea ice extent.* ANARE Research Notes. No 13. Australian Antarctic Division, Hobart.
Jacques G (1989) Primary production in the open Antarctic Ocean during the austral summer. A review. *Vie Milieu* **39**: 1–17.
Jacques G (1991) Is the concept of new production-regeneration production valid for the Southern Ocean? *Mar Chem* **35**: 273–86.
Jacques G, Panouse M (1991) Biomass and composition of size fractionated phytoplankton in the Weddell–Scotia confluence area. *Polar Biol* **11**: 315–28.
Johnson P W, Sieburth J M (1979) Chroococcoid cyanobacteria; a ubiquitous and diverse phototrophic biomass. *Limnol Oceanogr* **24**: 928–35.
Karentz D (1991) Ecological considerations of Antarctic ozone depletion. *Antarct Sci* **3**: 3–11.
Karentz D, Cleaver J E, Mitchell D L (1991) Cell survival characteristics and molecular responses of Antarctic phytoplankton to ultraviolet-B radiation. *J Phycol* **27**: 326–41.
Karl D M, Holm-Hansen O, Taylor G T, Tien G, Bird D F (1991*a*) Microbial biomass and productivity in the western Bransfield Strait, Antarctica during the 1986–87 austral summer. *Deep-Sea Res* **38**: 1029–55.
Karl D M, Tilbrook B D, Tien G (1991*b*) Seasonal coupling of organic matter and particle flux in the western Bransfield Strait, Antarctica. *Deep-Sea Res* **38**: 1097–126.
Koike I, Holm-Hansen O, Biggs D C (1986) Inorganic nitrogen metabolism by Antarctic phytoplankton with special reference to ammonium cycling. *Mar Ecol Prog Ser* **30**: 105–16.
Kottmeier S T, Sullivan C W (1990) Bacterial biomass and production in pack ice of Antarctic marginal ice edge zones. *Deep-Sea Res* **37**: 1311–30.
Lancelot C, Mathot S, Becquevort S, Dandois J-M, Billen G (1992) *Carbon and nitrogen cycling through the microbial network of the marginal ice zone of the Southern Ocean with particular emphasis to the northwestern Weddell Sea*. Final Report. University of Brussels. 196 pp.
Lessard E J, Rivkin R B (1986) Nutrition of microzooplankton and macrozooplankton from McMurdo Sound. *Antarct J US* **21**: 187–8.
Letelier R M, Karl D M (1989) Phycoerythrin-containing cyanobacteria in surface waters of the Drake Passage during February 1987. *Antarct J US* **24**: 185–8.
Lizotte M P, Sullivan C W (1992) Biochemical composition and photosynthate distribution in sea ice microalgae of McMurdo Sound, Antarctica: evidence for nutrient stress during the spring bloom, *Antarct Sci* **4**: 23–30.
Lubbers G W, Gieskes W W C, del Castillo P, Salomons W, Bril J (1990) Manganese accumulation in the high pH microenvironment of *Phaecocystis* sp. (Haptophyceae) colonies from the North Sea. *Mar Ecol Prog Ser* **59**: 285–93.
Lutjeharms J R E, Waters N M, Allanson B R (1985) Oceanic frontal systems and biological enhancement. In: Siegfried W R, Condy P R, Laws R M (eds) *Antarctic Nutrient Cycles and Food Webs.* Springer-Verlag, Berlin, pp 11–21.
Manahan D T, Shilling F M, Welborn J R, Colwell S J (1990) Dissolved organic material in seawater as a source of nutrition for invertebrate larvae from McMurdo Sound, Antarctica. *Antarct J US* **25**: 206–8.
Marchant H J (1993) Biological impacts of seasonal ozone depletion. In: Hempel G (ed) *Proceedings of*

the SCAR Antarctic Science Conference. Springer-Verlag, Berlin (in press).

Marchant H J, Davidson A T, Kelly G (1991) UV-B protecting pigments in the alga *Phaeocystis pouchetii* from Antarctica. *Mar Biol* **109**: 391–5.

Marchant H J, Davidson A T, Wright S W (1987) The distribution and abundance of chroococcoid cyanobacteria in the Southern Ocean. *Proc NIPR Symp Polar Biol* **1**: 1–9.

Marchant H J, Nash G V (1986) Electron microscopy of gut contents and faeces of *Euphausia superba* Dana. *Mem Natl Inst Polar Res Spec Issue* **40**: 167–77.

Marchant H J, Scott F J (1993) Uptake of sub-micrometre particles and dissolved organic material by Antarctic choanoflagellates. *Mar Ecol Prog Ser* (in press).

Marr J W S (1962) The natural history and geography of Antarctic krill (*Euphausia superba* Dana). *Discovery Rep* **32**: 33–464.

Martin J H, Gordon R M, Fitzwater S E (1990) Iron in Antarctic waters. *Nature* **345**: 156–8.

Miller D G M, Hampton I (1989) *Biology and Ecology of the Antarctic Krill (*Euphausia superba *Dana): a Review.* BIOMASS Scientific Series No 9. 166 pp. SCAR & SCOR Scott Polar Research Institute, Cambridge.

Mitchell B G, Holm-Hansen O (1991) Observations and modelling of the Antarctic phytoplankton crop in relation to mixing depth. *Deep-Sea Res* **38**: 981–1007.

Mullins B W, Priddle J (1987) Relationships between bacteria and phytoplankton in the Bransfield Strait and Southern Drake Passage. *Br Antarct Surv Bull* **76**: 51–64.

Murphy E J, Morris D J, Watkins J L, Priddle J (1988) Scales of interaction between Antarctic krill and the environment. In: Sahrhage D (ed) *Antarctic Ocean and Resources Variability.* Springer-Verlag, New York, pp 120–30.

Nöthig E-M, Bodungen B von (1989) Occurrence and vertical flux of faecal pellets of probably protozoan origin in the southeastern Weddell Sea (Antarctica). *Mar Ecol Prog Ser* **56**: 281–9.

Nöthig E-M, Bodungen B von, Sui Q (1991) Phyto- and protozooplankton biomass during austral summer in surface waters of the Weddell Sea and vicinity. *Polar Biol* **11**: 293–304.

Nöthig E-M, Gowing M M (1991) Late winter abundance and distribution of phaeodarian radiolarians, other large protozooplankton and copepod nauplii in the Weddell Sea, Antarctica. *Mar Biol* **111**: 473–84.

Owens N J P, Priddle J, Whitehouse M J (1991) Variations in phytoplanktonic nitrogen assimilation around South Georgia and in the Bransfield Strait (Southern Ocean). *Mar Chem* **35**: 287–304.

Paasche E (1988) Pelagic primary production in nearshore waters. In: Blackburn TH, Sorensen J (eds) *Nitrogen Cycling in Coastal Marine Environments.* John Wiley & Sons, Chichester, pp 33–57.

Painting S J, Lucas M I, Stenton-Dozey J M E (1985) The South African SIBEX I Cruise to the Prydz Bay region, 1984: X. Biomass and production of bacterioplankton in Prydz Bay, Antarctica, and phytoplankton, detritus and bacterial relationships. *S Afr J Antarct Res* **15**: 42–52.

Perrin R, Lu P, Marchant H J (1987) Seasonal variation in marine phytoplankton and ice algae at a shallow Antarctic coastal site. *Hydrobiologia* **145**: 33–46.

Peters R H (1983) *The Ecological Implications of Body Size.* Cambridge University Press. 329 pp.

Pomeroy L R (1974) The ocean's food web: a changing paradigm. *BioScience* **24**: 499–504.

Priddle J (1990) The Antarctic planktonic ecosystem. In: Medlin L K, Priddle J (eds) *Polar Marine Diatoms.* British Antarctic Survey, Cambridge, pp 25–34.

Priddle J, Croxall J P, Everson I, Heywood R B, Murphy E J, Prince P A, Sear C B (1988) Large-scale fluctuations in distribution and abundance of krill – a discussion of possible causes. In: Sahrhage D (ed) *Antarctic Ocean and Resources Variability.* Springer-Verlag, Berlin, pp 169–82.

Probyn T A, Painting S J (1985) Nitrogen uptake by size-fractionated phytoplankton populations in Antarctic surface waters. *Limnol Oceanogr* **30**: 1327–32.

Quetin L B, Ross R M (1985) Feeding by Antarctic krill, *Euphausia superba*: does size matter? In: Siegfried W R, Condy P R, Laws R M (eds) *Antarctic Nutrient Cycles and Food Webs.* Springer-Verlag, Berlin, pp 372–7.

Riebesell U, Schloss I, Smetacek V (1991) Aggregation of algae released from melting sea ice: implications for seeding and sedimentation. *Polar Biol* **11**: 239–48.

Rivkin R B, Anderson M R, Gustafson D E Jr (1990) Seasonal pattern of bacteria and net, nano- and pico-phytoplankton in McMurdo Sound, Antarctica, during the austral spring. *Antarct J US* **25**: 195–6.

Rivkin R B, Putt M, Alexander S P, Meritt D, Gaudet I (1989) Biomass and production in polar planktonic sea ice microbial communities: a comparative study. *Mar Biol* **101**: 273–83.

Ross R M, Quetin L B (1986) How productive are Antarctic krill? *BioScience* **36**: 264–9.

Sakshaug E, Holm-Hansen O (1984) Factors governing pelagic production in polar oceans. In: Holm-Hansen O, Bolis L, Giles R (eds) *Marine Phytoplankton and Productivity. Lecture Notes on Coastal and Estuarine Studies.* Springer-Verlag, New York, pp 1–18.

Sakshaug E, Skjoldal H R (1989) Life at the ice edge. *Ambio* **18**: 60–7.

Sanders R W (1991) Mixotrophic protists in marine and freshwater ecosystems. *J Protozool* **38**: 76–81.

Sanders R W, Porter K (1988) Phagotrophic phytoflagellates. In: Marshall K C (ed) *Advances in Microbial Ecology.* Plenum, New York, pp 167–92.

Satoh H, Fukami K, Watanabe K, Takahashi E (1989) Seasonal changes in heterotrophic bacteria under fast ice near Syowa Station, Antarctica. *Can J Microbiol* **35**: 329–33.

Satoh H, Watanabe K (1991) A red waterbloom caused by the autotrophic ciliate *Mesodinium rubrum*, in the austral summer in the fast ice area near Syowa Station, Antarctica, with a note on their photosynthetic rate. *J Tokyo Univ Fish* **28**: 11–17.

Sherman K, Gold B D (1990) Perspective: Large marine ecosystems. In: Shermann K, Alexander L M, Gold B D (eds) *Large Marine Ecosystems: Patterns, Processes and Yields*. American Association for the Advancement of Science, Washington, pp vii–xi.

Sherr E (1988) Direct use of high molecular weight polysaccharide by heterotrophic flagellates. *Nature* **335**: 348–51.

Sieburth J M (1960) Acrylic acid, an 'antibiotic' principle in *Phaeocystis* blooms in Antarctic waters. *Science* **132**: 676–7.

Siegel V (1988) A concept of seasonal variation of krill (*Euphausia superba*) distribution and abundance west of the Antarctic Peninsula. In: Sahrhage D (ed) *Antarctic Ocean and Resources Variability.* Springer-Verlag, Berlin, pp 219–30.

Smetacek V, Scharek R, Nöthig E-M (1990) Seasonal and regional variation in the pelagial and its relationship to the life history cycle of krill. In: Kerry K R, Hempel G (eds) *Antarctic Ecosystems: Ecological Change and Conservation.* Springer-Verlag, Berlin, pp 103–14.

Smith R C, Prezelin B B, Baker K S, Bidigare R R, Boucher N P, Coley T, Karentz D, MacIntyre S, Matlick H A, Menzies D, Ondrusk M, Wan Z, Waters K J (1992) Ozone depletion: ultraviolet radiation and phytoplankton biology in Antarctic waters. *Science* **255**: 952–9.

Smith S L, Schnack-Schiel S B (1990) Polar Zooplankton. In: Smith W O (ed) *Polar Oceanography.* Academic Press, London, pp 527–8.

Smith W O Jr (1987) Phytoplankton dynamics in marginal ice zones. *Oceanogr Mar Biol Ann Rev* **25**: 11–38.

Smith W O Jr, Keene N K, Comiso J C (1988) Interannual variability in estimated primary productivity of the marginal ice zone. In: Sahrhage D (ed) *Antarctic Ocean and Resources Variability.* Springer-Verlag, Berlin, pp 131–9.

Smith W O Jr, Nelson D M (1985) Phytoplankton bloom produced by a receding ice edge in the Ross Sea: spatial coherence with the density field. *Science* **227**: 163–7.

Smith W O Jr, Nelson D M (1986) Importance of ice edge phytoplankton production in the Southern Ocean. *BioScience* **36**: 251–7.

Smith W O Jr, Sakshaug E (1990) Polar phytoplankton. In: Smith W O Jr (ed) *Polar Oceanography.* Academic Press, London, pp 477–525.

Stoecker D K, Michaels A E, Davis L H (1987) Large proportion of marine planktonic ciliates found to contain functional chloroplasts. *Nature* **326**: 790–2.

Stretch J J, Hamner P P, Hamner W M, Michel W C, Sullivan C W (1988) Foraging behavior of Antarctic krill *Euphausia superba* on sea ice microalgae. *Mar Ecol Prog Ser* **44**: 131–9.

Sullivan C W, McCalin C R, Comiso J C, Smith W O Jr (1988) Phytoplankton standing crops within an Antarctic ice edge assessed by satellite remote sensing. *J Geophys Res* **93**: 12487–98.

Sullivan C W, Cota G F, Krempin D W, Smith W O Jr (1990) Distribution and abundance of bacterioplankton in the marginal ice edge zone of the Weddell–Scotia Sea during austral spring. *Mar Ecol Prog Ser* **69**: 239–52.

Tanoue E, Hara S (1986) Ecological implications of fecal pellets produced by the Antarctic krill *Euphausia superba* in the Antarctic Ocean. *Mar Biol* **91**: 359–69.

Thomsen H A, Buck K R, Coale S L, Garrison D L, Gowing M M (1990) Loricate choanoflagellates (Acanthoecidae, Choanoflagellida) from the Weddell Sea, Antarctica. *Zool Script* **19**: 367–87.

Tilzer M M, Elbrächter M, Gieskes W W C, Beese B (1986) Light-temperature interactions in the control of photosynthesis in Antarctic phytoplankton. *Polar Biol* **5**: 105–11.

Ulanowicz R E (1991) Formal agency in ecosystem development. In: Higashi M, Burns T (eds) *Theoretical Studies of Ecosystems: the Network Perspective.* Cambridge University Press, pp 58–70.

Vincent W F (1988) *Microbial Ecosystems of Antarctica.* Cambridge University Press. 304 pp.

Waterbury J B, Watson S W, Guillard R R L, Brand L E (1979) Widespread occurrence of a unicellular, marine, planktonic, cyanobacterium. *Nature* **277**: 293–4.

Weber L H, El-Sayed S Z (1987) Contributions of the net, nano- and picoplankton to the phytoplankton standing crop and primary productivity in the Southern Ocean. *J Plankt Res* **9**: 973–94.

Weber L H, El-Sayed S Z, Hampton I (1986) The variance spectra of phytoplankton, krill and water temperature in the Antarctic Ocean south of Africa. *Deep-Sea Res* **33**: 1327–43.

Wilson D L, Smith W O Jr, Nelson D M (1986) Phytoplankton bloom dynamics of the Eastern Ross Sea ice edge. I. Primary productivity and species-specific production. *Deep-Sea Res* **33**: 1375–87.

Trophic relationships and trends in population size and reproductive parameters in Antarctic high-level predators

K.-H. KOCK & Y. SHIMADZU

Introduction

THE PRINCIPAL objective of the BIOMASS Programme was 'to gain a deeper understanding of the structure and dynamic functioning of the Antarctic marine ecosystem as a basis for the future management of potential living resources' (BIOMASS, 1977). These potential resources included key organisms of the Southern Ocean's† food webs, such as krill, cephalopods, fish, birds, seals and whales. With the exception of krill all belong to the higher trophic levels with birds, seals and whales being top level predators.

The preparation of the BIOMASS Programme since 1974 and its final launching in 1977 has to be seen against the background of the history of exploitation in the Southern Ocean since the late eighteenth century. Soon after the discovery of a resource a major industry developed and the resource was depleted within decades, as in the case of fur seals and whales (Bonner, 1984), or even years, as in the case of the marbled notothenia, *Notothenia rossii* (Kock, 1991), without adequate measures for a rational management being discussed or developed for most species. The only industries for which regulations were implemented at an early stage were those for elephant seals in 1909 and shore-based whaling at South Georgia at about the same time (Headland, 1984). Despite much research, available information was inadequate for management decisions when exploitation of the next lower trophic level, i.e. of krill (*Euphausia superba*), started to develop (Miller & Hampton, 1989).

To achieve the very broad objective of the BIOMASS Programme, research had been specified into a number of fields:

1. Autecological studies on whales, seals, penguins, fish and squid; assessment of their standing stock/biomass and production in selected areas of the Southern Ocean.
2. Development of models to improve our understanding of the quantitative interaction between different elements of the ecosystem and the effects of climate, whaling and krill fishing on structure and efficiency of the food chain.
3. Compilation and analysis of data from exploratory and commercial fishing.
4. Provision of scientific information on the Antarctic ecosystem to the scientific community, governments, industries and other concerned bodies (BIOMASS, 1977).

Although it was part of the conceptual framework of the BIOMASS Programme, a large share of the research on higher trophic levels was carried out under the auspices of the International Whaling

† The Southern Ocean as recognized for the purposes of this article encompasses the area of the Convention on the Conservation of Antarctic Marine Living Resources (CCAMLR).

Commission (IWC) and the SCAR International Survey of Antarctic Seabirds (ISAS) or became an integral part of the fish stock assessment work and of the Ecosystem Monitoring Programme of the Commission for the Conservation of Antarctic Marine Living Resources (CCAMLR) before the termination of the BIOMASS Programme.

A wealth of scientific publications has arisen from this research in the past 15 years, which it is impossible to summarize adequately here. Thus, in reviewing what has been accomplished within the BIOMASS Programme on higher trophic levels in the food webs of the Southern Ocean, we have focused here on population sizes and trends and their possible causes, on changes in reproductive parameters, and on interactions within the higher trophic levels and with other components of the food webs.

Population sizes and trends

Cephalopods

There is still little information available on either the production or standing stock biomass of cephalopods in the Southern Ocean. Many of the *ca.* 50 cephalopod species known to occur south of the South Polar Front Zone reach their southernmost limit of distribution in the Ice-free Zone. This impoverishment of the squid fauna towards the south is paralleled by a similar reduction in the diversity of their main prey, mesopelagic fish (Kock, 1992). Fewer than 25 cephalopod species penetrate as far south as the Seasonal Pack-ice Zone and the High-Antarctic Zone in appreciable numbers (Lubimova, 1985). About 50% of them are benthic octopods.

A very crude estimate of cephalopod biomass extrapolated from annual squid consumption of 35 Mt by whales, seals and birds suggested a cephalopod biomass of more than 100 Mt. Although this biomass is partly made up by octopods the biomass estimate seemed to be very high by oceanic standards (Clarke, 1983, 1985). A production estimate of 17.6 Mt derived from consumption rates by predators (Everson, 1977) would result in a standing stock biomass of 5–9 Mt using the production: biomass ratio of 2–3.5 given by Soviet authors (Lubimova, 1985). Both biomass estimates appear to be little more than intelligent guesses. The only single species estimate available is one for the ommastrephid squid *Martialia hyadesi* in the Scotia Sea: about 330–380 kt have been estimated to be consumed annually by predators (primarily elephant seals) (Rodhouse, 1991).

Fish

Of *ca.* 270 fish species known from the Southern Ocean, 12 species (separated into 30 stocks) are or have been subject to regular commercial exploitation (Kock, 1991). Since 1984, this exploitation has been regulated by the Commission for the Conservation of Antarctic Marine Living Resources (CCAMLR). It is only for a number of these (abundant) stocks that population (stock) sizes and population trends are available.

Lanternfish (myctophid) biomass between 48 and 56°S and 8 and 48°W, which was mainly composed of *Electrona carlsbergi* but included also some other myctophids, was estimated to be 2 Mt in the summer–autumn period of 1987–9 (Filin *et al.*, 1991). The biomass of mesopelagic fish in the marginal ice zone of the Scotia Sea between the surface and 1000 m has been estimated to be 3.27 t km^{-2} (Torres *et al.*, 1984). The total biomass of Antarctic myctophids has been estimated to be 70–200 Mt (Lubimova *et al.*, 1987). However, the basis for this last figure is unclear and may also refer to myctophids that have their centre of abundance in or north of the Antarctic Convergence, such as *Electrona carlsbergi* (Kock, 1992).

There is evidence that a demersal fish biomass of *ca.* 750 kt had been present on the shelf around South Georgia (excluding Shag Rocks) prior to the beginning of commercial exploitation in 1969–70, with *Notothenia rossii* making up *ca.* 75% of it (Table 1; Fig. 1). This would correspond to a biomass of *ca.* 20–22 t km^{-2}. By 1975–6, fish biomass was only *ca.* 40% of this value owing to the removal of most *N. rossii* (Fig. 1). It would have been even less, if two strong year-classes of the mackerel icefish, *Champsocephalus gunnari*, had not been recruited to the stock. Other fish stocks which declined substantially after only a few years of fishing were the green notothenia, *Gobionotothen gibberifrons*, the Scotia Sea icefish, *Chaenocephalus aceratus*, and the South Georgia icefish, *Pseudoch-*

Table 1. *Sizes (kt) of individual fish stocks prior to commercial harvesting*

Species	Location	Pre-exploitation	1989–90
N. rossii	South Georgia	580	5–10
N. rossii	Kerguelen	180–290	25–30
L. squamifrons	Kerguelen	100–200	2–5
C. gunnari	S. Orkney Is/ Ant. Peninsula	280	8–10

Source: From Kock (1992).

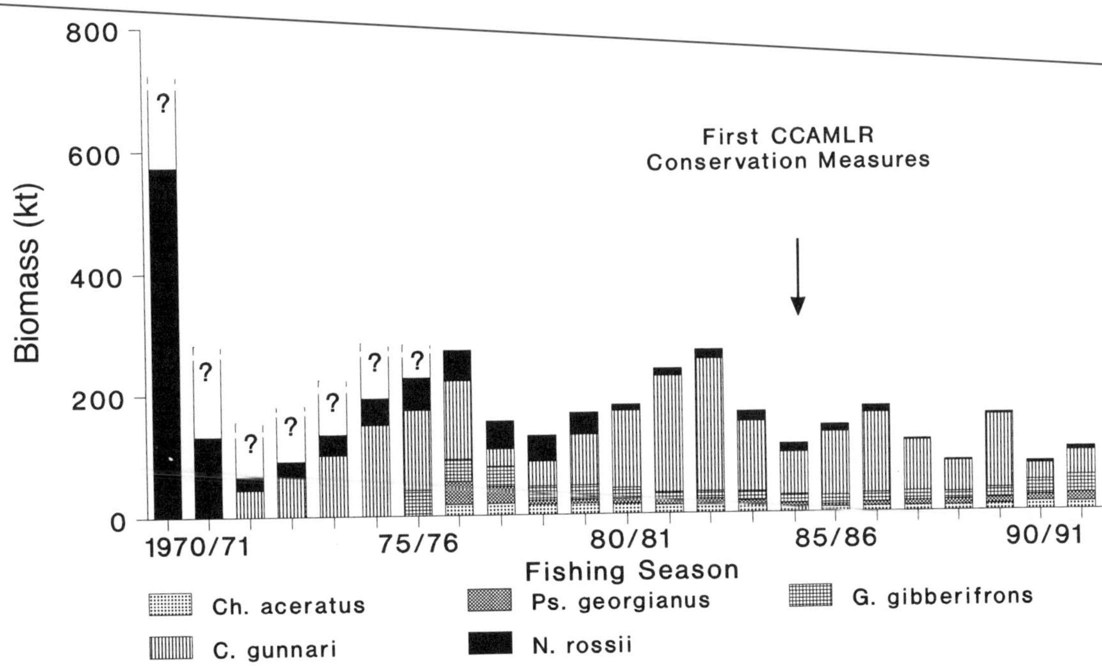

Fig. 1. Trends in biomass of the most abundant fish species around South Georgia.

aenichthys georgianus. At the end of the 1977–8 fishing season, standing stock biomass of fish around South Georgia was apparently less than 20% of its pristine size. Since then, it has fluctuated around 20–25% (Kock, 1991, 1992), largely dependent on the strength of the recruiting year-class of *Champsocephalus gunnari*. Stock size of *C. gunnari* at South Georgia was probably of the order of 50 kt when exploitation commenced. This was comparatively low in relation to stock sizes of 130–270 kt observed in years thereafter (Fig. 1) when strong year-classes recruited to the stock (Kock & Köster, 1989).

Fish biomass on the shelf of the South Orkney Islands should have been of the order of 250 kt prior to exploitation in 1977–8, of which at least 60–70% comprised *Champsocephalus gunnari* (Table 1). This would correspond to a biomass of 10–12 t km^{-2}. By 1979–80, fish biomass was probably less than 20% of the pristine biomass and has fluctuated around that value since then (Kock, 1991).

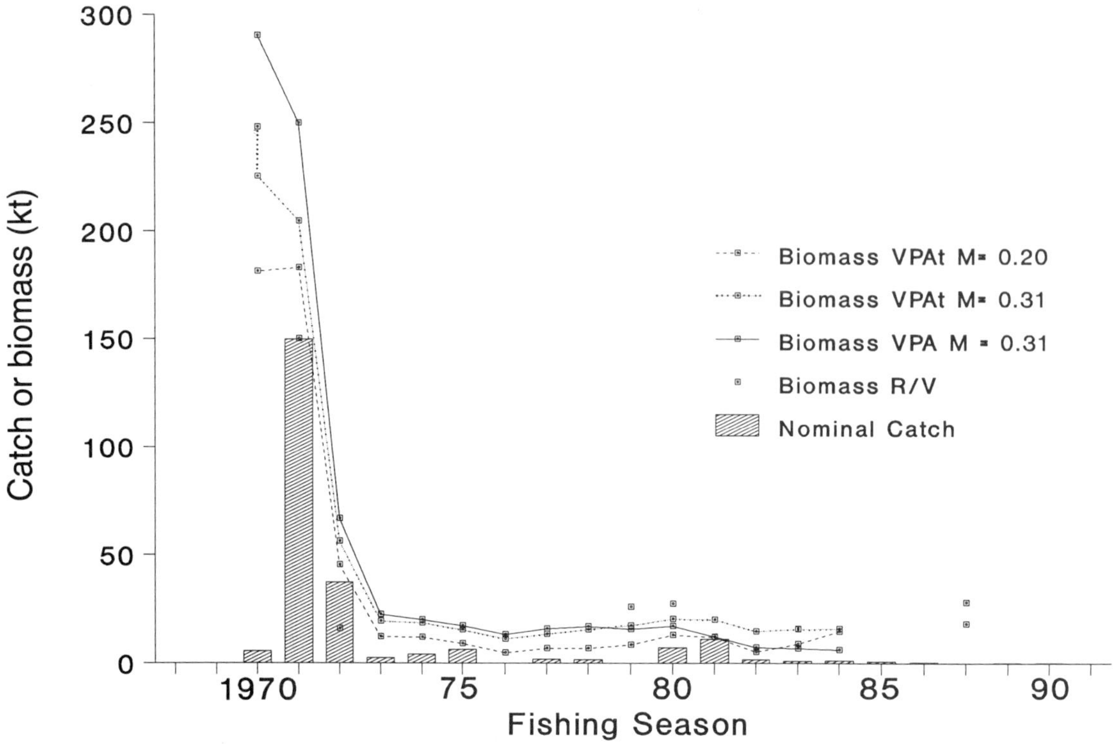

Fig. 2. Nominal catch and trends in biomass of *Notothenia rossii* around the Kerguelen Islands (from Kock, 1991). VPAt, virtual population analysis tuned to indices of biomass estimates; M, instantaneous rate of natural mortality; R/V, research vessel.

Demersal fish biomass on the Kerguelen shelf prior to exploitation in 1970–1 should have been of the order of 420–550 kt with *Notothenia rossii* and the grey notothenia, *Lepidonotothen squamifrons*, making up 50% and 25% of the biomass, respectively (Table 1). This would correspond to a biomass of 4–5 t km^{-2}. In the period 1974–6 the biomass may have been reduced to less than 200 kt mostly owing to a reduction in stock size of *N. rossii* (Fig. 2). Since the 1980s, the biomass is unlikely to have been much more than 120 kt owing to the low stock sizes of *N. rossii* and *L. squamifrons* (Duhamel, 1987; Kock, 1991).

Fish biomass in the High-Antarctic Zone appears to be one order of magnitude lower than in the Seasonal Pack-ice Zone and around islands north of it. It has been estimated to be 0.9 t km^{-2} on the shelf of the eastern Weddell Sea and only 0.3 t km^{-2} in its southern parts (Ekau, 1990). Stock size for the most dominant species in the Weddell Sea, the Antarctic silverfish *Pleuragramma antarcticum*, was estimated to be 600 kt (Hubold, 1992).

The trends in fish stock sizes primarily reflect the impact large-scale harvesting has had on most of these stocks. They were already overexploited only a few years after fishing had commenced. There is evidence that in at least three stocks, *Notothenia rossii* around South Georgia and the Kerguelen Islands and *Chaenocephalus aceratus* at South Georgia, stock sizes have been reduced to such a level that recruitment has been affected. In spite of the prohibition of directed fishing on *N. rossii* and a low by-catch provision, the stocks around South Georgia and Kerguelen Islands show little indication of a

Table 2. *Minimum breeding populations and biomass of penguins in the Southern hemisphere*[a]

Species	Minimum breeding population (pairs)	Minimum estimated total biomass (kg)	Estimated biomass, Southern Ocean only (kg)
Emperor	193 900	12 622 890	12 622 890
King	1 070 800	32 445 240	30 743 900
Adélie	2 465 800	21 700 800	21 700 800
Chinstrap	7 490 200	62 168 970	62 168 970
Gentoo	298 000	3 516 400	2 629 300
Macaroni	11 841 600	111 311 000	111 105 000
Rockhopper	3 687 600	19 175 500	3 765 850
Total	27 047 900	262 940 800	244 736 710

[a] Includes Iles St. Paul et Amsterdam, Macquarie Island, Auckland, Antipodes, Snares Islands, all South American colonies and the Falkland/Malvinas Islands.
Source: From Woehler (1992).

recovery from the heavy depletion in the 1970s (Duhamel, 1990; Kock, 1991). It has been hypothesized that increased predation by Antarctic fur seals on juvenile *N. rossii* in the fjords may have additional adverse effects on the recovery of the stock around South Georgia. Species which show an upward trend in stock size since the prohibition of bottom trawling around South Georgia are *Gobionotothen gibberifrons, Chaenocephalus aceratus* and *Pseudochaenichthys georgianus* (Kock, 1991, 1992). As with many other exploited fish populations, variability in the year-class strength has been identified as the second major variable affecting population sizes and trends in some fish species from the Southern Ocean. Recruitment (to the age at first capture) in *N. rossii* and *G. gibberifrons* at South Georgia has shown a high degree of stability (Anon, 1989; Kock & Köster, 1989). In contrast, year-class strength in *Champsocephalus gunnari* at South Georgia and on the Kerguelen Plateau varied significantly. Stock size is strongly dependent on the strength of the recruiting cohort (Duhamel, 1987; Kock & Köster, 1989). The absence of a clear stock–recruitment relationship in *C. gunnari* at South Georgia suggests that recruitment is probably largely determined by environmental factors and only loosely associated with parental stock size (Anon, 1989; Kock & Köster, 1989). In contrast, the two populations of *C. gunnari* on the Kerguelen Plateau exhibit clear stock–recruitment relationships in that a strong year-class is derived from the first spawning of the previous abundant cohort. It is unknown so far whether this is natural or whether this is a consequence of exploitation in the early 1970s (Duhamel, 1987; Duhamel & Hureau, 1990).

Seabirds

About 40 species of seabird are known to breed in the Southern Ocean or on islands near or in the Antarctic Convergence (Croxall, 1984). The most abundant species are chinstrap, *Pygoscelis antarctica,* macaroni, *Eudyptes chrysolophus*, and Adélie penguins, *Pygoscelis adeliae*, with breeding populations of 3–10 × 10^6 breeding pairs each (Table 2). More than 95% of the biomass of the first two penguin species is confined to the Atlantic Ocean sector. Breeding populations of the albatrosses and giant petrels, which are (with the exception of the southern giant petrel, *Macronectes giganteus*) confined to the peripheral, ice-free part of the Southern Ocean, are much smaller and commonly of the order of one to several tens of thousands of breeding pairs (Table 3). The total population sizes of smaller petrels, storm petrels and diving petrels are much less well known. Population sizes of breeding pairs in petrels, with Antarctic fulmar, *Fulmarus glacialoides*, and Antarctic prion, *Pachyptila desolata*, being the most abundant, are likely to be of the order of 10^8 breeding pairs, thus outnum-

Table 3. *Number of breeding pairs of albatrosses and giant petrels in the Southern Ocean*

Species	No. of breeding pairs (thousands)
Northern giant petrel *Macronectes halli*	5–6
Southern giant petrel *Macronectes giganteus*	35
Wandering albatross *Diomedea exulans*	20–25
Black-browed albatross *Diomedea melanophris*	100+
Grey-headed albatross *Diomedea chrysostoma*	60
Light-mantled sooty albatross *Phoebetria palpebrata*	27

Source: From Croxall (1987).

bering penguins. Storm petrels (the most prominent representative of which is Wilson's storm petrel, *Oceanites oceanicus*) and the diving petrels *Pelecanoides georgicus* and *P. urinatrix* are likely to account for 10^6–10^7 breeding pairs each. The remaining groups, shags, gulls and skuas and terns, are much less abundant with population sizes of 10^4–10^5 breeding pairs each (Croxall, 1987). The number of non-breeders in each of the species is unknown but may be of the same order of magnitude as the breeding population.

With the exception of penguins in some areas, overall population sizes even for species or regions considered to be well-censused still tend to tell us little about the populations status and trends. This refers in particular to species nesting in crevices, burrows and cliffs, such as petrels, or those breeding in more dispersed aggregations, such as skuas or terns. The improvement in logistic facilities in the past 10–15 years has led to the discovery of many previously unknown rookeries, and the accuracy of regional population estimates is improving continuously (Wilson, 1983; Poncet & Poncet, 1987; Shuford & Spear, 1988; Woehler, 1992).

Trends in population sizes are thus only detectable in rookeries that are regularly surveyed and censused. These trends, however, may not be consistent from colony to colony or region to region but may show considerable spatial and temporal variability. Wandering albatrosses, *Diomedea exulans*, southern giant petrels and Antarctic terns, *Sterna vittata*, have declined over all of their breeding range (Croxall *et al.*, 1990; Weimerskirch *et al.*, 1987; Rootes, 1988; Woehler, 1992). The rate of decline in wandering albatrosses, however, varied considerably between areas (islands) and colonies (Fig. 3). The breeding population of wandering albatrosses at Bird Island (South Georgia) declined by a mean rate of 1% per year between 1960 and 1988 although breeding success increased by about the same rate (1.2%) (Croxall *et al.*, 1990). On the Crozet Islands, the wandering albatross population was reduced by 50% during the 1970s, but has stabilized since 1977 on Possession Island (Weimerskirch *et al.*, 1987). Northern giant petrels, *Macronectes halli*, increased at South Georgia (Hunter, 1984), but decreased at other sites. Despite considerable short-term fluctuations populations of black-browed and grey-headed albatross, *Diomedea melanophris* and *D. chrysostoma*, and Gentoo penguin, *Pygoscelis papua*, appear to have been stable at South Georgia (Fig. 4) (Croxall *et al.*, 1988) while black-browed albatrosses on the Kerguelen Islands and the Crozet Islands declined (Weimerskirch *et al.*, 1987). Possible adverse effects of fishing operations on black-browed albatross at South Georgia may have been counterbalanced by increased breeding success due to the increased food availability from the fishery. At Beauchêne Island south of the Falkland/Malvinas Islands, for example, black-browed albatrosses obtain 10–15% of their food requirements between February and April from the fishery for the

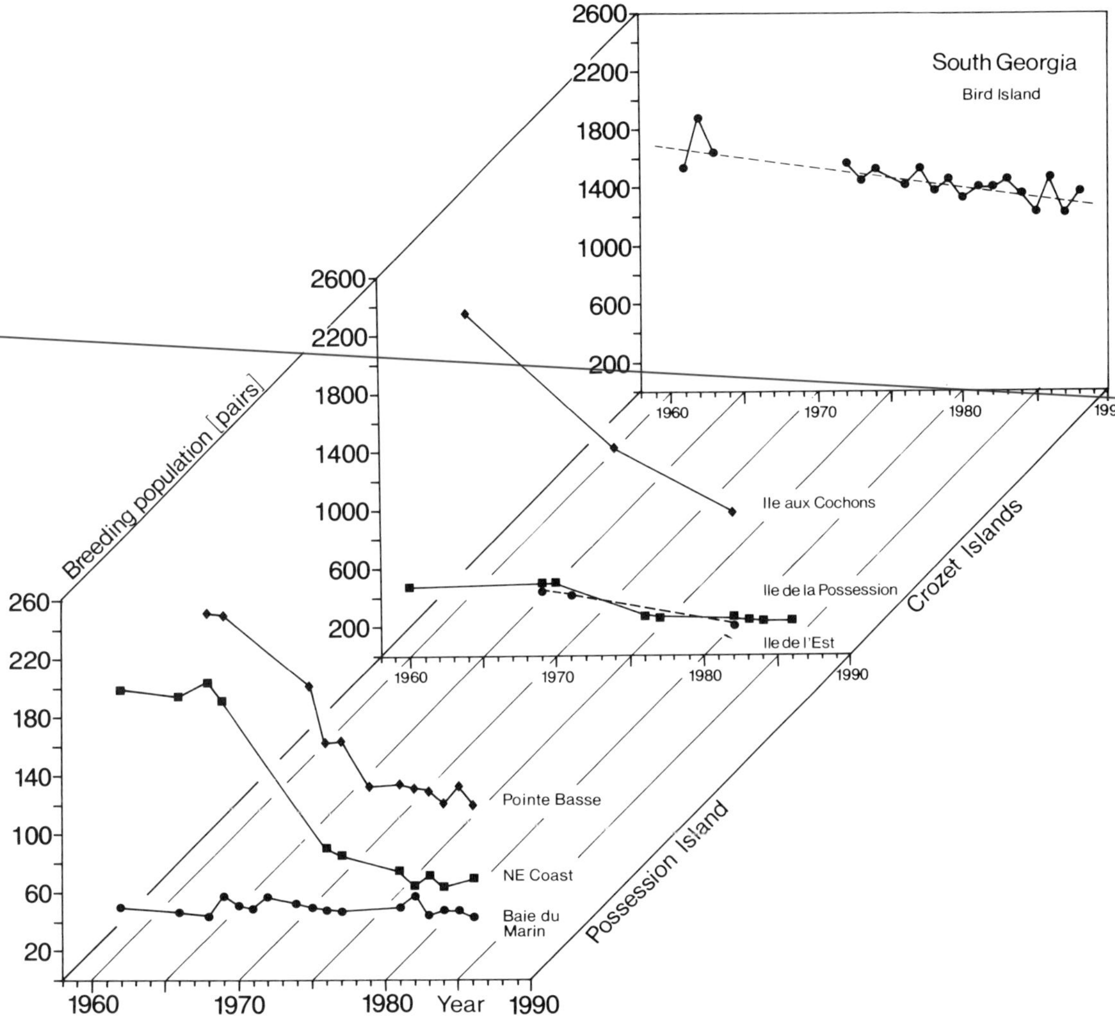

Fig. 3. Trends in breeding populations of wandering albatross in various localities (modified from Croxall *et al.*, 1990; Jouventin & Weimerskirch, 1990).

squid *Loligo gahi* (Thompson, 1990). King, chinstrap and Adélie penguins increased over almost their entire breeding range in the Southern Ocean.

Although precise data are lacking, populations of burrowing petrels (prions, storm petrels, diving petrels) on sub-Antarctic islands, such as South Georgia, Marion Island, Macquarie Island and the Crozet and the Kerguelen Islands, appear to have suffered severely from introduced predators, such as cats, rats, mice and, on Macquarie Island, weka rails (Brothers, 1984; Croxall, 1987). Totals of 440 000 and 1902 000 birds have been estimated to be killed on Marion Island in 1975 and 1982 by the feral cat population (Van Rensburg, 1985). Alien predators are believed to be responsible for the scarcity or even elimination of the common diving petrel, *Pelecanoides urinatrix* and storm petrels on Marion Island (Van Aarde, 1980) and the endemic

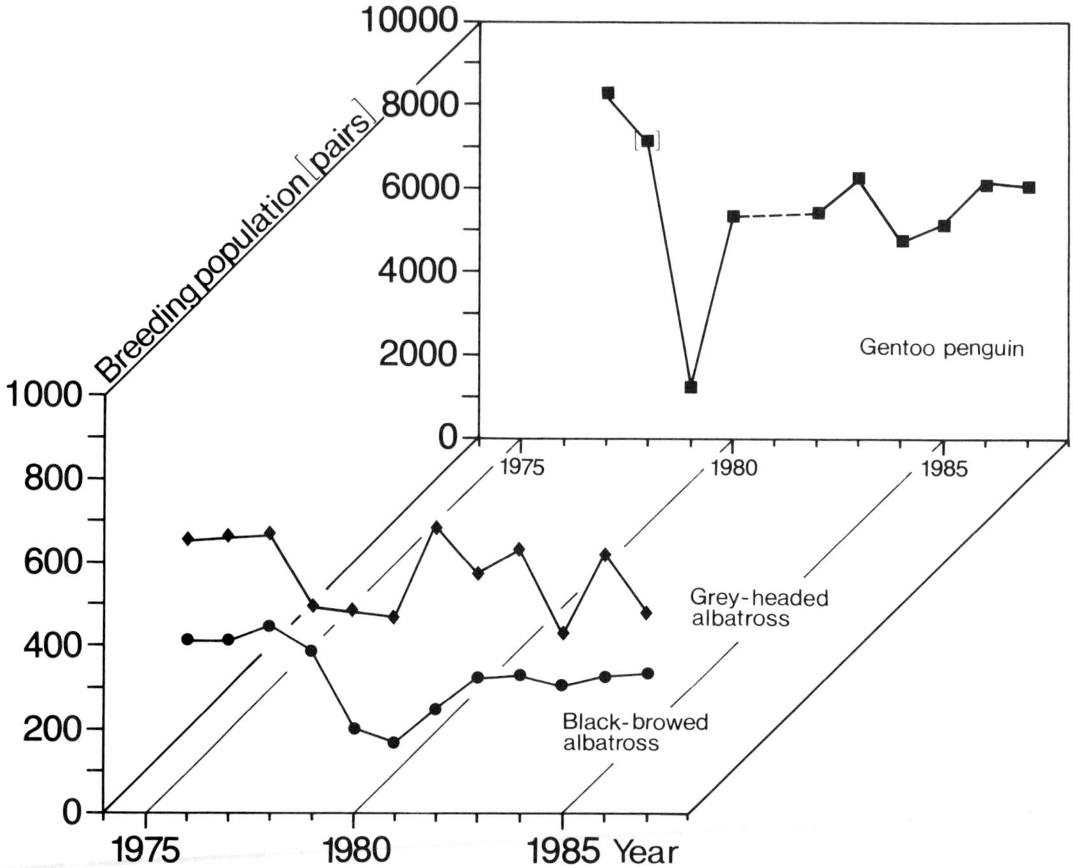

Fig. 4. Trends in breeding populations of gentoo penguin, black-browed albatross and grey-headed albatross at Bird island (South Georgia) (modified from Croxall *et al.*, 1988).

Antarctic pipit, *Anthus antarcticus*, from the mainland of South Georgia (Croxall, 1987). A number of factors have been identified or hypothesized to cause medium or long-term changes in seabird populations of the Southern Ocean. These may be very different between species or even within one species in that regions or even individual colonies are affected by different factors.

The ongoing expansion of king penguin colonies in many areas appears to be a response to exploitation to some extent (Rounsevell & Copson, 1982; Croxall *et al.*, 1988; Gales & Pemberton, 1988; Jouventin & Weimerskirch, 1990) although nineteenth-century exploitation does not fully explain the continuing increase in king penguins (JP Croxall, personal communication). Wandering albatross populations and probably also black-browed albatross and white-chinned petrel populations (*Procellaria aequinoctialis*) suffered from incidental mortality in longline fisheries on tuna in lower latitudes outside and on Patagonian toothfish (*Dissostichus eleginoides*) inside the Southern Ocean, and also from entanglement in net monitor cables and deliberate trapping and shooting during fishing operations (Weimerskirch & Jouventin, 1987; Croxall & Prince, 1990; Croxall *et al.*, 1990;

Anon, 1991*a*; Duhamel, 1991). This is evidenced both from recoveries of ringed wandering albatrosses from fishing operations (Croxall & Prince, 1990, Anon, 1991*a*) and from direct observations during longline fishing for tuna and Patagonian toothfish and pelagic trawling for mackerel icefish (Brothers, 1991; Anon, 1991*b*; Duhamel, 1991).

Human disturbance and incidental killing during fishing operations appear to be the major factors causing declines in southern giant petrels (*Macronectes giganteus*) and most northern giant petrel colonies. The increase of the northern giant petrel population at South Georgia has been attributed to a better survival of their chicks owing to the considerable expansion in breeding Antarctic fur seals, which provide an abundance of carrion for giant petrel nestlings (Hunter, 1984). Human disturbance is also the most likely cause for the decline of Antarctic terns (*Sterna vittata*) in the vicinity of shore stations (Rootes, 1988; Peter *et al.*, 1989).

Increases in chinstrap and Adélie penguin rookeries and the extension of the breeding range of chinstrap penguins along the west of the Antarctic Peninsula (Parmelee & Parmelee, 1987; Poncet & Poncet, 1987) have commonly been associated with a surplus of krill becoming available to these two penguin species after the dramatic reduction of the great baleen whales. This widely accepted 'whale reduction – krill surplus hypothesis' (Sladen, 1964) has been questioned recently by the presentation of equal evidence that increases in chinstrap colonies in the South Shetland Islands and Adélie penguins in the Ross Sea region may have been due to a gradual decrease in cold winters during which extensive ice covers had limited their access to food in winter (Taylor *et al.*, 1990; Fraser *et al.*, 1992).

Seals

Seven species of seal are found in the Southern Ocean. Four of them, crabeater seal, *Lobodon carcinophagus*, leopard seal, *Hydrurga leptonyx,* Weddell seal, *Leptonychotes weddellii*, and Ross seal, *Ommatophoca rossii*, are true Antarctic seals mostly confined to the pack- and fast-ice zones. The other three species occur abundantly on the sub-Antarctic islands (elephant seal, Antarctic fur seal) or are rare vagrants from north of the Antarctic Convergence (sub-Antarctic fur seal) (Laws, 1984).

Populations of the latter three species were depleted in the nineteenth century, those of the fur seals to very low levels. Ice seals have been taken regularly in small numbers to provide food for dog teams or irregularly during exploratory sealing operations in the pack-ice, such as in 1892–4 (32 558 seals in the Antarctic Peninsula region) (Bonner, 1985), in 1963–4 (861 seals in the South Orkney Islands region) (Oritsland, 1970), and in 1986–7 (4800 ice seals in the western Pacific Ocean sector) (Anon, 1988). It is unlikely that the *ca*. 39 000 seals taken during the sealing operations between 1892 and 1987 and those *ca*. 9200 ice seals killed for dog food between 1964 and 1985 (Laws, 1988), both taken over a wide geographical range, have affected any of the ice seal populations. Population sizes for the four true Antarctic seal species in the Southern Ocean are given in Table 4.

The population estimates for crabeater seals are considerably lower than a previous estimate. This down-scaling has been largely attributed to refined analytical procedures. However, it has been suggested from replicate censuses in the Pacific Ocean in 1972 and 1973 and in 1983 and in the western Weddell Sea in 1968 and 1969 and in 1983 that the crabeater seal populations in these two regions may have declined dramatically (Erickson & Hanson, 1990). Erickson & Hanson (1990) have speculated about a possible relationship between the depletion of most of the whale stocks by the end of the 1960s and initial recovery in the 1980s and a peak and subsequent decline of crabeater seals at about the same time (Erickson & Hanson, 1990). However, we are very reluctant to follow this speculation for a number of reasons:

1. Dispersal of crabeater seals may have been different in the census years (Erickson & Hanson, 1990).
2. None of the abundance estimates included a coefficient of variation allowing for some assessment of the degree of accuracy of the abundance estimates between areas and years.
3. Data to interpret the previous status of crabeater seals and competitive interactions with whales are absent. Theoretical arguments as to taxa advantaged in situations of competitive release suggest that benefits from the reduction in baleen whales are likely to occur to similar-sized predators with the broadest niche overlap. This in the marginal ice zone would be minke whales but is less likely to be seals or birds (Murphy *et al.*, 1988).

Table 4. *Estimates of stock sizes (in millions) of ice seals present in the Southern Ocean*

Species	Source of data	
	Erickson & Hanson (1990)	Laws (1984)
Crabeater seal	11–12	15–30
Weddell seal	0.9	0.8
Leopard seal	0.35	0.22–0.44
Ross seal	0.13	0.22

Until further evidence becomes available we feel unable to decide whether this change in abundance of crabeater seals has really occurred, or is related to an altered distribution or is primarily artefactual from the methods.

Recent (tentative) observations suggest a coupling of periodic fluctuations in crabeater, Weddell and leopard seal populations with large-scale oceanographic variations related to the El Niño–Southern Oscillation (ENSO) (Testa *et al.*, 1991). Although little is known on the relationship of ENSO to the Southern Ocean, among possible events are the disruption of mesoscale oceanic gyres and a resulting dispersal of high krill concentrations (Priddle *et al.*, 1988) or other effects on food availability of seals (Testa *et al.*, 1991). Strong year-classes of crabeater seals from the Antarctic Peninsula were separated by 4–5 year intervals. Dispersal of juvenile leopard seals and fluctuations in the reproductive rate of Weddell seals occurred at about the same interval. Events in Weddell seals were in general in phase with the Southern Oscillation Index whereas the relationship was less direct in the other two species. However, one would expect time lags and mediating oceanographic and/or weather effects due to the remoteness of the Southern Ocean from the tropics rather than direct couplings as in the case of Weddell seals. However, even in Weddell seals causal pathways will be very difficult to establish (Testa *et al.*, 1991).

Population numbers of elephant seals in the Atlantic Ocean sector are of the order of 300–400 $\times 10^3$ individuals (Laws, 1984). Populations in the Indian Ocean sector are probably at about the same level, while that in the Pacific Ocean sector (Macquarie Island) may be in the order of 50–60 $\times 10^3$ individuals. Elephant seals are declining over most of their distributional range (Anon, 1991*c*). The only population in the Southern Ocean which does not appear to decline is that at South Georgia. However, it should be noted that the suggestion of an apparently stable population at South Georgia is based upon only two censuses of pup production made 35 years apart (Anon, 1991*c*). Censuses on the Kerguelen Islands (Courbet Peninsula), for example, exhibit a considerable interannual variability (Jouventin & Weimerskirch, 1990). Despite these doubts of possible fluctuations in abundance of the South Georgia stock, it appears that this stock has not shown the long-term population decline of many other stocks. This may imply that the dynamics of the South Georgia stock are different from those of stocks elsewhere. After a decline prior to the 1980s, the stock on the Courbet Peninsula (Kerguelen Islands) appears to have stabilized at about two thirds of its initial size (Anon, 1991*c*; Jouventin & Weimerskirch, 1990).

Several hypotheses were formulated as offering possible explanations for the decline of southern elephant seal stocks (Anon, 1991*c*):

1. shortage of males in the population;
2. predation;
3. previous population overshoot over carrying capacity, followed by compensatory decline;
4. disease;
5. competition for prey with commercial fisheries;
6. broad ecological change in climate or in the ecosystem.

The first two factors are unlikely to play a major role in the general decline, given the social structure and small number of males necessary to maintain fertilization of the female population, the few natural predators, and the lack of evidence for any local or regional changes in their abundance. There is no real evidence available in respect of a compensatory decline or a decline due to disease (Anon, 1991*c*). Elephant seal–fishery interactions remain speculative in the absence of any reliable information on the proportion of fish in the diet of elephant seals, the main food of which around South Georgia is squid (Rodhouse *et al.*, 1992). Despite the fact that available food data may not be representative of the principal prey items during the year, there is yet no evidence to suggest that finfish exploitation in the CCAMLR Convention Area has played any causal role in the decline of stocks of elephant seals. The population at South Georgia appears to have remained stable despite its proximity to large-scale commercial finfishing over more than two decades (Anon, 1991*c*). It is thus likely that natural fluctuations, for example in prey abundance and availability, and also subtle changes in environmental conditions, about which we know little, have influenced the abundance of southern elephant seals. However, these changes must have been confined to or expressed principally at Indian Ocean sites (Anon, 1991*c*). It will be extremely difficult if not impossible to delineate causal relationships between these factors and shifts in the abundance of elephant seals at the present state of knowledge.

Antarctic fur seals probably numbered $1.5–2.0 \times 10^6$ individuals prior to exploitation. The first intensive period of sealing from 1790 to 1825 practically exterminated the fur seals in the Scotia Arc region. First indications of a recovery in the second half of the nineteenth century were soon thwarted by subsequent hunting; stock size obviously remained of the order of a few thousands or even only hundreds until the early 1950s. Thereafter, the population at South Georgia increased from a few hundreds to the current 1.5×10^6 individuals (Boyd *et al.*, 1990). In the late 1970s and during the 1980s, Antarctic fur seals started to recolonize their former area of distribution and rookeries in the South Orkney and the South Shetland Islands. Colonies on islands in the Indian Ocean sector are also increasing in size and numbers (Bengtson *et al.*, 1990; Jouventin & Weimerskirch, 1990). Population growth rate at Bird Island (South Georgia) was 17% annually from 1958 to 1972. It declined to 4–5% thereafter (Croxall & Prince, 1979) and was only 3% in 1988 (Fig. 5). However, there are few signs that Antarctic fur seals at South Georgia are close to their carrying capacity (Boyd *et al.*, 1990).

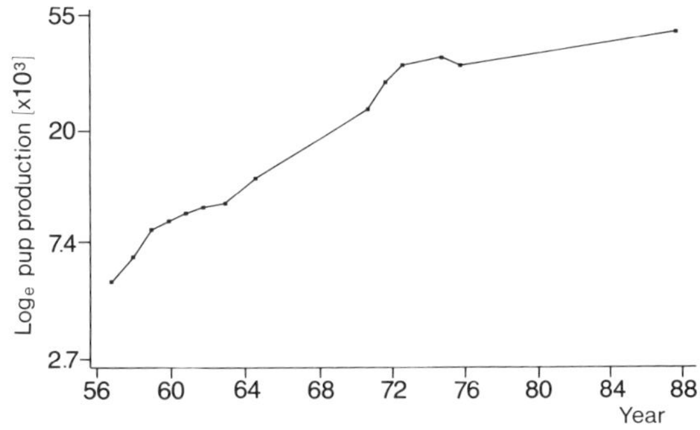

Fig. 5. Counts of fur seal pups at Bird Island (South Georgia) (from Boyd *et al.*, 1990).

The ongoing expansion of Antarctic fur seals is probably still a response following vigorous depletion in the course of the nineteenth century. Although we do not know whether fur seals and whales were competing for food, it is interesting to note that fur seals at South Georgia remained at a very low level for almost 70 years with little or no sealing occurring during that period. Their population started to increase only from the mid- or late 1950s onwards, concomitant with the heavy depletion of the whale stocks around the island which led subsequently to the cessation of land-based whaling on the island in the first half of the 1960s. The rate of increase of 14–17% was unusually high compared with the usual 6–10% for marine mammals recovering from overhunting (Laws, 1985).

Whales

Seven species or subspecies of baleen whale and eight species of toothed whale occur south of the South Polar Front Zone. All baleen whale species have been subject to extensive whaling. The only toothed whale taken regularly has been the sperm whale (*Physeter macrocephalus*). Killer whales (*Orcinus orca*) and southern bottlenose whales (*Hyperoodon planifrons*) were pursued irregularly and in small numbers only (Table 5).

Stock sizes prior to exploitation are not known, but have been inferred from the catch (Table 5).

Table 5. *Southern hemisphere whale population estimates, and accumulated catch and estimated mass of whales.*

Note that initial stock sizes should be regarded with caution as they are all biased in one way or another. The figures are used to illustrate the likely scale of changes rather than being figures approved by the Scientific Committee of the IWC.

Whale species	Stock size (thousands)		Catch		Catch prohibited after[b]
	Initial[a]	Present[a]	numbers	mass (kt)	
Blue and pygmy blue	220	0.7 and 5.5	331 644	27 530	1963–4
Fin	490	25	691 890	33 210	1976–7
Sei	200	38	149 594	2 620	1978–9
Humpback	130	12	68 294	1 840	1962–3
Right	100	3–4	?	?	1935
Minke	?	760	98 981	680	1985–6
Sperm	330	?	142 539	3 850	1979–80
Beaked whales (primarily Southern bottlenose whale)	?	170–320	43	?	1977–8
Killer whales	?	200	1 208	?	1979–80

[a] Initial stock sizes refer primarily to the recruited (i.e. sexually mature) part of the population, although it is known that juvenile whales have been taken to some extent in some species. Present stock sizes are based on sighting surveys and include all age classes of a population from age class 1+ onwards.
[b] It should be borne in mind that the dates of legal protection do not necessarily coincide with the actual cessation of catching.
Source: Gambell (1976, 1987); Kawamura (1978); Borchers (1991); Kasamatsu & Joyce (1991); J Kasamatsu (Personal communication).

The total reported catch of whales from 1904 to the moratorium at the beginning of the 1986–7 season was 1.483×10^6 individuals. An unknown portion of this catch, in particular of sperm whales, pygmy blue whales (*Balaenoptera musculus brevicauda*) and sei whales (*B. borealis*), has probably been taken north of the South Polar Front Zone, but it seems reasonable to assume that prior to Antarctic whaling 1–1.5 million whales may have penetrated the Antarctic Convergence in austral summer–autumn.

Most trends in whale stocks published over the years have eventually turned out to be largely biased in one way or other. However, the order of magnitude of the initial and current stock sizes (Table 5), and hence the scale of the changes which have been produced by whaling, should be correct (Gambell, 1987). With the exception of minke whales (*Balaenoptera acutorostrata*), and probably killer whales and southern bottlenose whales, all other species declined dramatically and are currently only fractions of their initial sizes, in particular those of blue (*B. musculus*), fin (*B. physalus*) and humpback whales (*Megaptera novaeangliae*). It has been suggested that the number of minke whales has increased. However, there are no observations directly supporting this increase. Of particular interest is the recent high estimate of the standing stock of beaked whales (primarily southern bottlenose whale, but also Arnoux's beaked whale, *Berardius arnouxi*, and *Mesoplodon* species) and killer whales in the Southern Ocean (Table 5) (Kasamatsu & Joyce, 1991; J Kasamatsu, personal communication). Southern bottlenose whales appear to be the second most abundant whale species in the Southern Ocean after the minke whale.

The decline or even collapse of the whale stocks is clearly associated with the whale fishery. Given the catch history (Fig. 6) this decline has obviously occurred successively. Right whales (*Eubalaena glacialis*) had apparently been largely depleted by nineteenth-century whalers on their breeding

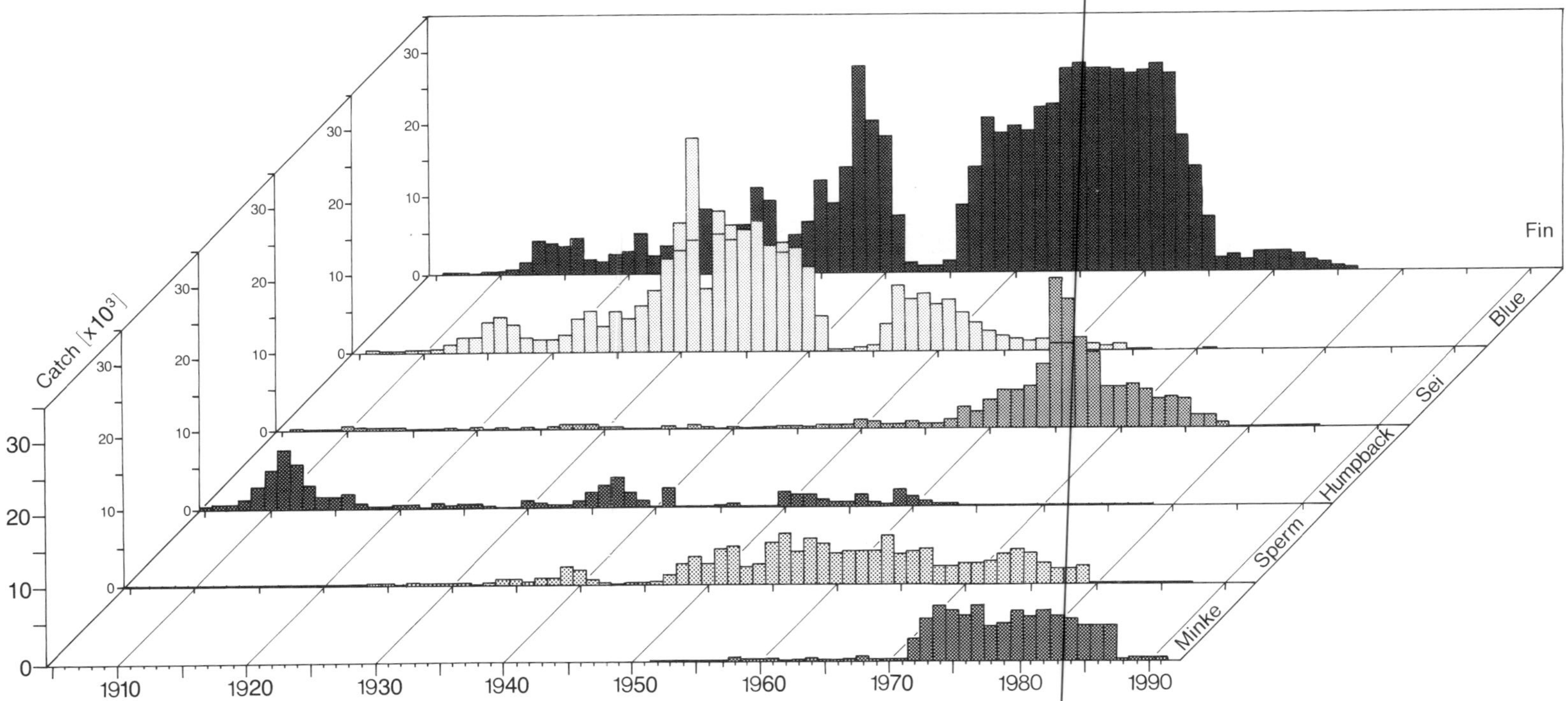

Fig. 6. Annual catches of whales in the Southern Ocean demonstrating the successive importance of each species in the fishery (modified from Gambell (1987) and International Whaling Statistics).

grounds off South America, South Africa and Australia (Best, 1970; Kawamura, 1978). Humpbacks formed an appreciable portion of the catch only until the beginning of the First World War and in the first half of the 1930s. Blue whales were depleted primarily between 1925 and 1940, while fin whale stock(s) collapsed in the late 1950s. Very few sperm and sei whales were taken prior to 1945 and 1955 owing to their comparatively smaller sizes (sei) and their chemically different oil (sperm). These stocks probably declined in the 1960s. Minke whales were not pursued in appreciable numbers before 1971 and owing to a prohibition of the take of other species in the 1970s were the target species until 1986–7. Since then, about 300 minke whales have been taken annually within a Japanese Scientific Whaling Programme (Fig. 6).

Among the baleen whales heavily depleted by hunting, only right whales and humpback whales show signs of recovery (IWC, 1990*b*). The number of blue whales is obviously still very low.

Changes in reproductive parameters

Food supply has been suggested as being an important determinant of maturity, i.e. the age at which sexual maturity is attained could decrease as the result of a higher level of nutrition stimulating growth (Laws, 1961). In fin whales, for example, energy requirements for lactating females are 60–70% higher than for other females (Lockyer, 1981). Increased food supply could also permit mature whales to reduce the time required to recover from a pregnancy and calf-rearing (Sampson, 1989), i.e. that females might be capable of breeding more rapidly than once in two years (Laws, 1961). It was thus tempting to associate reported changes in reproductive parameters with density-dependent responses to exploitation and a potential release of pressure on prey. However, we will show in the following that most of these reported changes do not stand critical examination and are inconclusive at best, primarily due to methodological problems.

Pregnancy rates in baleen whales

An increase in pregnancy rates has been recorded for blue and fin whales (Mackintosh, 1942; Laws, 1961; Doi *et al.*, 1969) and for sei whales (Gambell, 1973). The increased pregnancy rate in fin whales has been associated with an increase in the ovulation frequency during the normal breeding cycle (Gambell, 1973). These observations were primarily based on qualitative judgements. Two rigorous statistical analyses of the entire data set from 1930 to 1975 came to different conclusions. One (Mizroch, 1981; Mizroch & York, 1984) found a very high interseasonal and interannual variability in the data but no apparent trend to assert an increase in pregnancy rates or a shortened breeding cycle during the course of the exploitation. In contrast, Sampson (1989) observed that females of a given length were more frequently pregnant in more recent catches (Fig. 7) by including length of pregnant females more rigorously in his statistical models.

However, in addition to the fact that the recording of reproductive data, such as the presence of foetuses, became mandatory only in 1949, both studies tend to underestimate one aspect which alone could have accounted for most of the variance observed in Fig. 7: the recording of biological data was left to personnel of widely different qualification, from uninformed company personnel to graduate biologists. The quality of data could therefore be expected to vary considerably between seasons and expeditions (mother ships). After the establishment of the IWC, more qualified whaling inspectors and biologists became involved in biological sampling and this should have increased the likelihood of detecting foetuses. Any positive trend in the time series suggesting an increase in pregnancy rate during the course of exploitation has to be regarded as spurious if not accompanied by an interpretation of the associated levels of reporting efficiency (see also Best, 1989).

Recruitment rates in fin whales

Measurements of recruitment rates presented no evidence of any compensatory rise accompanying the decline of the whale stocks. Recruitment rates fell rather than rose during the 1930s (Clark, 1984). However, the method used (Cohort analysis) appears to be unsuitable for estimating recruitment rates in baleen whales as it is critically dependent on a number of assumptions, such as stable growth rates and the absence of size selection during exploitation which are unlikely to be met in reality.

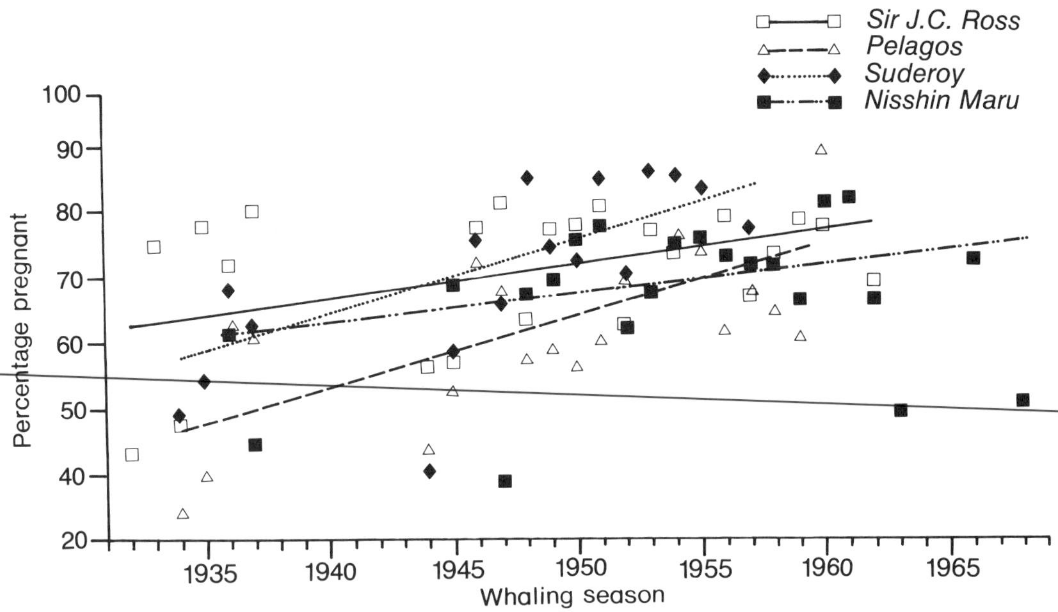

Fig. 7. Temporal trends in pregnancy rates of fin whales plotted by whaling season for four of the five major whaling expeditions that operated before and after World War II (modified from Sampson, 1989).

An increase in growth rate, for example, would lead to an erroneous decline in year-class strength.

Age at sexual maturity

Age at sexual maturity deduced from transition layer analysis in earplugs has been demonstrated to decline during whaling in fin and sei whales (Lockyer, 1972, 1974) and minke whales (Masaki, 1979; Kato, 1983). However, length at sexual maturity has not changed appreciably during the course of exploitation (Gambell, 1976). This would suggest that whales grew quicker and began to mature at an earlier age in more recent years.

This has been questioned by Cooke & de la Mare (1983, 1984) and Cooke (1985) who have suggested that the trends observed may have been artefacts from the data due to errors in age determination and under-representation of the most recent cohorts in the samples. Subsequent studies (Sakuramoto *et al.*, 1985; Kato & Sakuramotu 1991; Kato, 1987) have attempted to overcome some of these problems and tended to confirm that the observed decline in age at sexual maturity of minke whales from 12 to 13 years in the mid-1980s to 10 years in 1955 and a further decline to 7–8 years in the early 1970s was real (Fig. 8). However, the IWC Sub-Committee on Southern Hemisphere Minke Whales could not reach a unanimous view on the conclusiveness of these analyses, owing to methodological problems and the shortness of the time series (IWC, 1991).

A decline in the age at sexual maturity derived from back-calculating age from transition layer analysis has also been demonstrated in crabeater seals in the Marguerite Bay area. Age at sexual maturity had apparently dropped from 4.5 years in 1945 to 2.5 years in the late 1960s (Laws, 1977*a*). Age at sexual maturity deduced from ovarian material increased again from 2.5 years in the early 1960s to about 4.2 years in 1975 (Bengtson & Laws, 1985) (Fig. 9). Bengtson & Laws (1985), however, also demonstrated that the apparent decline in age at sexual maturity derived from transition layer read-

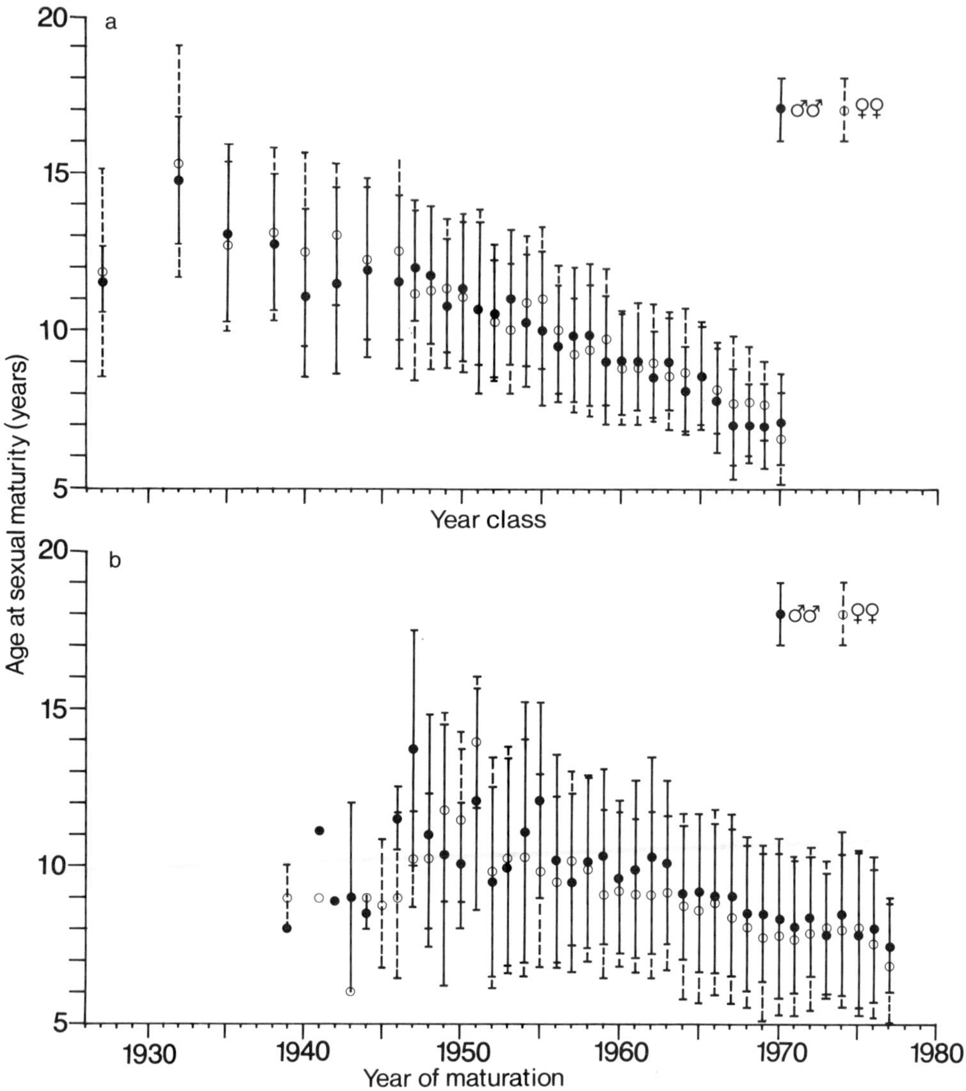

Fig. 8. Changes in mean age at sexual maturity (± one standard deviation) deduced from transition phase in earplug for males and females of the southern minke whales. Samples are grouped by year class (a) and year of maturation (b) (from Kato, 1987).

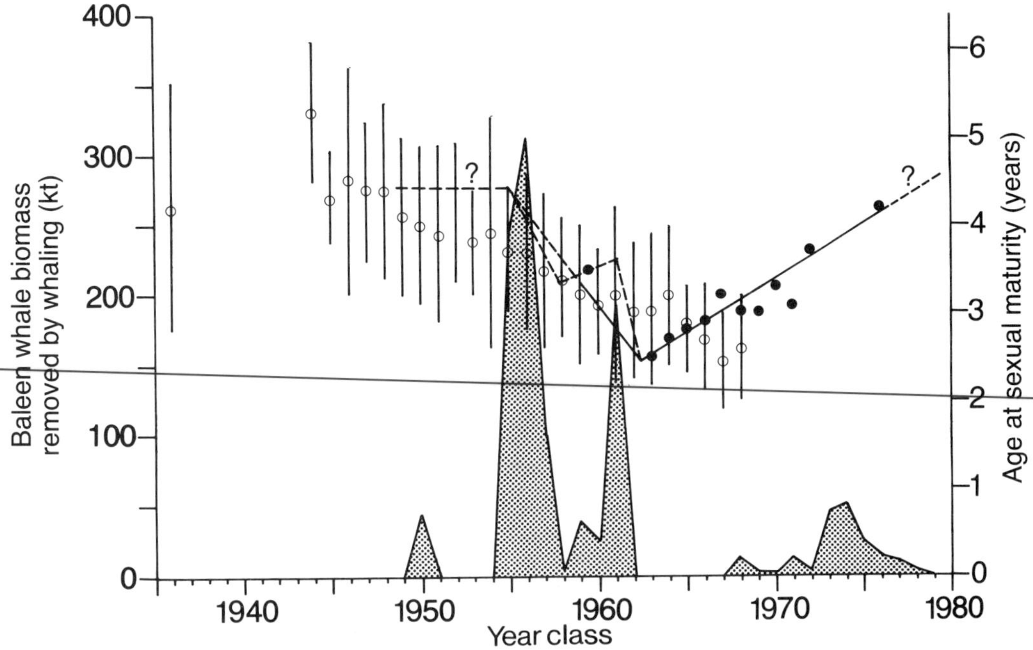

Fig. 9. Trends in age at sexual maturity of crabeater seals in the Antarctic Peninsula as estimated from ear-plug transition zone analysis (open circles) and estimated from ovarian material (solid circles) combined from Laws (1977*a*) and Bengtson & Laws (1985). Combined baleen whale biomass removed from Area I (Antarctic Peninsula, 120–60°W) (from Bengtson & Laws, 1985).

ings was an age-specific rather than a chronological trend, and hence was artefactual. A possible higher mortality rate in earlier-maturing individuals would bias transition layer readings upwards amongst older animals. The data (Fig. 9) would then suggest a more or less constant mean age over all year classes represented (Cooke, 1985).

Breeding success

Information on long-term changes in breeding success is scarce. Breeding, hatching and fledging success of wandering albatross at Bird Island (South Georgia) has increased by *ca*. 1% annually from 1976 to 1988 (Croxall *et al.*, 1990). No such trends have been observed for colonies of this species in the Crozet Islands (Weimerskirch & Jouventin, 1987). Wandering albatross at South Georgia may have benefitted from fish and offal discarded by the fishery (Kock, 1992). Black-browed albatrosses and white-chinned petrels, which are often seen in large numbers scavenging extensively behind fishing vessels, are other species which may have reacted by more successful chick-rearing to increased food availability due to fishing. However, data from South Georgia show no increase in population size of these species (JP Croxall, personal communication).

Diet composition and food consumption by high-level consumers

Within each of the five major groups of Antarctic high-level consumers most species rely on krill as their diet at least in summer and during their breeding season. The others depend mostly on squid and

Table 6. *Main prey species of Antarctic high-level consumers*

Predator species	Main prey species	Source
Blue, fin, minke, humpback whale	Krill	Kawamura, 1978, 1980; Bushuev, 1991
Sei, right whale	Copepods, *Themisto gaudichaudii*, *Thysanoessa* sp.	Kawamura, 1978
Sperm whale	Squid, fish	Berzin, 1971; Yukhov, 1982
Killer whale	Minke whale, seals, penguins, squid, fish	Ivashin, 1981; Mikhalev *et al.*, 1981
S. bottlenose whale	Squid	Frazer, 1945; Goodall & Galeazzi, 1985
Crabeater, fur seal	Krill, seasonally also fish	Laws, 1984
Leopard seal	Krill, penguins, seals, fish	Laws, 1984
Weddell seal	Fish, cephalopods	Laws, 1984
Ross, elephant seal	Squid, fish	Laws, 1984; Rodhouse *et al.*, 1992
Macaroni, Adélie, chinstrap penguin, black-browed albatross, snow petrel, Cape petrel, Antarctic prion	Krill, also fish (*Pleuragramma antarcticum*) in Adélie penguins	Croxall, 1984
Gentoo penguin, shags, Antarctic tern	Fish, also krill in Gentoo penguin and Antarctic tern	Croxall, 1984
King, emperor penguin, most albatross species	Fish, squid	Croxall, 1984
Most notothenioid fish species	Krill, other euphausiids, fish, *Themisto*	Kock, 1992
Mesopelagic fish species	Copepods, *Themisto*, euphausiids, pteropods	Kock, 1992
Squid	Mesopelagic fish, krill in some species	Lubimova, 1985; Nemoto *et al.*, 1985

fish or have significant components of other high-level consumers in their diet, as in killer whales and leopard seals (Table 6). The trophic relations among high-level consumers have been simplified in Fig. 10.

There appear to be two major systems of pelagic food webs in the Southern Ocean in respect to high-level predators.

(1) Northern slope water and oceanic food webs ranging over the Ice-free Zone and the oceanic parts of the Seasonal Pack-ice Zone in summer. This food web consists of copepods, mesopelagic fish and squid, right, sei and sperm whales, elephant seals, small petrels, wandering albatrosses and king penguins.

(2) Southern food webs with krill as a key food organism in the Seasonal Pack-ice Zone and at South Georgia. *Euphausia crystallorophias* and the nototheniid fish *Pleuragramma antarcticum* are partly replacing *Euphausia superba* in the high-Antarctic Zone. With regional differences in species

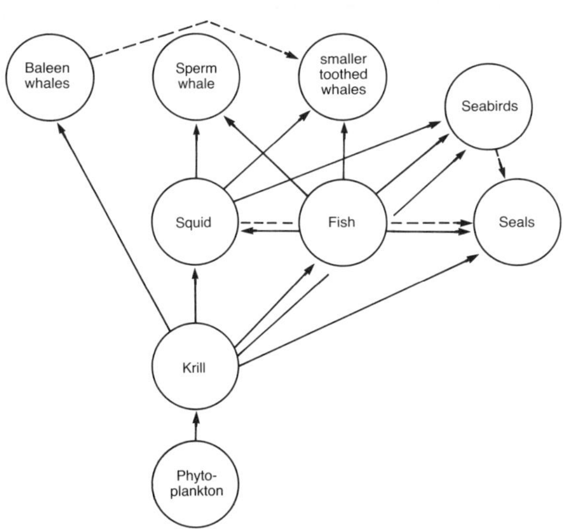

Fig. 10. Simplified trophic relations among high-level consumers (modified from Murphy *et al.*, 1988).

composition, these food webs consist (or consisted) mostly of blue, fin, minke and humpback whales, fur seals, crabeater and leopard seals, chinstrap, Adélie and macaroni penguins, petrels, notothenioid fish and juveniles of some squid species.

It was the food consumption and the trophic relations within these southern food webs on which the BIOMASS Programme focused. The northern oceanic food webs are still very imperfectly understood in quantitative terms.

Krill predators have been commonly ranked in the following order according to their importance as krill consumers.

Prior to whale exploitation	currently
baleen whales	seals
fish	seabirds
cephalopods	cephalopods
seabirds	fish
seals	baleen whales

This reversal in ranking reflects the removal of approximately 1.5 million whales between 1904 and 1985–6 and refined estimates of krill consumption by the major predator groups (see, for example, Everson, 1977; Kock, 1985, 1987; Laws, 1985). However, one should be aware that both rankings are still speculative and estimates of prey consumption are still subject to great uncertainties. Changing input parameters for the extrapolations, even if they appear to be minor, often change prey consumption estimates considerably. We show this by elucidating the role of squid as krill predators and by changing input variables in a recent estimate of krill consumption by minke whales.

Krill consumption by squid has always been a subject of extensive speculation. It has been estimated to be as high as 100 Mt annually, which is of the same order of magnitude as the estimated krill consumption of seals and seabirds (Laws, 1985). Interestingly, these extrapolations were based on the results of the analysis of hardly more than 200 squid stomachs from the entire Southern Ocean published until the mid-1980s (Clarke, 1980; Lubimova, 1985; Nemoto *et al.*, 1984, 1985). The number of squid stomachs analysed has now increased to more than 500 (Nemoto *et al.*, 1988; Rodhouse, 1991) but without clarifying the role of squid further.

Squid primarily inhabit the slope and oceanic waters which are under the influence of the Antarctic Circumpolar Current or the Circumpolar Warm Deep Water. The occurrence of krill, but also of other plankton organisms such as *Themisto gaudichaudii*, in the diet of a number of species suggests that these exhibit regular vertical migrations or live permanently in the epipelagic zone to exploit krill. The most common species with high proportions of *Euphausia superba* in their guts were *Kondakovia longimana, Gonatus antarcticus* and *Brachyteuthis picta*. However, one should be aware of the fact that most of the stomachs investigated originate from squid taken as by-catch in pelagic trawls targeted on krill (Nemoto *et al.*, 1984, 1985; Lubimova, 1985). Hence, it is not unlikely that their contents were biased towards the occurrence of krill and may not reflect the food composition of the species representatively. A comparison of the length compositions of these by-catch squid with those collected from sperm whale stomachs showed that squid feeding in the epipelagic zone were likely to be small-sized immatures (Nemoto *et al.*, 1985, 1988). Stomach contents of randomly collected squid, although their number is still small (<100–200), suggest that mesopelagic organisms, such as myctophids, paralepidids and squid, may play a much more important role in the diet of squid than does krill (Lubimova, 1985; Rodhouse, 1991). A larger number of squid which according to their frequent occurrence in sperm whale stomachs obviously represent a significant portion of the squid biomass in the Southern Ocean, such as *Mesonychoteuthis hamiltoni* or *Histioteuthis* species, have rarely been found above 500 m. It is unlikely that any of these species rely on krill as food. Among the squid species of the Southern Ocean hardly more than 10 species and in some of them only juveniles have been found to feed on krill in appreciable numbers. Hence there is evidence now that the role of squid as krill predators may have been overestimated in the past but that their true importance as predators in the food webs of the Southern Ocean still awaits elucidation.

Recently, Armstrong & Siegfried (1991) estimated euphausiid consumption by minke whales in the Southern Ocean south of 60°S to be 35.5 Mt annually. By changing mean daily consumption from 2.7% (females) and 3.6% (males) to 4% of body mass and a refinement of the average body mass of a minke whale in the Southern Ocean from 6.99 t (males) and 8.25 t (females) to 6.0 t based on results from Japanese Whaling under a Special

Scientific Permit in 1988–9 and 1989–90, Shimadzu (unpublished data) scaled Armstrong & Siegfried's estimate down to only 22 Mt annually.

We feel unable to decide which one of the estimates may be the more realistic. However, both examples illustrate the great uncertainties still surrounding overall abundance estimates including, for example, non-breeders in seabirds and seasonal variations in quantitative food composition and daily food consumption. Given these uncertainties we have refrained from extensively reviewing existing estimates of krill consumption here and adding new and revised estimates to the existing figures, which again may be subject to considerable revision in the near future. Furthermore we would strongly recommend to give up these whole-ocean calculations and carry out extrapolations of food consumption on a much smaller scale. Abundance estimates of krill predators as well as of krill within the limited foraging range of the predators and the estimation of food load sizes and feeding intervals are likely to be much more accurate and easier to obtain for individual colonies, groups of colonies, coastlines or islands during the breeding season. This approach, which is currently being pursued by the CCAMLR Ecosystem Monitoring Program (Anon, 1991*d*) offers among other things the opportunity to assess the impact a krill fishery might have on these predators by competing for the resource. Currently, 50–90% of the commercial krill catches are taken within the foraging range of krill predators during their breeding season (Anon, 1992). Rookeries of *Pygoscelis* species on King George Island and Nelson Island (South Shetland Islands), for example, collectively utilized *ca.* 24 kt of krill each summer to feed their chicks (Trivelpiece *et al.*, 1987).

Species interactions: the 'whale reduction–krill surplus' hypothesis and trends in environmental conditions

Despite a great deal of overlap in their main prey species (Table 5) there exists considerable ecological separation between predator species on both a spatial and a temporal scale. For example, balaenopterid and male sperm whale abundance in the Antarctic is very low in the winter months and at peak numbers in January–March when the retreat of the ice allows them to penetrate further south. Although their Antarctic distribution overlaps, each species occupies a characteristic latitudinal zone and varies in abundance longitudinally. Migratory patterns may be different by size and age class, sex and reproductive stage (Laws, 1977*a*). Fur seals are confined to the islands of the low Antarctic during their breeding season and the major part of the population probably migrates to the north of the Southern Ocean in winter. In contrast, the ice-inhabiting seals are resident in the Antarctic with crabeater and leopard seals preferring the looser pack-ice while Ross seals inhabit the dense pack-ice and Weddell seals predominate in the fast-ice regions. Beaked whales and a part of the minke whale population(s) are also resident in the Antarctic (Laws, 1977*b*).

Ecological segregation in the three pygoscelid species in the South Shetland Islands and the four species on Marion Island results from differences in breeding chronology, foraging behaviour and life history tactics rather than from competitive interaction (Trivelpiece *et al.*, 1987; Adams & Brown, 1989).

In spite of the ecological segregation of many species there exists a potential of competition for food. In the more northern food web sei whales and southern right whales are likely competitors for copepods over most of their distributional range. From anecdotal evidence, Kawamura (1978) has suggested that as long as the Southern Ocean ecosystem had been left untouched or little disturbed, southern right whales were at a high population size, whereas sei whales were much less abundant than in the twentieth century. The depletion of right whales from the early to mid-nineteenth century onwards may have led to the growth of the sei whale population(s). Right whales were still at a very low level for decades after whaling had ceased in 1935. They started to recover slowly after sei whales had been severely depleted in the period 1960–79.

Simultaneously with the decline of the great baleen whale stocks, considerable changes have been observed in population sizes and reproductive characteristics of krill and non-krill predators. With various reservations, numerous authors have suggested that these changes have been brought about by the availability of food released by the reduced number of whales. However, at the same time period, climatic changes in temperature, precipitation, snow cover and ice cover are apparent regionally. For example, there appears to have been a cold period during 1925–35 in the southern Scotia Arc region followed by a warming which amounted to 2 °C by 1940. This could have had considerable consequences on feeding and/or breeding success (Horwood, 1980).

Changes in reproductive characteristics have often been cited as an illustration of a density-dependent response to the reduction of inter- and intraspecific competition after the decline of the great baleen whales. However, as described earlier, in the past couple of years many of these changes have been demonstrated to be artefacts from the data rather than showing real trends, or have been found to be inconclusive at best. For sei whales, for example, which were unlikely to compete for food with other mysticetes in the Southern Ocean, comparable responses in reproductive parameters were manifest even before stocks of that species were significantly exploited in the 1960s (Gambell, 1973).

The example most often used to illustrate the 'whale reduction–krill surplus' hypothesis has been to associate increases in stock size of minke whales, crabeater seals and chinstrap and Adélie penguins with the decline of the great baleen whales. Blue whales, minke whales, crabeater seals, chinstrap and Adélie penguins are potential competitors for krill in the marginal ice zone in spring and early summer. Blue whales had been depleted to only a fraction of their initial stock sizes by 1940. Possible benefits from their reduction were likely to occur to similar-sized predators with the broadest niche overlap, which would probably be minke whales rather than seals or penguins (Murphy *et al.*, 1988). The number of minke whales may have increased although it is unknown whether it has and to what extent. Age at sexual maturity in minke whales has apparently declined from 12–13 years in the 1940s to 7 years in the 1970s. It is not known when the possible increase in the number of minke whales started. If we assume that the increase started with the decline in age at sexual maturity at about 1940, it would fit with the decline of the blue whale stocks. We do not know whether minke whales have been able to fill in the niche completely (this appears to be unlikely given the present estimate of krill consumption by minke whales) or whether part of the resource has become available to predators of the next size class, which would have been crabeater seals or even penguins. We do not even know whether krill availability to crabeater seals was ever limited by whales.

Antagonistic trends in age at sexual maturity and population sizes of crabeater seals and minke whales appear to be either largely biased (Bengtson & Laws, 1985) or still awaiting substantiation. Recent evidence indicates a coupling of fluctuations in ice-seal populations with large-scale oceanographic variation (Testa *et al.*, 1991).

Increases in the populations of chinstrap penguins in the South Shetland Islands and Adélie penguins in the Ross Sea area were likely to be associated with the gradual decrease of cold winters with extensive ice cover. This decrease would improve among other factors, overwinter survival and condition and early access to breeding grounds. Population increases are not likely to be associated with a pressure release on krill as such (Taylor *et al.*, 1990; Fraser *et al.*, 1992).

There is only some evidence so far that minke whales may have benefitted from the decline of the great rorquals. However, we should be aware that we know too little about trends in environmental conditions and their possible effects on minke whale populations in the same time period.

The recovery of Antarctic fur seals offers another example of a species that might have benefitted from the reduction of potentially competitive predators. Since the early 1950s the population has increased from a few hundreds to currently *ca.* 1.5×10^6 individuals. Recovery rates, at least from 1958 to 1972, were faster than expected from other marine mammal populations recovering from overhunting (Laws, 1977*a*; Payne, 1977). We do not know whether Antarctic fur seals and baleen whales were competing for krill. However, fur seals at South Georgia remained at a very low level from about 1870 to the early 1950s with very little or no sealing occurring during that time period (Bonner, 1984). Their increase only began with the severe reduction of baleen whales around the island.

Antarctic fur seal females, for example, apparently have a limited ability to increase the relative time foraging. Females obviously do not return to feed their pups until they have replenished their own reserves even though this may jeopardize the survival of their offspring (Costa *et al.*, 1989). One possible explanation for the unexpectedly high recovery rates may be that prey availability and hence breeding success increased appreciably in Antarctic fur seals after the depletion of the (potentially competitive) whale stocks around South Georgia.

It is obvious that many populations of Antarctic high-level consumers have undergone remarkable changes in abundance, some also in their biological parameters, in the last 200 years. Declines have commonly been attributed to direct and indirect effects of harvesting or the introduction of alien predators. This has been shown beyond doubt for the great baleen whales, sperm whales, fur seals, king penguins, some albatross species, small petrels on some sub-Antarctic islands, and a number of fish stocks. Increases in stock size and changes in reproductive parameters have heen responses to early exploitation or have often been associated with increased prey availability following the depletion of the great baleen whales. However, there appears to be much less direct evidence to support this 'whale reduction–krill surplus' hypothesis than previously thought. This is not to say that density-dependent effects have not occurred. However, biases in the data and trends in environmental conditions may have confounded or mediated trends in reproductive parameters and trophic interactions. Trends in environmental conditions presently offer equally or even more likely explanations for the observed trends in abundance or biological parameters, as has been the case in pygoscelid penguins.

Research in the framework of the BIOMASS Programme that has advocated direct causal links between krill availability and changes in high-level consumer populations may have been based on approaches that were too simplistic about species interactions in the ecosystems of the Southern Ocean (Fraser *et al.*, 1992). Experience of marine temperate systems, such as the North Sea, has shown that remarkable changes in upper trophic levels have occurred without direct effects of exploitation (Cushing, 1975). This emphasizes the need to include environmental parameters, such as the variability in the extent of sea-ice cover, in long-term monitoring programmes of the Southern Ocean (Croxall *et al.*, 1988; Anon, 1991*d*; Fraser *et al.*, 1992).

Acknowledgements

The manuscript has benefitted from comments provided by Drs JP Croxall, J Cooke and R Gambell. We are grateful to Ms U Harm for the preparation of the graphs and to Mrs J Arndt for typing the manuscript.

References

Adams N J, Brown C R (1989) Dietary differentiation and trophic relationships in the sub-Antarctic penguin community at Marion Island. *Mar Ecol Prog Ser* **57**: 249–58.

Anon (1988) *Report of the Meeting of the SCAR Group of Specialists on Seals.* Tasmania, Australia, 23–25 August 1988, Appendix 6, 2 pp.

Anon (1989) Report of the Working Group on Fish Stock Assessment. Hobart, Australia, 25 October to 2 November 1989. In: *Report of the 8th Meeting of the Scientific Committee.* CCAMLR, Hobart, Annex 6, pp 183–295.

Anon (1991*a*) *Report of the 10th Meeting of the Scientific Committee, Hobart, Australia, 21 to 25 October 1991.* CCAMLR, Hobart, 78 pp.

Anon (1991*b*) Information of interest to CCAMLR collected by MV Gondwana on Greenpeace's 1990/91 Expedition. CCAMLR–X/BG/18, 8 pp (mimeogr).

Anon (1991*c*) *Report of the Workshop on Southern Elephant Seals.* CCAMLR WG-CEMP-91/12, 28pp.

Anon (1991*d*) *Report of the Working Group for the CCAMLR Ecosystem Monitoring Program, Santa Cruz de Tenerife, Spain, 5 to 13 August 1991.* In: *Report of the Tenth Meeting of the Scientific Committee, Hobart, Australia, 21–25 October 1991.* CCAMLR, Hobart, Annex 7, pp 349–418.

Anon (1992) *CCAMLR, Report of the Eleventh Meeting of the Scientific Committee, Hobart, Australia, 26–30 October 1992.* CCAMLR, Hobart, 86 pp (mimeogr).

Armstrong A J, Siegfried W R (1991) Consumption of Antarctic krill by minke whales. *Antarct Sci* **3**(1): 13–18.

Bengtson J L, Ferm L M, Härkönen T J, Stewart BS (1990) Abundance of Antarctic fur seals in the South Shetland Islands, Antarctica, during the 1986/87 austral summer. In: Kerry K R, Hempel G (eds) *Antarctic Ecosystems. Ecological Change and Conservation.* Springer-Verlag, Berlin, pp 265–70.

Bengtson J L, Laws R M (1985) Trends in crabeater seal age at maturity: An insight into Antarctic marine interactions. In: Siegfried W R, Condy P M, Laws R M (eds) *Antarctic Nutrient Cycles and Food Webs.* Springer-Verlag, Berlin, pp 669–75.

Berzin A A (1971) *The Spermwhale.* Izd. Pishcheveya Promyshl, Moscow. (Transl.) Israel Progr for Scient Transl Jerusalem, 1972, 394 pp.

Best P B (1970) Exploitation and recovery of right whales, *Eubalaena australis* off the Cape Province. *Investl Rep Div Sea Fish S Afr* **80**: 1–20.

Best P B (1989) Some comments on the BIWS Catch Data Base. *Rep Int Whal Commn* **39**: 363–9.

BIOMASS (1977) *Biological Investigations of Marine Antarctic Systems and Stocks (BIOMASS).* Vol 1: *Research Proposals.* 79 pp.

Bonner W N (1984) Conservation and the Antarctic. In: Laws R M (ed) *Antarctic Ecology*, vol 2. Academic Press, London, pp 821–50.

Bonner W N (1985) Birds and mammals – Antarctic seals. In: Bonner W N, Walton D W H (eds) *Key Environments: Antarctica*. Pergamon Press, Oxford, pp 202–22.

Borchers D L (1991) Revised southern hemisphere whale abundance estimates south of 30° derived from IWC/IDCR survey and Japanese scouting vessel data (addendum to SC/24/SHMi 18). Paper presented at the Scientific Committee of the International Whaling Commission, May 1991, SC/42/Mi 34, 11 pp (mimeogr).

Boyd I L, Lunn N J, Rothery P, Croxall J P (1990) Age distribution of breeding female Antarctic fur seals in relation to changes in population growth rate. *Can J Zool* **68**: 2209–13.

Brothers N P (1984) Breeding, distribution and status of burrow-nesting petrels at Macquarie Island. *Aust Wildl Res* **11**: 113–31.

Brothers N P (1991) Approaches to reducing albatross mortality and associated bait loss in the Japanese longline fishery. *Biol Cons* **55**: 255–68.

Bushuev S G (1991) Distribution and feeding of minke whales in Antarctic Area I. *Rep Int Whal Commn* **41**: 303–12.

Clark W G (1984) Recruitment rates of Antarctic fin whales, *Balaenoptera physalus*, inferred from Cohort Analysis. *Rep Int Whal Commn Spec Issue* **6**: 411–15.

Clarke M R (1980) Cephalopoda in the diet of spermwhales of the southern hemisphere and their bearing on spermwhale biology. *Discovery Rep* **37**: 1–324.

Clarke M R (1983) Cephalopod biomass-estimation from predation. *Proc Vict Mus* **44**: 95–107.

Clarke M R (1985) Marine habitats – cephalopods. In: Bonner W N, Walton D W H (eds) *Key Environments: Antarctica*. Pergamon Press, Oxford, pp 193–200.

Cooke J G (1985) Has the age at sexual maturity of southern hemisphere minke whales declined? *Rep Int Whal Commn* **35**: 335–40.

Cooke J G, de la Mare W K (1983) The effects of variability in age data on the estimation of biological parameters in minke whales (*Balaenoptera acutorostrata*). *Rep Int Whal Commn* **33**: 333–8.

Cooke JG, de la Mare W K (1984) Some notes on the estimation of time trends in the age at sexual maturity in baleen whales from transition layer data. *Rep Int Whal Commn* **34**: 711–16.

Costa D P, Croxall J P, Duck C D (1989) Foraging energetics of Antarctic fur seals in relation to changes in prey availability. *Ecology* **70**(3): 596–606.

Croxall J P (1984) Seabirds. In: Laws R M (ed) *Antarctic Ecology*, vol 2. Academic Press, London, pp 533–619.

Croxall J P (1987) The status and conservation of Antarctic seals and seabirds: a review. *Environm Internat* **13**(1): 55–70.

Croxall J P, McCann T S, Prince P A, Rothery P (1988) Reproductive performance of seabirds and seals at South Georgia and Signy Island, South Orkney Islands, 1976–1987: Implications for Southern Ocean monitoring studies. In: Sahrhage D (ed) *Antarctic Ocean and Resources Variability.* Springer-Verlag, Berlin, pp 261–85.

Croxall J P, Prince P A (1979) Antarctic seabird and seal monitoring studies. *Polar Rec* **19**: 573–95.

Croxall J P, Prince P A (1990) Recoveries of wandering albatrosses *Diomedea exulans* ringed at South Georgia 1958–1986. *Ring Migr* **11**: 43–51.

Croxall J P, Rothery P, Pickering S P C, Prince P A (1990) Reproductive performance, recruitment and survival of wandering albatrosses *Diomedea exulans* at Bird Island, South Georgia. *J Anim Ecol* **59**: 775–96.

Cushing D H (1975) *Marine Ecology and Fisheries.* Cambridge University Press, 278 pp.

Doi T, Ohsumi S, Nasu K, Shimadzu Y (1969) Advanced assessment of fin whale stock in the Antarctic. *Bull Tokyo Reg Fish Res Lab* **60**: 95–141.

Duhamel G (1987) Ichtyofaune des secteurs indien occidental et atlantique oriental de l'ocean austral: biogéographie, cycles biologiques et dynamique des populations. Thèse de Doctorat d'Etat, Univ P et M Curie, Paris, 687 pp.

Duhamel G (1990) Supplementary data on exploited stocks in Division 58.5.1 (Kerguelen). *SC-CAMLR Select Sci Pap 1989:* 147–61.

Duhamel G (1991) Incidental mortality arising from fisheries activities around Kerguelen Islands (Division 58.5.1). SC-CAMLR-X/BG/14, 15 pp (mimeogr).

Duhamel G, Hureau J-C (1990) Changes in fish populations and fisheries around the Kerguelen Islands during the last decade. In: Kerry K R, Hempel G (eds) *Antarctic Ecosystems. Ecological Change and Conservation.* Springer-Verlag, Berlin, pp 323–33.

Ekau W (1990) Demersal fish fauna of the Weddell Sea, Antarctica. *Antarct Sci* **2**(2): 129–37.

Erickson A W, Hanson M B (1990) Continental estimates and population trends of Antarctic ice seals. In: Kerry K R, Hempel G (eds) *Antarctic Ecosystems. Ecological Change and Conservation.* Springer-Verlag, Berlin, pp 253–64.

Everson I (1977) *The Living Resources of the Southern Ocean.* FAO GLO/SO/77/1, 156 pp.

Filin A A, Gorchinsky K V, Kiselevy V M (1991) Biomass of myctophids in the Atlantic sector of the Southern Ocean as estimated by acoustic surveys. *SC-CAMLR Select Sci Pap 1990*: 417–29.

Fraser W R, Trivelpiece W Z, Ainley D G, Trivelpiece S G (1992) Increases in Antarctic penguin populations: Reduced competition with whales or a loss of sea ice due to environmental warming? *Polar Biol* **11**: 525–31.

Frazer F (1945) On a specimen of the Southern bottlenose whale, *Hyperoodon planifrons. Discovery Rep* **23**: 19–36.

Gales R, Pemberton D (1988) Recovery of the King penguin, *Aptenodytes patagonicus*, population on Heard Island. *Aust Wildl Res* **15**: 579–85.

Gambell R (1973) Some effects of exploitation on reproduction in whales. *J Reprod Fert (Suppl)* **19**: 531–51.

Gambell R (1975) Variation in reproduction parameters associated with whale stock size. *Rep Int Whal Commn* **26** (App V, Ann O): 182–9.

Gambell R (1976) World whale stocks. *Mammal Rev* **6**: 41–53.

Gambell R (1985) Birds and Mammals – Antarctic Whales. In: Bonner W N, Walton D W H (eds) *Key Environments: Antarctica.* Pergamon Press, Oxford, pp 223–41.

Gambell R (1987) Whales in the Antarctic ecosystem. *Environm Internat* **13**(1): 47–54.

Goodall R N P, Galeazzi A R (1985) A review of the food habits of the small cetaceans of the Antarctic and sub-Antarctic. In: Siegfried W R, Condy P R, Laws R M (eds) *Antarctic Nutrient Cycles and Food Webs.* Springer-Verlag, Berlin, pp 566–72.

Headland R K (1984) *The Island of South Georgia.* Cambridge University Press, 293 pp.

Horwood J W (1980) Competition in the Antarctic? *Rep Int Whal Commn* **30**: 513–17.

Hubold G (1992) Zur Ökologie der Fische im Weddellmeer. *Ber Polarforsch* **103**: 1–158.

Hunter S (1984) Breeding biology and population dynamics of giant petrels *Macronectes ssp.* at South Georgia *J Zool* **203**: 441–60.

International Whaling Commission (1990*a*) Chairman's Report of the 41st Annual Meeting. *Rep Int Whal Commn* **40**: 11–37.

International Whaling Commission (1990*b*) Report of the Scientific Committee. Annex F. Sub-Committee on Biological Parameters and MSY Rates, Appendix 6, Best P B Recovery rates in whale stocks that have been protected from commercial whaling for at least 20 years. *Rep Int Whal Commn* **40**: 129–30.

International Whaling Commission (1991) Report of the Scientific Committee. Annex E. Report of the Sub-Committee on Southern Hemisphere Minke Whales. *Rep Int Whal Commn* **41**: 113–31.

Ivashin M V (1981) USSR Progress Report on cetacean research June 1979–May 1980, Appendix I. *Rep Int Whal Commn* **31**: 221–6.

Jouventin P, Weimerskirch H (1990) Long-term changes in seabird and seal populations in the Southern Ocean. In: Kerry K R, Hempel G (eds) *Antarctic Ecosystems. Ecological Change and Conservation.* Springer-Verlag, Berlin, pp 208–13.

Kasamatsu J, Joyce G G (1991) Abundance of beaked whales in the Antarctic. IWC SC/43/012, 10 pp (mimeogr).

Kato H (1983) Some considerations on the decline in age at sexual maturity of the Antarctic minke whale. *Rep Int Whal Commn* **33**: 393–9.

Kato H (1987) Density-dependent changes in growth parameters of the Southern minke whale. *Sci Rep Whales Res Inst Tokyo* **38**: 47–73.

Kato H, Sakuramotu K (1991) Age at sexual maturity of Southern minke whales: a review and some additional analyses. *Rep Int Whal Commn* **41**: 331–7.

Kawamura A (1978) An interim consideration on a possible interspecific relation in Southern baleean whales from the viewpoint of their food habits. *Rep Int Whal Commn* **28**: 411–19.

Kawamura A (1980) A review of food of balaenopterid whales. *Sci Rep Whales Res Inst Tokyo* **32**: 155–98.

Kock K-H (1985) Marine habitats – Antarctic fish. In: Bonner W N, Walton D W H (eds) *Key Environments: Antarctica.* Pergamon Press, Oxford, pp 173–92.

Kock K-H (1987) Marine consumers: Fish and squid. *Environm Internat* **13**(1): 37–45.

Kock K-H (1991) The state of exploited fish stocks in the Southern Ocean – a review. *Arch FischWiss* **41**(1): 1–66.

Kock K-H (1992) *Antarctic Fish and Fisheries.* Cambridge University Press, 359 pp.

Kock K-H, Köster F W (1989) The state of exploited fish stocks in the Atlantic sector of the Southern Ocean. *Mitt Inst Seefisch Hamburg* **46**: 1–73.

Laws R M (1961) Reproduction and growth of southern fin whales. *Discovery Rep* **31**: 327–485.

Laws R M (1977*a*) The significance of vertebrates in the Antarctic marine ecosystem. In: Llano G A (ed) *Adaptations within Antarctic Ecosystems.* Smithsonian Institution, Gulf Publishing Company, Houston, pp 411–38.

Laws R M (1977*b*) Seals and whales of the Southern Ocean. *Phil Trans R Soc Lond B***279**: 81–96.

Laws R M (1984) Seals. In: Laws R M (ed) *Antarctic Ecology*, vol 2. Academic Press, London, pp 621–715.

Laws R M (1985) The ecology of the Southern Ocean. *Am Scient* **73**: 26–40.

Laws R M (1988) Number of seals killed and captured in the Antarctic 1964–85. In: *CCAS Report of the 1988 Meeting to review the operation of the Convention for the Conservation of Antarctic Seals, London, 12–16 September 1988,* Appendix F, 4 pp.

Lockyer C (1972) The age at sexual maturity of the southern fin whale (*Balaenoptera physalus*) using annual layer counts in the ear plug. *J Cons Perm Int Explor Mer* **34**: 276–94.

Lockyer C (1974) Investigations of the ear plug of the southern sei whale (*Balaenoptera borealis*) as a valid means of determining age. *J Cons Perm Int Explor Mer* **36**: 71–81.

Lockyer C (1981) Growth and energy budgets of large baleen whales from the southern hemisphere. *FAO Fish Ser 5 (Mammals in the Sea)* **3**: 379–487.

Lubimova T G (1985) Results of Soviet investigations of distribution and ecology of pelagic squids (Oegopsida) in the Southern Ocean. *SC-CAMLR Select Sci Pap 1985*: 79–111.

Lubimova T G, Shust K V, Popkov V V (1987) Ecology of mesopelagic fishes, family Myctophidae, in the Southern Ocean. In: Skarlato O A, Alekseev A P, Lubimova T G (eds) *Biological Resources of the Arctic and Antarctic.* Nauka, Moscow, pp 320–337. (In Russian.)

Mackintosh N A (1942) The southern stocks of whalebone whales. *Discovery Rep* **22**: 197–300.

Masaki Y (1979) Yearly change of the biological parameters for the Antarctic minke whale. *Rep Int Whale Commn* **29**: 375–96.

Mikhalev Yu A, Ivashin M V, Savusin V P, Zelenaya F E (1981) The distribution and biology of killer whales in the southern hemisphere. *Rep Int Whal Commn* **31**: 551–66.

Miller D G M, Hampton I (1989) Biology and ecology of the Antarctic krill. *BIOMASS Sci Ser* **9**: 1–166.

Mizroch S A (1981) Further notes on southern hemisphere baleen whale pregnancy rates. *Rep Int Whal Commn* **31**: 629–33.

Mizroch S A, York A E (1984) Have pregnancy rates of southern hemisphere fin whales, *Balaenoptera physalus,* increased? In: *Whales, Dolphins and Porpoises.* International Whaling Commission, Cambridge. (Spec Issue 6), pp 401–10.

Murphy E J, Morris D J, Watkins J L, Priddle J (1988) Scales of interaction between Antarctic krill and the environment. In: Sahrhage D (ed) *Antarctic Ocean and Resources Variability.* Springer-Verlag, Berlin, pp 120–130.

Nemoto T, Okiyama M, Iwasaki N, Kikuchi T (1988) Squid as predators on krill (*Euphausia superba*) and prey for sperm whales in the Southern Ocean. In: Sahrhage D (ed) *Antarctic Ocean and Resources Variability.* Springer-Verlag, Berlin, pp 292–6.

Nemoto T, Okiyama M, Takahashi M (1984) Aspects of the role of squid in the food chains of the Antarctic marine ecosystems (extended abstract). *Mem Natl Inst Polar Res Spec Issue* **32**: 89–92.

Nemoto T, Okiyama M, Takahashi M (1985) Aspects of the roles of squid in food chains of marine Antarctic ecosystems. In: Siegfried W R, Condy P R, Laws R M (eds) *Antarctic Nutrient Cycles and Food Webs.* Springer-Verlag, Berlin, pp 415–20.

Oritsland T (1970) Sealing and seal research in the south-west Atlantic pack ice, Sept–Oct 1964. In: Holdgate MW (ed) *Antarctic Ecology*, vol 1. Academic Press, London, pp 367–76.

Parmelee D F, Parmelee J M (1987) Revised penguin numbers and distribution for Anvers Island, Antarctica. *Br Antarct Surv Bull* **76**: 65–73.

Payne M R (1977) Growth of a fur seal population. *Phil Trans R Soc Lond* B**279**: 67–80.

Peter H-U, Bannasch R, Bick A, Gebauer A, Kaiser M, Mönke R, Zippel D (1989) Bestand und Reproduktion ausgewählter antarktischer Vögel und Robben im Südwestteil von King George Island, South Shetland Islands. *Wiss Ztschr Friedrich Schiller Univ Jena, Naturwiss* R, **38**(4/5): 645–57.

Poncet S, Poncet J (1987) Censuses of penguin populations of the Antarctic Peninsula 1983–87. *Br Antarct Surv Bull* **77**: 109–29.

Priddle J, Croxall J P, Everson I, Heywood R B, Murphy E J, Prince P A, Sear C B (1988) Large-scale fluctuations in distribution and abundance of krill – A discussion of possible causes. In: Sahrhage D (ed) *Antarctic Ocean and Resources Variability.* Springer-Verlag, Berlin, pp 169–82.

Rodhouse P (1991) Towards an assessment of the stock of ommastrephid squid *Martialia hyadesi* in the Scotia Sea: data from predators (abstract only). *SC-CAMLR Select Sci Pap 1990:* 479–80.

Rodhouse P G, Arnbom T R, Fedak M A, Yeatman J, Murray A W A (1992) Cephalopod prey of the southern elephant seal, *Mirounga leonina L. Can J Zool* **70**: 1007–15.

Rootes D M (1988) The status of birds at Signy Island, South Orkney Islands. *Br Antarct Surv Bull* **80**: 87–119.

Rounsevell D E, Copson G R (1982) Growth rate and recovery of a King penguin *Aptenodytes patagonicus* population after exploitation. *Aust Wildl Res* **9**: 519–25.

Sakuramoto K, Kato H, Tanaka S (1985) A simulation study about several factors influencing the estimates of a trend in age at sexual maturity of the Antarctic minke whale. *Rep Int Whal Commn* **35**: 285–90.

Sampson D B (1989) Pregnancy rate versus length in southern fin whales. *Rep Int Whal Commn* **39**: 255–65.

Shuford W D, Spear L B (1988) Surveys of breeding penguins and other seabirds in the South Shetland Islands, Antarctica, January–February 1987. NOAA Techn Memorandum NMFS-F/NEC 59, 27 pp.

Sladen W J L (1964) The distribution of the Adélie and chinstrap penguins. In: Carrick R, Holdgate M W, Prevost J (eds) *Biologie Antarctique.* Hermann, Paris, pp 359–65.

Taylor R H, Wilson P R, Thomas B W (1990) Status and trends of Adélie penguin populations in the Ross Sea region. *Polar Rec* **26**(159): 293–304.

Testa J W, Oehlert G, Ainley D G, Bengtson J L, Siniff D B, Laws R M, Rounsevell D (1991) Temporal variability in Antarctic marine ecosystems: periodic fluctuations in the phocid seals. *Can J Fish Aquat Sci* **48**: 631–9.

Thompson K (1990) Calamares and mollyhawks. *Newsl Falkland Is Found* **10**: 5–8.

Torres J J, Lancraft T M, Weigle B M, Hopkins T L, Robison B H (1984) Distribution and abundance of fishes and salps in relation to the marginal ice-zone of the Scotia Sea, November and December 1983. *Antarct J US* **19**: 117–19.

Trivelpiece W Z, Trivelpiece S G, Volkman N J (1987) Ecological segregation of Adélie, gentoo and chinstrap penguins at King George Island, Antarctica. *Ecology* **68**(2): 351–61.

Van Aarde R J (1980) The diet and behaviour of feral cats, *Felis catus*, at Marion Island. *S Afr J Wildl Res* **10**: 123–8.

Van Rensburg P J J (1985) The feeding ecology of a decreasing feral house cat, *Felis catus*, population at Marion Island. in: Siegfried W R, Condy P R, Laws R M (eds) *Antarctic Nutrient Cycles and Food Webs.* Springer-Verlag, Berlin, pp 620–4.

Weimerskirch H, Clobert J, Jouventin P (1987) Survival in five southern albatrosses and its relationship with their life history. *J Anim Ecol* **56**: 1043–55.

Weimerskirch H, Jouventin P (1987) Population dynamics of the wandering albatross, *Diomedea*

exulans, at the Crozet Islands: causes and consequences of the population decline. *Oikos* **49**: 315–22.
Wilson G J (1983) Distribution and abundance of Antarctic and sub-Antarctic penguins: a synthesis of current knowledge. *BIOMASS Sci Ser* **4**: 1–46.
Woehler E J (1992) *The Distribution and Abundance of Antarctic and sub-Antarctic Penguins.* SCAR, Cambridge.
Yukhov V L (1982) *The Antarctic Toothfish.* Nauka, Moskow, 113 pp. (In Russian.)

The history of the BIOMASS Data Centre and lessons learned during its lifetime

M. R. THORLEY & P. N. TRATHAN

Introduction

BEFORE THE BIOMASS Data Centre was set up there was little or no previous experience of organizing large-scale biological databases and little experience of how an Antarctic biological data centre should be run. This paper is intended to provide an overview of how the BIOMASS Data Centre was developed, as well as to highlight a number of insights and lessons which have been learned and which may be applied in the future. As the initial objectives for the BIOMASS Data Centre were revolutionary in aspect, this paper is not offering criticism, but considers the lessons learned and how they may be applied to future SCAR information management systems.

The history of the BIOMASS Data Centre

The experiences of the Data Centre are very closely related to the experiences of the BIOMASS Programme. This section reviews the history of the BIOMASS Data Centre, from its inception to the present day.

Inception

The need for a Data Centre to be made available for the duration of BIOMASS was identified from the outset of the programme (BIOMASS, 1979). However, it was some time before the BIOMASS Data Centre was finally established. It was agreed that the initial responsibilities of the Data Centre (BIOMASS, 1979) should include the following:

- the provision of advice and guidance to data collectors on data exchange formats and mechanisms;
- the receipt and storage of BIOMASS data sets and the creation of a data inventory;
- the support of data interpretation workshops;
- answering requests regarding BIOMASS data holdings;
- liaison with the appropriate archiving agencies, for example the World Data Centres for oceanography, and transfer of data to these agencies;
- the preparation of standard data products.

The first data management systems

The first data management system was established at the Fachbereich Informatik, University of Hamburg, in 1981. Fachbereich Informatik was a research group looking at methods of data handling; it made its experimental data management system available to BIOMASS, free of charge, as a practical test and demonstration of the methods being developed. It was never intended that Fachbereich Informatik should provide more than a temporary home for the Data Centre and only the first FIBEX data interpretation workshop was held there (BIOMASS, 1980, 1981).

In 1984, the database group of the Fachbereich Informatik of the University of Hamburg moved to the Computer Science Department at the University of Frankfurt. As a consequence of this, the FIBEX database was also moved to the University of Frankfurt. The task of implementing the database at Frankfurt involved some restructuring of the database; this task was hindered by the poor quality of the documentation, a consequence of the haste with which the database was developed for the 1981 FIBEX data interpretation workshop (BIOMASS, 1984*a*).

A couple of subsidiary, subject-specific databases were also set up to support individual workshops. In 1982 an oceanographic database was set up at the Institut für Seefischerei, Hamburg, using the data handling system of the institute. This was to support the FIBEX hydrographic workshops held in 1982 and 1983 (BIOMASS, 1982). In 1985, for the FIBEX seabird data interpretation workshop, a database was set up in Cape Town, using a subset of the data from the main database (BIOMASS, 1985).

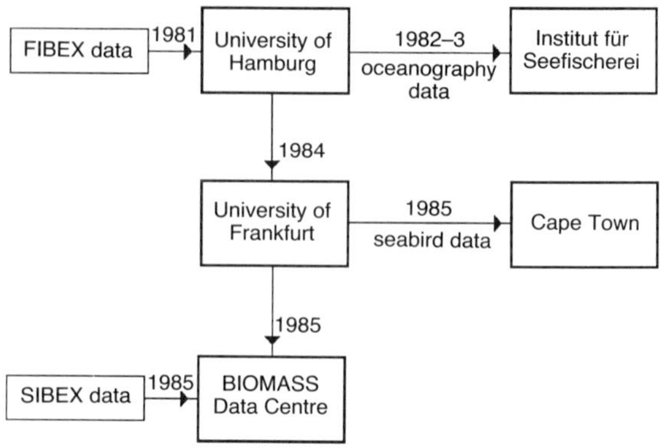

Fig. 1. Relationships between the various BIOMASS databases.

The move to Cambridge

In 1984 the BIOMASS Executive agreed that a permanent home for the BIOMASS database should be set up at Cambridge following an offer to host a BIOMASS Data Centre by the British Antarctic Survey (BAS). The Executive agreed that the BAS offer provided the kind of facilities necessary for a successful data centre. These included adequate analytical computing and data management facilities as well as infrastructure support. A similar offer was also made by the Alfred-Wegener-Institut, Bremerhaven; however, the Executive agreed to accept the BAS offer in view of the urgency with which the facility was required and the more advanced state of the BAS system (BIOMASS, 1984*b*). Figure 1 illustrates the relationship between the various BIOMASS databases.

The early BAS system was run on a Honeywell computer using the G-EXEC database system. In 1986 the system was transferred to a VAX computer using the Oracle database management system. Oracle is a relational system which uses Structured Query Language (SQL), the internationally accepted standard for relational query languages. Throughout the time that the Data Centre has been at Cambridge, BAS have provided the funds to cover the employment of the Data Centre staff and for a large proportion of the computing resources required.

The workshops

The BIOMASS data validation and analysis workshops which have made use of the various BIOMASS databases are detailed in Table 1. There have been other workshops held to look at BIOMASS data, but they have not made direct use of any of the BIOMASS databases. Initially the amount of work required to load and validate data for workshops was underestimated. For example, for the first FIBEX data interpretation workshop not all the planned analyses could be completed, as many data only arrived with the workshop participants, and most of the workshop was spent in loading and validating data (BIOMASS, 1981).

More recently, data validation has been carried out prior to the data analysis workshops. For example, before the SIBEX fish data evaluation workshop there was a separate data validation workshop (BIOMASS, 1986) and for the SIBEX non-acoustic krill workshop data were validated by Data Centre and BAS staff, in correspondence with other BIOMASS scientists (BIOMASS, 1991). This validation of data before the main analysis workshops enabled many more analyses to be carried out and increased the success of these workshops.

Table 1. *BIOMASS workshops that have utilized the BIOMASS database*

Year	Workshop title	Venue[a]
1981	Post-FIBEX Data Interpretation Workshop	UH
1982	First Post-FIBEX Hydrographic Data Interpretation Workshop	H
1983	Second Post-FIBEX Hydrographic Data Interpretation Workshop	H
1984	Post-FIBEX Acoustic Workshop	F
1985	FIBEX Seabird Data Interpretation Workshop	PF
1986	Post-SIBEX Fish Data Evaluation Workshop: Validation	C
1987	SIBEX Physical Oceanography Workshop	C
1987	Post-SIBEX Fish Data Evaluation Workshop: Evaluation	C
1988	SIBEX Acoustic Data Validation and Analysis Workshop	C
1991	SIBEX Non-Acoustic Krill Data Workshop	C

[a] Venues: C, BIOMASS Data Centre, Cambridge; F, Fachhochschule, University of Frankfurt; H, Institut für Seefischerei, Hamburg; PF, Percy FitzPatrick Institute of African Ornithology, Cape Town; UH, Fachbereich Informatik, University of Hamburg.

Current position

Following an extensive review of the BIOMASS data holdings and database in 1990 the database was rationalized and reorganized, with a new database structure being defined and the data transferred from the old database. The Data Centre now holds a data set comprising the majority of the data that have been submitted. There are a few data sets which the Data Centre has, as yet, been unable to load. It is difficult to say how complete the data set now is, as there are BIOMASS data which have never been submitted to the Data Centre. Some subject areas, for example phytoplankton species composition, were very poorly represented in the data submitted.

The FIBEX and SIBEX data analysis workshops have been responsible for validating the data that they used. However, the validation of the chlorophyll and nutrient data still needs to be completed, though this work has been started by Priddle and co-workers (see Priddle *et al.*, this volume).

Availability of the BIOMASS data set

The BIOMASS Executive has decided that the BIOMASS data set, with its supporting documentation, will be made available to any organization or individual that requests it (BIOMASS, 1992). It is intended that BIOMASS national representatives will receive copies free of charge, while others will be charged a nominal fee to cover the cost of materials and postage. Requests for copies of the data set will initially be handled by the BIOMASS Data Centre, though in future other arrangements may be made.

Data life cycle, database access and structure, and current problems

This section describes the life cycle of the data held within the Data Centre, reviews the structure of the database and access to it, and also discusses some of the problems currently facing the Data Centre.

The data life cycle

All BIOMASS data follow a similar path before they become available for use in queries within the BIOMASS Data Centre.

Field data, whether from individual manual observations or from sophisticated capture methods utilizing computer logging and reduction, must first be transferred to a permanent machine-readable format.

The machine-readable data must be validated by the group of scientists in their own national institution.

After conversion to BIOMASS format the data must be sent to the BIOMASS Data Centre where they are loaded into a temporary area of the database. This transfer is normally done using magnetic tape, though other media have been used, for example punched cards or personal computer diskettes.

The data are again validated by both Data Centre staff and the scientists involved, often in the context of other BIOMASS data sets.

Validated data can finally be incorporated into the BIOMASS database and deleted from the temporary area.

The data are now of scientific value and can be integrated with other data on the database, to support the queries that are made against the database.

It is important to note that the database can only be as good as the validated data that it holds.

Database access

The BIOMASS database is designed around a two-tier level of access. The top level of access allows any item of data to be added, deleted or modified and is only accessible to the staff of the Data Centre. The second level of access, open to any user, allows read-only access of the data and is used for querying the database. Access to the database is normally made from the Data Centre, although it is possible to have access from a remote site by using international data networks. However, only limited use has been made of this facility.

The specialized, high-level database query language, SQL, can be difficult for non-specialists to use. The syntax and grammar of the language, though English-like, are not always easy to assimilate and various queries can be ambiguous. The structure of the database can also be difficult for a non-specialist to undersand. Other systems available for querying databases can be easier to use, but do not offer the power or flexibility of SQL. Once data have been extracted from the database there are few defined interfaces to statistical or graphical packages. This paucity of interfaces is a further deterrent experienced by many scientists. For these reasons, scientists have not directly accessed the database, relying instead on the specialist staff of the Data Centre, who are able to extract and analyse the data for them.

The current database structure

The database is divided into a number of distinct subject areas:

cruise details, including stations and ships track;
oceanographic data, including CTD and chlorophyll data;
net haul data, comprising krill length-frequency and ichthyoplankton data;
krill acoustic data;
bird observation data.

These major areas are in turn made up from a number of different collections of data known as tables. These tables have specific and logical interrelationships and are provided with keys enabling the data to be selected in a precise and logical manner. Figure 2 shows the relationship between the tables in the BIOMASS database.

Some BIOMASS data will be excluded from the final database, as they are either incomplete or have not been validated. The data to be excluded are phytoplankton species composition and zooplankton data. Phytoplankton data, apart from chlorophyll values, have only been supplied by one country, while the zooplankton data, excluding those for krill or ichthyoplankton, are impossible to validate at this late stage of the programme.

Problems with the current BIOMASS database

The BIOMASS Data Centre holds data for only a subset of the BIOMASS cruises, and for those cruises, only a subset of the available data. There are other data sets which are known to exist but they are not currently available to the Data Centre. Some of the large-scale questions which BIOMASS was set up to investigate can only be adequately

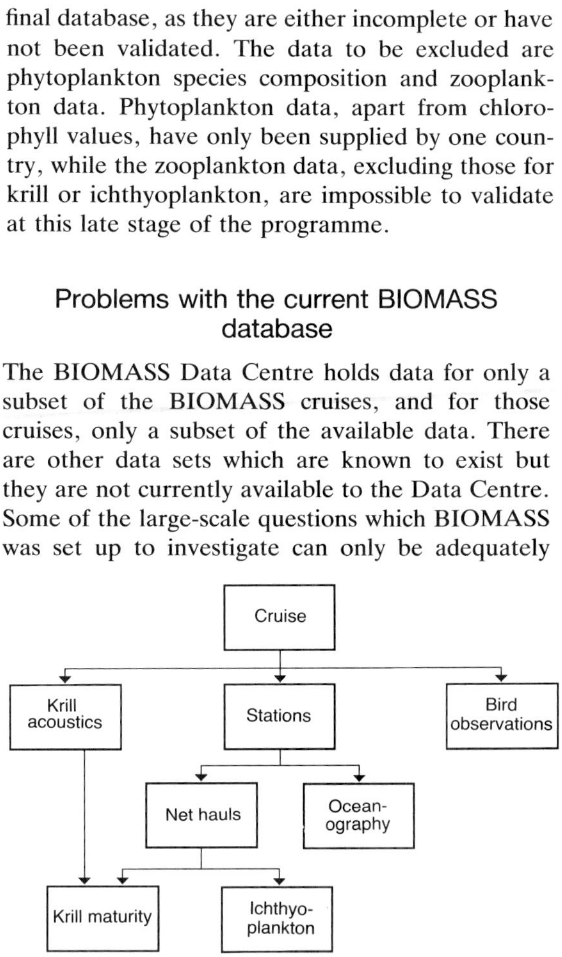

Fig. 2. Relationships between tables in the BIOMASS database.

investigated if all the data sets collected during the BIOMASS Programme are available for analysis. Proper validation after a period of ten years is extremely difficult, especially when original data and the original scientists are no longer available for consultation. Therefore, even if these data sets were to be supplied today it would prove difficult to incorporate them into the database. Notwithstanding this, is it vital that every effort should be made to make the BIOMASS database as complete as possible.

Key insights and their application to a future data centre

By carrying out a review of the development and running of the BIOMASS Data Centre it has been possible to describe a number of activities that with hindsight could have been carried out differently. These 'lessons' learned from the running of the BIOMASS Data Centre ought not to be forgotten and they should be taken into account when planning any future SCAR Information Management Strategy. This section reviews these lessons, and recommends actions that must be undertaken by future programmes.

Integration of information management and science

A majority of the information management problems suffered by BIOMASS were due to the lack of integration of data management within the science programmes and also to insufficient account being taken of the objectives for the BIOMASS Data Centre.

The need for a BIOMASS Data Centre was identified in 1979, yet it was not until 1985 that the Data Centre was finally established at Cambridge. Since then, the Data Centre has been concerned with collating and integrating data into the central database and then using these data within data analysis workshops. The work of the Data Centre has been reactive and driven by the immediate scientific needs, without sufficient consideration being given to the long-term requirements of the BIOMASS Programme and the initial objectives of the Data Centre (see section on Inception, above).

In future research programmes, a better integration of information management and science will be required, with information management being planned into the programmes from the beginning. This planning must include estimates of the costs of information management and the identification of the sources of funding which will be required to meet these costs. The work of the information management facilities (information centres, data networks, etc.) must be reviewed at regular intervals to ensure that they still meet the required objectives. It is reassuring to note that a key part of the planning for the International Geosphere–Biosphere Programme (IGBP) is the information management facilities that will be required; for example, see IGBP (1990).

Choice of hardware and software

The BIOMASS database was initially implemented using an experimental relational database system, PASCAL/R, at the Fachbereich Informatik, University of Hamburg, which was then moved to the Computer Science Department at the University of Frankfurt. On transfer to the BIOMASS Data Centre at BAS, it was installed using G-EXEC, before being transferred to the widely used commercial relational database system, Oracle. Each transfer stage took time and resources. This time used for transfer and revalidation was not time spent doing science. The transfer also created the potential to corrupt the database. For example, the Australian FIBEX krill length–frequency data have been lost from the BIOMASS database between Hamburg and Cambridge.

Future programmes, operating within the remit of the SCAR–IGBP, have an obligation to make their data available for the next 10–20 years. Whatever data management systems are used, they must be based on industry standard hardware and software and should also be upgraded to keep abreast with advances in computing technology. However, even with recent advances in open systems technology, making the transfer of databases between different systems much easier to carry out, transfers still offer the potential for data corruption and use valuable resources. For this reason, transfer to new systems should not be undertaken without good reason. Since 1986, BAS have taken this approach, with the BIOMASS database being based on Oracle, one of the most widely used relational database management systems, and Digital Equipment VAX hardware.

A data centre should not devote resources to developing its own 'in-house' software unless absolutely necessary. Full use should be made of commercially produced and supported software for graphical and statistical analyses. Obviously, given the nature of the work being carried out, there will be a requirement for a data centre to develop certain software, but when this is done, all software should be fully validated and documented.

The structure of the databases should be well thought out, and full use should be made of the commercially available tools and techniques for the design and documentation of databases. Too often the structure of the BIOMASS database has reflected the complexity and variability of the data that were supplied, rather than the underlying generic model of the data.

Transfer of data to a data centre

The BIOMASS Programme expended considerable resources in the design and development of data transfer formats. Data contributors had to develop their own computer programs for converting their data into the BIOMASS format, and the BIOMASS Data Centre had to develop software to convert data from the BIOMASS format into that used within the database. However, though difficult for scientists, it was an advantage to the BIOMASS Data Centre to have all data arriving in the same format, as only one set of data load software had to be produced. In order to cope with the variability of the data the BIOMASS formats were extremely complex and this was reflected in the difficulty scientists experienced in using them. This is also true of other complex data formats, for example GF3, the general format for geo-referenced data used for data exchange by the World Data Centres for oceanography.

The use of the BIOMASS data transfer formats was a disincentive to the transfer of data to the Data Centre. To convert data to BIOMASS format took time and resources which could be better spent doing science. The scientist was required to make a significant investment without any immediate quantifiable return. For this reason in the future, data transfer needs to be made as easy as is possible. A data centre should be able to accept data in the format used by scientists and on a variety of media, for example, floppy disks, tapes or network transfer. The format and media of the data are not important, but it is important that the data are documented. It is essential that data contributors supply all the required supporting information.

An innovative idea, developed by M. Reinke (Alfred-Wegener-Institut, Bremerhaven) was an IBM PC-based data entry and storage system, which was designed for use within the home institutions of data contributors. Using such a system, data could be entered and validated on a local database and then a copy sent to the Data Centre. Had this approach been adopted within BIOMASS, it could well have improved the quality of the data received by the Data Centre. Such a method forces data contributors to enter the supporting information that is required by the Data Centre.

Scientists must also know why they are transferring data to a data centre. They must be aware of the data sharing and integration requirements of the programme and must appreciate the long-term benefits of having such data available through a central source. These benefits include being able to make more of their own data by being able to combine it with other related data sets, as these combined data sets have a greater value than the individual components.

Data validation

Data validation must be given a high priority. It should be carried out when data sets are initially loaded and not ten years later, as has been the case with the BIOMASS chlorophyll data. Validation requires the full participation of the scientists who collected the data, with the aim of making sure that the data that are on the database are the data that were collected.

Unlike some forms of data which have largely automated capture and validation stages, for example remote sensing or geophysical data, the biological data handled by the BIOMASS Data Centre have required a large amount of manual processing from the scientists who collected them. For example, krill length and maturity stage classification, or ichthyoplankton species determinations, have to be done by hand. The accuracy of these data depends on the technical skill and taxonomic confidence of the scientist. Some data sets were supplied without sufficient supporting information. For example, chlorophyll data were supplied without any information on the methods used to determine the chlorphyll concentration, or whether the figure included phaeopigments. Data were supplied where the position of the station did not correspond, even approximately, with the BIOMASS standard station being visited, and at least one data set was supplied without any positional information at all!

In future, minimum acceptable quality standards for data must be defined and adhered to. It should be a requirement to supply with these data some indication of the level of familiarity of the scientists with the material being investigated. Ideally, this should also include the names of the scientists responsible for ensuring the quality of the data, allowing queries regarding the data to be directed to the correct person. When a data centre responds to requests for data, this supporting information must be supplied along with the data that are required.

Data validation is the responsibility of the scientists contributing the data and it requires their active participation. Without validation, data will be of no use or interest to any other scientists and will not be worth holding in a data centre. However, data validation takes both time and money. For example, the most successful way that was found of validating BIOMASS data has been through a series of data validation workshops. These workshops, held at the Data Centre, were attended by many BIOMASS scientists, but were expensive to organize and run. For a data centre to hold such workshops in future would require a substantial source of funding to enable scientists to attend.

Data access protocols

The BIOMASS Executive had initially proposed an open access protocol for data in the BIOMASS database, but this was restricted at the insistence of some of the BIOMASS nations. This restricted access protocol was in operation for most of the life of the Data Centre and made it difficult for data to be accessed outside the analysis workshops. The original protocol was replaced with a new open protocol defined at the BIOMASS Executive meeting of January 1990 (BIOMASS, 1990). However, by then the damage had been done as the scientific work that could have been carried out was reduced because the Data Centre was unable to satisfy requests for data from BIOMASS scientists.

Restrictive data access protocols will stop a data centre from working effectively. Any new data centre should aim for open access to all its data holdings. The ideal protocol would be similar to that of the World Data Centre System, where data held by a World Data Centre must be completely accessible by scientists in all countries, upon written request or personal visit (Anon, 1987). With a future data centre, emphasis must be given to the task of allowing scientists immediate access to the data they require, without having to ask the data centre staff to prepare the data for them. The SCAR Seismic Data Library system on CD-ROM is a good example of how this might be achieved (SCAR, 1993). Given a physically open system, it would be difficult to implement a restrictive protocol, as scientists familiar with the benefits of an open system would become frustrated with a protocol that was too restrictive and would demand one that was more open.

Management of a data centre

The management of the BIOMASS Data Centre was overseen by two groups, the British Antarctic Survey and the Data Centre Advisory Group. Unfortunately, there was no clearly defined set of responsibilities for each group, with the result that some management tasks were never carried out. For example, no quality assurance of the type and amount of documentation within the Data Centre, and no review of the validity of the database structures, was undertaken. In this instance, each party thought it was the other's responsibility. Another major problem was the period of 18 months between the resignation of the first Data Centre Manager and the appointment of his replacement. During this time, both the BIOMASS Executive and the British Antarctic Survey seem to have been unsure of what future, if any, the Data Centre should have.

Any future data centres will require a clearly defined management structure with allocated roles and responsibilities. The management of a data centre must not fall into a void between its host institution and the scientific community it is serving, and to a large extent it is the responsibility of this scientific community to provide the management structures. They must ensure that standards of both software and documentation are given a high priority and that their requirements, as users, are being met. At the same time, the work of the data centre must be fully integrated with the research programme which it is serving.

Responding to the needs of the user community

It has been said that the BIOMASS Data Centre became a 'black hole' for data: once data had been contributed they never saw the light of day again! This was due partly to the very real problems that the restrictive data access protocol caused. Even though there is now a much freer access protocol in force, this impression of the Data Centre has stuck. The image that a data centre presents to its users is very important. If scientists feel that they can work with a data centre and that it is doing the kinds of things they want it to do, it will be much more successful than one that they feel uninterested in, or remote from.

A data centre requires some form of 'User Group' to represent the interests of those using it. The data centre is a facility for the users and should be responsive to their needs. This requires communication between the data centre, its management group, and the user community.

Resources required for a data centre

Running a successful data centre is an expensive operation; the financial costs of providing the information management facilities for future programmes are likely to be too great for any single institution to cover. For example, the cost of the computer resources used by the BIOMASS Data Centre in 1991 was in excess of US$100 000. This is on top of salaries and other expenses. Since 1985 funding for the Data Centre has been provided by BAS, with a contribution of about US$130 000 from the BIOMASS special fund.

In future, funding arrangements must be defined at the outset of a programme. It is likely that the costs would have to be shared more evenly than has been done with BIOMASS. This could possibly be implemented by organizations paying a subscription to use the facilities.

The Data Centre staff should ideally have a scientific background in the subject area and a thorough appreciation of the scientific objectives of the programme they are supporting. They will also require support from scientists who understand the type of data held and the scientific objectives behind the collection and analyses of these data. This could, in the first instance, be provided by the scientific staff within the host institution.

Summary of recommendations

This section summarizes the recommendations made above. It is important that they should be incorporated into the design of future programmes.

In future research programmes, a better integration of information management and science will be required, with information management planned into programmes from the beginning.

The information management strategy and its implementation must be reviewed at regular intervals throughout the lifetime of a programme to ensure that scientific objectives are met.

Although information management systems must be upgraded to keep abreast with advances in computing technology, migration to new systems should only be undertaken with good reason. A data centre must be based on industry standard hardware and software and not on technologically obsolete systems.

The structure and implementation of any database must be well thought out and full use should be made of commercially available tools and techniques for its design, maintenance and documentation.

Transfer of data to a data centre must be made as easy as is possible, with data being accepted in the format used by scientists.

Scientists must know why they are transferring data to a data centre and must be aware of the data integration and sharing requirements of the programme.

Data validation must be given a high priority and carried out as data are loaded. Validation will require the full participation of the scientists who collected the data.

Full supporting documentation must be supplied with all data.

Minimum acceptable quality standards for data must be defined and adhered to.

There should be no restrictions on access to data.

A future data centre will require a clearly defined management structure with allocated roles and responsibilities. Ultimately, the scientific community will be responsible for ensuring that their requirements are met, that standards are adhered to and that the overall management of the data centre is being carried out in accordance with agreed plans.

A data centre should be responsive to the needs of the user community, with liaison through a 'User Group' representing the interests of scientists who use the data centre.

Before a data centre is established, adequate resources must be made available to ensure that it will be able to carry out its agreed role.

Conclusions

It must be stressed that this chapter has not been an attempt to level criticism at any individual or organization, but it has sought, with the benefit of hindsight, to point out where mistakes were made and to show how we can learn from these mistakes in the future. This chapter has been an attempt to review the operation of the BIOMASS Data Centre, looking at some of the problems associated with data management for large-scale programmes and some of the lessons learned. Over the past six years the Data Centre has built up a wealth of valuable knowledge on the management of large-scale data sets, which will be of great relevance to future information management initiatives within the Antarctic scientific community. Overall, the BIOMASS Data Centre has played a valuable role within this community, both as a key resource for the BIOMASS Programme and also as a source of knowledge on data management. The BIOMASS data themselves may now be of less interest, but the skills and knowledge within the Data Centre are of great relevance to the future.

Acknowledgements

The authors thank the following people for their help and support during their involvement with the BIOMASS Data Centre: Andy Clarke, Barry Heywood, Bruce Lamden, David North, Julian Priddle and Andy Wood from the British Antarctic Survey, Manfred Reinke from the Alfred-Wegener-Institut, and Sayed Z El-Sayed, Chairman of the BIOMASS Executive.

References

Anon (1987) *Guide to the World Data Center System*, Part 1, *The World Data Centers.* ICSU.

BIOMASS (1979) Technical group on data, statistics and resource evaluation, report of the first meeting. *BIOMASS Rep Ser* **4.**

BIOMASS (1980) Technical group on data, statistics and resource evaluation, report of the second meeting. *BIOMASS Rep Ser* **16**.

BIOMASS (1981) Post-FIBEX data interpretation workshop, report of the workshop. *BIOMASS Rep Ser* **20**.

BIOMASS (1982) First Post-FIBEX hydrographic data interpretation workshop, report of the workshop. *BIOMASS Rep Ser* **30**.

BIOMASS (1984*a*) Acoustic working party, report of the meeting to prepare for the Post-FIBEX Acoustic Workshop. *BIOMASS Rep Ser* **37**.

BIOMASS (1984*b*) BIOMASS Executive, report of the 1984 meeting. *BIOMASS Rep Ser* **38**.

BIOMASS (1985) Working party on bird ecology, report of the FIBEX sea-bird data interpretation workshop. *BIOMASS Rep Ser* **44**.

BIOMASS (1986) Post-SIBEX fish data evaluation workshop: Phase 1, validation, report of the workshop. *BIOMASS Rep Ser* **52**.

BIOMASS (1990) BIOMASS Executive, report of the 1990 meeting. *BIOMASS Rep Ser* **61**.

BIOMASS (1991) SIBEX non-acoustic krill data analysis workshop, report of the workshop. *BIOMASS Rep Ser* **66**.

BIOMASS (1992) BIOMASS Executive, report of the 1991 meeting. *BIOMASS Rep Ser* **67**.

IGBP (1990) Data and information systems for the IGBP. In: *The International Geosphere-Biosphere Programme: A Study of Global Change, The Initial Core Projects.* Report Number 12, IGBP Secretariat, Royal Swedish Academy of Sciences, Stockholm.

SCAR (1993) A SCAR seismic data library system for cooperative research: Summary report of the international workshop on Antarctic seismic data, Oslo, Norway, April 11–15, 1991. *SCAR Rep Ser* (in press).

Discussant's Report: Antarctic marine systems

R. M. LAWS

Introduction

THE GENERAL OBJECTIVE of the BIOMASS Programme was 'to gain a deeper understanding of the structure and dynamic functioning of the Antarctic Marine Ecosystem as a basis for the future management of potential living resources' (Anon, 1977). However, ecosystem studies were not given priority in the planning of BIOMASS, although there was a BIOMASS Working Party on Ecosystem Modelling. Ecosystem modelling had been attempted in the 1970s but we still have only very general conceptual models; as the extent of the qualitative and quantitative uncertainties has become more apparent, there has been relatively little recent modelling work.

Types of Antarctic marine ecosystems

Although the Southern Ocean ecosystem has often been assumed to be a single homogeneous ecosystem, it contains a number of different ecosystems that can be classified in various ways. Thus, Marchant & Murphy (this volume) consider the smaller organisms and define four ecosystem types, three of which are based on zones: a Northern or Ice-Free Zone, an Intermediate or Seasonal Pack-Ice Zone and a Southern or Permanent Pack-Ice Zone. A fourth category includes ecosystems typical of various Zone-Independent Regions, associated with enhanced production: fronts, upwelling, eddies, islands and shelf regions. The breeding colonies of higher predators are often associated with such regions of high production.

When higher-level predators are considered, there appear to be two major food webs superimposed on the systems characterized by the lower trophic levels. These food webs (Kock & Shimadzu, this volume) are more conspicuous to the observer and may be characterized as: a northern slope water and oceanic food web of the Ice-Free Zone and the oceanic parts of the Seasonal Pack-Ice Zone in summer, comprising copepods, mesopelagic fish and squid, small petrels, wandering albatrosses, macaroni and king penguins, elephant and fur seals, right whale, sei whale and sperm whale; and a southern food web with krill (*Euphausia superba*) as the key food organism in the Seasonal Pack-Ice Zone, partly replaced by *Euphausia crystallorophias* and the fish *Pleuragramma antarcticum*, in the High-Antarctic and Shelf Zones, also comprising juveniles of some squid species, notothenioid fish, petrels, chinstrap and Adélie penguins, blue, fin, minke and humpback whales, and southern bottlenosed whales, *Berardius arnouxii*.

In addition to these broad circumpolar classifications there are regional ecosystems. Examples of these are the Antarctic Peninsula and Bransfield Strait, South Georgia waters, Prydz Bay and the

Weddell and Ross Seas. There are interactions between some of these ecosystems, for example the Weddell Sea and Bransfield Strait, and the ocean current 'conveyor belt' between the Bellingshausen Sea and South Georgia, which maintains the vast breeding colonies of birds and seals on that island.

Marine ecosystems are three-dimensional, with vertical as well as lateral components, influenced, for example, by the current systems at different depths and the association of bird feeding concentrations with different water masses.

In addition, there have been perturbations associated with commercial exploitation over long time scales (in the case of seals, over two hundred years, and for whales approaching a hundred years). More recently commercial fisheries for krill, fish, and squid have developed. The perturbations of baleen whale stocks were associated with the 'krill surplus' theory, although increases in other predators such as penguins and seals, whether by competitive release, or owing to large-scale environmental change, probably obviated the accumulation of any real surplus. This emphasizes the importance of studying trends with time.

Energy fluxes

Most of the available data are from the summer months and the rest of the year has been neglected, but there are major seasonal changes in the balance between processes. Thus, a striking feature of Antarctic systems is the seasonal light change which results in a pulse of production in the summer. Winter studies in the pack-ice zone have led to a reappraisal of the biology of krill, *Euphausia superba* (Smetacek *et al.*, 1990).

Another recent interest has been in the direct effects of atmospheric forcing on ecosystem dynamics, most notably with the El Niño–Southern Oscillation (ENSO) phenomenon associated with year-to-year fluctuations over periodicities of 4–6 years. Thus, fluctuations in year class abundance of several seal species (Laws, 1984) may be causally related, although a more recent analysis was inconclusive (Testa *et al.*, 1991); krill year-class strength is critically dependent on water temperature and sea ice distribution during larval development, with up to 50% annual post-larval mortality.

Sakshaug (this volume) emphasized the importance of fluxes in ecosystem studies, but there is still a need to define the energy 'boxes' and to decide the connections. In this respect, it is very difficult to obtain data on the food of higher predators. Diet differs regionally (for example the Weddell seal in the Weddell, Bellingshausen and Ross Seas) and seasonally (for example, fur seals switch between krill and squid; leopard seals switch between crabeater seals, penguins and krill). There are also changes within species over the life history and between sexes.

The nature of Antarctic ecosystems

Lower trophic levels

When the BIOMASS Programme began, textbooks still described a short Antarctic food chain: diatoms→ krill→whales, which was highly productive, because it contained few component groups. Formerly the focus of primary production studies was on the relatively large diatoms, but in the past decade research has also encompassed the smaller picoplankton, nanoplankton, bacteria and protozoa. Marchant & Murphy (this volume) have summarized the lower trophic levels of the Southern Ocean, based on a generalized relationship between organism size (0.1 μm to 10 cm) and generation time (hours to years), from bacteria to krill. One of the most important recent developments has been the re-evaluation of the microbial components of marine ecosystems and the development of the 'microbial-loop' hypothesis.

Thus, the *new production system* is essentially the traditional view of the Southern Ocean ecosystem, with nitrate-based production dominated by large diatoms, grazed by the large herbivorous plankton, which are in turn taken by the larger predators at higher trophic levels. The *regeneration systems* involve the microzooplankton groups, such as ciliates, which consume the smaller heterotrophic and autotrophic groups and themselves provide a food source for the larger zooplankton. Much of the autotrophic production by microorganisms is based on ammonia produced by larger animals in the ecosystem. These regeneration systems provide energy flow pathways which transfer energy up the food web in areas where new production is low.

Phaeocystis has recently received attention as a

producer of dimethylsulphate, providing cloud seeding nuclei and therefore potentially affecting global climate. In a higher ultra-violet radiation regime, promoted by the thinning of the ozone layer, *Phaeocystis* may have a competitive advantage as it appears more UV-B-tolerant than other groups.

The seasonal pack-ice zone is a highly variable environment in time and space, ranging from interannual changes and seasonal changes in sea-ice area and the ice edge position (hundreds of kilometres), to the effect of individual floes on local light input (minutes to hours and tens of metres). Patchiness is very pronounced, and pathways of energy flux show high variability. The highest (new) production occurs in blooms coinciding with the recent retreat of the ice edge, recycled production probably increasing in amount with increasing distance from the bloom regions. Regeneration systems are likely to be important in the permanent pack-ice zone, although new production by diatoms is seasonally important within the sea-ice community. In the northern zone new production is low because of wind-induced turbulence, low light, lowered trace elements, and grazing pressure. Copepods are the dominant zooplankton there and much of the production is probably initiated and recycled by the small organisms, though low ammonium levels may be restricting.

In some regions other zooplankton may be dominant. For example, salps are very abundant and show high productivity in the Pacific sector (Foxton, 1966). Environmental changes may also have widespread effects on marine systems.

Higher levels

The higher predators of the Antarctic are conspicuous, but their role in terms of energy fluxes is uncertain. First, the role of squid may have been overestimated as a predator of krill; their true significance in the food webs is unknown. Squid biomass estimates are controversial, ranging from 5 to 100 Mt. There is little information on the fifty species found south of the Polar Front; most have southern limits in the ice-free zone, north of the main krill concentrations. Recent work suggests that mesopelagic organisms such as myctophid and paralepid fish, and other squid, may feature more in squid diets than krill.

Secondly, this latitudinal decline in squid is matched by a parallel reduction in the diversity of their prey, the mesopelagic fish. Thus, the fish biomass in higher latitudes is an order of magnitude lower than in the seasonal pack-ice zone and the northern islands. Some 270 fish species are known, of which 12 are commercially exploited in the ice-free zone on a regular basis. *Notothenia rossii*, for example, was drastically depleted within a few years (Kock, 1991). For only a few species are population sizes and trends known, but current estimates of the biomass at South Georgia, South Orkney Islands and Kerguelen Islands are of the order of 20% of their pristine abundance.

The ecological interactions of these two groups remain speculative and poorly understood.

Thirdly, there are some 40 species of breeding seabirds in the Southern Ocean, dominated in biomass terms by the penguins, the most abundant of which are the macaroni, chinstrap and Adélie, with populations estimated at 12 million, 7.5 million and 2.5 million breeding pairs, respectively, and perhaps as many again in the non-breeding populations. More than 95% of the biomass of the first two species occurs in the Atlantic sector. Although the ecology, physiology and behaviour of Antarctic seabirds is becoming increasingly better known, relatively little is known of their overall population status or trends. A number of species have increased (including most penguins); others have declined, some such as wandering albatross as a result of incidental mortality from the commercial fisheries.

Fourthly, the seven species of seals have also been intensively studied in respect of their ecology, physiology and behaviour, but again their overall population status and trends are relatively poorly known. Research has advanced, but in some respects knowledge of the population status of several species is more uncertain than it was; this uncertainty is due to the high costs of logistic operations in the pack-ice zone and the fact that it is nine years since the most recent major ice seal census was carried out in 1983. Conversely, knowledge of land-breeding species has greatly improved and the population trends of most stocks are known, although a satisfactory explanation for the sustained decline of the Indian Ocean stocks of the elephant seal is lacking (SCAR Group of Specialists on Seals, 1991).

However, the foraging behaviour of seabirds and seals is becoming increasingly well known, as more sophisticated telemetry is applied (geolocation, time-depth recorders, physiological monitors). This is revealing astonishing physiological feats and quantifying energetic processes, which can be related to the food resources, and so to ecosystem structure and function. However, our understanding is still very far from complete.

Fifthly, the BIOMASS Programme did not address cetacean research, which is coordinated by the Scientific Committee of the IWC. The biology of six species of great whales and eleven species of smaller toothed whales is becoming better known, but with the ending of the commercial whaling industry knowledge of the population sizes and trends of the great whales is very uncertain and research on other aspects of their biology has stopped, although direct observations on some species have increased. Even though whaling has virtually ended, it is not known whether species are increasing, stable or decreasing, except for the right whale and humpback whale (which can be counted in coastal waters in low latitudes) and the minke whale, which has been studied through specialized sightings cruises in the Antarctic which show it to have increased, to 760 000 by 1985. Knowledge of the smaller toothed whales is still very limited, although the southern bottlenosed whale, *Berardius arnouxii*, with populations estimated at 170 000–320 000 appears to be the second most abundant cetacean in the Southern Ocean after the minke whale. The killer whale, *Orcinus orca*, is known to have trophic links within the Antarctic food web with fish, birds, seals and other whales.

Future research needs

Despite the undoubted success of the BIOMASS Programme, there is still little coordination between workers on the different groups of organisms in the Southern Ocean. There is an urgent need to promote this by interdisciplinary studies within regions; integrated studies of interactions between and within trophic levels should be promoted more actively (examples are the EASIZ and GLOBEC programmes). There will be increased understanding of the fundamental importance of the seasonal pack-ice zone for primary and secondary production.

There are serious gaps in knowledge within taxonomic groups. For example there was little direct work on squid biology in the BIOMASS Programme, but it led to the production of an *Identification Key and Species Description for Antarctic Squids* (BIOMASS Handbook No. 21). This is a very ecologically significant and unusual group of animals, with a very short lifespan, high growth rate and high population turnover; it is a food resource for higher predators, and an important predator itself on mesopelagic fish and krill.

Two views have been expressed in the BIOMASS Colloquium on the role of krill in the ecosystem. Some confirmed existing ideas; others have suggested that krill might not be as significant as formerly thought, or important only when it occurs as swarms, which are the points for significant transfer of energy and material to higher levels. Where is the krill when it is not found in swarms? What about the vast areas between krill swarms, the apparent lacunae in its distribution, where are found copepods, salps, fish and squid larvae?

More work will be done on exciting aspects of the alternative krill biology, winter grazing on ice algae coating the underside of the pack-ice. There is also a need for further work to attempt to resolve the controversy surrounding the idea that the decline of the great whales through overfishing resulted in density-dependent, krill-abundance-related, compensatory increases in populations of penguins, crabeater seals and minke whales. What role have climatic and physical oceanographic change played in the dynamics of the systems as compared with trophodynamics? What effects can be expected if global warming develops? Much more needs to be discovered about the krill standing biomass, krill production and its consumption by predators, and krill energetics.

Then there is a new research topic, resulting from the discovery of the thinning of the ozone layer, causing enhanced penetration of UV radiation. Could this change the nature of ecosystems at lower trophic levels, owing to differential responses by organisms to changes in UV levels, and lead to perturbations as profound as (or more so than) may have been caused by the decline in abundance of whales?

Increasingly we have become aware of the need to identify fluxes not just between 'labelled boxes' within ecosystems, but also between ecosystems. For example, the geographical distribution of elephant seals during the breeding and non-breeding seasons shows that they regularly cross boundaries between the Antarctic ecosystems listed earlier, and even migrate north of the Antarctic Polar Front, as do the baleen whales.

Conclusions

Comprehensive understanding of Antarctic marine ecosystems is still remote. To what extent has BIOMASS influenced thinking about Antarctic marine systems? There has certainly been a dramatic increase in knowledge at all levels of the ecosystems, and BIOMASS has been the trigger for much of this work. However, comprehensive ecosystem studies are just beginning; exciting times are ahead. Understanding is more necessary than ever if CCAMLR is to succeed in its task of managing the Antarctic marine ecosystems and ensuring their survival.

References

Anon (1977) Biological investigations of marine Antarctic systems and stocks. *BIOMASS Sci Ser* **1** (El-Sayed SZ, ed). SCAR/SCOR, Cambridge, 79 pp.

Foxton P (1966) The distribution and life history of *Salpa thomsoni* Foxton with observations on a related species *Salpa gerlachei* Foxton. *Discovery Rep* **34**: 1–116.

Kock K-H (1991) The state of exploited fish stocks in the Southern Ocean – a review. *Arch FischWiss* **41**: 1–66.

Laws R M (1984) Seals. In: Laws R M (ed) *Antarctic Ecology.* Academic Press, London, pp 621–715.

SCAR Group of Specialists on Seals (1991) *Report of the workshop on southern elephant seals.* SCAR, 29 pp.

Smetacek V, Scharek R, Nöthig E-M (1990) Seasonal and regional variation in the pelagial and its relationship to the life history of krill. In: Kerry K R, Hempel G (eds) *Antarctic Ecosystems: Ecological Change and Conservation.* Springer-Verlag, Berlin, pp 103–14.

Testa J W, Oehlert G, Ainley D G, Bengtson J L, Siniff D B, Laws R M, Rounsevell D (1991) Temporal variability in marine Antarctic ecosystems: periodic fluctuations in the phocid seals. *Can J Fish Aquat Sci* **48**: 631–9.

Future developments

The potential contribution of the BIOMASS Programme to global change research, especially the JGOFS core project of IGBP

J. PRIDDLE, M. R. THORLEY & P. N. TRATHAN

Introduction

THE BIOMASS PROGRAMME was initiated in 1977 against a background of increasing concern that a new episode of exploitation of marine living resources in the Southern Ocean would have serious environmental and ecological effects. As such, it had a structure and approach which were defined by its goals. Fogg (this volume) examines the degree to which the BIOMASS Programme has achieved the goals which it set itself. In this paper, however, we seek to assess the utility of the BIOMASS data and other products for purposes other than those for which they were intended, so it will not be surprising if only a small proportion of the BIOMASS Programme is of relevance to global change studies.

We start with brief reviews of the BIOMASS Programme and its evolution, and consideration of global change research. We will then identify the general ways in which these two areas of science could be linked. These set the scene for a detailed consideration of the BIOMASS data and other products which could feed into global change research. We conclude with recommendations for the most likely ways in which the BIOMASS Programme can contribute to global change research. We should emphasize that the views expressed here are our personal opinions; any critical observations should not be taken as a reflection on the programme as a whole.

Background

The establishment and structure of the BIOMASS Programme

As already noted, BIOMASS was established in response to concern over possible serious environmental and ecological consequences of developing exploitation of Southern Ocean marine living resources. The aims of BIOMASS were:

> To contribute to mankind's understanding of the Southern Ocean and its biota.
>
> To develop a sound ecological strategy for the exploitation of the living resources and for the conservation of the Antarctic marine ecosystem. (BIOMASS, 1977)

The ten-year programme, which included two major field campaigns, has made a significant contribution to Southern Ocean ecology. It has achieved a significant degree of coordination, both within the field programme and in subsequent data analysis.

The concern over exploitation within the Southern Ocean that prompted the joint SCOR–SCAR symposium in 1976 in Woods Hole, USA, which resulted in the genesis of the programme, has of course not diminished (BIOMASS, 1977, 1986). Since then, the Antarctic Treaty nations have framed the Convention on the Conservation of Antarctic Marine Living Resources (CCAMLR). This has addressed the management of existing and future Southern Ocean fisheries on the innovative basis of minimal impact on the ecosystem as a whole. CCAMLR undertakes its own scientific assessments, and also has strong links with the BIOMASS Programme, the BIOMASS scientific community, and the BIOMASS database. Croxall (this volume) demonstrates clearly both the close relationship between areas of science in CCAMLR and BIOMASS and the crucial role which the BIOMASS community had in establishing CCAMLR. Likewise, in a discussion document considering the future role and possible evolution of the BIOMASS Programme (BIOMASS, 1986), Professor El-Sayed rightly emphasized the links with CCAMLR. He focused on CCAMLR and related programmes of research into living resources in the Southern Ocean as likely successors to BIOMASS.

Global change research and its Antarctic component

During the lifetime of the BIOMASS Programme, other environmental concerns have also achieved prominence. With growing worldwide awareness, the Antarctic became the flagship of global conservation in the 1980s. At the same time, several aspects of global problems have focused attention on Antarctica. A good example is the diminution of stratospheric ozone (Farman *et al.*, 1985). This is a phenomenon of global significance in which Antarctica plays a key role. Such major problems cannot be tackled in isolation, and the international research community began to organize major collaborative programmes on global change. These focus on the impact of a wide range of anthropogenic perturbations, including the increase in atmospheric concentrations of several radiatively active ('greenhouse') gases, on the planet as a whole. These research programmes indicate areas where our understanding of the natural system is insufficient to explain what is happening at present, let alone to predict what might happen in the future. For instance, only about half of the known emissions of CO_2 remain in the atmosphere, but our present knowledge of the magnitude of ocean carbon fluxes is not yet good enough to determine whether oceanic sinks are currently much more, or much less, important than terrestrial ones (SCOR, 1990). Antarctica is an area of key importance in many aspects of global change, yet at the same time is one which is badly understudied in several respects. Global change research achieves a higher level of integration. This is accomplished through two umbrella schemes: the World Climate Research Programme (WCRP) and the International Geosphere-Biosphere Programme (IGBP). SCAR has elected to organize a programme of research within the framework of IGBP, under the title *The role of Antarctica in global change* (SCAR, 1989).

We can narrow down the scope of the following discussion to certain aspects of global change research which concern the pelagic marine ecosystem. The IGBP encompasses a very broad range of global change research. This is summarized in a recent publication (IGPB, 1990), where the major topics are identified, and each addressed by one or more core projects. Here we will concentrate almost exclusively on the Joint Global Ocean Flux Study (JGOFS), an existing international study of the role of the oceans in the global carbon cycle (see also SCOR, 1990). Other IGBP core projects, such as LOICZ (Land Ocean Interactions in the Coastal Zone), have more limited potential overlap with studies undertaken during BIOMASS. It should also be noted that SCAR has initiated a regional study within IGBP, and that the organization of the science areas differs slightly from that adopted by IGBP as a whole.

The scientific aims of JGOFS are defined as follows.

> To determine and understand on a global scale the processes controlling the time-varying fluxes of carbon and associated biogenic elements in the ocean, and to evaluate the related exchanges with the atmosphere, sea-floor, and continental boundaries.

> To develop a capability to predict on a global scale the response of oceanic biogeochemical processes to anthropogenic perturbations, in particular those related to climate change.

An Antarctic regional study within JGOFS is already being planned and involves several groups that also participated in BIOMASS. The study will concentrate on several features of particular importance to global ocean carbon flux, including the potential for enhanced carbon fixation by high-latitude phytoplankton (see, for example, Martin *et al.*, 1990), and the climatically sensitive sea-ice zone (see Sullivan *et al.*, 1988).

What can BIOMASS offer to global change research?

In assessing the possible contribution of BIOMASS to projects such as JGOFS, we can identify two ways in which such an interaction could take place. Firstly, information obtained by the BIOMASS Programme may be useable in JGOFS. This could either be by the direct use of data held on the BIOMASS database or available elsewhere, or through the benefit of data synthesis. In the latter category, we will allow the BIOMASS Programme to speak for itself through the proceedings of the evaluation colloquium.

Secondly, there are more intangible but none the less important contributions which the BIOMASS Programme can offer in the form of what could be termed 'infrastructure'. Here we include aspects of BIOMASS which act as foundations for JGOFS and similar projects, and the value of the experience of an internationally coordinated oceanographic programme.

The value of the BIOMASS data

A great deal of international effort was invested in the BIOMASS field programmes and subsequent work-up and data analysis. Obviously, this has made an impressive contribution to the original aims of the BIOMASS Programme, but does this information also relate to global change? Can we make direct use of some or all of the data in future studies, or must we content ourselves with a foundation of increased knowledge of the Southern Ocean? Certainly, the great advances in Southern Ocean ecology promoted by the BIOMASS Programme and several contemporary, and often related, projects cannot be overestimated. We will take it as self-evident that the community of Southern Ocean marine scientists has benefitted greatly by the studies which have taken place. This view is supported by the comprehensive reviews of Southern Ocean ecology in this colloquium volume, and by numerous publications and meetings which preceded it. An enhanced understanding of the Southern Ocean ecosystem will be a clear legacy of the BIOMASS Programme. For instance, the participating scientists have provided a description of the distribution and biology of krill which could probably not have been obtained without the coordinated efforts and standard techniques of the SIBEX cruise series. However, for us the key area for assessment must be the data themselves, and their direct application to problems of global change.

Data holdings

As part of a planned ten-year programme, BIOMASS participants undertook two main campaigns of fieldwork. In FIBEX (First International BIOMASS Experiment), the accent was on spatial variation on a circumcontinental scale. By contrast, the SIBEX (Second . . .) programme was more restricted in coverage but sampled at (supposedly) standard stations during two consecutive seasons.

The BIOMASS Programme included 33 cruises, sixteen in FIBEX and an approximately even division of the remainder between the two years of SIBEX (eight for SIBEX 1 and nine for SIBEX 2). For the SIBEX cruises there was considerable replication at standard stations. The BIOMASS Data Centre holds results from all of these cruises. The total number of stations worked in BIOMASS was 1987, with between 194 and 23 stations worked per cruise. In the two SIBEX areas of Bransfield Strait and Prydz Bay, there were 53 and 127 standard stations, respectively. This is a large dataset: as a comparison, the Southern Ocean Atlas of Gordon *et al.* (1982) drew on data from approximately 6300 oceanographic stations, selected from over 30 000 stations undertaken south of 30°S up until *ca.* 1980. Thus the BIOMASS data set is of considerable significance in terms simply of size.

Data quality varies from measurement to measurement and from cruise to cruise, and data submission to the database was patchy. Oceanography is well represented, with station coverage ranging from approximately 50 to 100% in the 25 cruises covered in the database. Fifteen cruises have provided zooplankton net haul data, and 22 provide data on acoustic estimation of krill biomass. Twenty-one cruises reported bird observations, although the number of records varied widely, from 17 to 618 per cruise. Only two cruises' data on phytoplankton species composition were deposited with the Data Centre, although it is known that more analyses were undertaken. Before considering the complete suite of BIOMASS data, we should establish the criteria which must be satisfied for their use in IGBP projects such as JGOFS.

Data requirements for global change studies

Data requirements in global change programmes such as those within IGBP relate to the measurement or understanding of change, and the prediction of future events. Change may take place rapidly, at rates which can be measured on relatively short time scales (such as change in atmospheric ozone or carbon dioxide), or very much more slowly. In order to be used in future climate-change and other 'global' programmes, data must satisfy one of the following general criteria:

Do they provide an adequate baseline against which to evaluate future change?
Do they provide evidence of contemporary change: the so-called 'fingerprint of global change'?
Do they provide information on present or past trends, or large-scale natural variability which might distort such trends?
Do they provide means of predicting the effects of future changes?

Furthermore, there are important and basic criteria of data quality. These are already explored to some extent within the BIOMASS Programme: indeed, BIOMASS made exemplary attempts to define measurement protocols for several key variables. The BIOMASS Handbook series sets out methods for population estimation, species identification, measurement of rate processes and data recording and submission. Of the 25 volumes, two cover phytoplankton biomass and productivity, ten deal with investigation of krill and other euphausiids, and a further ten cover investigations of squid, fish, seabirds and seals. JGOFS will also standardize methodology, by defining measurement protocols for properties and processes which are seen to be crucial to biogeochemical cycling (SCOR, 1989). Clearly, the topics crucial to JGOFS studies of carbon cycling differ in their emphasis from the BIOMASS investigations centred on krill.

Quality control becomes crucial where there is an intention to incorporate data in long-term monitoring, or to use them in other comparisons with additional data. In particular:

Have the data been properly checked and validated?
Are the data documented so that methods and sampling techniques can be detailed or evaluated?
Are data consistent within a sampling campaign?
Are the data available in an accessible form?
Does the sampling programme allow the suggested application of the data?
Is there an indication of how representative, or variable, the measurements are within a given domain in time and space?

In this chapter, we concentrate on the JGOFS project, which examines the processes of carbon fluxes between atmosphere and ocean. A key aspect of these carbon transfer studies is an understanding of the so-called 'biological pump', whereby primary production enhances the flux of carbon to deep water where it is temporarily isolated from the atmosphere. The Southern Ocean is one of the few areas of the world ocean where the activity of the biological pump could increase. Underutilization of nutrients, especially nitrogen, points to at least the potential for an increase in the biological fixation of carbon (Martin *et al.*, 1990). There remains debate over why the Antarctic marine phytoplankton underperforms, whether any enhanced production would indeed be effective in arresting or reversing the current increase in atmospheric CO_2, and how the Southern Ocean ecosystem might respond to likely environmental change.

Utility of the BIOMASS data

Are the data from the BIOMASS Programme useful for JGOFS studies? First, we have already acknowledged that the two programmes lay different emphases on key variables and parameters. JGOFS is particularly concerned with the processes of particle production and transformation, whereas BIOMASS took a broader view of the ecosystem, with an emphasis on species at higher trophic levels which might be affected by fisheries. However, the BIOMASS database represents a rich archive of information on plankton abundance, distribution and variation, and a detailed framework for the oceanography of specific key areas. A more general description of the dataset and its future availability is presented by Thorley & Trathan (this volume). In the following sections we concentrate on two main areas which are most likely to be of importance to JGOFS. We will attempt to evaluate the data using the quality criteria described above.

'Oceanographic' data

Data on the physical, chemical and microbial variables measured during BIOMASS are all potentially important to JGOFS, although many of the measurements prescribed by JGOFS were not undertaken in BIOMASS, or are not incorporated in the database. Primary production measurements were made on only a few cruises and are not archived.

Have the data been properly checked and validated?

Physical oceanographic data have been checked and validated during data analysis workshops. To a more limited extent, the same workshops checked through data on nutrient chemistry. Phytoplankton biomass data were not analysed during such workshops. Priddle *et al.* (this volume) provide an assessment of chlorophyll data for the FIBEX and SIBEX cruises in the southwest Atlantic and southeast Pacific regions, but no similar evaluation has been undertaken for the Indian Ocean sector cruises.

Are the data documented so that methods and sampling techniques can be detailed and evaluated?

The BIOMASS standard data set contains only limited quality control and methodological information on some physical oceanographic measurements. Details on the methodology used in determinations of nutrient concentrations and phytoplankton biomass can be obtained in most cases from the publications arising from the cruises, but are not documented in the database. In many cases, the data in the BIOMASS database are by no means comparable to those collected by JGOFS process or time-series studies and cannot be incorporated with them.

Are data consistent within a sampling campaign?

Different methods were commonly employed by the different countries and groups involved in the BIOMASS cruises, even within the standardized SIBEX series.

Are the data available in an accessible form?

Oceanographic data holdings will form part of the standard BIOMASS data set, which will be distributed to BIOMASS participating countries when documentation of the database is complete (see Thorley & Trathan, this volume).

Does the sampling programme allow the suggested application of the data?

Large data sets such as those in BIOMASS, which contain relatively low-grade information on phytoplankton biomass, nutrients and water column structure, can be used to explore functional relations and to test model conclusions. The designation of standard stations in the SIBEX cruises, and their repeat sampling over two summer seasons, is a particular strength of the BIOMASS data set.

This data area is the one which has already been addressed by enquiries from outside the BIOMASS community, following relaxation of data access restrictions. Principal interest has been in the chlorophyll data for retrospective validation of satellite ocean colour data.

Is there an indication of how representative, or variable, the measurements are within a given domain in time and space?

Physical oceanographic data from the BIOMASS cruises have been used to provide effective summaries of mesoscale and larger variability in the SIBEX areas, where repeat cruises at standard stations allowed the identification of the positions of water mass boundaries (BIOMASS, 1990).

Priddle *et al.* (this volume) have used phytoplankton biomass and nutrient data from several BIOMASS cruises in the Antarctic Peninsula area to analyse the possible causes of patterns of spatial and temporal variation in these variables. By consolidating the full set of measurements from SIBEX cruises, they were able to generate a crude seasonal pattern of phytoplankton growth and nutrient utilization for different water masses or regions. However, they found that, even by combining all cruises in the area, the spatial and temporal patchiness of phytoplankton production was not resolved adequately by the BIOMASS station grid. On present evidence, it indicates that measurements of phytoplankton biomass using conventional hydrographic stations will not offer any clear indication of long-term trends which can be resolved over the very variable background.

Zooplankton data

Research on zooplankton, especially krill, formed a major component of the BIOMASS Programme. A significant innovation was the use of acoustic techniques to undertake a systematic large-scale survey of krill abundance and distribution. In combination with the net sampling and the ecological and physiological data from the BIOMASS and other cruises (see, for instance, Siegel & Kalinowski, this volume), we have unparalleled data on the grazing community and its variability.

Have the data been properly checked and validated?

Data for both net-derived and acoustic estimation of krill biomass have been validated during a series of data analysis workshops (see Thorley & Trathan, this volume, their Table 1). Apart from these, only data on ichthyoplankton abundance and distribution have been validated, again in a series of workshops. Data on the remainder of the zooplankton may be published for individual cruises, but the BIOMASS data set has not been checked or cross-referenced as a whole.

Are the data documented so that methods and sampling techniques can be detailed and evaluated?

The BIOMASS data set contains some documentation on sampling methods and the acquisition of krill acoustic biomass. Further data are in the appropriate workshop reports and in individual scientific papers.

Are data consistent within a sampling campaign?

Despite a major attempt to standardize on zooplankton sampling, there remain inconsistencies in sampling methods. For krill biomass data, different echo-integrators, depth strata and reset distances were used. Further analysis of these data should resolve most problems. A range of different nets were used for zooplankton sampling.

Are the data available in an accessible form?

The data on zooplankton abundance and species distribution will form part of the standard BIOMASS data set, which will be distributed to BIOMASS participating countries when documentation of the database is complete (see Thorley & Trathan, this volume).

Does the sampling programme allow the suggested application of the data?

To be of use for JGOFS (for incorporation in models, for instance) these data need further analysis. Work is already in progress on multivariate correlations with hydrography, which not only reinforce observational detail but also allow more general relationships to be identified. A great deal of information on swarm structure and variability could also be extracted from the BIOMASS data set.

Is there an indication of how representative, or variable, the measurements are within a given domain in time and space?

The BIOMASS data provide an excellent body of information on the spatial and temporal variability of zooplankton, especially krill and to a lesser extent ichthyoplankton. The FIBEX data give large-scale variability, whilst the SIBEX cruises provide repeat sampling along standard transects and standard stations, which can be related to oceanographic variables.

Overall strengths and weaknesses of data of potential use to JGOFS

We note above the limits of the utility of two main areas of BIOMASS data, in the context of application to a Southern Ocean JGOFS programme. In particular, the differing data requirements for the two programmes imply that BIOMASS data will not be capable of standing alongside the data from JGOFS studies: the scope and standards of measurement differ too widely. None the less, the BIOMASS data represent an important resource which offers the opportunity to derive parameters for models, to assess spatial and temporal variability and to validate remote sensing information. The strong points of the BIOMASS data include:

- the use of standard methods for some measurements, allowing intercomparison, especially in zooplankton acoustics;
- the comprehensive spatial coverage in FIBEX, providing an extensive data set for modelling and remote sensing calibration applications;
- the repeat sampling of standard stations in the SIBEX cruises, providing limited data on spatial and temporal variability.

However, there are some specific drawbacks to the BIOMASS data, apart from the general criticisms of data quality and coverage:

- absence of information on ice-associated systems;

hydrographic complexity of the SIBEX areas, making regional environmental correlations difficult;
poor resolution of microbial processes, especially the heterotrophic components.

Data on higher predators, which could have provided a unique data set pertinent to a possible key issue of Southern Ocean carbon cycling (Huntley *et al.*, 1991), are too sketchy to be useable for this purpose.

Ways in which the BIOMASS infrastructure can benefit global change research

The establishment and organization of the BIOMASS Programme was complex, and continues an established pattern of Antarctic research (Fogg, this volume). BIOMASS not only addressed the logistic problems of elaborate field campaigns, but also covered methodology, data management and analysis. In these cases, BIOMASS encountered exactly the same problems with which JGOFS and the other IGBP projects are faced at present. BIOMASS can now transmit its experience to IGBP, and to SCAR's own regional study.

The community

BIOMASS drew in participation from a wide range of nations. It is unclear how many will join IGBP projects, including SCAR's regional programme, but it is obvious that some of this community is not incorporated into Southern Ocean JGOFS at present. We should look to increase involvement.

The intiative for a Southern Ocean regional study within JGOFS has differed from the establishment of BIOMASS. Southern Ocean JGOFS grew out of a number of more-or-less independent research cruises by national groups, some of whom were already committed to JGOFS studies in other oceans. When it became clear that a coordinated study would be needed, drawing these planned cruises into a more formal framework, it was relatively simple to introduce the minor changes and additions needed. This is clear evidence of the effective way in which the Antarctic marine biological community has come to work together in BIOMASS, and in similar international efforts such as the European EPOS programme.

The experience of running multiship experiments

The scale of operations in both FIBEX and SIBEX probably represents one of the largest oceanographic campaigns in history. It is important that the lessons learned in this should be carried forward. However, although an international programme, BIOMASS organized its field campaigns as an agglomeration of national programmes. This was not a problem, because all participants in BIOMASS were capable of undertaking the full suite of activities with equal competence. However, with programmes such as JGOFS there are sometimes serious deficiencies in resources, be it equipment or personnel. Southern Ocean JGOFS includes in its remit the bringing together of scientists and ships from different nations to optimize the deployment of expertise and logistics. Cruises will also need to be better coordinated in that several ships are likely to be deployed in the same area at the same time.

The intention to standardize methods, and the subsequent publication of protocols, sets an excellent precedent for future collaborative ventures. The only element which was missing was the means to ensure that investigators adhered to the protocols, which sometimes they were either unable or disinclined to do. The SIBEX experiments demonstrated the value of coordinated studies at standard stations. Similar work in future, often linked to sites of long-term measurements such as the French ANTARFIX mooring at Kerguelen, will be needed within JGOFS.

Data management

The lessons from the BIOMASS experience of data management are spelled out in detail by Thorley & Trathan (this volume). It suffices merely to say that global change programmes in Antarctica will face problems similar in character but vastly more complex. Again, BIOMASS was correct in principle in addressing its data management problems early on.

Data analysis

For a collaborative venture such as BIOMASS or JGOFS, the linking and analysis of numerous disparate data sets is a crucial activity if the information is to be used to its maximum potential: 'the whole must be greater than the sum of its parts'. It has to be said that the BIOMASS community selected a data analysis procedure, based on greatly restricted access and poor-quality data submission, which fell well short of this ideal. It is also greatly to be regretted that analysis of some data sets was never undertaken within the BIOMASS Programme. In some other cases, such as the nutrient and photoplankton-biomass data, which would be of crucial relevance to JGOFS, much data analysis was carried out only to present data to the final BIOMASS colloquium.

Conclusions and recommendations

We have attempted to evaluate the BIOMASS Programme in the context of present and planned global change research, especially the JGOFS core project of IGBP. Although some BIOMASS data are of potential use for such a project, their utility is reduced by the absence of certain measurements as standard (e.g. primary productivity), lack of standardization, and the overall sampling strategy imposed by a living resources focus.

We would make a series of recommendations, in order that global change research derives the maximum benefit from the BIOMASS Programme.

The general availability of the BIOMASS database should be publicized extensively so that maximum use can be made of these data.

An urgent attempt should be made to complete the documentation of the data, especially with regard to methods, before the information enters the public domain.

These data should continue to be analysed. In particular, the overall synthesis which an ecosystem model can provide should be attempted (Fogg, this volume).

Sites for JGOFS process studies should be selected to capitalize on BIOMASS studies if possible.

The community that participated in BIOMASS should be encouraged to contribute to JGOFS where appropriate, with emphasis on truly multinational field programmes.

The lessons of BIOMASS data management should be transmitted to relevant experts in IGBP and SCAR.

We would conclude that the BIOMASS Programme could make a significant contribution to global change research. This is not only in the sense of results – both as data and as information synthesis – but also in the valuable development of an international community with useful facilities, which could usefully be transferred to appropriate global change studies.

References

BIOMASS (1977) *Biological Investigations of Marine Antarctic Systems and Stocks.* Volume 1: *Research Proposals.* SCAR, Cambridge.

BIOMASS (1986) *Biological Studies of the Southern Ocean: Research Plans and International Coordination.* SCAR/SCOR, College Station, Texas.

BIOMASS (1990) SIBEX physical oceanography workshop. *BIOMASS Rep Ser* **62**: 1–37.

Farman J C, Gardiner B G, Shanklin J D (1985) Large losses of total ozone in Antarctica reveal ClO_x/NO_x interaction. *Nature* **315**: 207–10.

Gordon A L, Molinelli E J, Baker T N (1982) *Southern Ocean Atlas.* Columbia University Press, New York.

Huntley M E, Lopez M D G, Karl D M (1991) Top predators in the Southern Ocean: a major leak in the biological carbon pump. *Science* **253**: 64–6.

IGBP (1990) *The International Geosphere-Biosphere Programme: a study of global change. The initial core projects.* IGBP report 12. IGBP, Stockholm.

Martin, J H, Gordon R M, Fitzwater S E (1990) Iron in Antarctic waters. *Nature* **345**: 156–8.

SCAR (1989) *The role of Antarctica in global change. Scientific priorities for the International Geosphere-Biosphere Programme (IGBP).* ICSU Press/SCAR, Cambridge.

SCOR (1989) *Joint Global Ocean Flux Study: core measurement protocols. Reports of the core measurement working groups.* JGOFS report 6. SCOR, Halifax, Nova Scotia.

SCOR (1990) *Joint Global Ocean Flux Study: science plan.* JGOFS report 5. SCOR, Halifax, Nova Scotia.

Sullivan C W, McClain C R, Comiso J C, Smith W O (1988) Phytoplankton standing crops within an Antarctic ice-edge assessed by satellite remote sensing. *J Geophys Res* **93**: 12487–8.

BIOMASS–CCAMLR relations: past, present and future

J. P. CROXALL

Introduction

THE SCIENTIFIC COMMITTEE on Antarctic Research (SCAR) programme on Biological Investigations of Marine Antarctic Systems and Stocks (BIOMASS), for which planning began in 1972, was the largest multinational, multidisciplinary programme of marine research in the Antarctic. The Convention on the Conservation of Antarctic Marine Living Resources (CCAMLR), for which preparations started in 1977 leading to ratification in 1980 and entry into force in 1982, has been the Antarctic Treaty Consultative Parties' single most important and influential initiative that deals with the Antarctic marine environment. This chapter briefly reviews:

1. the historical links between BIOMASS and CCAMLR, over the period when the active BIOMASS Programme contributed significantly to the development and operation of the Scientific Committee for the Conservation of Antarctic Marine Living Resources (SC-CAMLR);
2. the present situation, with a very active programme of research, evaluation and management within CCAMLR, coinciding with the conclusion of the BIOMASS Programme; and
3. the most practical and productive future scientific relations between SCAR and CCAMLR.

Because the rest of this volume has dealt almost exclusively with BIOMASS, this chapter treats CCAMLR in more detail. It focuses on the South Atlantic sector of the Southern Ocean, because this is where most of the fisheries regulated by CCAMLR occur, and also because it was also the principal area of BIOMASS research.

The past

The BIOMASS Programme arose from discussions in SCAR in the early 1970s, a time of expanding interest in exploiting the living resources of the Southern Ocean. These discussions concerned the need for increased knowledge of the Southern Ocean system, in particular the relations between potentially exploitable species and other elements of the marine ecosystem.

The publication of the research proposals (BIOMASS, 1977), which included outlines for multinational, multidisciplinary field programmes in 1980–1 (FIBEX) and 1983–4 (SIBEX; later implemented in both 1983–4 and 1984–5) marked the effective start of the programme's implementation phase. The research proposals also stated the principal objective of the BIOMASS Programme: 'to gain a deeper understanding of the structure and dynamic functioning of the Antarctic marine ecosystem as a basis for the future management of potential living resources'.

Similar general interest and concern also stimu-

Table 1. *The role of the Commission of CCAMLR (Article IX of the Convention)*

The Commission shall:	
(a)	facilitate research into and comprehensive studies of Antarctic marine living resources and of the Antarctic marine ecosystem;
(b)	compile data on the status of and changes in population of Antarctic marine living resources and on factors affecting the distribution, abundance and productivity of harvested species and dependent or related species or populations;
(c)	ensure the acquisition of catch and effort statistics on harvested populations;
(d)	analyse, disseminate and publish the information referred to in sub-paragraphs (b) and (c) above in the reports of the Scientific Committee;
(e)	identify conservation needs and analyse the effectiveness of conservation measures;
(f)	formulate, adopt and revise conservation measures on the basis of the best scientific evidence available, subject to the provisions of paragraph 5 of this Article;
(g)	implement the system of observation and inspection established under Article XXIV of this Convention;
(h)	carry out such other activities as are necessary to fulfil the objectives of this Convention.

lated the Antarctic Treaty Consultative Parties in 1977 to start negotiating a wide-ranging conservation convention which would deal not only with the direct effects of harvesting specific organisms but also with the indirect effects of exploitation on other species. There is no doubt that the major contributions of resources and support to BIOMASS by international agencies (particularly SCAR but including SCOR, IABO, ACMRR and several others), and by participating nations and scientists, played an important role in expediting progress towards the Convention which, following several diplomatic and scientific conferences, was drawn up in 1980 and came into force in 1982.

In particular, the scope and content of the developing BIOMASS Programme was very influential in ensuring that the Convention maintained an ecosystem perspective, requiring that its area of jurisdiction extended beyond that of the Antarctic Treaty to encompass the whole of the Southern Ocean south of the Antarctic Polar Front. The scientific principle of considering this whole area as comprising a complex suite of fundamentally interlinked units and systems extended to the principles of management of commercial harvesting. Thus Article II of the Convention includes the statement that:

Any harvesting and associated activities shall be conducted in accordance with the following principles of conservation:

a) prevention of decrease in the size of any harvested population to levels below those which ensure its stable recruitment.
b) maintenance of the ecological relationships between harvested, dependent and related populations of Antarctic marine living resources; restoration of depleted populations to the levels ensuring stable recruitment.
c) prevention of changes or minimization of the risk of changes in the marine ecosystem which are not potentially reversible over two or three decades.

This commitment to consider interspecies effects in regulating fisheries was well ahead of its time, and the concern for dependent species unique in any comparable fishery management legislation.

Implementation of CCAMLR was to proceed via a Commission and Scientific Committee (Fig. 1) and Article IX of the Convention provided for the duties of the Commission. These are extremely wide-ranging (Table 1), including roles in facilitating comprehensive research programmes, compiling data on status and trends of marine living resources and

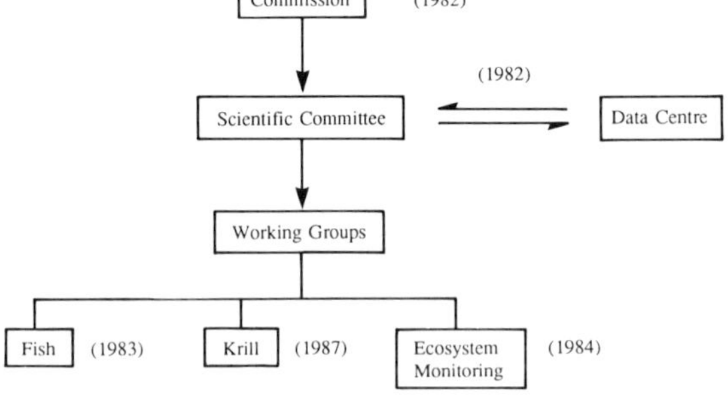

Fig. 1. Structure of CCAMLR, with the dates of establishment of the various groups.

dependent species, identifying conservation needs, promulgating conservation measures and operating a system of observation and inspection. The Scientific Committee (aided by the data management group within the CCAMLR secretariat) was charged with the conduct of the data acquisition, assessment and analysis operations necessary to provide advice to the Commission on all relevant scientific and management matters. The Scientific Committee was also required to formulate proposals for the conduct of international and national programmes of research into Antarctic marine living resources.

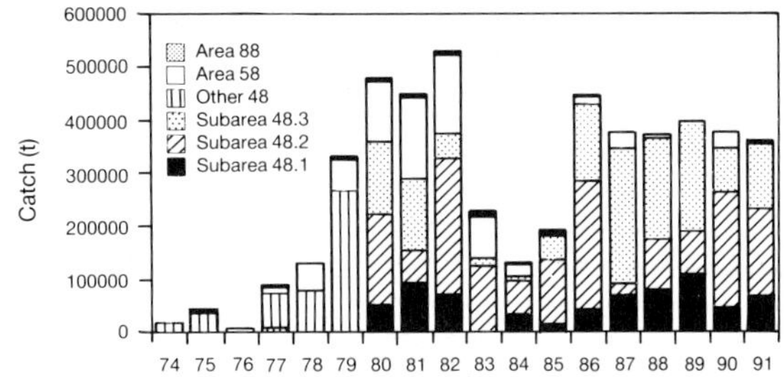

Fig. 2. Annual catches of Antarctic krill within the convention area, as reported by commercial vessels.

Such broad terms of reference obviously had considerable potential overlap with the marine research being conducted by SCAR and particularly within BIOMASS. However, in 1982, when SC-CAMLR met for the first time, the reported commercial harvest of krill (Fig. 2) was 500 kt (having risen from zero ten years earlier) and the commercial fish harvest had exceeded 100 kt in several previous years (Fig. 3) with some species, particularly *Notothenia rossii*, showing clear signs of overexploitation (Kock *et al.*, 1985; Kock, 1992). This ensured that attention was focused initially and urgently on resource assessment for exploited fish species and on attempts to develop appropriate conservation measures. Nearly all the initial assessments and advice came in the form of documentation which had been prepared in the planning and implementation of BIOMASS; many of the key scientists in BIOMASS participated in the early work of SC-CAMLR.

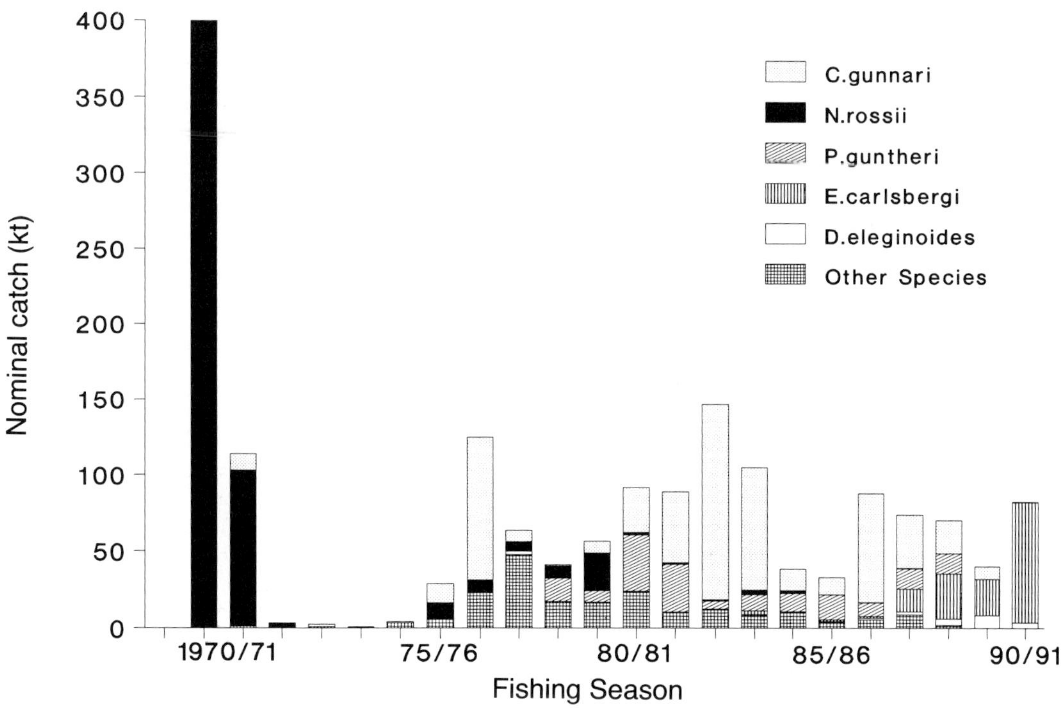

Fig. 3. Annual reported catches of the main species of fish for subarea 48.3 (South Georgia).

However, the need to request, compile, analyse and evaluate data from the commercial finfish fisheries rapidly led CCAMLR to establish in 1983 its own group of experts to undertake these tasks. This group, the *ad hoc* Group on Fish Stock Assessment, became the Working Group on Fish Stock Assessment (WG-FSA) in 1987. Subsequent BIOMASS fish research concentrated on ichthyoplankton and general biology and ecology of fish species, and direct links between BIOMASS and CCAMLR diminished.

For several more years, except for the summary of commercial catches, most information on Antarctic krill (*Euphausia superba*) was provided directly to SC-CAMLR by scientists engaged in the BIOMASS Programme, for which krill research was the central theme. Annual commercial catches had stabilized at 300–500 kt and were at a level accepted as very low in relation to stocks (Everson, 1988; Miller, 1990). CCAMLR, therefore, generally accorded krill lower priority than fish. However, the success of WG-FSA, the slow progress in developing useful catch per unit effort (CPUE) indices from krill fishery data and the requirements elsewhere in CCAMLR for information on krill, led to the establishment in 1987 of an expert group, which became the Working Group on Krill (WG-Krill) in 1988, to review current and past research on krill and to evaluate its applications in meeting the Convention's objectives.

The work of WG-Krill initially relied very extensively on data collected and techniques used (and standardized) during the BIOMASS Programme, especially with respect to survey design, acoustic techniques for stock estimation, quantitative use of nets, length–frequency and analyses, etc. (see Everson & Miller, this volume). In reviewing and utilizing information on krill growth rates, biochemistry and physiology, longevity, age and sex structure of populations, structure, size and dynamics of swarms, WG-Krill depended on research by scientists who initiated such work as part of, or in response to, the BIOMASS objectives (see Siegel & Kalinowski, this volume; Quetin *et al.*, this volume).

Simultaneously with the above initiatives, SC-CAMLR was also considering how best to treat the requirement of the Convention to investigate and evaluate relations between harvested and dependent species. SC-CAMLR noted that such work required specifically designed studies on dependent species. It therefore set up the Working Group for the CCAMLR Ecosystem Monitoring Programme (WG-CEMP) in 1985 (an *ad hoc* Group having been formed in 1984) to consider the use of indicator species within a coordinated scheme addressing issues of ecosystem monitoring. The objectives of this programme were subsequently defined as:

1. to detect and record significant changes in critical components of the ecosystem, to serve as a basis for the conservation of Antarctic Marine Living Resources; and
2. to distinguish between changes due to the harvesting of commercial species and changes due to environmental variability, both physical and biological.

Critical consumer groups were deemed to be seabirds, seals and whales. Priority species were selected for monitoring chiefly on the basis of primarily consuming exploited resources. Because most dietary data at that time came from species or sites where krill predominated in the diet, and because known piscivores took mainly non-harvested species, initial attention focused on species eating krill and consequently on the need to monitor krill stocks themselves. A network of predator monitoring sites was established and three priority research areas (the South Georgia, Antarctic Peninsula and Prydz Bay Integrated Study Regions) selected for the development of integrated predator–prey (krill) – environment monitoring (Fig. 4). Standard methods were drawn up for acquiring annual indices relating to the condition (mass), population size, reproductive performance (breeding success, foraging trip duration, offspring growth rate, etc.), diet, and survival of the priority predator species (Fig. 5). Programmes of directed research to support interpretation of the predator indices and to investigate the utility of other indices were also started, as were initiatives to obtain relevant environmental data to aid interpretation of events, especially within Integrated Study Regions (ISRs). Special attention was directed to the appropriate design of surveys to estimate krill biomass.

Because SCAR was already coordinating existing national programmes monitoring reproductive performance of seabirds (through its Bird Biology Subcommittee, which doubled as the BIOMASS Bird Ecology Working Group from 1980 to 1986; see Croxall, 1990*a*) and was coordinating relevant

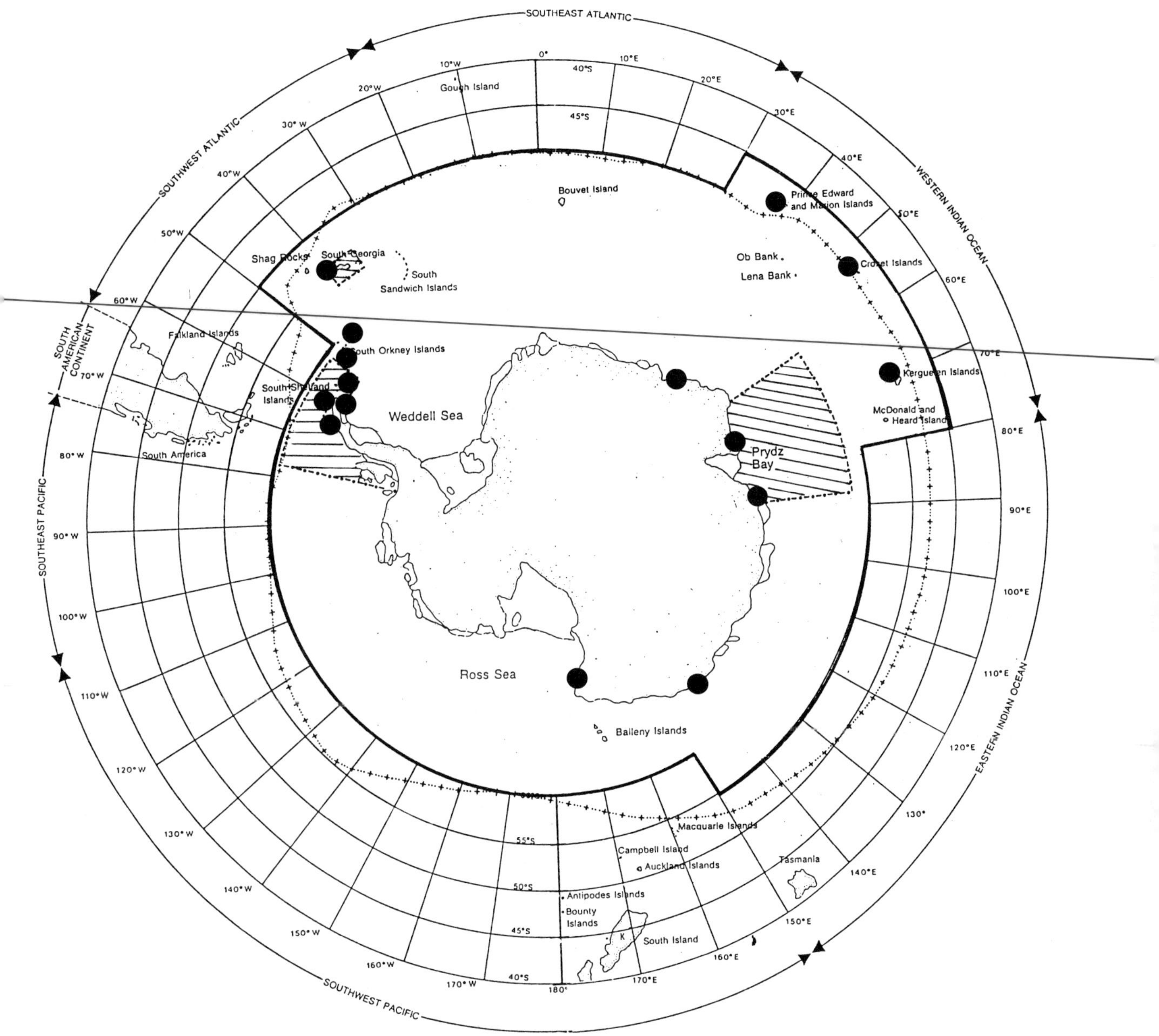

Fig. 4. The CCAMLR convention area with the Ecosystem Monitoring Programme integrated study regions (hatched) and sites at which monitoring studies are currently underway.

research on seals through its Group of Specialists on Seals, these groups and their constituent scientists provided the main sources of advice and data guiding the initiation and development of WG-CEMP programmes on dependent species.

Meanwhile, the BIOMASS Programme was successfully undertaking its major collaborative field programmes, FIBEX and SIBEX, the largest oceanographic research campaigns in history, consisting of 31 cruises (14 in FIBEX, 8 in SIBEX 1 and 9 in SIBEX 2) involving 12 nations. Since 1984 efforts have been directed at submitting data from these cruises to the BIOMASS Data Centre, checking and validating data and holding data analysis workshops.

Ultimately the BIOMASS Data Centre received data from 26 of the 31 cruises, for a total of 1917 standard stations. Oceanographic data (coupled with meteorological and sea state information) were the most extensive, being available for 25 cruises and most of the standard stations. Acoustic data estimating krill biomass were also available for 25 cruises and zooplankton (including krill and ichthyoplankton) net-haul data from 15. Observations of seabirds at sea, concurrent with acoustic transects, derived from 13 cruises; data on phytoplankton species composition were submitted from 2 cruises.

Workshops have been held and publications produced in the BIOMASS literature based on analysis

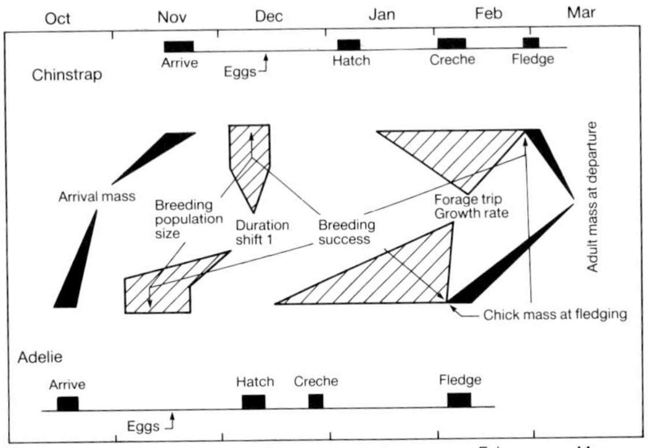

Fig. 5. The timing and duration of several of the parameters measured within the CCAMLR Ecosystem Monitoring Programme for Adélie and Chinstrap penguins.

of most of these data sets, although attempts to analyse simultaneously and interactively data from different disciplines (e.g. oceanography, krill, seabirds) have been very limited to date. However, over the same period many participants independently published results of analyses of their own FIBEX and SIBEX data. These have included several publications relating physical oceanography (e.g. water-mass characteristics) to the distribution of zooplankton and seabirds and examining correlations between krill distribution and simultaneous predator observations (see Hunt *et al.*, this volume; Cooper & Woehler, this volume). A full review of the achievements of BIOMASS is, of course, provided by the earlier chapters in this volume.

The present

BIOMASS

An important aim of this whole volume is to assess the achievements of the BIOMASS Programme. It is sufficient here to note that it:

1. formed a milestone in the development of international cooperation in biological investigations of the Southern Ocean, particularly through the simultaneous multiship surveys in FIBEX and SIBEX;
2. established an international Antarctic marine science community, without which the practical development of the work of CCAMLR would have been impossible and to which many of the present initiatives in Antarctic marine science owe their origin;
3. generated immense scientific (and political) interest in the Antarctic marine ecosystem, creating a climate in which many research programmes addressing elements of the overall and specific objectives of BIOMASS were able to flourish, and leading to an explosion of marine research in the region, including research in many fields not tackled in FIBEX and SIBEX;
4. made possible major advances in the development, standardization and use of new methods and equipment, particularly in relation to field research on krill;
5. collected within the BIOMASS Data Centre invaluable oceanographic and zooplankton data, particularly for the Antarctic Peninsula and Prydz Bay areas.

CCAMLR

In reviewing the achievements and current status of CCAMLR it should be remembered that the Commission can only adopt decisions, including and especially conservation measures, by consensus, and depends on voluntary compliance. The annual report of the Scientific Committee of CCAMLR is the principal source of advice to the Commission on scientific and management matters. SC-CAMLR annually receives and publishes the deliberations and advice from each of its three specialist working groups (Fig. 1), together with additional information on important issues such as squid, status and trends of bird and seal (both via SCAR) and whale (via the International Whaling Commission) populations and incidental mortality (of seabirds and seals).

Table 2. *Conservation measures adopted by CCAMLR for management of Antarctic fin-fisheries*

Action	First enacted
Mesh regulations	1985
Catch limits	1987
Closed seasons	1987
Closed areas	1990[a]
Protected species	1986[b]
Finescale reporting	1985

[a] At the level of a complete subarea. A 12 mile no-fishing zone around South Georgia was declared in 1985 (to protect fish nursery grounds).
[b] *Notothenia rossii* in subarea 48.3; subsequently extended to whole of Area 48 in 1987.

Fish

WG-FSA annually reviews all relevant data on each exploited fish stock in the Southern Ocean. This involves undertaking (with the help of the secretariat's data management group) a range of standard analyses of the data submitted by the fishery and from research vessels and then performing appropriate assessments and evaluations, which are summarized in standard fashion (Fig. 6). WG-FSA has also guided the Scientific Committee and Commission in developing an overall strategy for the management of Southern Ocean fisheries. As this policy has been implemented, conservation measures have been necessary for most of the exploited stocks. Regulations are currently in force concerning the net mesh size, closed seasons, closed areas and catch limits (Table 2). These are now accompanied by strict reporting requirements both for the annual submission of finescale data and for the fishery while it is in progress. Thus, the catches of several species of fish are reported to the secretariat every five days through the open season, allowing monitoring of catch trajectories (reported back to the fishery by the secretariat), ensuring (in theory) that catch limits are adhered to.

Despite the Convention's ecosystem perspective and its emphasis on rational management and care for depleted and dependent species, progress in fisheries management has invariably been too slow and insufficient to prevent consistent over-exploitation of stocks. This results from the need to obtain consensus on each issue, the difficulty of overcoming the objections of fishing nations to performing other than single-species assessments using traditional methods, and the problems in reconciling two opposing lines of argument. The first, from fishing nations, is that where comprehensive and conclusive scientific information is lacking, no management action should be taken. The second is the contention of non-fishing nations that the absence of relevant data (which could and should be provided from the fishery), or uncertainty over the correct interpretation of existing data, indicates a particular need for conservative management measures (Fig. 7). The development of principles for precautionary measures is now under active debate within CCAMLR, particularly in respect of their potential application to new and developing fisheries.

Squid

Catches of squid in the Convention area have so far totalled only a few tonnes. However, there is a major squid fishery on *Nototodarus sloani* in subAntarctic waters off New Zealand and another, only about 500 km outside the Convention area, at the Falkland Islands. The squid species which is third in importance to this latter fishery is *Martialia hyadesi*, known also to be common within the CCAMLR area in the South Georgia region (Rodhouse *et al.*, 1993) and important there in the diet of several sea-

Assessment Summary: *Champsocephalus gunnari*, Subarea 48.3

Source of Information: This Report

Year	1986	1987	1988	1989	1990	1991	Max[2]	Min[2]
Recommended TAC			31500	10200	12000	·		
Agreed TAC			35000	- [4]	8000	26000		
Landings	11107	71151	34619	21359	8027	92	128194[6]	7592
Survey Biomass		159283	15716	22328[5]	149598[a] 442168[b]	26204[a] 192144[b]		
Surveyed by		Spain	USA/POL	UK/POL	UK/POL[a] USSR[b]	UK[a] USSR[b]		
Sp. Stock Biomass[3]								
Recruitment (age...)								
Mean F (.....)[1]								

Weights in tonnes

1 ... weighted mean over ages (...)
2 Over period 1981 to 1991
3 From VPA using (..........)
4 Prohibition from 4 November 1988
5 Standard estimate from Appendix D
6 Maximum catch in 1983

Conservation Measures in Force: 19/IX, 20/IX, 21/IX, 25/IX

Catches:

Poland	41 tonnes (commercial)
UK	3 tonnes (research)
USSR	49 tonnes (research)

Data and Assessment: Commercial length data in WG-FSA-91/36. VPA assessments tuned to commercial effort and survey abundance indices in WG-FSA-91/27 and WG-FSA-91/15.

Fishing Mortality: Very low **F** in 1990/91.

Recruitment: Uncertainty regarding current strength of 1987/88 year class. Survey reported in WG-FSA-91/14 shows high proportion of 1 year olds. Subject to significant uncertainty, indications of a large decline since 1989/90.

State of Stock:

Forecast for 1991/92:

Option Basis	1991 F	1991 Stock	1991 Catch	1992 F	1992 Stock	1992 Catch	Implications/ Consequences
CPUE tuned $\mathbf{F}_{0.1}$ $\mathbf{M}$ = 0.48				0.39	236779	61870	Could lead to serious over-exploitation if stock status is accurately assessed by survey tuned VPA
Survey tuned $\mathbf{F}_{0.1}$ $\mathbf{M}$ = 0.48				0.39	41834	9672	

Weights in '000 tonnes

Fig. 6. Example of CCAMLR fish stock assessment summary sheet: *Champsocephalus gunnari*, subarea 48.3, 1991.

Table 3. *Summary of known and potential consumption of the squid* Martialia hyadesi *(tonnes per year) by vertebrate predators in the Scotia Sea*

	Known predators	Possible predators
Seabirds		
Grey-headed albatross	14 214–18 746	
Black-browed albatross	3496	
Wandering albatross	43–55	
Light-mantled sooty albatross	23	
Southern giant petrel	300	
Northern giant petrel	6	
Others		?
Seals		
Southern elephant seal	308 016	
Leopard seal		?
Fur seal		48 100?
Whales		
Sperm whale		<3300?
Other toothed whales		?
Baleen whales		?
Total	326 098–330 642	<51 400?

Source: Rodhouse *et al.* (1993).

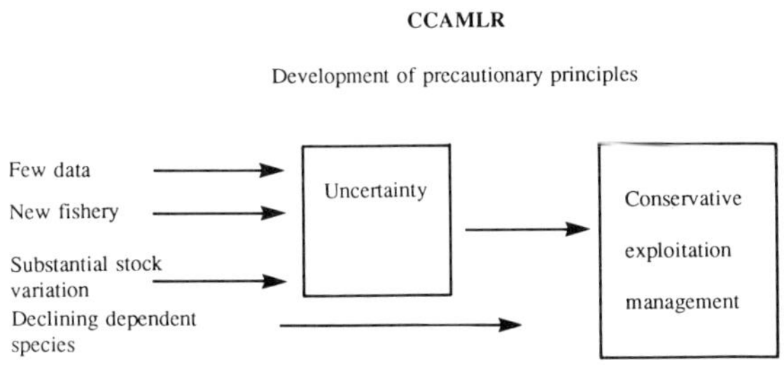

Fig. 7. Examples of circumstances in which conservative management action in respect of fisheries would be appropriate.

bird and seal species (Table 3). Most squid live for only one year and are therefore exceptionally easy to overexploit. Accordingly, CCAMLR is using the data from predators to assess potential stock size as a first step towards defining appropriate precautionary catch limits, in the event that any fishery develops.

Krill

The role of WG-Krill in the provision of management advice has been complicated by the contention of fishing nations that it is unnecessary and inappropriate to manage the krill fishery at the present level of catches, which is agreed to be at least two orders of magnitude smaller than the total Southern Ocean krill stock. However, analyses of the finescale data for krill catches (Everson & Goss, 1991) has revealed that not only does the bulk of the fishery come from Area 48 (South Atlantic sector) but also, at certain times, the fishery is consistently located in the same places (Fig. 8). Thus, in the winter months the fishery operates over the South Georgia shelf (subarea 48.3), whereas in summer it operates over the shelf to the north of the South Shetland Islands (subarea 48.1) and South Orkney Islands (subarea 48.2). Furthermore, most of the fishing is close to land and well within the foraging range of large concentrations of breeding penguins (and other seabirds) and fur seals at the time of year when those in subareas 48.1 and 48.2 are rearing offspring (Fig. 9). Therefore earlier beliefs that the current level of krill fishing is unlikely to pose significant problems for dependent species have been replaced by recognition that, at least in some local areas, there may be significant potential for competition between fisheries and natural predators.

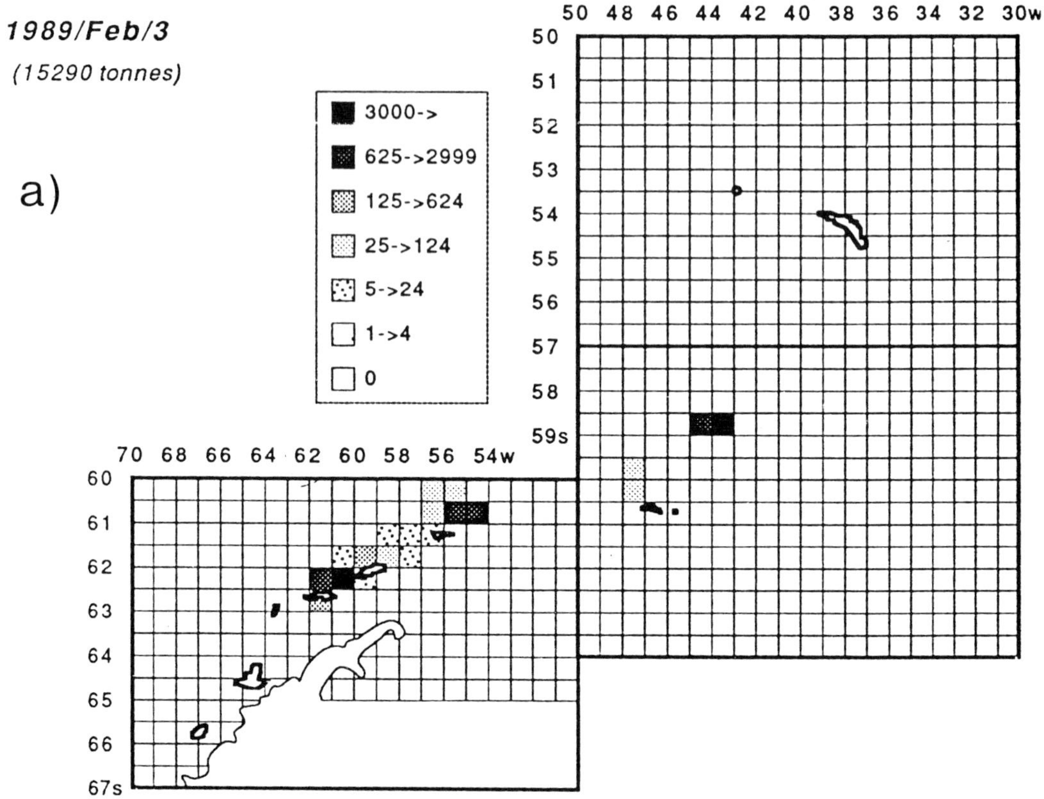

Fig. 8. Distribution of reported catches of Antarctic krill in area 48: (a) 21–28 Feb 1989; (b) (opposite page) 11–20 May 1989.

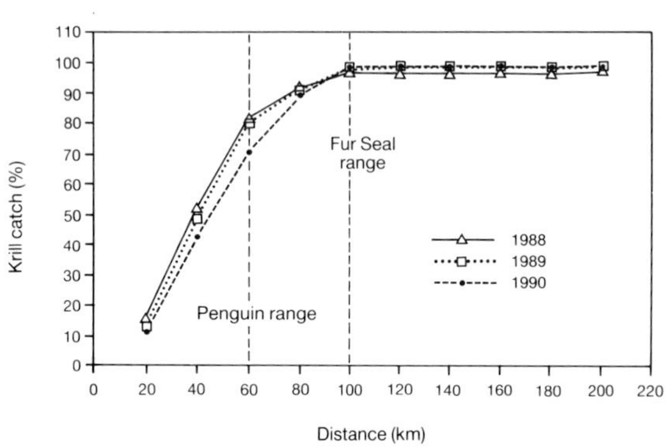

Fig. 9. Cumulative percentage catches of Antarctic krill in subarea 48.1 (Antarctic Peninsula) in December–March of 1988–90 in relation to distance to the nearest penguin and fur seal colonies. Mean maximum foraging ranges of penguins and fur seals at this time of year are indicated.

This has greatly strengthened the case for the rapid development of appropriate management measures, possibly involving some combination of precautionary catch limits (currently being developed on the basis of estimates of potential yield derived from the FIBEX survey data), together with restrictions on the location and/or time of fishing to protect dependent species. It has also highlighted the need to estimate krill stocks, their yield (involving establishment of production-to-biomass ratios) and their residence times. The latter requires knowledge of transport rates, which depend on the position and dynamics of local currents, eddies, gyres, and other physical processes and events.

An additional major role of WG-Krill has been to coordinate the development of research on topics of relevance to management and monitoring. Thus

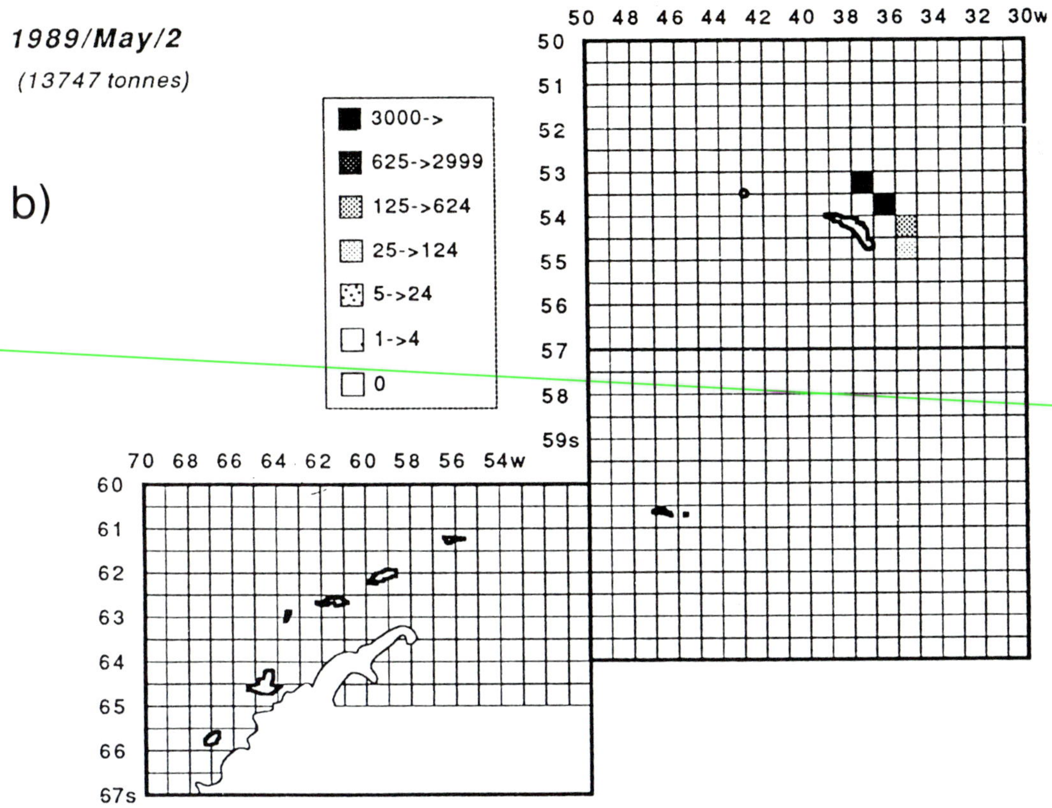

it has specialist *ad hoc* sub-groups, in which most CCAMLR nations are active participants, investigating survey design, improved acoustic techniques for biomass estimation, demography, swarm characteristics, etc. WG-Krill is also still trying to extend earlier CCAMLR initiatives (Butterworth, 1989*a, b*; Mangel, 1989) and develop approaches based on CPUE data. Progress has been hampered by the fisheries' apparent inability to provide the essential haul-by-haul data and by the difficulty of quantifying search time. However, recent Chilean research, using data from their fishery, demonstrates that haul-by-haul data is not necessarily difficult to collect and that its analysis can yield results with important management potential (Marin *et al.*, 1991). Some progress has also been made in developing a composite index, using data on krill concentrations from USSR surveys and on krill abundance within concentrations from Japanese fishing vessels (CCAMLR, 1989).

Monitoring

WG-CEMP has developed what is currently the major collaborative international field research programme under the auspices of CCAMLR. At present 8 nations, operating at 12 sites, most within the ISRs, are collecting data on reproductive performance and related topics of various krill-eating seabirds and seals. Summarized data are transmitted annually to the CCAMLR data centre (Fig. 10) where they are analysed according to agreed methods in order to provide standardized comparisons within and between species and sites and across years. The data and analyses are evaluated

CCAMLR Data submission

CEMP METHOD C2 Fur seal pup growth Procedure B | Form E/C2/B V2

CATEGORY A - ALL ITEMS MUST BE COMPLETED

1. Standard Method Version C2 • __
2. Date of preparation (Y/M/D) __ __ __ __ / __ __ / __ __
3.* Member __ __ __
4. Researcher contact __ __ __ __ __ __ __ __ __ __ __ __ __ __
5.* Area/Subarea/Division __ __ __ __
6. Split year __ __ __ __
7. Site designator __ __ __
8. Colony designator __ __ __ __
9. Sex: Male ☐ Female ☐ Unknown/Unsexed ☐

CATEGORY B - COMPLETE AS APPROPRIATE

PROCEDURE B ONLY

10. Date of first observation (Y/M/D) __ __ / __ __ / __ __
11. Date of last observation (Y/M/D) __ __ / __ __ / __ __
12. Date of first birth (Y/M/D) __ __ / __ __ / __ __
13. Date of last birth (Y/M/D) __ __ / __ __ / __ __

SUMMARY STATISTICS

14. Regression coefficient (kg/day) 0 • __ __ __
    Standard error __ • __ __ __
15. Intercept (kg) __ __ • __ __
    Standard error __ • __ __ __
16. F ratio for regression __ __ • __ __
17. Number of measurements __ __

	Date (Y/M/D)	Mean weight (Kg)	Std. deviation (S_{n-1})	N
18.	__ __ / __ __ / __ __	__ __ __ • __ __	__ __ • __ __	__ __ __
19.	__ __ / __ __ / __ __	__ __ __ • __ __	__ __ • __ __	__ __ __
20.	__ __ / __ __ / __ __	__ __ __ • __ __	__ __ • __ __	__ __ __
21.	__ __ / __ __ / __ __	__ __ __ • __ __	__ __ • __ __	__ __ __
22.	__ __ / __ __ / __ __	__ __ __ • __ __	__ __ • __ __	__ __ __
23.	__ __ / __ __ / __ __	__ __ __ • __ __	__ __ • __ __	__ __ __
24.	__ __ / __ __ / __ __	__ __ __ • __ __	__ __ • __ __	__ __ __
25.	__ __ / __ __ / __ __	__ __ __ • __ __	__ __ • __ __	__ __ __

26. Comments: ______________________________

Use capital letters and dark pencil or black ink

E/C2/B V2 page ____ of ____

*For appropriate codes see the document "CCAMLR Data Codes"

Fig. 10. Example of a CEMP data submission sheet.

annually by WG-CEMP. In many cases the data from the earlier monitoring initiatives coordinated by SCAR are entirely congruent with the CEMP; thus data for more than a decade are already available for a number of sites (e.g. Croxall *et al.*, 1988). Formats and requirements for collecting environmental data (weather, sea-ice etc.) have also been prepared. The development by WG-Krill of recommended survey designs for estimating krill biomass is the first step towards standardizing monitoring of krill stocks in the ISRs.

WG-CEMP has three other main initiatives at present. First, it is encouraging research directed at identifying aspects of predator biology and ecology which will provide additional (and hopefully better) insights into predator–prey relations. To this end it is planning a workshop to investigate the utility of data on foraging performance of seabirds and seals currently being collected by remote-recording and transmitting devices. Second, it is coordinating the synthesis of data on predator distribution, abundance, biomass and diet within ISRs as a key contribution to investigating relationships between krill stocks and the timing, location and magnitude of commercial harvests and the requirements of krill-dependent predators (Croxall *et al.*, 1991). Third, and perhaps most important, WG-CEMP is trying to create the mechanisms by which the annual results of the predator monitoring studies, and the scientific advice developed on their basis, can be formally incorporated into the fishery management deliberations of the Scientific Committee and Commission (Croxall, 1990*b*).

The future

From this review it will be evident that many of the original aims of the BIOMASS Programme are now being actively pursued within CCAMLR. Perhaps one of the most notable recent achievements of SCAR is to have provided the inspiration for CCAMLR itself and, through the BIOMASS Programme, the means by which CCAMLR has achieved scientific credibility. This is leading, however gradually and imperfectly, to significant achievements and improvements in the conservation of Antarctic marine living resources.

However, it is important to recognize that there are several research fields, geographical areas and topics, important in Antarctic marine science, which formed part of the overall objectives of BIOMASS, and which are hardly, or not at all, covered by CCAMLR at present. Major research fields include benthos, physicochemical interactions, oceanic biogeochemical cycling, and modelling. Important topics and areas omitted or underemphasized reflect the present focus on exploited fish stocks and krill and on krill-dependent predators. Thus, areas of the Southern Ocean where *Euphausia superba* is rare or absent have not yet received serious attention within the scope of CEMP. This effectively excludes the sub-Antarctic region of the Indian and Pacific Ocean sectors and the Ross Sea, areas where myctophids, *Pleurogramma antarcticum* and *Euphausia crystallorophias* are important in food webs.

Similarly, the lack of attention to predators not dependent on krill excludes from detailed consideration significant species such as elephant and Weddell seals, emperor and king penguins and many albatrosses and petrels. This not only prevents consideration of data derived from important long-term studies, such as those in Adélie Land and the Prince Edward, Kerguelen, Crozet and Macquarie Islands (Watkins, 1987; Whitehead *et al.*, 1990; Jouventin & Weimerskirch, 1991) but reduces attention on important prey species of these predators which are (myctophids) or may become (squid) the targets of commercial fishing. For CCAMLR to develop programmes of monitoring and directed research in areas where species other than *Euphausia superba* are important in food webs will require not only the provision of appropriate scientific information and advice but also increased participation in the work of CCAMLR by scientists interested in these species, areas and systems.

Furthermore, progress in the areas where CCAMLR is already active requires the continuing provision of data not just from the fisheries but also from directed research. Some of the most important information required by CCAMLR to improve its management of Southern Ocean fisheries is summarized in Table 6. Many of these data are difficult to obtain and will require multinational collaboration sustained over several years. What, if any, role will SCAR play in such research, which is essentially the direct continuation of the BIOMASS Programme?

As Priddle *et al.* (this volume) notes, during the lifetime of the BIOMASS Programme several other

Table 4. *Priority areas for data acquisition, analysis and research to maintain and improve current resource management operations within the Working Groups of the Scientific Committee of CCAMLR*

Fish	Krill	Ecosystem Monitoring
Fishery: Finescale catch data Length, age and general biological data Fish by-catch data Incidental catch of seabirds and seals	Fishery: Finescale catch data Biological data from catches CPUE data Fish by-catch data	Monitoring: More sites Indices of foraging performance Indices of condition Automated data collection Relations between indices and krill stocks Relations between indices and physical environment Develop approaches for fish- (and squid-) eating species
Directed research: Biomass/abundance surveys Identification of nursery grounds Recruitment surveys	Directed research: Standing stock biomass Production: biomass ratios Transport rates/residence times Swarm characteristics Stock size/production in relation to e.g. ice distribution/extent, chlorophyll concentrations Stock identity Acoustic target identification Frequency-dependence of target strength	Directed research: Foraging areas and feeding sites while breeding Prey capture rates Activity-specific energy budgets Distribution, movements and diet outside breeding season Characteristics of prey concentrations utilized by predators

aspects of the Antarctic environment have risen to global prominence. While some of these issues have focused exclusively on atmospheric events (e.g. diminution of stratospheric ozone and increases in 'greenhouse' gases), others have substantial marine connections, e.g. the role of the Southern Ocean in carbon fluxes. The increasing attention given to global change has stimulated numerous attemps to produce global scale models. For two of the biggest initiatives, the World Climate Research Programme (WCRP) and the International Geosphere–Biosphere Programme (IGBP), Antarctica is an area of major importance.

SCAR is currently organizing a research programme entitled *The Role of Antarctica in Global Change*, within the framework of IGBP. In this context, the future of SCAR's research programme in marine ecology has been under review by the Group of Specialists on Southern Ocean Ecology (SCAR, 1988, 1989, 1992*a, b*). At present there are two major initiatives. The Southern Ocean–Joint Global Ocean Flux Study (SO-JGOFS), field programmes for which start in 1992–93, is a study of biogeochemical fluxes at lower trophic levels and has little direct relevance to CCAMLR. The Southern Ocean–Global Ocean Ecosystems Dynamics Research (SO-GLOBEC) Programme, whose formal field programmes are not planned to start until 1996, is broadly complementary to SO-JGOFS. SO-GLOBEC aims to study the role of physical and biological processes influencing the dynamics of marine animal populations within the context of global change. Research on this theme, which is of great relevance to CCAMLR, is at present planned for two main study areas: the Bellingshausen Sea (extending east to the Antarctic Peninsula) and in the Indian Ocean (and likely to encompass the Prydz Bay region).

Thus, extensive coordinated information on the physical and biological environment in relation to animal population dynamics and relevant to CCAMLR ISRs is likely to become available from 1996 to the end of the century. CCAMLR itself would have much to contribute to, as well as to gain from, such a programme, so it is important to develop complementary links of mutual benefit.

However, relatively short term multinational research programmes, even of the potential scope and objectives of SO-GLOBEC, can only make a limited direct contribution to many of the more specific management objectives of CCAMLR. There is also a clear need for SCAR to continue to coordinate and promote research of the type that formed the core of the BIOMASS Programme and which led to such major advances in Antarctic marine ecology and to the application of this knowledge to the rational management and conservation of Southern Ocean systems.

Acknowledgements

I thank Inigo Everson, Karl-Hermann Kock and Denzil Miller for helpful advice and comment, three anonymous referees for their suggestions, David Agnew and Karl-Hermann Kock for providing illustrations, and Julie Thomson for typing the manuscript.

References

BIOMASS (1977) *Biological Investigations of Marine Antarctic Systems and Stocks.* Volume 1: *Research Proposals*. SCAR/SCOR, Cambridge, 79 pp.

Butterworth D S (1989*a*) A simulation study of krill fishing by an individual Japanese trawler. *CCAMLR Select Pap* 1988: 1–108.

Butterworth D S (1989*b*) Some aspects of the relation between Antarctic krill abundance and CPUE measures in the Japanese krill fishery. *CCAMLR Select Pap 1988*: 109–26.

CCAMLR (1989) Report of the Workshop on the krill CPUE simulation study. *Report of the Eighth Meeting of the Scientific Committee of CCAMLR*, pp. 81–137.

Croxall J P (1990*a*) Report of the meeting of the SCAR Bird Biology Subcommittee, Sao Paulo, Brazil, 1990. *Mar Ornithol* **18**: 79–89.

Croxall J P (1990*b*) Use of indices of predator status and performance in CCAMLR fishery management strategies. *CCAMLR Select Pap 1989*: 353–65.

Croxall J P, McCann T S, Prince P A, Rothery P (1988) Reproductive performance of seabirds and seals at South Georgia and Signy Island 1976–1986: implications for Southern Ocean monitoring studies. In: Sahrhage D (ed) *Antarctic Ocean and Resources Variability.* Springer-Verlag, Berlin, pp 261–85.

Croxall J P, Ricketts C, Wood A G (1991) Food consumption by predators in CCAMLR Integrated Study Regions. *CCAMLR Select Pap 1990*: 489–519.

Everson I (1988) Can we satisfactorily estimate variation in krill abundance? In: Sahrhage D (ed) *Antarctic Ocean and Resources Variability.* Springer-Verlag, Berlin, pp 199–208.

Everson I, Goss C (1991) Krill fishing activity in the southwest Atlantic. *Antarct Sci* **3**: 351–8.

Jouventin P, Weimerskirch H (1991) Changes in the population size and demography of southern seabirds: management implications. In: Perrins C M, Le Breton J D, Hirons G M (eds) *Bird Population Studies: Relevance to Conservation and Management.* Oxford University Press, pp 297–316.

Kock K-H (1992) *Antarctic Fish and Fisheries.* Cambridge University Press.

Kock K-H, Duhamel G, Hureau J-C (1985) Biology and status of exploited Antarctic fish stocks: a review. *BIOMASS Sci Ser* **6**: 1–143.

Mangel M (1989) Analysis and modelling of the Soviet Southern Ocean krill fleet. *CCAMLR Select Pap 1988*: 127–236.

Marin V H, Mujica A, Eberhard P (1991) Chilean krill fishery: analysis of the 1991 season. SC-CAMLR, Working Group on Krill 91/39, 8 pp.

Miller D G (1990) Commercial krill fisheries in the Antarctic 1973 to 1988. *CCAMLR Select Pap 1989*: 229–82.

Rodhouse P, Croxall J P, Prince P A (1993) Towards an assessment of the stock of the ommastrephid squid *Martialia hyadesi* in the Scotia Sea. In: Okutani T, Kubodera T (eds) *Recent Advances in Cephalopod Fishery Biology.* Tokai University Press, Tokyo (in press).

SCAR (1988) Report on meeting of SCAR/SCOR Group of Specialists on Southern Ocean Ecology, Paris, 1987. SCAR Report No. 3, 20 pp.

SCAR (1989) SCAR/SCOR Group of Specialists on Southern Ocean Ecology. Report on meeting at Hobart, Tasmania, 1988. SCAR Report No. 5, pp 21–8.

SCAR (1992*a*) SCAR/SCOR Group of Specialists on Southern Ocean Ecology. Report of meeting held in Trondheim, Norway, 1990. SCAR Report No. 8, pp 1–22.

SCAR (1992*b*) SCAR/SCOR Group of Specialists on Southern Ocean Ecology. Report of meeting held in Bremerhaven, Germany, 1991. SCAR Report No. 8, pp 23–40.

Watkins B P (1987) Population sizes of king, rockhopper and macaroni penguins and wandering albatrosses at the Prince Edward Islands and Gough Island, 1951–1986. *S Afr J Antarct Res* **17**: 155–62.

Whitehead M D, Johnstone G W, Burton H R (1990) Annual fluctuations in productivity and breeding success of Adélie penguins and fulmarine petrels in Prydz Bay, East Antarctica. In: Kerry K R, Hempel G (eds) *Antarctic Ecosystems: Ecological Change and Conservation.* Springer-Verlag, Berlin, pp 214–23.

BIOMASS in relation to the sea-ice zone

G. HUBOLD

The Antarctic Sea-Ice Zone

MORE THAN HALF of the Southern Ocean is ice-covered by the end of winter; only one tenth remains under sea-ice during summer (Zwally *et al.*, 1983). Sea-ice growth and retreat in conjunction with the seasonal light regime are the major environmental signals in the Antarctic Ocean. The seasonal ice cover influences the water column and benthos by modifying the processes of primary production and sedimentation. Therefore, the large-scale bio-zonation of the Antarctic Ocean is related to the ice cover (Hempel, 1985).

Besides the modifying effect which the ice cover has on the water column and the bottom below, the sea-ice is a habitat in itself. Specialized taxa live on, in and under the sea-ice (Garrison *et al.*, 1986). The fact that some of its fauna, e.g. the crab-eater seals (*Lobodon carcinophagus*) and their principal prey, krill (*Euphausia superba*) are among the most abundant Antarctic species is indicative of the high productivity of the sea-ice habitat.

The sea-ice habitat changes seasonally. Maximum extent and stability occurs in winter, when the ice provides a substratum for adapted plants and animals. Satellite observations show that a dramatic decrease in sea-ice extent occurs within a few weeks in spring and summer (Comiso *et al.*, 1990). Crab-eater seals concentrate on the remaining ice areas on drifting floes everywhere in the seasonal sea-ice zone in summer and autumn. The sea-ice microbes and planktonic organisms, however, are released into the water, where they are occasionally mixed down into the deeper layers of the water column.

The considerable amount of meltwater produced in a relatively short period in spring and summer stabilizes the sea surface layer. At low temperatures between -2 and $+2$ °C, the density of seawater is determined primarily by salinity and not by temperature (Ostapoff, 1965). Of the 20×10^6 km² sea-ice which covers the Southern Ocean by the end of winter, about 15×10^6 km² melt annually (Zwally *et al.*, 1983). Considering a mean ice thickness of 1–1.5 m, melting of the ice results in 1–1.5 m³ of low-salinity ($4–6 \times 10^{-3}$ psu) water at the sea surface. Under turbulent conditions, the meltwater will be gradually mixed into the water column. One cubic metre of surface water mixed into a 50 m³ water column (average salinity of 34.5×10^{-3} psu) can reduce salinity to $< 33.9 \times 10^{-3}$ psu, low enough to provide a significant stratification.

In the oceanographic literature, values of low surface salinity or density have rarely been described because such values are often disregarded as surface 'noise' in the databases. A few published data refer to low surface density, for example, in lenses of meltwater near Mawson Station in December and January (Bunt, 1966), in the vicinity of melting ice edges in Lützow–Holm Bay, where salinity was as low as 30.5×10^{-3} psu (Fukuchi *et al.*, 1985), and in the southern Weddell Sea (Gammelsröd & Slotsvik, 1981).

The low-salinity meltwater which is released during a relatively short period in austral summer thus stabilizes the water column and reduces the amount of downmixing for the sea-ice biota (Fig. 1). Owing to low wind fetch between ice floes, the stratification of the water column becomes particularly strong in the ice edge zone and in leads within the pack ice. In leads and in coastal polynyas, surface water can be advected over large distances according to the predominant wind conditions, i.e. from north-east to south-west in the East-Wind Zone. The considerable surface stratification at high latitudes (78°S) in the southern Weddell Sea (Gammelsröd & Slotsvik, 1981) (Fig. 2) in an area where air temperatures do not rise above freezing point and *in situ* ice melting does not take place can only be explained by large-scale advection of meltwater from the north-east.

The large-scale dynamics of the Antarctic Ocean surface layer are determined by year-round ice formation in the southern continental areas and ice melting at the northern ice edge. This results in a northward transport during autumn and winter, when freezing is most pronounced. An example of this transport is shown by the parallel tracks of the frozen-in ships *Deutschland* and *Endurance* in the Weddell Sea (Fig. 3). Starting in February, the ice-floe of the *Deutschland* drifted over a distance of approximately 3000 km in 230 days, i.e. 13 km d^{-1} (Brennecke, 1921). Observations with drifting Argos buoys revealed drift velocities of ice floes of 17 km d^{-1} off Coats Land in the Weddell Sea (Kottmeier & Hartig, 1988). Recent models confirm the importance of advective effects in the seasonal sea-ice dynamics in the Weddell Sea (Lemke *et al.*, 1990; Owens & Lemke, 1990).

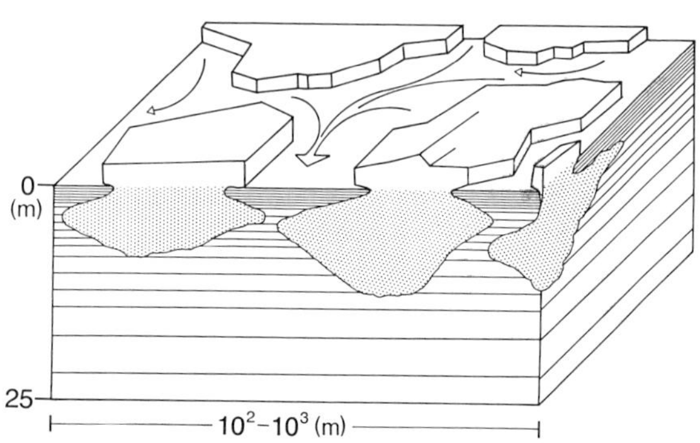

Fig. 1. Conceptual model of the Antarctic sea-ice and meltwater system. Low-salinity water (indicated by narrowly spaced lines) spreads horizontally between ice floes and provides a stratified surface habitat for sympagic microorganisms during the meltwater phase of the sea-ice system.

In spring, the northern ice edge begins to retreat. When the melting zone comes under the relatively calm conditions of the Antarctic Divergence, the thin surface layer spreads between the melting floes to the south and under the regime of the East Winds to the west into the southern Weddell Sea. Similar surface advection patterns can be observed in the southern Ross Sea and Prydz Bay (Littlepage, 1965; Smith *et al.*, 1984). The surface advection of meltwater contributes to a very rapid decay of the ice cover in an asymmetric fashion in the eastern Weddell Sea during December, when the larger part of the ice melts in about one month (Comiso *et al.*, 1990) (Fig. 3). The directed and regular surface advection of surface water to the south compensates for the northward ice transport in the Weddell Gyre (Lemke *et al.*, 1990). As a result, a predictable pattern of circular surface motion provides a retention system for organisms which are able to stay in the surface layer all year round. Due to its vertical stratification and predictable horizontal dynamics, sea-ice *and* meltwater can thus be considered as two phases of the same surface ecosystem. Both phases are characterized by relatively high stability and sufficient light in an otherwise turbulent and dark Antarctic environment.

Many taxa of zooplankton are found in the surface layer in summer. These are mainly the early stages of copepods, euphausiids and fish both of the oceanic and the neritic pelagic communities. However, the juveniles of these pelagic copepods, euphausiids and fish leave the surface after a relatively short time before the summer stratification is disturbed (Hubold *et al.*, 1988; Schnack-Schiel *et al.*, 1991). Only specialized sea-ice taxa remain in the surface layer in autumn and thus qualify as truly ice-adapted (sympagic) taxa. Krill (*Euphausia superba*) is one such taxon.

It is well known that krill live in the stratified surface layer *and* in sea-ice (e.g. Hamner *et al.*, 1983; Spiridonov *et al.*, 1985; Garrison *et al.*, 1986). The ice phase of krill is an integral part of its life

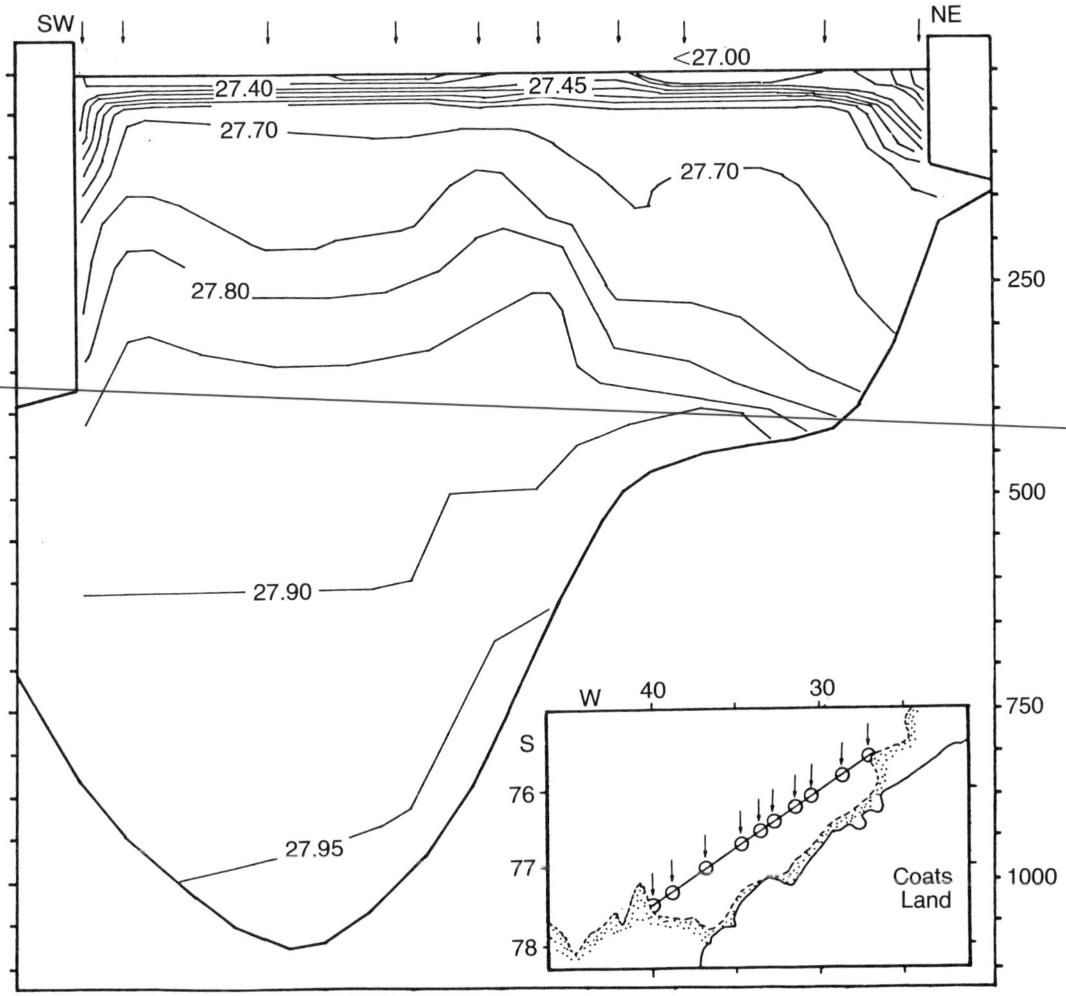

Fig. 2. Water column stratification (in density units) in the southern Weddell Sea in January 1980. (Data from Gammelsröd & Slotsvig (1981 and unpublished) and Hubold (1992).

cycle and not just a marginal aspect (summarized by Smetacek *et al.*, 1990). In the Weddell Sea, the krill life cycle can be conceptually connected to the seasonal sea-ice and meltwater cycle.

Three major systems can be distinguished within the Antarctic sea-ice zone:

1. the pelagic oceanic community;
2. the neritic community;
3. the ice and meltwater layer.

Systems (1) and (2) correspond to the seasonal and permanent pack-ice zones as defined by Hempel (1985), which are hydrographically linked to the deep-water circulation of the oceanic gyres and coastal current. The overlaying ice layer is characterized by a distinct surface circulation pattern (Fig. 4). During summer, members of the three communities can be mixed in the liquid phase of the sea-ice layer, but they remain separated during most of the year. Copepods and other mesoplankton dominate

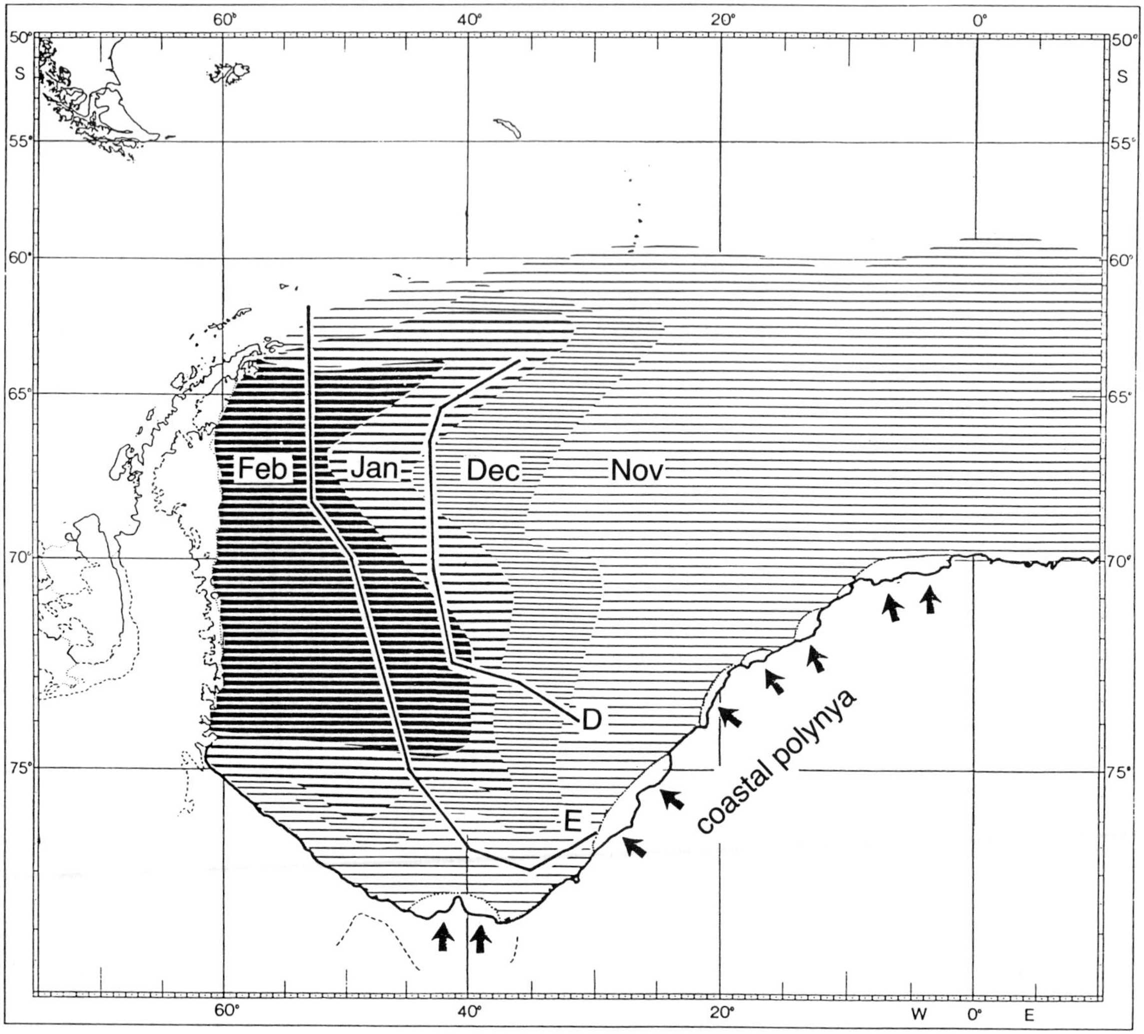

Fig. 3. Minimum sea-ice extension during summer months (Nov–Feb) in the Weddell Sea and drift paths of frozen-in ships *Endurance* (E) and *Deutschland* (D) during the winters of 1911 and 1913, respectively (Hubold, 1992). Arrows indicate locations of all-year open water (coastal polynyas) along the south-eastern coast.

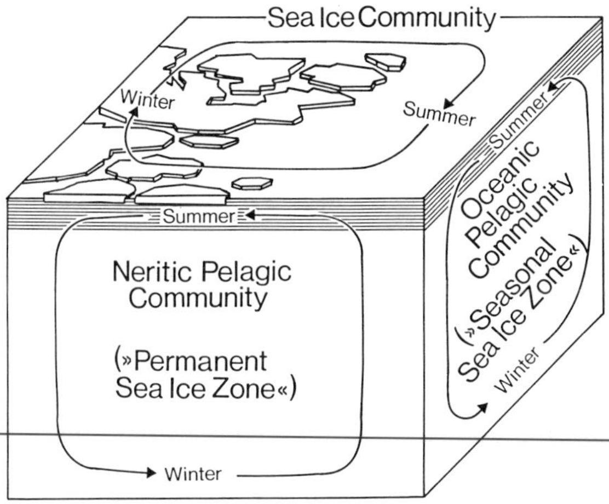

Fig. 4. Conceptual model of seasonal cycles of zooplankton taxa in pelagic oceanic, neritic and surface sea-ice systems in the Antarctic Ocean.

the pelagic communities, and the megaplankton species, krill, dominates in the ice (Siegel *et al.*, 1992).

BIOMASS, sea-ice and krill

Man has experienced the ice as a hostile environment, and from an anthropocentric point of view it was concluded that the ice must be a major threat to Antarctic biota. Owing to the fact that very few of the ships available were ice-breakers or ice-strengthened, the BIOMASS investigations concentrated on working in open water in summer in areas which, historically, have been known for heavy krill concentration. Therefore, the sampling strategies and hypotheses of BIOMASS for several reasons had to aim at krill in its pelagic summer phase in its northern distribution areas.

The results of many plankton investigations show that in spite of locally high concentrations in 'superswarms' the average krill biomass in net catches, even during maximum summer concentration in the Antarctic Peninsula region, is of the order of 6–8 g m^{-2} (Siegel, 1986), which is less than that of meso- and microplankton in the same area (Boysen-Ennen *et al.*, 1991). Towards the south, this relation is even more in favour of the smaller plankton and other euphausiid species. Even the most recent acoustic survey results of 25–90 g m^{-2} krill swarm biomass (Anon, 1991) are of the same order of magnitude as the mesoplankton net biomass.

Given the 4–5 times higher P:B ratio of the copepods compared with krill (Voronina *et al.*, 1980; Everson, 1977; Siegel, 1986), it becomes clear that in the pelagic realm outside the ice, mesoplanktonic copepods contribute most to the total zooplankton production of the Southern Ocean. In the *pelagic* Antarctic environment *krill*, therefore, *is not* the absolutely dominant zooplankton species, and the estimated pelagic biomass and production hardly can account for the food demands of large populations of krill consumers.

The large krill production which is necessary to meet the food demand of the historic whale and the present-day seal stocks could, therefore, probably be located in and under the sea-ice (together with the meltwater layer in summer). In pack-ice areas of the northern Weddell Sea, 90–98% of the net plankton biomass in the surface layer is made up of krill (Siegel *et al.*, 1992). Ice-inhabiting crabeater seals of the pack-ice zone obviously rely on the presence of their prey (krill) in their icy habitat all year round.

Thus it can be concluded that the paradigmatic Antarctic marine food chain of diatoms→krill→birds and mammals, if it exists, is located in the sea-ice zone and in the sea-ice itself, and for this reason it has to be investigated in the sea-ice in much more detail in future.

The BIOMASS Programme treated the Antarctic ecosystems and krill from a 'cryophobic' position, and it was centred around the hypothesis that krill is a dominating *pelagic* species (in spite of its unusual mass and high energetic cost of swimming). Sampling effort was directed toward those areas and times where pelagic krill indeed occurred in masses. The ecology of plankton, fish, birds and mammals was then related to this mass presence of krill, and a high degree of correlation was derived. The postulated strong trophic relations of many Antarctic taxa and krill, derived from findings in areas and times of high krill density, in my view contain a considerable amount of bias.

In order to evaluate the true significance of the results of BIOMASS one has to be aware of this bias and to consider as an alternative the possibility of the existence of krill stocks in relation to the seasonal and permanent sea-ice areas on one hand, and the existence, on the other hand, of non-krill-dominated pelagic ecosystems over most of the Antarctic ocean and over most of the time.

In the light of the above, our view of the manifold ecological features of the Antarctic sea-ice and the sea-ice zone should therefore be developed and extended toward a more encompassing ecological approach in the sea-ice zone. The newly developed SCAR programme on the Ecology of the Antarctic Sea-ice Zone (EASIZ) could be the logical successor to BIOMASS.

Given that the Antarctic sea-ice zone is a key component of the Antarctic Ocean and has the capacity for significant interactive change with global climate, the overall objective of the EASIZ research programme is 'to determine the role of the Antarctic sea-ice zone in Antarctic marine systems and in the control of global biogeochemical and energy exchanges' (Anon, 1992). Three topics in the Antarctic sea-ice zone system which need priority attention are:

the factors controlling the life cycles and survival of the biota;
the impact of sea-ice and ice biota on ocean-atmosphere exchanges;
the nature of biogeochemical cycles in the water column and benthos.

The BIOMASS community which has successfully grown during the past ten years is encouraged to join and contribute to the EASIZ programme and further develop the concept of a multinational, interdisciplinary ecological research effort in Antarctica as an essential part of SCAR's contribution to IGBP.

References

Anon (1991) CCAMLR WG CEMP 91/4.

Anon (1992) SCAR/SCOR Group of Specialists on Southern Ocean Ecology: Report of the meeting held in Trondheim, Norway, 21–23 May 1990. SCAR Report No. 8.

Boysen-Ennen E, Hagen W, Hubold G, Piatkowski U (1991) Zooplankton biomass in the ice-covered Weddell Sea, Antarctica. *Mar Biol* **111**: 227–35.

Brennecke W (1921) Die ozeanographischen Arbeiten der Deutschen Antarktischen Expedition 1911–1912. *Arch Deut Seewarte* **39**(1): 1–215.

Bunt J S (1966) Microalgae of the Antarctic pack ice zone. In: Currie I (ed) *Symposium on Antarctic Oceanography.* Scott Polar Inst., Cambridge, pp 198–218.

Comiso J C, Maynard N G, Smith W O, Sullivan C W (1990) Satellite ocean color studies of Antarctic ice edges in summer and autumn. *J Geophys Res* **95**(C6): 9481–96.

Everson I (1977) The living resources of the Southern Ocean. FAO GLO/SO/77/1: 1–156.

Fukuchi M, Tanimura A, Ohtsuka H (1985) Marine biological and oceanographical investigations in Lützow-Holm Bay, Antarctica. In: Siegfried W R, Condy P R, Laws R M (eds) *Antarctic Nutrient Cycles and Food Webs.* Springer-Verlag, Berlin, pp 52–9.

Gammelsröd T, Slotsvik N (1981) Physical oceanography of the Weddell Sea. Some results from the German Expedition 1979–80. *Polarforschung* **51**(1): 101–11.

Garrison D L, Sullivan C W, Ackley S F (1986) Sea-ice microbial communities in Antarctica. *BioScience* **36**(4): 243–50.

Hamner W M, Hamner P P, Strand S W, Gilmer R W (1983) Behavior of Antarctic krill *Euphausia superba*: Chemoreception, feeding, schooling and molting. *Science* **220**: 433–5.

Hempel G (1985) On the biology of Polar Seas, particularly the Southern Ocean. In: Gray J S, Christiansen M E (eds) *Marine Biology of Polar Regions and Effects of Stress on Marine Organisms.* Wiley & Sons, New York, pp 3–330.

Hubold (1992) Zur Ökologie der Fische im Weddellmeer. *Rep Polar Res* **103**: 1–157.

Hubold G, Hempel I, Meyer M (1988) Zooplankton communities in the southern Weddell Sea (Antarctica). *Polar Biol* **8**: 225–33.

Kottmeier Ch, Hartig R (1988) Drifting Buoy Data. In: Fahrbach E (ed) *Meteorological and Oceanographic Data of the Winter-Weddell-Sea Project 1986* (ANT V/3). *Rep Polar Res* **46**: 5–38.

Lemke P, Owens W B, Hibler W D (1990) A coupled sea-ice – mixed layer – pycnocline model for the Weddell Sea. *J Geophys Res* **95**(C6): 9513–25.

Littlepage J L (1965) Additional oceanographic studies in McMurdo Sound, Antarctica. In: *Biology of the Antarctic Seas II. Am Geophys Union Antarct Res Ser* **5**: 1–37.

Ostapoff F (1965) Antarctic Oceanography. In: Mieghem J van, Oye P van (eds) *Biogeography and Ecology in Antarctica*. Junk, The Hague, pp 97–126.

Owens W B, Lemke P (1990) Sensitivity studies with a sea-ice – mixed layer – pycnocline model in the Weddell Sea. *J Geophys Res* **95**(C6): 9527–38.

Schnack-Schiel S, Hagen W, Mizdalski E (1991) Seasonal comparison of *Calanoides acutus* and *Calanus propinquus* (Copepoda: Calanoida) in the southeastern Weddell Sea. *Mar Ecol Prog Ser* **70**: 17–27.

Siegel V (1986) Structure and composition of the Antarctic krill stock in the Bransfield Strait (Antarctic

Peninsula) during the Second International BIOMASS Experiment (SIBEX). *Arch FischWiss* **37**: 51–72.
Siegel V, Skibowski A, Harm U (1992) Community structure of the epipelagic zooplankton community under the sea-ice of the northern Weddell Sea. *Polar Biol* **12**: 15–24.
Smetacek V, Scharek R, Nöthig E-M (1990) Seasonal and regional variation in the pelagial and its relationship to the life history of krill. In: Kerry K R, Hempel G (eds) *Antarctic Ecosystems: Ecological Change and Conservation.* Springer-Verlag, Berlin, pp 103–14.
Smith N R, Zhaogian D, Kerry K R, Wright S (1984) Water masses and circulation in the region of Prydz Bay, Antarctica. *Deep-Sea Res* **31**(9): 1121–47.
Spiridonov V A, Gruzov E N, Pushkin A F (1985) Investigations of schools of Antarctic *Euphausia superba* (Crustacea, Euphausiacea) under the ice. *Zool Zh* **64**(11): 1655–60.
Voronina N M, Menshutkin K V, Tseytlin V B (1980) Production of the common species of antarctic copepods *Calanoides acutus. Oceanol Acad Sci USSR* **20**(1): 90–3.
Zwally H J, Comiso J C, Parkinson C L, Campbell W J, Carsey F D, Gloersen P (1983) Antarctic Sea-Ice, 1973–76: Satellite Passive Microwave Observations. *Nat Aeronaut Space Admin Spec Publ* **459**: 206 pp.

The Antarctic marine ecosystem in global perspective

K. SHERMAN

Introduction

THE ANTARCTIC MARINE ecosystem is the largest global area where focused scientific efforts are under way in support of the conservation and management of living marine resources from an ecosystems perspective. In the 1980s the fisheries for krill and finfish raised questions concerning the long-term sustainability of Antarctic living marine resources. The concerned nations established the Commission for the Conservation of Antarctic Marine Living Resources (CCAMLR). The CCAMLR is developing and implementing a programme of research, monitoring, and management from a holistic ecosystem perspective. The Scientific Committee of CCAMLR evolved from an earlier scientific programme, the SCAR/SCOR BIOMASS Programme, which was directed to improve the understanding of the key role of krill in the Antarctic ecosystem. At present, research scientists from 21 countries are involved in the activities of the Scientific Committee.

A Krill Working Group of CCAMLR is focused on attempting to better define the spatial and temporal ranges of variability in the abundance and distribution of krill. Several excellent syntheses of information available from the efforts to learn more about the dynamics of krill in the Antarctic marine ecosystem can be found in: the two volumes on Antarctic ecosystems edited by Laws (1984*a, b*); a description of the BIOMASS Programme by El-Sayed (1988); in the fine volume edited by Sahrhage (1988) on Antarctic Ocean resources and variability; and in the annual reports of the CCAMLR Scientific Committee (1990). Even now, following a decade of intensive scientific investigation, estimates of krill abundance remain as a significant uncertainty in Antarctic ecosystem studies. The uncertainty is based in the difficulties of acoustic back-scatter calibration and interpretation, and the biases associated with avoidance and extrusion problems of net sampling for krill. In consideration of the problems associated with estimating krill abundance and variability, the CCAMLR Working Group on Krill recognized the need for a two-fold approach to the problem: (1) the implementation of coordinated surveys using standardized techniques and methodologies; and (2) continued refinement of krill biomass estimates using survey and commercial catch data.

The need for improvement of estimates of krill abundance and distribution problems in relation to the variability in the dynamics of water movements have been underscored by several investigators (Everson, 1984; Makarov *et al.*, 1988; Murphy *et al.*, 1988; Priddle *et al.*, 1988). Actions to promote the recovery of depleted fish populations have been taken by CCAMLR, and studies designed to measure the effects of variability in krill abundance on dependent predator species have been initiated by the CCAMLR Ecosystem Monitoring Programme.

Table 1. *National krill landings (t) since 1982–3*

	Split-year[a]							
Member	1983	1984	1985	1986	1987	1988	1989	1990
Chile	3 752	1 649	2 598	3 264	4 063	5 938	5 329	4 527[b]
GDR	0	0	50	0	0	0	0	396
Japan	42 282	49 531	38 274	61 074	78 360	73 112	78 928	62 179[b]
Republic of Korea	1 959	5 314	0	0	1 527	1 525	1 779	4 040
Poland	360	0	0	2 065	1 726	5 215	6 997	1 275
Spain	0	0	0	0	379	0	0	0
USSR	180 290	74 381	150 538	379 270	290 401	284 873	301 498	302 376
Total	228 643	130 875	191 460	445 673	376 456	370 663	394 531	374 793

[a] The Antarctic split-year begins on 1 July and ends on 30 June. The 'split-year' heading refers to the calendar year in which the split-year ends (e.g. '1989' refers to the 1988–9 split-year).
[b] From catch data tabled during the Meeting.

The krill catch for 1989–90 was 375 kt, approximately 5% lower than for the 1988–9 season. Since 1983, the annual catch has not exceeded 450 kt (Table 1), suggesting that the krill fishery is evolving slowly, with sufficient lead time to apply appropriate conservation strategies. In contrast, the fish stocks have been depleted with catches at a two-decade peak for one of the principal fishing areas, South Georgia, reaching a total of 399 kt in 1970, followed by a variable but declining trend to 40 kt for 1990 (Table 2).

Table 2. *Catch of finfish from South Georgia Island, a major fishing ground within the CCAMLR area*

Split-year	Total catch (t)
1970	399 704
1971	113 713
1972	3 351
1973	2 595
1974	747
1975	4 053
1976	28 732
1977	124 611
1978	37 626
1979	24 705
1980	56 664
1981	9 167
1982	89 036
1984	104 742
1985	38 517
1986	32 723
1987	87 882
1988	73 794
1989	70 160
1990	40 148

The directed research and monitoring activities of the Scientific Committee of CCAMLR are consistent with the management approach of the Commission which seeks to:

1. prevent any harvested population from falling below the level that ensures the greatest net annual increment to stable recruitment;
2. maintain the ecological relationships between harvested, dependent, and related populations of Antarctic marine living resources;
3. restore depleted populations; and
4. prevent or minimize the risk of changes in the marine ecosystem that are not potentially reversible over two or three decades.

The CCAMLR approach to marine ecosystems management can be considered as a model framework for supporting the management of living marine resources from an ecosystems perspective. Support for ecosystem management of marine resources can be found under the terms of the United Nations Convention on the Law of the Sea (UNCLOS). Under the terms of UNCLOS coastal nations assume responsibility for: (1) ensuring through proper conservation and management measures, that the maintenance of the living resources in the Exclusive Economic Zone (EEZ) is not endangered by overexploitation; (2) promoting the optimal utilization of the living marine resources in the EEZ without prejudice to the need for conservation of those resources; and (3) the obligation to protect and preserve the marine environment and to take all measures necessary to ensure that activities under their jurisdiction or control are so conducted as not to cause damage by pollution from any source to other states and their environment (Juda & Burroughs, 1990).

Antarctic ecosystem

For the Antarctic ecosystem, significant progress has been made in applying science to mitigate the stress imposed on fish stocks through heavy exploitation. Conservation measures have been taken by CCAMLR to close areas to fishing, adjust mesh sizes of fishing trawls, and limit catch levels (CCAMLR, 1990). Other issues being addressed by the CCAMLR Scientific Committee and Commission include:

1. development of approaches to manage the krill fishery;
2. management advice to enhance the recovery of depleted fish stocks, and maintain the long-term sustainability of healthy fish stocks;
3. consideration of methods for reducing the mortality of fish larvae and juvenile fish caught in the krill fishery;
4. assessment of the status of squid resources with regard to dependent predator populations;
5. development and implementation of a study of the krill-dependent predator population as part of an ecosystem monitoring programme;
6. acquisition of information on the status and trends of marine mammal and bird populations with the SCAR Group of Specialists on Seals and the SCAR Bird Biology Subcommittee;
7. assessment and monitoring of the effects of marine debris and incidental catch on marine mammal and bird populations.

On the question of how best to develop approaches to the conservation of Antarctic marine living resources from a holistic perspective, CCAMLR (1990) has identified two areas for concentrating effort:

1. improving assessments in key areas involving coordination and integration of studies which would enable definition of appropriate management options. It was suggested that the fundamental problem is how to deal with the uncertainty in the assessments. An example would be investigation of the krill flux in the South Shetlands/Peninsula area combined with determination of the impact of predators on stocks, leading to drawing up a budget of predator–prey interactions; and
2. the wider task of evaluating the effectiveness of approaches to management adopted by the Commission in the light of the objectives of the Convention.

Two other related issues have been directed to the Working Groups of the Scientific Committee:

1. operational definitions of depletion and target levels for recovery of depleted species; and
2. the ability of the CCAMLR Ecosystem Monitoring Programme to detect changes in ecological relations and to recognize the effects of simple dependencies between species, including distinguishing between the effects of natural fluctuations and those induced by fisheries.

Other large marine ecosystems

The large-scale CCAMLR ecosystems approach to marine resources research, monitoring, and management has provided a useful model for other regions of the globe. The temporal and spatial scales influencing important processes in biological production in the sea have been the topic of a number of studies. The selection of scale in any study is related to the process under investigation. An excellent treatment of this topic can be found in Steele (1988). He indicates that in relation to general ecology, the best known work in fish population dynamics is represented by the early studies of Schaefer (1954) and Beverton & Holt (1957). Evolution to more holistic ecological models is evident in the energy flow approach of Steele (1965), following the early pioneering approach of Lindeman (1942). However, as noted by Steele (1988), such models are unsuitable for consideration of temporal or spatial variability in the ocean. The concept of large marine ecosystems (LMEs) defines the unit of study on the order of thousands of kilometres in scale, with regard to fish and fisheries yields, and represents consideration of spatial and temporal energy flow in relation to both human and natural factors which contribute to variability (Sherman & Alexander, 1986).

Large marine ecosystems (LMEs) are extensive areas of ocean space of over 200 000 km^2, characterized by distinct hydrographic regimes, submarine topography, productivity, and trophically dependent populations. The ecological concept that critical processes controlling the structure and function of biological communities can best be addressed on a regional basis (Ricklefs, 1987), is consistent with the LME approach to research on living marine resources and with management measures designed to mitigate stress and enhance long-term sustainability of living marine resources.

Living marine resources represent a major component of the community structure and dynamics of LMEs. Changes in the abundance levels of millions of tonnes (Mt) of fish, molluscs, and crustaceans through human intervention by fishing or from natural environmental perturbations can alter the structure and dynamics of LMEs, generating cascading effects up the food chain to predators (including cetaceans, pinnipeds, and sea birds) and down the food chain to the plankton. The cascading effects of krill removals are exacerbated through fishing at a relatively lower level within the Antarctic ecosystem. In an effort to provide a series of readily available case studies of the effects of fishery removals and other human-induced and natural perturbations, a series of workshops and symposia on the changing states of LMEs have been convened.

A list of reports describing the effects of biological and physical perturbations on the fisheries biomass yields of large marine ecosystems is given in Table 3. The questions generally posed by these investigations are not dissimilar from those posed a few years ago by Beddington (1984):

> There are a number of scientific questions which are central to the rational management of marine communities, but all revolve around the question of sustainability.
>
> What levels of mortality imposed by a fishery will permit a sustainable yield? Are there levels below which a fish population will not recover? Can judicious manipulation of the catch composition of the fishery alter the potential of the community to produce yields of a particular type, e.g., high value species? Can a community be depleted to a level where its potential for producing a harvestable resource is reduced?
>
> With the exception of the first question, these questions and others like them are rarely explicitly addressed in the scientific bodies of the various fisheries' organizations. Instead, such bodies concentrate on the estimation of stock abundance and the calculation of allowable catch levels, although often implicit in the advice given by these bodies to management are a set of beliefs about the answers to such questions.

Given the increasing number of responsibilites of government agencies for: (1) managing fisheries for longer-term sustainability, (2) protecting endangered species, (3) mitigating pollution, (4) reducing environmental stress, and (5) restoring lost habitat, it is not surprising that interest is growing to formulate solutions to resource management problems from an ecosystems approach.

The topic of change and persistence in marine communities and the need for multispecies and ecosystem perspectives in fishery management were reviewed at the Dahlem Conference on Exploitation of Marine Communities in 1984 (May, 1984). Examples of the changing states of marine ecosystems were summarized at the conference by Sugihara *et al.* (1984). Collapses of the Pacific sardine in the California Current ecosystem, the pilchard in the Benguela Current ecosystem, and the anchovy in the Humboldt Current ecosystem, are but a few examples of cascading effects on other ecosystem components including marine birds (MacCall, 1986; Croxall, 1987; Burger, 1988; Crawford *et al.*, 1989).

Progress in fisheries research and management can be enhanced by comparing the changing states among LMEs with regard to the cause of stress or perturbation on the system and the feedback of the system to stress.

The concept of management to maximize benefits, taking account of the long-term value of learning about factors affecting renewable resources, has been described as adaptive management strategy by Walters (1986), Walters *et al.* (1988) and Collie (1991). The key to their strategy is to recognize and model alternative hypotheses about factors affecting fish production. There are two approaches. The passive approach is based on identifying optimal management strategy based on average conditions across different stable-states of the ecosystem. The active approach manipulates the population size to distinguish among alternative hypotheses. This is necessary in order to interpret the principal forces driving changes in abundance levels in fish produc-

Table 3. *List of large marine ecosystems for which syntheses relating to principal, secondary, or tertiary driving forces controlling variability in biomass yields have been reported in Conference and Symposium volumes up to October 1990*

Large marine ecosystem	Volume No.[a]	Authors
U.S. North-east Continental Shelf	1	M Sissenwine
	4	P Falkowski
U.S. South-east Continental Shelf	4	J Yoder
Gulf of Mexico	2	Richards & McGowan
	4	Brown *et al.*
California Current	1	A MacCall
	4	M Mullin
	5	D Bottom
Eastern Bering Sea	1	Incze & Schumacher
West Greenland Sea	3	Hovgaard & Buch
Norwegian Sea	3	Ellertsen *et al.*
Barents Sea	2	Skjoldal & Rey
	4	V Borisov
North Sea	1	N Daan
Baltic Sea	1	G Kullenberg
Iberian Coastal	2	Wyatt & Perez-Gandaras
Adriatic Sea	5	G Bombace
Canary Current	5	C Bas
Gulf of Guinea	5	Binet & Marchal
Benguela Current	2	Crawford *et al.*
Patagonian Shelf	5	A Bakun
Caribbean Sea	3	Richards & Bohnsack
Gulf of Thailand	2	T Piyakarnchana
Yellow Sea	2	Q Tang
Sea of Okhotsk	5	V V Kusnetsov
Humboldt Current	5	Alheit & Bernal
Banda Sea	3	Zijlstra & Baars
Bay of Bengal	5	S N Dwivedi
Antarctic Marine	1,5	Scully *et al.*
Weddell Sea	3	G Hempel
Kuroshio Current	2	M Terazaki
Oyashio Current	2	T Minoda
Great Barrier Reef	2	Bradbury & Mundy
	5	G Kelleher

[a] Vol. 1: Sherman K, Alexander LM (eds) (1986) *Variability and Management of Large Marine Ecosystems* (AAAS Selected Symposium 99). Westview Press, Boulder, Colorado.
Vol. 2: Sherman K, Alexander LM (eds) (1989) *Biomass Yields and Geography of Large Marine Ecosystems* (AAAS Selected Symposium 111). Westview Press, Boulder, Colorado.
Vol. 3: Sherman K, Alexander LM, Gold BD (eds) (1990) *Large Marine Ecosystems: Patterns, Processes, and Yields*. AAAS, Washington, D.C.
Vol. 4: Sherman K, Alexander LM, Gold BD (eds) (1991) *Food Chains, Yields, Models, and Management of Large Marine Ecosystems*. Westview Press, Boulder, Colorado.
Vol. 5: Sherman K, Alexander LM, Gold BD (eds) (1993). *Stress, Mitigation, and Sustainability of Large Marine Ecosystems*. AAAS, Washington, D.C.
Source: Sherman (1991).

tion among environmental, biological, and fishing effects. Sissenwine & Cohen (1991) have also suggested that adaptive management strategy should be considered for the US North-east Shelf ecosystem to test hypotheses about species interactions and to develop strategies to return the system to a more favourable state that will enhance long-term sustainability of biomass yield levels. In the application of adaptive management, options for mitigating the results of overexploitation or environmental perturbation appear to hold promise, but it is necessary to move with extreme caution given the uncertainties of multispecies interactions in LMEs. Populations of adaptive management measures will require the ability to monitor the response variable as well as the control variable to provide the basis for evaluation of the mitigation measure. Although discussion of mitigation is speculative, those projects with potential for adaptive management should be carefully evaluated and tested.

Ecosystems management

The designation and management of LMEs is, at present, an evolving scientific and geopolitical process (Morgan, 1988; Alexander, 1989). Sufficient progress has been made to allow for useful comparisons to be made of the different processes influencing large-scale changes in the biomass yields of LMEs (Bax & Laevastu, 1990). For example, a comparison of the sources of major mortalities of fish in five different ecosystems shows clearly that fish–fish predation is the largest single cause of mortality. Exploitation by humans approaches the level of fish predation in the North Sea, where considerable fishing effort is directed to species underutilized in other ecosystems (Fig. 1).

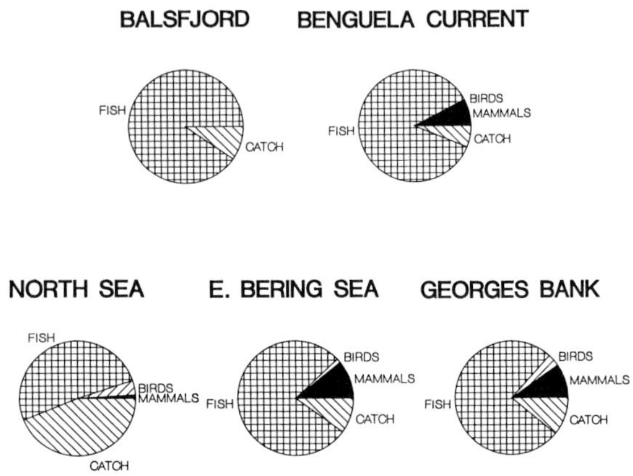

Fig. 1. Sources of major mortalities of fish in five different ecosystems (from Bax & Laevastu, 1990).

In addition to the Antarctic marine ecosystem, under the CCAMLR management regime (Scully *et al.*, 1986; Sherman & Ryan, 1988), other ecosystems that are being managed from a more holistic perspective are: the Yellow Sea ecosystem, where the principal effort is being made by the People's Republic of China (Tang, 1989); the multispecies fisheries of the Benguela Current ecosystem under the management of the government of South Africa (Crawford *et al.*, 1989); and the Great Barrier Reef ecosystem (Bradbury & Mundy, 1989) and the Northwest Australian Continental Shelf ecosystem (Sainsbury, 1988) under management by the state and federal governments of Australia. Within the EEZ of the United States, the state governments of Washington and Oregon have developed a comprehensive plan for the management of marine resources within the Northern California Current ecosystem (Bottom *et al.*, 1989).

Global Strategy

There is a growing awareness of the utility of the large marine ecosystem (LME) approach to resource management among marine scientists, geographers, economists, government representatives, and lawyers (Byrne, 1986; Christy, 1986; Alexander, 1989; Belsky, 1989; Crawford *et al.*, 1989; Morgan, 1989; Prescott, 1989). Effective stress mitigation and management strategies will be contingent on the identification of the major, secondary, and tertiary driving forces causing large-scale changes in biomass yields. Management of species responding to strong environmental signals will be enhanced by improving the understanding of the physical factors forcing biological changes, whereas in other LMEs, where the prime driving force is predation, options can be explored for implementing adaptive management strategies. Remedial actions are required to ensure that the 'pollution', or eutrophication, of the coastal zone of LMEs is reduced and does not become a principal driving

Table 4. *Key spatial and temporal scales and principal elements of a systems approach to the research and management of large marine ecosystems*

1. *Spatial–temporal scales*	Spatial	Temporal	Unit
1.1	Global (World Ocean)	millennia–decadal	pelagic biogeographic
1.2	Regional (Exclusive Economic Zones)	decadal–seasonal	large marine ecosystems
1.3	Local	seasonal–daily	subsystems
2. *Research elements*			
2.1	Spawning strategies		
2.2	Feeding strategies		
2.3	Productivity, trophodynamics		
2.4	Stock fluctuations/ recruitment/mortality		
2.5	Natural variability (hydrography, currents, water masses, weather)		
2.6	Human perturbations (fishing, waste disposal, petrogenic hydrocarbon impacts, aerosol contaminants, eutrophication effects)		
3. *Management elements – options and advice – international, national, local*			
3.1	Bioenvironmental and socioeconomic models		
3.2	Management to optimize fisheries yields		
4. *Feedback loop*			
4.1	Evaluation of ecosystem status		
4.2	Evaluation of fisheries status		
4.3	Evaluation of management practices		

Source: Sherman (1991).

force in any LME. Concerns remain regarding the socioeconomic and political difficulties in management across national boundaries, as in the case of the Sea of Japan ecosystem where the fishery resources are shared by five countries (Morgan, 1988), or the North Sea ecosystem, or the Caribbean Sea ecosystem where 38 nations share the resources.

The LMEs for which syntheses have been completed regarding the principal forces driving variability in biomass yields are shown in Fig. 2. The seven LMEs for which effort is under way to manage fisheries biomass yields from an ecosystems perspective within the United States Exclusive Economic Zone are depicted in Fig. 3. Within each of the LMEs research is conducted on principal ecosystem components including the environment (e.g. temperature, regional climate, salinity, O_2 concentrations, variability in water movements), assessments of fish, molluscan, and crustacean resources, zooplankton, ichthyoplankton, cetaceans, pinnipeds, sea turtles, and habitat protection of the wetlands and coastal zones, including protected areas.

A systems approach to the management of LMEs is depicted in Table 4. The system allows for the

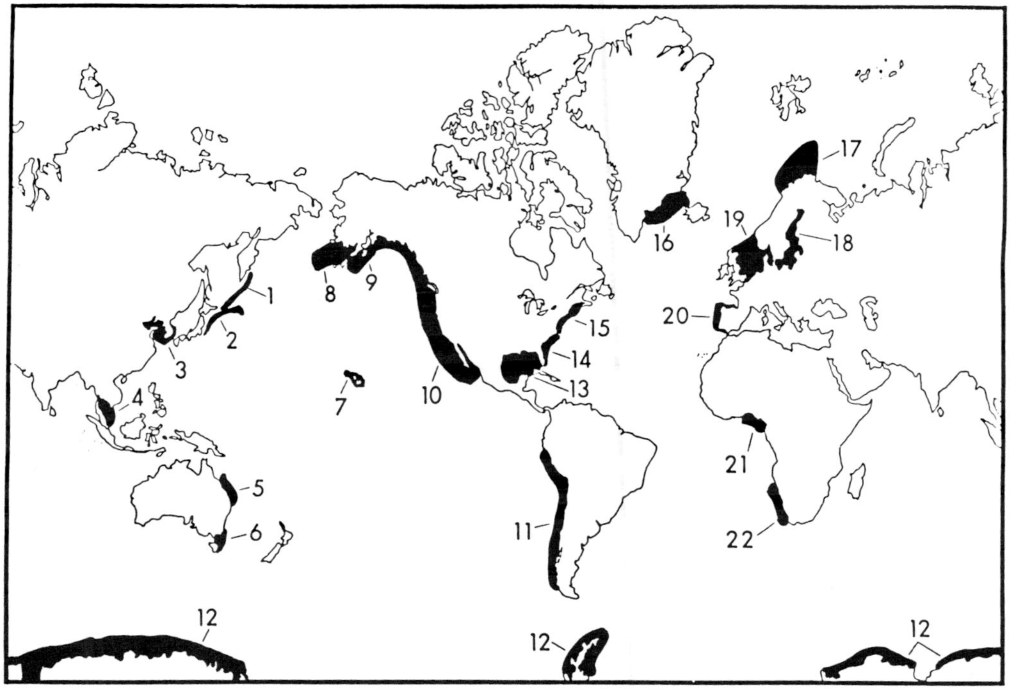

1 OYASHIO CURRENT ECOSYSTEM (O)
2 KUROSHIO CURRENT ECOSYSTEM (O)
3 YELLOW SEA ECOSYSTEM (X)
4 GULF OF THAILAND ECOSYSTEM (X)
5 GREAT BARRIER REEF ECOSYSTEM (X)
6 TASMAN SEA ECOSYSTEM (+)
7 INSULAR PACIFIC ECOSYSTEM (+)
8 EAST BERING SEA ECOSYSTEM (+)
9 GULF OF ALASKA ECOSYSTEM (+)
10 CALIFORNIA CURRENT ECOSYSTEM (O)
11 HUMBOLDT CURRENT ECOSYSTEM (O)
12 ANTARCTIC ECOSYSTEM (+)
13 GULF OF MEXICO ECOSYSTEM (+)
14 SOUTHEAST CONTINENTAL SHELF ECOSYSTEM (+)
15 NORTHEAST CONTINENTAL SHELF ECOSYSTEM (X)
16 EAST GREENLAND SEA ECOSYSTEM (+)
17 BARENTS SEA ECOSYSTEM (O)
18 BALTIC SEA ECOSYSTEM (P)
19 NORTH SEA ECOSYSTEM (+)
20 IBERIAN COASTAL ECOSYSTEM (O)
21 GULF OF GUINEA ECOSYSTEM (+)
22 BENGUELA CURRENT ECOSYSTEM (O)

Fig. 2. Predominant variables influencing changes in fish species biomass in large marine ecosystems. Predominant variable: predation (X); environment (O); pollution (P); inconclusive information (+) (Sherman, 1991).

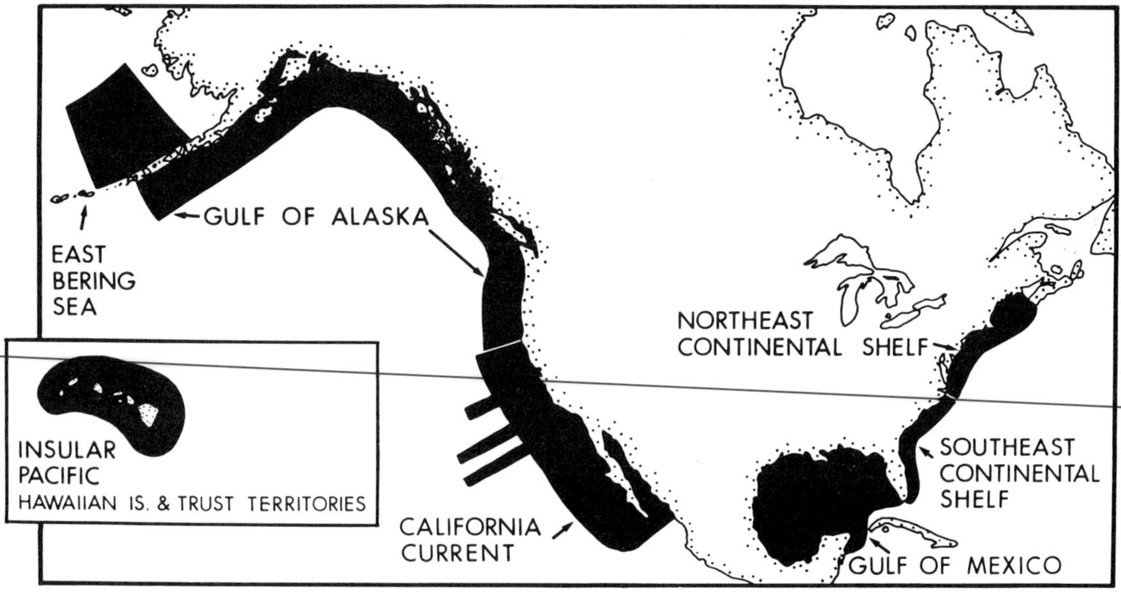

Fig. 3. Large marine ecosystems of the United States where the National Marine Fisheries Service of NOAA is conducting research. Projection based on NOAA folio map no. 7 (NOAA, 1988).

LMEs to serve as the link between local events (e.g. fishing, pollution, storms) occurring on the daily-to-seasonal temporal scale and their effects on living marine resources, and the more ubiquitous global effects of climate changes on the multidecadal time scale. The regional and temporal focus of season to decade is consistent with the evolved spawning and feeding migrations of the fishes, the keystone species of most large marine ecosystems. These migrations are seasonal, and occur over distances of hundreds to thousands of kilometres within the unique physical and biological characteristics of the regional LME to which they have adapted. As the fisheries represent most of the useable biomass yield of the LMEs, and fish populations consist of several age classes, it follows that measures of variability in growth, recruitment, and mortality should be conducted over multiyear time scales. This is necessary in order to interpret the effects of environmental, biological, and fishing effects on changing abundance levels of the year class to the populations of the species constituting the fish community, their predators and prey, and physical environment.

Consideration of the naturally occurring environmental events and the human-induced perturbations affecting demography of the populations within the ecosystem is necessary. Based on a firm, scientific understanding of the principal causes of variability in abundance, and with due consideration to socio-economic needs, management options can be considered for implementation from an ecosystems perspective. The final element in the systems approach is the feedback loop that allows for evaluation of the effects of mitigation of stress, and management actions at the fisheries level (single species, multispecies) and the ecosystem level, with regard to the concept of resource maintenance and sustained yield. It will be necessary to conduct supportive research on the processes controlling sustained productivity of LMEs. Within several of the LMEs, important hypotheses concerned with the growing impacts of pollution, overexploitation, and environmental changes on sustained biomass yields are under investigation (Table 5). By comparing the results of research among the different systems, it should be possible to accelerate an understanding

Table 5. *Selected hypotheses concerning variability in biomass yields of large marine ecosystems*

Note that references can be found in Table 3.

Ecosystem	Predominant variables	Hypothesis
Oyashio Current, Kuroshio Current, California Current, Humboldt Current, Benguela Current, Iberian Coastal	Density-independent natural environmental perturbations	*Clupeoid population increases*: Predominant variables influencing changes in biomass of clupeoids are major increases in water-column productivity resulting from shifts in the direction and flow velocities of the currents and changes in upwelling within the ecosystem.
Yellow Sea, U.S. Northeast Continental Shelf, Gulf of Thailand	Density-dependent predation	*Declines in fish stocks*: Precipitous decline in biomass of fish stocks is the result of excessive fishing mortality, reducing the probability of reproductive success. Losses in biomass are attributed to excesses of human predation expressed as overfishing.
Great Barrier Reef	Density-dependent predation	*Change in ecosystem structure*: The extreme predation pressure of crown-of-thorns starfish has disrupted normal food chain linkage between benthic primary production and the fish component of the reef ecosystem.
East Greenland Sea, Barents Sea, Norwegian Sea	Density-independent natural environmental perturbations	*Shifts in abundance of fish stock biomass*: Major shifts in the levels of fish stock biomass within the ecosystems are attributed to large-scale environmental changes in water movements and temperature structure.
Baltic Sea	Density-independent pollution	*Changes in ecosystem productivity levels*: The apparent increases in productivity levels are attributed to the effects of nitrate enrichment resulting from elevated levels of agricultural contaminant inputs from the bordering land masses.
Antarctic Marine	Density-dependent perturbations	*Status of krill stocks*: Annual natural production cycle of krill is in balance with food requirements of dependent predator populations. Surplus production is available to support economically significant yields, but sustainable level of fishing effort is unknown.
	Density-independent natural environmental perturbations	*Shifts in abundance in krill biomass*: Major shifts in abundance levels of krill biomass within the ecosystem are attributed to large-scale changes in water movements and productivity.

Source: Sherman (1991).

of how the systems respond and recover from stress. The comparisons will allow for narrowing the context of unresolved problems and for capitalizing on research efforts in the different ecosystems.

GLOBEC and the Antarctic ecosystem

Progress in the research, monitoring, and assessment of LMEs in relation to stress mitigation and resource sustainability is being accelerated through a series of recent actions. The Global Ocean Ecosystems Dynamics (GLOBEC) Programme of the US National Science Foundation has been developing an implementation plan for North-west Atlantic ecosystems, in cooperation with Canadian scientists, as a contribution to the Cod and Climate Change Programme of the International Council for the Exploration of the Sea (ICES) for North Atlantic ecosystems (ICES, 1990). The development and rationale of GLOBEC are presented in the GLOBEC Initial Science Plan (GLOBEC, 1991*a*). The general objectives of GLOBEC are to determine how (a) changes in ocean physics affect biological processes and population dynamics, and (b) population-level responses to physical variables affect structure and stability of ocean ecosystems. These main objectives will underlie all GLOBEC studies and will guide the development of the NW Atlantic Implementation Plan. A workshop in Halifax, Nova Scotia, during 19–21 June 1990 produced recommendations for a Georges Bank field study, technology developments, field programme logistics, and data management (GLOBEC, 1991*b*). These recommendations were taken into account by the Implementation Committee when the plan was written.

In addition, GLOBEC will contribute significantly to understanding the controls over ocean variability and feedback mechanisms from expected changes in LMEs and ocean ecology from large-scale global climate changes (Reeve, 1993).

The GLOBEC Programme will support a Southern Ocean Programme that proposes to investigate the linkages between climate and ocean dynamics, and population dynamics of krill (*Euphausia superba*) and the salp *Salpa thompsoni*, benthos, fish predators including *Champsocephalus gunnari, Pleuragramma antarcticum, Notothenia neglecta*, penguin species, and crabeater and fur seals. The GLOBEC study should contribute significantly to an understanding of the population dynamics of the Antarctic marine ecosystem, and serve to complement the more applied scientific efforts under way by the Scientific Committee of CCAMLR and the contributions made to the effort by the 21-nation membership of CCAMLR.

Acknowledgements

I am pleased to acknowledge the contributions made by Antarctic marine scientists to the development of the large marine ecosystem concept during the many provocative discussions of the CCAMLR Scientific Committee. Special thanks are extended to Jennie Dunnington of the Narragansett Laboratory of the National Marine Fisheries Service for her preparation of the manuscript and fine editorial assistance.

References

Alexander LM (1989) Large marine ecosystems as global management units. In: Sherman K, Alexander LM (eds) *Biomass Yields and Geography of Large Marine Ecosystems* (American Association for the Advancement of Science [AAAS] Selected Symposium 111). Westview Press Inc, Boulder, Colorado pp 339–44.

Bax NJ, Laevastu T (1990) Biomass potential of large marine ecosystems: a systems approach. In: Sherman K, Alexander LM, Gold BD (eds) *Large Marine Ecosystems: Patterns, Processes and Yields.* American Association for the Advancement of Science, Washington, DC, pp 188–205.

Beddington JR (1984) The response of multispecies systems to perturbations. In: May RM (ed) *Exploitation of Marine Communities.* Springer-Verlag, Berlin, pp 209–55.

Belsky MH (1989) The ecosystem model mandate for a comprehensive United States ocean policy and Law of the Sea. *San Diego Law Rev* **26**(3): 417–95.

Beverton RJH, Holt SJ (1957) On the dynamics of exploited fish populations. *Fishery Investigations* Series II. *Marine Fisheries* **19**: 1–533. UK Ministry of Agriculture, Fisheries, and Food.

Bombace G, Fabi G, Fiorentini L (1989) Preliminary analysis of catch data from artificial reefs in Central Adriatic. *Fish Rep* N**428**: 86–98. Fisheries and Agriculture Organization, Rome.

Bottom D L, Jones K K, Rodgers J D, Brown R F (1989) *Management of living resources: A research plan for the Washington and Oregon continental margin.* National Coastal Resources Research and Development Institute, Newport OR. Publication No NCRI-T-89–004, 80 pp.

Bradbury R H, Mundy C N (1989) Large-scale shifts in biomass of the Great Barrier Reef ecosystem. In: Sherman K, Alexander L M (eds) *Biomass Yields and*

Geography of Large Marine Ecosystems (AAAS Selected Symposium 111). Westview Press, Boulder, Colorado, pp 143–67.

Burger J (1988) Interactions of marine birds with other marine vertebrates in marine environments. In: Burger J (ed) *Seabirds and Other Marine Vertebrates.* Columbia University Press, New York, pp 3–28.

Byrne J (1986) Large marine ecosystems and the future of ocean studies. In: Sherman K, Alexander L M (eds) *Variability and Management of Large Marine Ecosystems* (AAAS Selected Symposium 99). Westview Press, Boulder, Colorado, pp 299–308.

Christy F T Jr (1986) Can large marine ecosystems be managed for optimum yields? In: Sherman K, Alexander L M (eds) *Variability and Management of Large Marine Ecosystems* (AAAS Selected Symposium 99). Westview Press, Boulder, Colorado, pp 263–7.

Collie J S (1991) Adaptive strategies for management of fisheries resources in large marine ecosystems. In: Sherman K, Alexander L M, Gold B D (eds) *Food Chains, Yields, Models and Management of Large Marine Ecosystems* (AAAS symposium proceedings volume). Westview Press, Boulder, Colorado, pp 225–42.

Commission for the Conservation of Antarctic Marine Living Resources (CCAMLR) (1990) Report of the ninth meeting of the scientific committee. Hobart, Australia, 22 October – 2 November, 1990. SC-CCAMLR-IX.

Crawford R J M, Shannon L V, Shelton P A (1989) Characteristics and management of the Benguela as a large marine ecosystem. In: Sherman K, Alexander L M (eds) *Biomass Yields and Geography of Large Marine Ecosystems* (AAAS Selected Symposium 111). Westview Press, Boulder, Colorado, pp 169–219.

Croxall J P (ed) (1987) *Seabirds: Feeding Ecology and Role in Marine Ecosystems.* Cambridge University Press, 408 pp.

El-Sayed S Z (1988) Seasonal and interannual variabilities in Antarctic phytoplankton with reference to krill distribution. In: Sahrhage D (ed) *Antarctic Ocean and Resources Variability.* Springer-Verlag, Berlin, pp 101–19.

Everson I (1984) Marine interactions. In: Laws R M (ed) *Antarctic Ecology*, vol 2. Academic Press, London, pp 783–819.

Global Ocean Ecosystems Dynamics (GLOBEC) (1991*a*) *Report Number 1. Initial science plan.* February 1991. Produced by Joint Oceanographic Institutions Inc, Washington, DC.

GLOBEC (1991*b*) *Report Number 2.* GLOBEC: Northwest Atlantic Program, GLOBEC Canada/US Meeting on NW Atlantic Fisheries and Climate. February 1991. Produced by Joint Oceanographic Institutions Inc, Washington, DC.

International Council for the Exploration of the Sea (ICES) (1990) Report of the ICES Study Group on Cod Stock Fluctuations. Cod and Climate Change (CCC), framework for the study of global ocean ecosystem dynamics. ICES CM 1990/G:50, Ref C+L.

Juda L, Burroughs R H (1990) The prospects for comprehensive ocean management. *Mar Pol* **14**: 23–35.

Laws R M (ed) (1984*a*) *Antarctic Ecology*, vol 1. Academic Press, London.

Laws R M (ed) (1984*b*) *Antarctic Ecology*, vol 2. Academic Press, London.

Lindeman R L (1942) The trophic dynamic aspect of ecology. *Ecology* **23**: 399–418.

MacCall A D (1986) Changes in the biomass of the California Current System. In: Sherman K, Alexander L M (eds) *Variability and Management of Large Marine Ecosystems* (AAAS Selected Symposium 99). Westview Press, Boulder, Colorado, pp 33–54.

Makarov R R, Maslennikov V V, Solynkin E V, Spiridonov V A, Yakovlev V N (1988) Variability in population density of Antarctic krill in the Western Scotia Sea in relation to hydrological conditions. In: Sahrhage D (ed) *Antarctic Ocean and Resources Variability.* Springer-Verlag, Berlin, pp 231–6.

May R M (ed) (1984) *Exploitation of Marine Communities.* Springer-Verlag, Berlin, 366 pp.

Morgan J R (1988) Large marine ecosystems: An emerging concept of regional management. *Environment* **29**(10): 4–9; 26–34.

Morgan J R (1989) Large marine ecosystems in the Pacific Ocean. In: Sherman K, Alexander L M (eds) *Biomass Yields and Geography of Large Marine Ecosystems* (AAAS Selected Symposium 111). Westview Press, Boulder, Colorado, pp 377–94.

Murphy E J, Morris D J, Watkins J L, Priddle J (1988) Scales of interaction between Antarctic krill and the environment. In: Sahrhage D (ed) *Antarctic Ocean and Resources Variability.* Springer-Verlag, Berlin, pp 120–30.

National Oceanic and Atmospheric Administration (NOAA) (1988) *A National Atlas: Health and Use of Coastal Waters, United States of America.* Office of Oceanography and Marine Assessment, National Ocean Service, National Oceanic and Atmospheric Administration, Department of Commerce, Washington, DC.

Prescott J R V (1989) The political division of large marine ecosystems in the Atlantic Ocean and some associated seas. In: Sherman K, Alexander L M (eds) *Biomass Yields and Geography of Large Marine Ecosystems* (AAAS Selected Symposium 111). Westview Press, Boulder, Colorado, pp 395–442.

Priddle J, Croxall J P, Everson I, Heywood R B, Murphy E J, Prince P A, Sear C B (1988) Large-scale fluctuations in distribution and abundance of krill – a discussion of possible causes. In: Sahrhage D (ed) *Antarctic Ocean and Resources Variability.* Springer-Verlag, Berlin, pp 169–82.

Reeve M (1993) Application of international global change research programs to long-term LME management. In: Sherman K, Alexander L M, Gold B D (eds) *Stress, Mitigation, and Sustainability of Large*

Marine Ecosystems. AAAS, Washington, DC, pp. 30–55.
Ricklefs R E (1987) Community diversity: relative roles of local and regional processes. *Science* **235**(4785): 167–71.
Sahrhage D (ed) (1988) *Antarctic Ocean and Resources Variability.* Springer-Verlag, Berlin, 304 pp.
Sainsbury K J (1988) The ecological basis of multispecies fisheries, and management of a demersal fishery in tropical Australia. In: Gulland J A (ed) *Fish Population Dynamics*, 2nd edn, John Wiley & Sons, New York, pp 349–82.
Schaefer M B (1954) Some aspects of the dynamics of populations important to the management of the commercial marine fisheries. *Inter-American Tropical Tuna Comm Bull* **1**: 27–56.
Scully R T, Brown W Y, Manheim B S (1986) The Convention for the Conservation of Antarctic Marine Living Resources: A model for large marine ecosystem management. In: Sherman K, Alexander L M (eds) *Variability and Management of Large Marine Ecosystems* (AAAS Selected Symposium 99). Westview Press, Boulder, Colorado, pp 281–6.
Sherman K (1991) The large marine ecosystem concept: Research and management strategy for living marine resources. *Ecol Applic* **1**(4): 349–60.
Sherman K, Alexander L M (eds) (1986) *Variability and Management of Large Marine Ecosystems* (AAAS Selected Symposium 99). Westview Press, Boulder, Colorado, 319 pp.
Sherman K, Ryan A F (1988) Antarctic marine living resources. *Oceanus* **31**(2): 59–63.
Sissenwine M P, Cohen E B (1991) Resource productivity and fisheries management northeast shelf ecosystem. In: Sherman K, Alexander L M, Gold B D (eds) *Food Chains, Yields, Model and Management of Large Marine Ecosystems* (AAAS symposium proceedings volume). Westview Press, Boulder, Colorado, pp 107–23.
Steele J H (1965) Some problems in the study of marine resources. *Int Comm Northwest Atl Fish Spec Publ* **6**: 463–76.
Steele J H (1988) Scale selection for biodynamic theories. In: Rothschild B J (ed) *Toward a Theory on Biological-Physical Interactions in the World Ocean* (NATO ASI Series C: Mathematical and Physical Sciences, vol 239). Kluwer Academic Publishers, Dordrecht, The Netherlands, pp 513–26.
Sugihara G, Garcia S, Gulland J A, Lawton J H, Maske H, Paine R T, Platt T, Rachor E, Rothschild B J, Ursin E A, Zeitzschel B F K (1984) Ecosystem dynamics: group report. In: May R M (ed) *Exploitation of Marine Communities.* Springer-Verlag, Berlin, pp 130–53.
Tang Q (1989) Changes in the biomass of the Yellow Sea ecosystems. In: Sherman K, Alexander L M (eds) *Biomass Yields and Geography of Large Marine Ecosystems* (AAAS Selected Symposium 111). Westview Press, Boulder, Colorado, pp 7–35.
Walters C J (1986) *Adaptive Management of Renewable Resources.* Macmillan, New York, 374 pp.
Walters C J, Collie J S, Webb T (1988) Experimental designs for estimating transient responses to management disturbances. *Can J Fish Aquat Sci* **45**: 1848–54.

Discussant's Report: The legacy of BIOMASS

G. HEMPEL

The scene at the onset of BIOMASS

BIOMASS WAS a mission-oriented programme of regional marine ecology whose main objective was to gain 'a deeper understanding of the structure and functioning of the Antarctic marine ecosystem as a basis for the future management of potential living resources'. It had been triggered by an overall interest in the Southern Ocean's living resources. Meadows *et al.* (1972) pointed to the 'limits of growth', in general, but had singled out the Southern Ocean as a potential source of much food. Attempts were made by FAO to estimate the productivity and the fishing potential of the Southern Ocean at the various levels. In 1977 I. Everson produced for FAO a comprehensive review of the state of knowledge of the living resources of the Southern Ocean. Distant water fleets of various countries, particularly of the Eastern bloc, had lost their historic fishing grounds off foreign coasts, and turned to the Southern Ocean.

The whaling industry had succeeded in decimating the large baleen whales one species after the other, but the stocks of minke whales were on the rise. In addition, the populations of the formerly exploited Antarctic fur seals and elephant seals recovered at an astonishing rate and there was evidence for advanced growth and reproduction in marine mammals. These observations gave rise to the 'krill surplus hypothesis': almost 200 million tonnes of krill formerly consumed by large baleen whales every year would now be available to other consumers, and could hence provide a substantial fisheries resource.

In the early 1960s Ryther's (1963) primary productivity estimate of the Antarctic Ocean of 100 g C m^{-2} y^{-1} was still much quoted, although Currie (1964) had already reduced this figure by half. The hypothesis of the richness of 'These Great Waters' dates back to Captain James Cook's and George Forster's reports of the first exploration in 1773, and had been propagated by Hardy & Gunther (1935), but contested by T.J. Hart (1942), both using evidence from the *Discovery* Investigations. S. Z. El-Sayed and others in their early measurements of primary production of Antarctic waters concentrated on highly productive regions, but showed that primary production varies greatly within the Southern Ocean. Holm-Hansen *et al.* (1977) had demonstrated that primary production was often rather low in spite of generally high nutrient concentrations. For the sustainable exploitation of Antarctic krill and fish, multispecies assessment was obviously needed, taking into account also the various natural consumers of krill.

Energy flux models of ecosystems as developed by Odum (1971) were in vogue in the 1970s. The Southern Ocean was considered a rather simple ecosystem, being a circumpolar continuum with some zonation, very rich in nutrients, highly productive on the phytoplankton level, but food limited on the krill consumer's level. This system seemed to lend itself to straightforward modelling as attempted very optimistically by K. Green Hammond (1981) for the Ross Sea. Further models by Beddington & May (1977) centred mainly around the krill and its consumers. BIOMASS kept in mind the multispecies approach but concentrated mainly on stock assessments rather than being able to go deeply into the questions of structure and function of the Antarctic marine ecosystems.

Regarding living resources other than whales, the future looked rather promising when BIOMASS was launched in 1976. The Soviet finfish fishery had already started in the early 1970s off South Georgia and Kerguelen Islands and moved towards the South Orkney and South Shetland Islands as well as to the banks south of Kerguelen Islands. After the very rapid decline of the exploited stocks due to some years of heavy fishing, we now know that the sustainable yield of notothenoid fish population was very limited.

For krill production Soviet estimates of 2 billion tonnes were reported, while Hempel's (1971) assessments for FAO amounted to 150 million tonnes annually. The expeditions by RV *Walther Herwig* in 1975–6 had demonstrated the feasibility of locating and harvesting large concentrations of krill by using conventional methods of echosounding and midwater trawling. On the other hand, the technical problems of processing large quantities of krill for human consumption became obvious.

The child BIOMASS was conceived in 1972 and was born in 1976. Three parents adopted it: SCAR, SCOR and FAO. The interest of SCAR was in the overall importance of BIOMASS to Antarctic research, while SCOR considered biological studies in the Southern Ocean as a contribution to biological oceanography in general. FAO's interest was in the assessment and management of the Antarctic fisheries resources. As it turned out, BIOMASS over its 15 year lifespan has served the interests and expectations of all three sponsoring organizations.

The scene at the end of BIOMASS

At the termination of BIOMASS, the Southern Ocean scene has greatly changed. We are far more cautious now with regard to broad statements on the productivity of the Southern Ocean, in general, and its potential resources, in particular. Such caution is largely based on BIOMASS-related studies, but also on the surveys of the pack-ice regions, and on a change in the public perception in large parts of the world regarding issues involving Man's rights and obligations in relation to the Antarctic creatures and biota.

From the organizational viewpoint, all important elements of BIOMASS have been taken over by new organizations. Croxall (this volume) discusses in detail the BIOMASSES–CCAMLR relationships.

The BIOMASS main objective mentioned at the outset '. . . for the future management of potential living resources . . .' has been fully taken up by CCAMLR. The key words of its goals are 'sustainable yield', maintenance of 'ecological relationships', avoidance of 'long-term upsetting' of any part of the system (every impact has to be reversible within 2–3 decades). For several years, because of Soviet denial, CCAMLR suffered from poor fisheries statistics, which are the most important source of information for any stock and yield assessment. Furthermore, instead of being able to monitor the stocks of krill directly through regular quantitative surveys, CCAMLR had to rely mostly on monitoring breeding populations of krill consumers in the hope that they would inform us about variations and trends in krill populations. The recent introduction of a precautionary catch limit for krill and its division into quotas by areas reflects the concern about potential effects on krill consumers rather than on the krill stocks themselves. CCAMLR is forced to look at the krill not only as a prospective resource but also as a stock which has to be protected as much as possible for the sake of its consumers. The variations of krill abundance and distribution from year to year, and place to place have been detected including total failure of certain year classes. Relations to the dynamics of sea-ice have been described in the course of regular studies in certain areas, but there is still the need for synoptic surveys of the entire current systems year after year. The data on krill distribution collected during FIBEX should be reassessed in light of present knowledge of echo-target strength and of the structure of krill swarms (Everson *et al.*, 1990). The basic question of a reliable age determination for krill still awaits an answer (Quetin *et al.*, this volume).

We now know that krill may concentrate and thrive in the underside of sea-ice, but the overall importance of the sea-ice for the krill populations at large remains to be studied by new quantitative under-ice survey methods, which have yet to be developed. Siegel & Kalinowski (this volume) in their review of krill demography gave an outlook on those future questions too.

The potential side-effects of krill fishing on other species are of particular concern to CCAMLR as to other organizations. Thus the system of monitoring sites on seal and penguin rookeries has been introduced. The selection of these sites has been done on an opportunistic basis by choosing old study areas near overwintering stations for which time series of observations already existed. The very valuable data sets on the variations in abundance and population dynamics will have to be augmented in the future by more year-round environmental data including indices of sea-ice distribution and of krill abundance outside each site. Whether or not remote sensing will be useful in estimating krill distribution and abundance is still as uncertain as it was at the inception of BIOMASS.

For the analysis of the exploitable Antarctic fish stocks, we are on much safer grounds as demonstrated by Kock & Shimadzu (this volume) in their review on the higher trophic levels. For future management, mainly straightforward fishery biology and monitoring are needed. As to the squid resources of the Southern Ocean, little progress has been made so far but new research has just started particularly by the British Antarctic Survey.

As was pointed out by several contributors to this volume, fishing operations for catching finfish, krill and squid may have adverse effects on other species, that is to say the krill fishery takes many young fish. Bottom trawling affects the benthos on the narrow shelves and slopes. Longlining is dangerous for sea birds like the albatrosses and petrels. Driftnetting is already prohibited by CCAMLR. The side-effects of fishing operations on non-target species and communities are also the concern of the SCAR Group of Specialists on Environment and Conservation (GOSEAC) and IUCN who have benefitted from the early interest of the BIOMASS Programme in ecosystem conservation matters.

Seals and whales were not covered by the BIOMASS Programme. Major breakthroughs in remote tracking of seals and birds opened a wide spectrum of studies in migration routes as well as feeding and diving ecology. The observations of year class strength, possibly influenced by El Niño–Southern Oscillation (ENSO) phenomenon, as stated by Laws (this volume), as well as the decline in the elephant seals, point to the need for studies in remote interactions of species with long distance feeding and breeding migrations.

In terms of basic research in Antarctic ecosystems, there is not much belief any more in global figures on total biomass, production and energy budget, but rather in the detailed analysis of metabolism on both the organismic and community levels. The new lines of interest call for studies of:

rates rather than of stocks;
fluxes rather than of compartments;
variations rather than average steady states.

New research strategies and methods are needed, including new forms of international cooperation, such as the EPOS cruises in 1988–9 where experimentalists as well as traditional 'samplers' of many European laboratories worked together on RV *Polarstern* in the Weddell Sea.

The concept of Large Marine Ecosystems (LME) has been applied to the Antarctic by K. Sherman (this volume). Of course, there is not one single LME in the Southern Ocean, but several distinct systems within the three major zones determined by the current and ice regimes. Hempel (1990) described the Weddell Sea as an LME consisting of parts of the seasonal and permanent ice zones.

There is a certain geographical shift in the area of interest. BIOMASS has mainly focused on the southern part of the ice-free zone of the West Wind Drift and on the summer conditions in the seasonal ice zone (SIZ) in the Weddell Sea and Prydz Bay area. Nowadays interest is growing in the study of the permanent sea-ice zone and in the sea-ice aspects of the SIZ.

Also for Antarctic marine biologists, 'Global Change' has become a key phrase. Changes in the world climate system are brought about by regional processes, just as the effects of global change are measured on a regional or local scale. The Southern Ocean is often taken as a particular example both for cause and effects of global change.

The importance of the sea-ice to the overall primary productivity, carbon flux and CO_2 budget is a matter of heated debate at present. Under the umbrella of IGBP, the role of the sea-ice zone as a source or sink of atmospheric CO_2 becomes an important topic of international programmes, particularly JGOFS. In the framework of large-scale biogeochemical studies, biologists are indispensable for contributing basic data on the primary production by phytoplankton and ice algae and on the remineralization in the microbial loop and by herbivores. I do not expect many new high-ranking biological insights from the programme itself, which is basically driven by geochemical questions. Nevertheless, JGOFS provides opportunities for the study of biological processes against the background of broad physical, chemical and sedimentological information. Participation in Southern Ocean JGOFS would bring very important features of BIOMASS into Global Change research as outlined in the presentation of Priddle, Thorley & Trathan, (this volume).

Hubold (this volume) discussed the importance of the sea-ice for the overall understanding of the ecology of the high antarctic zone. In contrast to the permanently ice-free parts, the surface layer is rather stabilized by the sea-ice, or, in summer, by meltwater. Rapid surface currents and wind drift transport surface communities, particularly ice biota, over large distances, e.g. in the Weddell Gyre. Some organisms are highly adapted to live exclusively in the sea-ice while others, like krill, can live both on ice and in the open waters. The stabilization of the surface layer and the sea ice, as inverted sea bottom in the euphotic zone, make the sea-ice zone fairly prolific in terms of primary production. This is reflected in the high abundance of krill, Adélie penguins, crabeater seals and detritophagous benthos.

The organization's future

The unique blending of applied and pure ecological work under BIOMASS will not be continued. Its organizational 'umbrella' with an infrastructure of workshops, data centre, symposia, scientific publications and newsletter, all steered by the Executive and with some central funding, will be missed by the scientific community.

The resource-oriented objectives of BIOMASS are now well taken care of by CCAMLR which became far more ecologically minded than originally anticipated. Other organizations like GOSEAC and IUCN took up the conservation aspects of the BIOMASS Programme. There is no uniform organizational framework for basic research in biological oceanography of the Southern Ocean.

For the study of the sea-ice zone, SCAR has proclaimed the programme Ecology of the Antarctic Sea Ice Zone (EASIZ) as *the* major multinational interdisciplinary biological research effort in the Southern Ocean for the years to come. With the increasing number of ice-bound research vessels in the Southern Ocean and the good data on distribution and pattern of sea ice and its polynas, the chances are good for getting meaningful answers to the burning questions on, for example, the overall importance of krill in the ice, on the role of the polynyas and small leads in the transfer of CO_2 and other gases and on primary production of the region.

In order to get off the ground and to coordinate with the programmes initiated by CCAMLR, WOCE and JGOFS, there is a need for a secretariat for EASIZ and hopefully some central funds. With that kind of support, similar to the BIOMASS system, EASIZ might develop into the most effective follow-up programme to BIOMASS in the terms of basic biological science.

Basic research in biological oceanography of the Southern Ocean goes in two directions:

studies of processes and their variations in space and time;
studies of the biological factors of carbon dioxide budgets and other elements of Global Change.

Both kinds of study shift partly from the open waters into the sea ice. For the process-oriented work the EASIZ programme, possibly in combination with GLOBEC, may provide an adequate organizational home, while the global change-related work will be carried out within the well organized framework of IGBP.

An important dividend of BIOMASS lies in the increase in the number of marine biologists from all parts of the world cooperating in the Southern Ocean. In the course of the years a certain regionalization of the global BIOMASS community became noticeable. In particular the Latin American, but also the European, marine biologists working in the Antarctic developed specific contacts within their respective continents. Most likely this will continue and can be useful as long as strong interaction between the regional groups is ensured by a multitude of individual contacts, joint actions and continuous communication.

References

Beddington J R, May R M (1977) Harvesting populations in a randomly fluctuating environment. *Science* **197**: 463–5.

Currie R (1964) Environment features in the ecology of Antarctic seas. In: Carrick R, Holdgate M W, Prévost J (eds) *Biologie Antarctique*. Paris, Herman, pp 87–94.

Everson I (1977) The living resources of the Southern Ocean. FAO GLO/SO 77/**1**: 1–156.

Everson I, Watkins J L, Bone D G, Foote K G (1990) Implications of a new acoustic target strength for abundance estimates of Antarctic krill. *Nature* **345**: 338–40.

Green Hammond K A (1981) Modelling of Antarctic ecosystems. In: El-Sayed S Z (ed) *Biological Investigations of Marine Antarctic Systems and Stocks* (*BIOMASS Sci Ser* No 2). SCAR and SCOR, Scott Polar Research Institute, Cambridge, pp 23–9.

Hardy A C, Gunther E R (1935) The plankton of the South Georgia whaling grounds and adjacent waters, 1926–27. *Discovery Rep* **11**: 1–456.

Hart T J (1942) Phytoplankton periodicity in Antarctic surface waters. *Discovery Rep* **21**: 261–365.

Hempel G (1971) Southern Ocean. In: Gulland J A *The Fish Resources of the Ocean.* Fishing News (Books) Ltd, West Byfleet, UK.

Hempel G (1990) The Weddell Sea, a high polar ecosystem. In: Sherman K, Alexander L M, Gold B D (eds) *Large Marine Ecosystems.* AAAS, Washington, DC, pp 5–18.

Holm-Hansen O, El-Sayed S Z, Franceschini G A, Cuhel R (1977) Primary production and the factors controlling phytoplankton growth in the Antarctic Seas. In: Llano G A (ed) *Adaptations within Antarctic Ecosystems* (Proceedings of SCAR Symposium in Antarctic Biology). Gulf Publishing Co, Houston, Texas, pp 11–50.

Meadows D, Meadows D, Zahn E, Melling P (1972) *The Limits of Growth. Report to the Club of Rome.* Universe Books, New York.

Odum H T (1971) *Environment, Power, and Society*. New York, Wiley Interscience, 332 pp.

Ryther J H (1963) Geographic variations in productivity. In: Hill M N (ed) *The Sea*, vol 2. John Wiley & Sons, New York, pp 347–80.

Critical appraisal of the BIOMASS Programme

G. E. FOGG

Introduction

MY REMIT IS to give a critical appraisal of BIOMASS, but I must say at the outset that my feeling is predominantly one of admiration for what it has achieved. It has been a venture in search of understanding, carried out by some dozen nations in as many ships over the best part of ten years in the world's most remote and hostile seas. Rather than being criticized at this stage, it deserves to be celebrated in poetry or music. With some elements both of the *Odyssey* and *Der Fliegende Holländer* it calls for a Homer or a Wagner to do it justice. Alternatively, a performance of Vaughan Williams' *Sea Symphony* might have been an appropriate conclusion to this meeting. Vaughan Williams, who, of course, wrote the *Symphonia Antarctica*, composed his *Sea Symphony* as a setting of verses by Walt Whitman which have as their themes the brotherhood of the seafaring nations and the living entity of the ocean.

The basic concepts of BIOMASS

The two themes – of international cooperation and the need to look at the marine ecosystem as a whole – are basic to the BIOMASS Programme and are well established and well tested in Antarctic research. Professor El-Sayed has given us an admirable account of the origins of BIOMASS but he might have gone further back. We already find the beginnings of the same collaboration between nations and holistic outlook 150 years ago in the great French, American and British expeditions to the Southern Ocean. A main objective of these voyages was geomagnetic survey, and all three of them followed similar protocols and internationally agreed-upon term days for making observations. That this happened was largely due to the efforts of one man, Alexander von Humboldt, who also pressed for similar systematic and widely based studies on marine life. In a letter to the British Admiralty in 1839, he urged that observations on the colour of the sea, surface slicks and the microscopic animals which could be collected by filtration through fine linen should be made at different latitudes and longitudes. Such a comprehensive survey was not carried out, but two discoveries of central importance to BIOMASS were made. One was the description of *Euphausia superba* by James Dwight Dana of the US expedition; the other was the recognition of phytoplankton as the primary producers in the oceans by Joseph Dalton Hooker of the British expedition.

The detailed investigation that Humboldt envisioned had to wait for another eighty years until the British government initiated the *Discovery* Investigations. It may be noted that data and collections from these Investigations are still of value in the study of the population dynamics of krill (Basson &

Beddington, 1989). In fact, the essential objectives of the *Discovery* Investigations and of BIOMASS are remarkably similar. Both set out to provide a scientific basis for the management of stocks of economically important animals such as whales and krill, and both recognized that this could not be done without knowledge of the physicochemical factors controlling the fertility of the sea and of interactions with other organisms in the ecosystem. The working hypothesis of the sea's fertility which provided the basis of the *Discovery*'s work and which is still generally accepted today was that proposed by Karl Brandt of the University of Kiel.

Thus, the BIOMASS Programme did not introduce any fundamentally new scientific concepts. What was new was a practical approach to the immense task of making observations over extensive sea areas over a sufficient span of time to yield significant results. It was recognized that this could be done only by pooling the resources of a number of nations and by using the latest technology for gathering and processing data. The general scientific objectives were outlined succinctly so as to allow the inclusion of almost any kind of biological oceanography or marine biology. The emphasis in the BIOMASS organization has been on facilitating the exchange of information, selection and standardization of techniques, collection of information and group discussion of results, rather than on laying down detailed programmes of research. The individual national Antarctic research organizations were thus able to adapt their existing programmes rather than sacrifice them to contribute to BIOMASS. The outcome has been diffuse: in preparing this paper I have often been quite unable to decide whether a particular piece of work belonged to BIOMASS or not. This does not really matter; the organization has proved flexible and has allowed evolution of the programme as more information became available and as new techniques appeared. There is no doubt that the result has been a great stimulus to Antarctic marine biology and has led to greater public awareness of the interest and value of the ecosystem of the Southern Ocean.

The coverage of BIOMASS

Has this approach meant that important areas have been neglected or missed altogether? Resources have not been unlimited, and I think it was realistic to concentrate on krill as the main subject of the study. It is inevitable that specialists will complain that their particular fields have received insufficient attention, but I do not think that any important aspect of the marine ecosystem has been inadvertently missed. However, I might perhaps be allowed to consider briefly how two of my own particular interests in biological oceanography have fared.

The ultraplankton

In the past ten years there has been increasing recognition by marine microbiologists in general that the pelagic microbial community – which I like to call the ultraplankton and which includes protozoans, nanoflagellates, picophytoplankton, bacteria and viruses – forms a highly dynamic equilibrium system within which elements such as carbon, nitrogen and phosphorus are rapidly cycled. El-Sayed (1983, 1984) pointed out that the existence of this 'microbial loop' complicates the previously simple picture of the food chain in the Southern Ocean and that its further study might resolve some of the perplexing problems of primary production in these waters. Components of the ultraplanktonic community such as bacteria (Hanson *et al.*, 1983) and picophytoplankton (Marchant *et al.*, 1987) occur in Antarctic seas in concentrations similar to those found in temperate waters, and the heterotrophic activity appears to be comparable with that in warmer waters (Hodson *et al.*, 1981; Cota *et al.*, 1990). Cycling in the microbial loop seems to be as active in Antarctic waters as it is elsewhere. The consensus among marine microbiologists at present tends to the opinion that little of the biomass of the ultraplanktonic community is used as food by higher trophic levels (see Fogg, 1991), and this seems to be true in some coastal waters in Antarctica (Cota *et al.*, 1990). However, there is some evidence that protozoans, such as choanoflagellates and tintinnids, which graze on bacteria and picophytoplankton may sometimes form an appreciable proportion of the diet of krill (Vincent, 1988). The microbial community of sea-ice may be of more importance in the nutrition of these animals. Sea-ice provides microhabitats in which heterotrophic production by bacteria, fungi, foraminifera, ciliates and flagellates is high. The carbon source for this production may come from diatom photosynthesis or from the dissolved organic matter of seawater. An estimate of 35 mg C m^{-2} day^{-1} by Kottmeier & Sullivan (1987) may be taken as indicative of the upper range of primary production by sea-ice algae. The dissolved organic matter normally present in seawater will be concentrated during ice formation, and if we take 4 mg C l^{-1} as an average value for its initial concentration (Sugimura & Suzuki, 1988) and assume that it is concentrated proportionately with sodium chloride, then the average salt flux during the ice-growth season of 1–5 kg m^{-2} mo^{-1} (Gow & Tucker, 1990) would indicate a supply of 4–20 mg C m^{-2} day^{-1} from dissolved organic matter. Not all of this is necessarily available to bacteria but, nevertheless, it seems that this source of substrate for heterotrophic growth may be appreciable relative to that provided by photosynthesis. More information about the ultraplanktonic community, especially at the ice-edge, where it seems to be particularly active (Buck & Garrison, 1983), is essential for proper understanding of the Antarctic marine ecosystem and one could wish that BIOMASS had promoted more research on this.

Physical oceanography

The second particular aspect that I should like to touch on is that of physical oceanography in relation to the biology. Biological oceanographers are absolutely dependent on their physical colleagues, as was recognized from the outset in BIOMASS. Expert advice was taken and routine physical and chemical observations have been carried out on all cruises. An innovation introduced by the British Antarctic Survey has been to use a microprocessor to display temperature and salinity data in real time so that water masses can be identified on the spot. Such things are of great convenience to the biologist. Polish, Soviet and British workers have paid particular attention to the relations of plankton development and distribution to hydrographical features. However, the value of the physical and chemical data will fully emerge only when full use is made of the data bank. This is all highly commendable, but I have one reservation. The physical oceanography has been planned with expert advice, but largely carried out on a do-it-yourself basis by biologists. As far as I know, there has been little innovative input from professional physical oceanography, perhaps because its two eminent representatives, Professor ESW Simpson and Sir George Deacon, both of whom had great interest in BIOMASS, died in 1983 and 1984, respectively.

The life of planktonic organisms is profoundly influenced by water movements on all scales, from molecular motion to oceanic currents. In a hazy way we begin to see how currents, fronts, gyres, eddies and small-scale turbulence have varying effects on different types of organism (Fogg, 1991), but real advances in our understanding of plankton will only come when physical oceanographers and fluid dynamicists with an appreciation of the biological problems give precision to these ideas. This is something that BIOMASS has lacked.

BIOMASS as collectivized research

In addition to incomplete coverage in its research programme, it is possible that BIOMASS had actually had negative effects on Antarctic marine science by diverting resources from other projects. For example, one has heard grumbling within the British Antarctic Survey that development of its offshore research programme in response to BIOMASS has meant reducing support for inshore marine biology. This is one instance of growing problems as science becomes increasingly collectivized and less of an individualistic activity. Innovative science is essentially an individual matter, and by devolving more of its organization and allocation of resources onto committees we may be cutting science off from the source of its inspiration. Ziman (1983), who considered this problem in his Bernal Lecture to the Royal Society, put it in a nutshell when he asked whether, in the highly collectivized and rationalized system that we are moving towards, an absurd idea, against all common sense and contrary to authoritative opinions (such as was Alfred Wegener's theory of drifting continents), would have any chance of survival. Anything similar today would be unlikely to get support. Who knows what initiatives that might have led to significant advances have been stifled because a young scientist has been directed into counting krill instead of pursuing his own ideas, or because a project has failed to get support because BIOMASS had priority? I do not think, however, that BIOMASS has been too inhibitory in this way. From the beginning, Antarctic science has been 'big science', requiring government funding and organization by committees and therefore collective in Ziman's sense. Nevertheless, once in the field and, until recently, out of the reach of bureaucracy, organization has been flexible and individuals, besides doing the work required of them, have had some scope for following up their own ideas. The 'big science' has usually carried some 'little science' along with it. Thus, as the early magnetic survey expeditions took naturalists along with them, the *Discovery* Investigations, although primarily biological, provided opportunity for work on physical oceanography which was to prove of the greatest importance, and the International Geophysical Year (IGY) has helped many biologists to do work in the Antarctic. BIOMASS inherited this tradition of flexibility and its committees, which have had a minimum of bureaucratic interference, have been largely concerned with ways and means rather than with detailed planning of the scientific programme. BIOMASS has succeeded, I think, by providing a framework and standardized techniques, in enabling the nations just entering into Antarctic research to obtain useful information straightaway without waste of effort. At the same time, the creative scientist was given scope within broad limits to make his full contribution.

Ziman (1983) suggested that a way to counteract the inhibitory tendency of collectivization would be to fund university research departments on the basis of their past performances and without regard for what projects they planned. This might not work very well if they wished to work in the Antarctic, since the imperatives of logistics would still, under present circumstances at least, force them into the collective system.

Against the disadvantage of collectivization there are some very real advantages which have all been manifest in the BIOMASS Programme. Apart from the fact that it is the only feasible way of carrying out an intensive research programme on anything approaching the necessary scale, one may list the following.

1. Individual research is placed in a wider context and acquires more significance.
2. It is an effective means of establishing fruitful contacts between different laboratories and different disciplines.
3. It provides good-quality background data for all the component projects.
4. It gives broader opportunity and training to young scientists.

Answers to a questionnaire distributed in 1991 to participants in the UK Biogeochemical Ocean Flux Study (BOFS) indicated general approval of the organization and research coordination and appreciation of the advantages, such as those just listed, of working in a community without the barriers of language and distance which hamper interchange of ideas in its international counterpart, the Joint Global Ocean Flux Study (JGOFS), or in BIOMASS. Nevertheless, I get the impression that participants in BIOMASS have found the same advantages and derived satisfaction from working together with a common purpose. That this has happened must surely be due to the system of workshops, involving a few participants coming together to deal with a specific task, and the admirable *Newsletter* which has kept everyone informed of what has been going on.

The results achieved by BIOMASS, and their practical value

The results of BIOMASS have been summarized in the chapters of this volume and it would not be appropriate here to discuss them in detail. We now have a much more complete picture of the Antarctic marine ecosystem and of some of its component organisms. We have some idea of the relative importance of the different krill predators, in particular the massive and evidently increasing roles of seals and birds. However, Miller & Hampton (1989) concluded that neither the estimates of stocks and production of krill nor the models attempting to describe the effects of predation and the mechanisms determining distribution were adequate for setting optimum harvesting levels. The inadequacy was very evident in the gap between the direct estimates of krill abundance from acoustic surveys and those based on annual production–lifespan values, which were an order of magnitude higher. The finding that the target strengths used in the acoustic estimations were too high has brought the two figures nearer together (Everson *et al.*, 1990). Another major advance has been the confirmation by thorough and detailed observation of the anecdotal evidence (which has existed for some time) of the presence in quantity of grazing krill immediately under sea-ice during autumn, winter and spring (Marschall, 1988). These observations were made in the Weddell Sea and it remains to be seen whether they apply generally, but they have focussed attention on the importance of primary production in sea-ice (see section on ultraplankton, above). Nevertheless, in spite of these advances, Miller & Hampton's (1989) conclusion that our knowledge still falls short of what CCAMLR requires for management of commercial fisheries remains true. We are all convinced that the ecosystem approach to management is the only sound one, but are still hazy as to how it can be applied.

Many of the BIOMASS data have been incorporated into the database, and it is one of the great strengths of BIOMASS that every effort has been made to ensure that data from different sources are comparable and that the material incorporated is valid. The importance of this cannot be overemphasized: it is a notorious fact of science that even with an apparently simple technique, different laboratories manage to produce widely different results, and that once it has been loaded into a computer a number is accepted as beyond question. The issue of handbooks on methods and the interchange of information in the Working Parties will have gone some way to minimize laboratory-made variation, and the validation procedure which must be gone through before data are loaded into the database eliminates at least the obvious uncertainties and inconsistencies. The BIOMASS database is exemplary, not only for the soundness of its material, but in being multisubject, international, interactive and, it is hoped, continuing.

The data relate mainly to the two field exercises FIBEX and SIBEX and thus, since they cover only two or three years, are in themselves inadequate to provide a baseline against which to assess the effects of krill harvesting. Natural variations in ecosystems are so great that it requires many years (two sunspot cycles?) of continuous records before perturbations due to human activities can be distinguished against the background noise. Paine (1986) showed how correlations between changes brought about by El Niño in marine communities and water column characteristics appeared statistically significant when examined over six years, but disappeared when sought for over a longer period. Some national programmes, that of the USSR for example, extend observations of the type made during FIBEX over a stretch of years for certain areas. BIOMASS chlorophyll data can be supplemented with observations such as those of SZ El-Sayed to cover the period from 1962 to 1987, and are being examined by Dr J Priddle as a possible means of providing a baseline (Trathan, 1991).

The database will be an invaluable source of facts, but if it is used only for this, it will be a sad waste. One hopes that the database will provide the means for the elaboration and testing of hypotheses, not only on the dynamics of krill populations, but on the functioning of the whole Antarctic marine ecosystem. It is here that the individual scientist will have an opportunity for innovation. If, as has been proposed in the BIOMASS Data Centre manager's 'Discussion Document' (Thorley, 1990), it is incorporated by SCAR into an Antarctic data centre, its capacity to contribute to global change research will be greatly enhanced. Programmes such as the International Geosphere–Biosphere Programme (IGBP) and the Joint Global Ocean Flux Study (JGOFS), now under way, could obviously make excellent use of BIOMASS data and experience.

BIOMASS and the understanding of ecosystem functioning

I began by referring to the traditional holistic approach of Antarctic science. The weakness of this approach has been its lack of the rigour and precision which we have come to expect from reductionist science. It seems to me that a principal way to overcome this weakness must be through mathematical modelling. This was envisaged for BIOMASS from the beginning (Green Hammond, 1981). Mathematical modelling, however, will only be of advantage if we take a realistic view of its limitations as well as of its capabilities. Even with the capacity of present-day computers, it is necessary to simplify the functions relating variables, often to an extent that the physiologist knows to be incorrect, and to ignore some variables of potential importance altogether. Furthermore, as we are beginning to learn, there is always a lurking element of chaos in complex systems. In marine science, at least, modelling is, at present, of most use in testing hypotheses rather than in making predictions, for which it is only of value in limited circumscribed situations. A simple observation at the natural history level (for example, that krill derived a substantial amount of sustenance from ice algae during the winter) may call for reorganization of a model, with vastly increased complication of the mathematical formulation.

I do not altogether share Miller & Hampton's (1989) conclusion that it is premature to attempt a global production model for krill. Admittedly, some of the necessary information is as yet of dubious value if not altogether missing, but if we wait for everything to be complete and accurate, nothing will get done. It can be instructive to do what one can with what is available and see what new ideas emerge. Baird *et al.* (1991) have recently carried out a comparative network flow analysis of six different marine ecosystems and have come to the conclusion that the aggregate amount of cycling of materials, such as carbon, is not an indication of system maturity but rather of stress. It would be of great interest to extend this analysis to the BIOMASS data which, I suspect, are no worse, and perhaps better, than those used by Baird *et al.*, and gain some idea of how the Antarctic ecosystem compares with those in temperate and tropical regions. A model of the Antarctic marine food system, which again makes some bold assumptions and relies on some questionable data, suggests that the Southern Ocean, contrary to the generally accepted view, is remarkably inefficient as a carbon sink (Huntley *et al.*, 1991). This inefficiency arises from the large flux of carbon respired to the atmosphere by air-breathing birds and mammals. This conclusion may be erroneous, but it is thought-provoking to say the least, and it may be of great significance for predicting the course of carbon dioxide accumulation in the atmosphere.

In stressing the importance of mathematical modelling, I must add the reservation that I hope that continuation of BIOMASS-inspired work will not be entirely on computers in comfortable home laboratories. Only the acute observation and intuitions of the biologist actually working with living organisms in their natural habitats can ensure the sound development of our science.

Acknowledgements

My gratitude must be expressed to Professor Sayed Z El-Sayed for his encouragement and help in many ways in the preparation of this paper and to the following who provided me with advance copies of reports on national contributions: Dr F Brandini (Brazil), Dr C de Broyer (Belgium), Prof Dong Zhaoqian (China), Dr RB Heywood (UK), Dr T Hoshiai (Japan), Dr RR Makarov (USSR), Dr HJ Marchant (Australia), Dr G di Prisco (Italy), Dr S Rakusa-Suszczewski (Poland) and Dr Schnack-Schiel (Germany).

References

Baird D, McGlade JM, Ulanowicz RE (1991) The comparative ecology of six marine ecosystems. *Phil Trans R Soc Lond B***333**: 15–29.

Basson M, Beddington JR (1989) Preliminary analysis of growth and mortality rates of Antarctic krill (*Euphausia superba*) from length frequency samples. In: Heywood RB (ed) *University Research in Antarctica.* British Antarctic Survey, Cambridge, pp 51–5.

Buck KR, Garrison DL (1983) Protists from the ice-edge region of the Weddell Sea. *Deep-Sea Res* **30**: 1261–77.

Cota GF, Kottmeier ST, Robinson DH, Smith WO Jr, Sullivan CW (1990) Bacterioplankton in the marginal ice zone of the Weddell Sea: biomass, production and metabolic activities during austral autumn. *Deep-Sea Res* **37**: 1145–67.

El-Sayed SZ (1983) Biological productivity of the Antarctic waters: present paradoxes and emerging paradigms. In: El-Sayed SZ, Tomo AP (eds) *Antarctic Aquatic Biology (BIOMASS Sci Ser* No 7). SCAR and SCOR, Scott Polar Research Institute, Cambridge, pp 1–22.

El-Sayed S Z (1984) Productivity of Antarctic waters – A reappraisal. In: Holm-Hansen O, Bolis L, Gilles R (eds) *Marine Phytoplankton and Productivity.* Springer-Verlag, Berlin, pp 19–34.

Everson I, Watkins JL, Bone DG, Foote KG (1990) Implications of a new acoustic target strength for abundance estimates of Antarctic krill. *Nature* **345**: 338–40.

Fogg GE (1991) The phytoplanktonic ways of life. *New Phytol* **118**: 191–232.

Gow AJ, Tucker WB III (1990) Sea ice in the polar regions. In: Smith WO Jr (ed) *Polar Oceanography*, part A (*Physical Science*). Academic Press, San Diego, pp 47–122.

Green Hammond KA (1981) Modelling of Antarctic ecosystems. In: El-Sayed SZ (ed) *Biological Investigations of Marine Antarctic Systems and Stocks* (*BIOMASS Sci Ser* No 2). SCAR and SCOR, Scott Polar Research Institute, Cambridge, pp 23–9.

Hanson RB, Lowery HK, Shafer D, Sorocco R, Pope DH (1983) Microbes in Antarctic waters of the Drake Passage: vertical patterns of substrate uptake, productivity and biomass in January 1980. *Polar Biol* **2**: 179–88.

Hodson RE, Azam F, Carlucci AF, Fuhrman JA, Karl DM, Holm-Hansen O (1981) Microbial uptake of dissolved organic matter in McMurdo Sound, Antarctica. *Mar Biol* **61**: 89–94.

Huntley M E, Lopez M D G, Karl D M (1991) Top predators in the Southern Ocean: a major leak in the biological carbon pump. *Science* **253**: 64–6.

Kottmeier ST, Sullivan CW (1987) Late winter primary production and bacterial production in sea ice and seawater west of the Antarctic Peninsula. *Mar Ecol Prog Ser* **36**: 287–98.

Marchant HJ, Davidson AT, Wright SW (1987) The distribution and abundance of chroococcoid cyanobacteria in the Southern Ocean. *Proc NIPR Symp Polar Biol* **1**: 1–9.

Marschall H-P (1988) The overwintering strategy of Antarctic krill under the pack-ice of the Weddell Sea. *Polar Biol* **9**: 129–35.

Miller DGM, Hampton I (1989) Biology and ecology of the Antarctic krill (*Euphausia superba* Dana): a review. (*BIOMASS Sci Ser* No 9). SCAR and SCOR, Scott Polar Research Institute, Cambridge. 166 pp.

Paine R T (1986) Benthic community – water column coupling during the 1982–1983 El Niño. Are community changes at high latitudes attributable to cause or coincidence? *Limnol Oceanogr* **31**: 351–60.

Sugimura Y, Suzuki Y (1988) A high-temperature catalytic oxidation method for the determination of non-volatile dissolved organic carbon in seawater by direct injection from a liquid sample. *Mar Chem* **24**: 105–31.

Thorley M (1990) The BIOMASS Data Centre: What's in store after 1991? *BIOMASS Newsl* **12**(1): 3.

Trathan P (1991) News from the BIOMASS Data Centre. *BIOMASS Newsl* **13**(1): 7.

Vincent W F (1988) *Microbial Ecosystems of Antarctica.* Cambridge University Press, 304 pp.

Ziman J M (1983) The Bernal Lecture, 1983. The collectivization of science. *Proc R Soc Lond* B**219**: 1–19.

Index

Note: Page numbers in italic refer to charts, diagrams and maps. References to land masses and islands include waters in the region of those locations.